Metal Interactions with Boron Clusters

MODERN INORGANIC CHEMISTRY

Series Editor: John P. Fackler, Jr.
Texas A&M University

METAL INTERACTIONS WITH BORON CLUSTERS
Edited by Russell N. Grimes

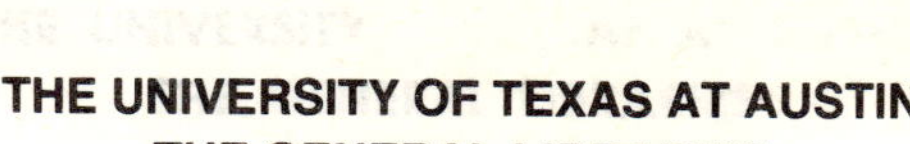

Metal Interactions with Boron Clusters

Edited by
Russell N. Grimes
University of Virginia
Charlottesville, Virginia

PLENUM PRESS • NEW YORK AND LONDON

Library of Congress Cataloging in Publication Data

Main entry under title:

Metal interactions with boron clusters.

(Modern inorganic chemistry)
Includes bibliographies and index.
1. Borane. 2. Metal-metal bonds.
3. Reactivity (Chemistry) I. Grimes, Russell N., 1935–   . II. Series.
QD181.B1M47 1982   546'.67159   82-9068
ISBN 0-306-40933-X   AACR2

© 1982 Plenum Press, New York
A Division of Plenum Publishing Corporation
233 Spring Street, New York, N.Y. 10013

Printed in the United States of America

To the memory of our friend and colleague, Ralph Rudolph

Contributors

Silvano Bresadola • Istituto di Chimica, Universitá di Trieste, Trieste, Italy, and CNR–Centro di Studio sulla Stabilità e Reattività dei Composti di Coordinazione, Padova, Italy

Donald F. Gaines • Department of Chemistry, University of Wisconsin, Madison, Wisconsin

William E. Geiger, Jr. • Department of Chemistry, University of Vermont, Burlington, Vermont

Norman N. Greenwood • Department of Inorganic and Structural Chemistry, University of Leeds, Leeds LS2 9JT, England

Russell N. Grimes • Department of Chemistry, University of Virginia, Charlottesville, Virginia

Steven J. Hildebrandt • Minerals and Chemicals Division, Engelhard Minerals and Chemicals Corporation, Edison, New Jersey

John D. Kennedy • Department of Inorganic and Structural Chemistry, University of Leeds, Leeds LS2 9JT, England

Marion E. O'Neill • Chemistry Department, Durham University, South Road, Durham OH1 3LE, England

Lee J. Todd • Department of Chemistry, Indiana University, Bloomington, Indiana

Kenneth Wade • Chemistry Department, Durham University, South Road, Durham OH 1 3LE, England

Preface

Molecular clusters, in the broad sense that the term is commonly understood, today comprise an enormous class of species extending into virtually every important area of chemistry: "naked" metal clusters, transition metal carbonyl clusters, hydrocarbon cages such as cubane (C_8H_8) and dodecahedrane $(C_{20}H_{20})$, organometallic cluster complexes, enzymes containing Fe_4S_4 or $MoFe_3S_4$ cores, high polymers based on carborane units, and, of course, the many kinds of polyhedral borane species. So large is the area spanned by these diverse classes that any attempt to deal with them comprehensively in one volume would, to say the least, be ambitious—and also premature. We are presently at a stage where intriguing relationships between the various cluster families are becoming apparent (particularly in terms of bonding descriptions), and despite large differences in their chemistry an underlying unity is gradually developing in the field. For example, structural changes occurring in Fe_4S_4 cores as electrons are pumped in and out, in some measure resemble those observed in boranes and carboranes. The cleavage of alkynes via incorporation into carborane cages and subsequent cage rearrangement, a sequence familiar to boron chemists, is a thermodynamically favored process which may be related to the behavior of unsaturated hydrocarbons on metal surfaces; analogies of this sort have drawn attention from theorists and experimentalists.

The perception of such relationships between boranes and other cluster types arises partly from new discoveries, but is also largely a matter of tying together facts that have been around for some time. Yet most chemists today are specialists; few are expert in separate fields as seemingly disparate as, for example, metal carbonyl clusters and metallaboranes. Bridges, therefore, are much in need, and the present volume is a modest attempt in that spirit. This book deals with the interface between metal chemistry and the prototype cluster family, the boranes. In labeling the boranes and their derivatives "prototype", I rest partly on the historic truth that the cage-like boron hydrides discovered by Alfred Stock early in this century were the first true discrete, covalent cluster species (though their structures were not elucidated until some years later). I also draw attention to the fact that, as Kenneth Wade has noted, the boranes are "pattern-makers" for clusters in general. The widely cited electron-counting rules for polyhedral cages were inspired by the boron systems; moreover, the boranes, carboranes, and their metal complexes form by far the most extensive and structurally varied class of molecular clusters in existence, ranging over dozens of different cage geometries and thousands of individual compounds. No other cluster family is really comparable, either in terms of extensively developed chemistry or in the central role occupied by the boron clusters in the theory of bonding in electron-delocalized polyhedra.

The selection of topics for this book was conducted with two things in mind: first, to provide selective and up-to-date coverage of the fascinating chemistry that results when polyhedral boron compounds and metal reagents get together, and second, to deal explicitly with certain aspects of this field that have not previously been reviewed *per se*. To my knowledge there has been no *specific* treatment of $B_3H_8^-$-metal complexes, metallaborane and metallacarborane electrochemistry, polyhedral boranes with metal–hydrogen bonds, or metallaboron compounds of the main group metals; until now, anyone interested in any of these important topics has had to comb the literature for scattered references. The subjects of the three remaining chapters, dealing with cluster bonding, σ-bonded metal complexes of boranes and carboranes, and transition-metal metallaboranes, have been reviewed before but not recently; the present treatments are intended to bring the reader abreast of these rapidly developing areas.

Finally, it is important to stress the *interface* aspect of this book; the content is as much "metal chemistry" as "boron chemistry". It is to be hoped that the contribution of this volume to both fields will be significant.

Russell N. Grimes
Charlottesville, Virginia

Contents

3. Interactions of Metal Groups with the Octahydrotriborate (1–) Anion, $B_3H_8^-$

Donald F. Gaines and Steven J. Hildebrandt

4. Metallaboron Cage Compounds of the Main Group Metals

Lee J. Todd

5. *closo*-Carborane–Metal Complexes Containing Metal–Carbon and Metal–Boron σ-Bonds

Silvano Bresadola

6. Electrochemistry of Metallaboron Cage Compounds

William E. Geiger Jr.

7. Boron Clusters with Transition Metal–Hydrogen Bonds
Russell N. Grimes

1

Structural and Bonding Features
of Metallaboranes
and Metallacarboranes

Marion E. O'Neill and Kenneth Wade

1. INTRODUCTION

Metallaboranes and metallacarboranes have played an important role in the development of cluster chemistry. They provided the vital link by which boron clusters were seen to be related to metal clusters and metal–hydrocarbon π-complexes, paving the way for general theories of cluster bonding. They also provided, and continue to provide, unprecedented networks of atoms that prompted researchers to look for new approaches to cluster bonding, or to refine existing approaches. This chapter is concerned with these related themes.

Before mixed clusters with both metal and boron atoms in their skeletons were known, the higher boranes tended to be regarded as chemical curiosities, compounds whose peculiar structures, apparently without parallel elsewhere in chemistry, required their own bonding rules. However, with the discovery of carboranes and their metalla derivatives, it became evident that carbon and metal atoms could participate in the same type of bonding as the boron atoms of boranes, and that the bonding between metal atoms and borane or carborane ligands had much in common with that between metal atoms and unsaturated hydrocarbons in organometallic compounds such as ferrocene. As a result, boron clusters and metal clusters, metal–hydrocarbon π-complexes, and aromatic ring systems, even certain small-ring hydrocarbons and a few carbo-cations, were seen

Marion E. O'Neill and Kenneth Wade • Chemistry Department, Durham University, South Road, Durham OH1 3LE, England. Work supported by the Science and Engineering Research Council.

1

to be members of the same family of cluster systems, all conforming to a common structural and bonding pattern. This pattern has proved to be quite useful for rationalizing or predicting the structures of many individual clusters, though recent developments in the chemistry of metal clusters and of metal–boron clusters have shown that the pattern has many more ramifications than were first apparent.

In the following pages, we first consider how the simple structural and bonding pattern that holds for most metal–boron clusters can be used for a general theory of cluster shapes, allowing useful analogies to be drawn and generalizations and predictions to be made. In doing so, we consider the relative merits of localized bond and molecular orbital treatments of the bonding. We then examine the implications of a number of exceptions to the simple pattern and indicate what alternative approaches are available to accommodate them. Our intention is to supplement an earlier review[1] written by one of us illustrating the structural and bonding pattern more comprehensively than would be appropriate here. Further discussion of these themes is to be found in Refs. 2–10.

2. STRUCTURES AND BONDING OF BORANES AND THEIR METALLA DERIVATIVES

2.1. The Borane-Carborane Structural Pattern

The way that the structures of the higher boranes and carboranes reflect the numbers of electrons available to hold their skeletal atoms together is a familiar feature of boron cluster chemistry, and needs only to be briefly summarized here to provide a basis for the consideration of metallaboranes and metallacarboranes.

The key structural feature of boranes that provided the basis for the bonding treatments now used was recognized by Lipscomb[11] from his early structural studies. He noted that, in the molecules of typical boranes B_pH_{p+q} such as B_4H_{10}, B_5H_9, B_5H_{11}, B_6H_{10}, and $B_{10}H_{14}$, the atoms effectively lie on two concentric spherical surfaces. The p boron atoms and q hydrogen atoms lie on the inner spherical surface, while the remaining p hydrogen atoms lie on the outer one, each eclipsing one boron atom. The molecules can, therefore, be thought of as clusters of p B–H units, in which the B–H bonds point radially outwards away from the cluster center, held together by the electrons they contribute (two apiece) for skeletal bonding, together with q further electrons from the q hydrogen atoms on the inner sphere. The bonding problem these clusters pose, thus reduces to explaining how the $(2p + q)$ skeletal bonding electrons hold the inner sphere boron and hydrogen atoms together, while at the same time rationalizing the precise way they are arranged on the inner sphere.

Figure 1 illustrates some of the spatial arrangements that require explanation. The polyhedra on which all the structures are based are shown in Figure

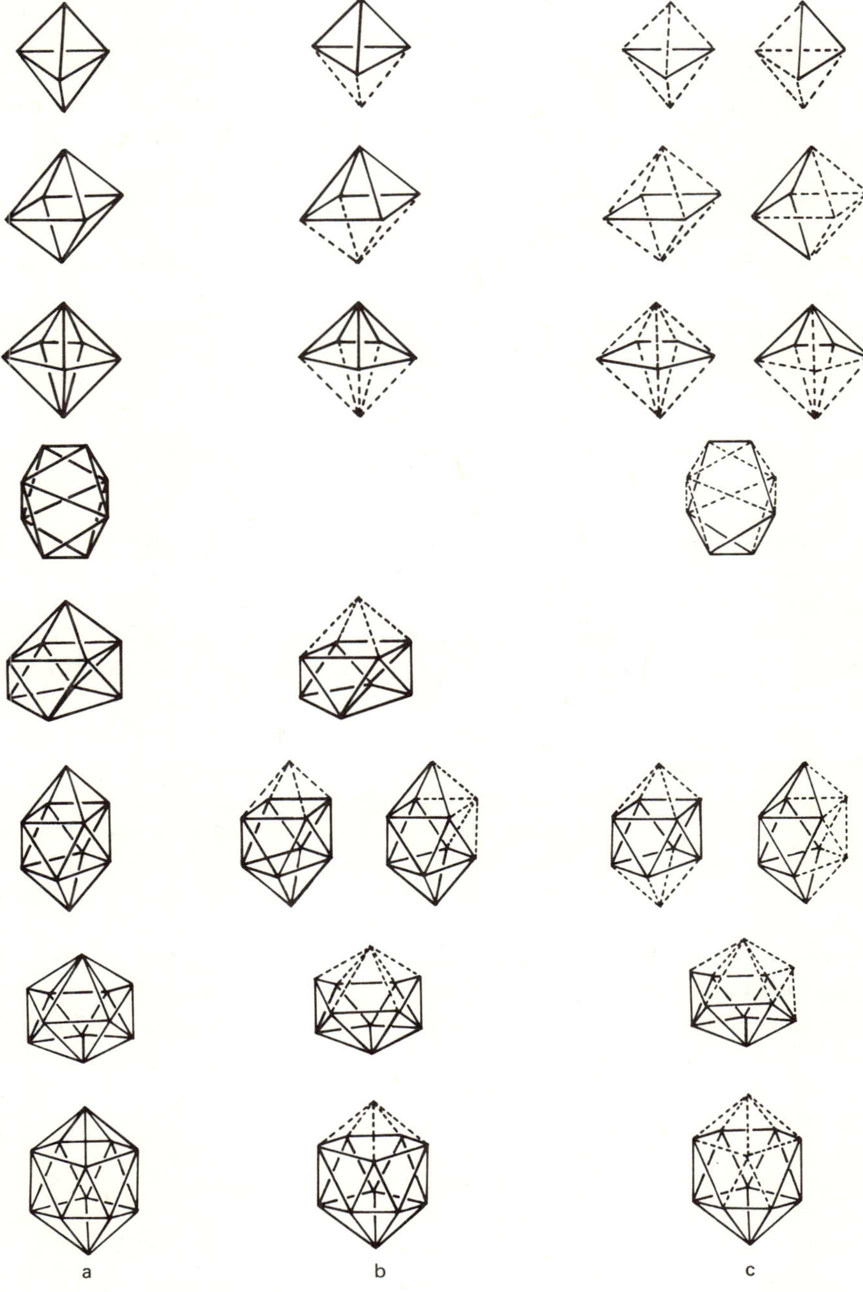

Figure 1. Polyhedral skeletons of: (a) *closo*-borane anions $B_pH_p^{2-}$ and carboranes $C_2B_{p-2}H_p$; (b) *nido*-boranes B_pH_{p+4} and carboranes $C_2B_{p-2}H_{p+2}$; (c) *arachno* boranes B_pH_{p+6} and carboranes $C_2B_{p-2}H_{p+4}$.

1(a).[1,3,12,14] The polyhedra are complete, each vertex being occupied by a boron (or carbon) atom, in the series of *closo*-borane anions $B_pH_p{}^{2-}$ and isoelectronic carboranes $C_2B_{p-2}H_p$, which contain **p** skeletal boron (or carbon) atoms held together by (**p**+1) skeletal bond pairs. One polyhedron vertex is left vacant in the *nido* series [Figure 1(b)] of formulas B_pH_{p+4}, $C_2B_{p-2}H_{p+2}$, etc., which contain **p** skeletal atoms, held together by (**p**+2) skeletal bond pairs. Two vertices are left vacant in the *arachno* series [Figure 1(c); B_pH_{p+6}, etc.] in which (**p**+3) skeletal bond pairs hold together **p** skeletal atoms. A few *hypho* species, of formulas derivable from hypothetical boranes B_pH_{p+8} are also known, containing (**p**+4) skeletal bond pairs. These have yet more open structures.

For all these structural types: *closo, nido, arachno,* and *hypho,* the structures are thus based on that polyhedron of Figure 1 with one vertex fewer than the number of skeletal bond pairs.

2.2. The Significance of the Numbers of Skeletal Bond Pairs

2.2.1. Localized 2- and 3-Center Bonds and Styx Numbers

The connectivities of the skeletal atoms of boranes B_pH_{p+q} and related carboranes can be rationalized by the elegant and ingenious topological treatment devised by Lipscomb.[11] This approach, which allocates the skeletal bonding electrons to 2- and 3-center bonds, has become the standard method for indicating the electron distribution in *nido*- and *arachno*-boranes and carboranes, although it has disadvantages when applied to *closo* species and is of limited value for predicting the polyhedral shapes of borane clusters. Since it can be applied to metalla derivatives, it is worth outlining the salient features here.

The (2**p**+**q**) skeletal electrons of boranes B_pH_{p+q} are allocated to the 2- and 3-center bonds that are assumed to hold the **p** inner-sphere boron atoms and **q** inner-sphere hydrogen atoms together. These consist of **s** BHB and **t** BBB 3-center bonds, and **y** BB and **x** BH 2-center bonds. Each of the **p** outer-sphere (*exo*) hydrogen atoms is assumed to be linked to its neighboring boron atom by a normal 2-center BH bond. Each inner-sphere (*endo*) hydrogen atom is assumed to be bound either to one neighboring boron atom by one of the **x** 2-center BH bonds, or to two neighboring boron atoms by one of the **s** 3-center BHB bonds. Each boron atom is assumed to use all four of its valence shell orbitals, and so to participate in four bonds, one of which is to the *exo*-hydrogen atom. The remaining three (skeletal) bonds link it to the neighboring inner-sphere boron and hydrogen atoms.

Since each boron atom is involved in three skeletal bonds, but furnishes only two skeletal bonding electrons (the other valence shell electron being assigned to the *exo*-BH bond), the number of 3-center bonds (**s**+**t**) must equal the number of boron atoms (**p**). Because the **q** *endo*-hydrogen atoms are involved in either 2-center BH bonds or 3-center BHB bonds, then the total number of these (**s**+**x**) must equal **q**. Moreover, the number of skeletal bond pairs (2**p**+**q**)/2 must equal the number of skeletal bonds $(s+t+y+x)$. Hence, the following equations relate s, t, y, and x to p and q:

$$s = p - t \qquad x = q - s \qquad 2y = s - x$$

Using these equations, possible values of s, t, y, and x can be deduced from the formulas of boranes $B_p H_{p+q}$. Unique solutions are not to be expected, since there are four unknowns and only three equations, but the equations can nevertheless be used to deduce what bond networks are, in principle, capable of holding the skeletal boron and *endo* hydrogen atoms together. The two bond networks (corresponding to **styx** numbers 4012 and 3103) compatible with the formula B_4H_{10} are shown in Figure 2(a). The 4012 topology is that adopted by

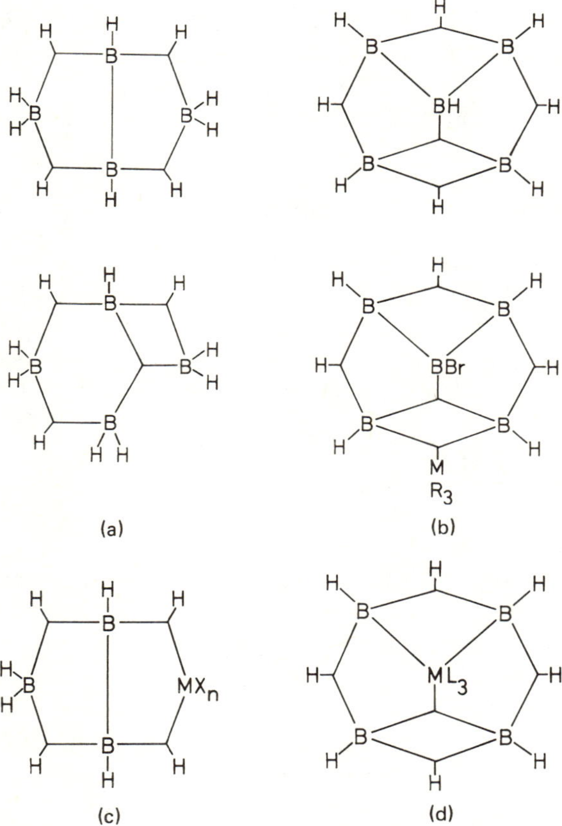

Figure 2. Localized 2- and 3-center bond schemes for: (a) B_4H_{10} with **styx** numbers 4012 and 3103; (b) B_5H_9 and $B_5H_8MR_3$ (M = Si, Ge, Sn, or Pb); (c) $B_3H_8MX_n$ $\{MX_n = Cu(PR_3)_2$ or $[Cr(CO)_4]^-\}$; (d) $B_4H_8ML_n$ $[(ML_n = Fe(CO)_3$ or $Co(\eta^5-C_5H_5)]$.

B_4H_{10}. The 3103 topology represents a less symmetrical and so less satisfactory alternative in which one of the bridging hydrogen atoms of B_4H_{10} has been converted into an *endo* terminal hydrogen atom, a species through which hydrogen scrambling reactions may well occur.

Latterly, the topological approach has undergone considerable refinement[15-17] and has been applied to carboranes as well as boranes. Its success lies more in the way it indicates the electron distribution in boranes and carboranes of known structure than in its capacity to predict the three-dimensional arrangement of their skeletal atoms. Nevertheless, when applied to metallaboranes or metallacarboranes in which metal atoms or units replace hydrogen atoms or BH, CH, or BH_2 units of boranes or carboranes, the bonding environment of the replaced atom or group can be assumed to be taken up by the metal concerned. For example, monovalent MR_3 units (M = Si, Ge, Sn, or Pb) can replace the bridging or terminal hydrogen atoms of B_5H_9 [Figure 2(b)].[18] A BH_2 unit of B_4H_{10}, effectively a source of one electron and two AO's, can be replaced by other sources of one electron and two AO's like $Cu(PR_3)_2$[19] or $[Cr(CO)_4]^-$[20] [Figure 2(c)] while a source of two electrons and three AO's, like a BH unit of B_5H_9, can be replaced by other sources of two electrons and three AO's such as $Fe(CO)_3$[21] or $Co(C_5H_5)$.[22,27]

2.2.2. The Polyhedral Skeletal Electron Pair Theory (PSEPT)[1-9]

Like Lipscomb's topological approach, the polyhedral skeletal electron pair theory (PSEPT) focuses attention on the number of electrons associated with the skeletal bonding. However, instead of allocating these electrons to localized 2- and 3-center bonds, it simply uses their total number as a guide to the shape of the skeleton, exploiting (and attempting to explain) the link between electron numbers and shape noted at the end of Section 2.1 above, i.e., that the polyhedra on which the structures are based have one vertex fewer than the number of skeletal bond pairs available.

Rationalization of this link between shape and electron numbers is simplest for the *closo* series of borane anions $B_pH_p^{2-}$ and carboranes $C_2B_{p-2}H_p$. Their closed polyhedral structures [Figure 1(a)] clearly represent the arrangements of **p** BH (or CH) units that make the most effective use of their **(p+1)** skeletal bond pairs. This is not at all obvious from localized bond treatments, which require these **(p+1)** bond pairs to be allocated to three 2-center BB (or BC or CC) bonds and to **(p−2)** 3-center BBB (or BBC or BCC) bonds, numbers ill-suited to match appropriate proportions of the 3**(p−2)** polyhedral edges and 2**(p−2)** polyhedral faces among which they must be distributed. However, molecular orbital (MO) treatments have shown that the symmetries of these polyhedra are appropriate to generate **(p+1)** bonding MO's, with a sizeable energy gap between HOMO and LUMO, from the basis set of 3**p** AO's available. Each BH or CH unit supplies three AO's; an *sp* hybrid AO pointing towards the polyhedron center, and a pair of **p** AO's perpendicular to this, orientated tangentially with respect to the pseudospherical surface of the polyhedron—see Figure 3(a). Of these $(p+1)$ bonding MO's, one unique MO, of A symmetry, results from a fully in-phase combination of the **p** radially oriented AO's. The remaining **p** bonding MO's result from combinations of the 2**p** tangentially orientated AO's, with occasional stabilizing contributions from the radial AO's where the symmetry permits this. This is illustrated in Figure 3(b) for the octahedral $B_6H_6{}^{2-}$.[23] The unique A_{1g} MO arises from the in-phase combination of the inward-pointing sp hybrid AO's. The T_{2g} MO's are derived solely from "σ-bonding" combinations of **p** AO's around squares of boron atoms. The T_{1u} MO's correspond to fully in-phase "π-bonding" combinations of **p** AO's around such squares, stabilized by contributions from the sp hybrid AO's of the remaining pairs of boron atoms located above and below the centers of these squares.

In the case of the polyhedral-fragment *nido, arachno,* and *hypho* species, the symmetries of the fragments are such as to generate the same numbers of skeletal bonding MO's as the complete parent polyhedra, provided that the **q** inner-sphere hydrogen atoms, whether terminal or bridging, are situated around the open face in such a manner as to match the symmetry of the skeletal atoms, so stabilizing some of the skeletal bonding MO's without adding to their number. For example, the square pyramidal arrangement of the five BH units of B_5H_9 and the butterfly-shaped arrangement of the four BH units of B_4H_{10}, like the

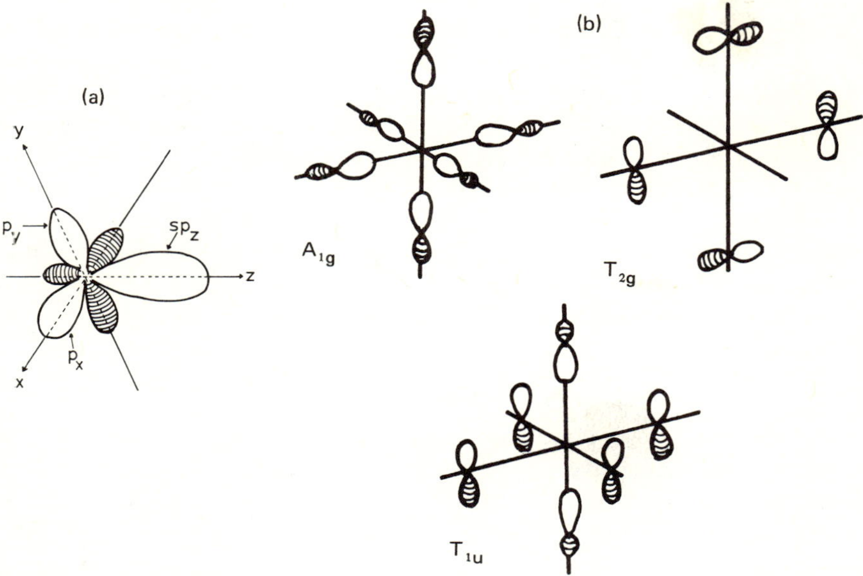

Figure 3. Orbitals used in skeletal bonding: (a) The radially orientated sp hybrid AO, and tangentially oriented p AO's that a BH unit can use for skeletal bonding; (b) the skeletal bonding MO's of $B_6H_6{}^{2-}$.

octahedral arrangement of the six BH units of $B_6H_6{}^{2-}$, are appropriate to generate seven skeletal bonding MO's in each case. Again, the shapes adopted are those that evidently make best use of the electrons available.

Given that the structures of the *closo-*, *nido-*, and *arachno*-boranes and carboranes represent those most appropriate for clusters of **q** atoms held together by **(p+1)**, **(p+2)**, and **(p+3)** skeletal bond pairs, respectively, then similar structures may be expected for other clusters with similar atom:electron ratios, provided that their cluster atoms, like the boron and carbon atoms of boranes and carboranes, can supply a radially orientated orbital of cylindrical symmetry and a pair of tangentially orientated orbitals of π-symmetry.

It is now evident that a wide range of other atoms and groups can replace the BH or CH groups of boranes and carboranes. The systems that first illustrated this were the metallaboranes and metallacarboranes whose structural types and electron numbers are discussed in the following section.

2.2.3. Skeletal Electron Numbers in Metallacarboranes

The structures of the first metallacarboranes prepared (during K. P. Callahan and M. F. Hawthorne's pioneering researches[24] in the 1960's) and the route by which they were synthesized showed how boron atoms of carborane clusters could be replaced by metal atoms, and indicated the relationship between metallacarboranes and metal–hydrocarbon π-complexes, although further work was necessary before the full implications of these relationships became apparent.

These first metallacarboranes were derivatives of the *closo*-icosahedral carborane $C_2B_{10}H_{12}$, base degradation of which removed a boron atom, leaving the *nido*-icosahedral-fragment carborane anion $C_2B_9H_{11}{}^{2-}$. Reaction of this anion with suitable metal cations regenerated the icosahedron by placing a metal atom where originally there had been a boron atom (Figure 4):[24]

$$C_2B_{10}H_{12} \xrightarrow[\text{(ii) NaH}]{\text{(i) NaOR/ROH}} C_2B_9H_{11}{}^{2-} \begin{cases} \xrightarrow{Fe^{2+} + C_5H_5{}^-} [(C_5H_5)Fe(C_2B_9H_{11})]^- \\ \xrightarrow{\frac{1}{2}Fe^{2+}} [Fe(C_2B_9H_{11})_2]^{2-} \end{cases}$$

Conceptually, this work was important in three respects. It showed that a BH unit of a carborane could be replaced by metal units {in this case the anionic units $[Fe(C_5H_5)]^-$ or $[Fe(C_2B_9H_{11})]^{2-}$}; it showed that a metal ion could bond to the three boron and two carbon atoms of the open face of the *nido* anion $[C_2B_9H_{11}]^{2-}$ in the same way that it could bond to the cyclopentadienide anion $[C_5H_5]^-$, forming metallocene-like products; and it showed how a metal atom could link two polyhedral clusters together by occupying a common vertex of both. Although the analogy with cyclopentadienyl complexes prompted a

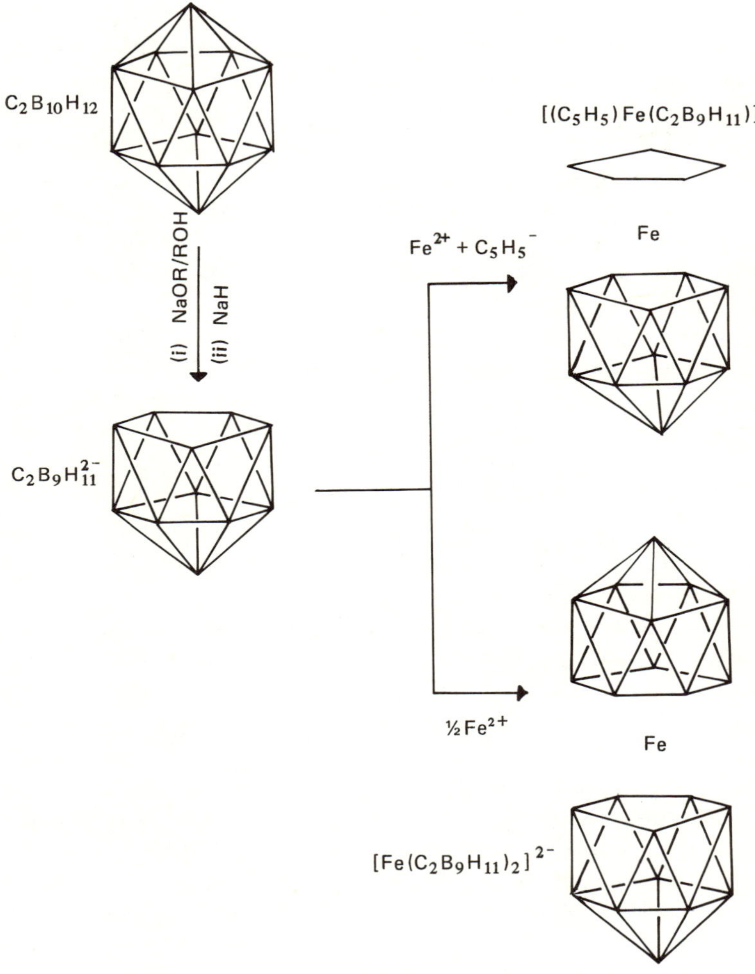

Figure 4. The formation of the *closo*-metallacarboranes $[(\eta^5-C_5H_5)Fe(C_2B_9H_{11})]^-$ and $[Fe(C_2B_9H_{11})_2]^{2-}$ from the *closo*-carborane $C_2B_{10}H_{12}$.

coordination chemist's view of the bonding, in that the $[C_2B_9H_{11}]^{2-}$ ligand was seen to be able to function as a six-electron ligand to a suitable metal acceptor, the complementary cluster chemist's interpretation, that the anionic units $[Fe(\eta^5-C_5H_5)]^-$ and $[Fe(\eta^5-C_2B_9H_{11})_2]^{2-}$ could function, like a BH unit, as sources of two electrons and three AO's for use in cluster bonding, was also apparent.

It soon emerged that there was a wide range of metal units capable of formally replacing the BH units of carboranes, and many *closo*- and several *nido*-metallacarboranes were isolated. Neutral BH analogs included $Fe(CO)_3$,

$Co(\eta^5-C_5H_5)$, $Ni(PR_3)_2$, $BeNMe_3$, AlR, GaR, Sn, and Pb, and species isoelectronic with these. What they had in common was that, like a BH unit, they contained four electrons too few to fill the valence shell of their metal atoms. In the terminology of transition metal coordination chemistry, the transition metal units found capable of replacing a BH unit were typically (although not exclusively) 14-electron species. The main group units contained four valence shell electrons rather than the octet expected for a closed shell configuration. If required to make available three AO's for cluster bonding, both types would necessarily function as sources of two electrons. Units capable of taking the place of CH units in carborane clusters, and also functioning as sources of three electrons and three AO's, included $Co(CO)_3$, $Ni(\eta^5-C_5H_5)$, $Cu(PR_3)_2$, BCO, P, As, and Sb. The formulas of some representative examples of metallaboranes and metallacarboranes with *closo* structures are given in Table 1;[25-74] Table 2 lists examples with *nido* structures.[21,22,28,75-94]

The examples listed in Tables 1 and 2, like the majority of metallaboranes and metallacarboranes, have the structures expected if their metal atoms conform with the 18-electron or octet rules. Exceptions that do not conform with these rules are quite common (see the following), but nevertheless the assumption that all the available metal orbitals are filled allows the expected structure of a metallacarborane to be deduced from its molecular formula. For example, species $C_2B_{p-2}H_p(ML_n)_x$, where the unit ML_n, like a BH unit, can contribute two skeletal bonding electrons, are expected to have *closo* structures based on polyhedra with $(p+x)$ vertices. Species $C_2B_{p-2}H_{p+2}(ML_n)_x$, having $(p+x+2)$ electron pairs to bond $(p+x)$ skeletal atoms, are expected to have *nido* structures, as are species $C_2B_{p-2}H_p(M'L_n)_2$, in which the unit $M'L_n$ is a source of three skeletal electrons. More generally, the number of electrons available for skeletal bonding can be deduced by adding the number of electrons in the valence shell of each skeletal atom to the number of electrons furnished by the ligands and subtracting two for each main group metal and twelve for each transition metal. Comparison of the total number of skeletal electrons with the number of skeletal atoms then allows the structural type (*closo, nido, arachno*) and fundamental polyhedron (one vertex fewer than the number of skeletal bond pairs) to be deduced.

Tables such as Tables 3 and 4[1] have been constructed to facilitate electron counting, and structural predictions, by listing the number of skeletal bonding electrons contributed by the various likely types of cluster unit. We have already noted that units $Fe(CO)_3$ or $[Fe(C_5H_5)]^-$ can contribute two skeletal electrons. Table 4 shows that neutral units $M(CO)_2$, $M(\eta^5-C_5H_5)$, $M(CO)_3$, and $M(CO)_4$, where $M = Fe$, Ru, or Os, are formally capable of contributing 0, 1, 2, and 4 skeletal electrons, respectively. Similar units containing cobalt-group metal atoms contribute an extra electron apiece.

Although Tables 3 and 4 facilitate electron counting, they need to be used with caution. For example, although a unit $Co(CO)_4$ may formally be a source

Table 1. Some Metallaboranes and Metallacarboranes with *closo* Structures

Number of skeletal atoms	Shape	Examples
6	Octahedron	$B_5H_3(CO)_2Fe(CO)_3$;[25] $B_4H_6Co_2(\eta^5-C_5H_5)_2$;[26,27] $B_5H_5Co_3(\eta^5-C_5H_5)_3$;[27–29] $B_3H_3Co_3(\eta^5-C_5H_5)_3BH$;[26–28] $C_2B_3H_5Co(\eta^5-C_5H_5)$;[30] $C_2B_3H_5Fe(CO)_3$ [30]
7	Pentagonal bipyramid	$C_2B_4H_6Fe(CO)_3$;[31] $C_2B_4H_6MMe$ (M=Ga, In);[32] $C_2B_4H_6Ni(PPh_3)_2$;[30] $C_2B_3H_5Fe_2(CO)_6$;[30] $C_2B_3H_5Co_2(\eta^5-C_5H_5)_2$;[30,33] $C_2B_4H_6Pt(PEt_3)_2$;[34] $C_2B_3Me_2H_3CoFeH(\eta^5-C_5H_5)_2$;[35] $C_2B_4Me_2H_3RCo(C_2B_3H_5Me_2)$;[36] $(C_2B_4Me_2H_4)_2FeH_2$;[37] $C_3B_3H_5MeMn(CO)_3$;[38] $C_4BH_3EtPhMn_2(CO)_6$ [39]
8	Dodecahedron	$C_2B_4Me_2H_4SnCo(\eta^5-C_5H_5)$ [40]
9	Tricapped trigonal prism	$C_2B_6H_8Co(\eta^5-C_5H_5)$;[41,42] $[C_2B_6H_8Mn(CO)_3]^-$;[43] $[(C_2B_6H_8)_2Co]^-$;[42] $C_2B_6H_8Pt(PMe_3)_2$;[44] $C_2B_5H_7Co_2(\eta^5-C_5H_5)_2$;[45] $[CB_7H_8Co(\eta^5-C_5H_5)]^-$ [46]
10	Bicapped square antiprism	$[B_9H_9Ni(\eta^5-C_5H_5)]^-$;[47] $CB_7H_8CoNi(\eta^5-C_5H_5)_2$;[48] $[(C_2B_7H_9)_2Co]^-$;[49] $C_2B_7H_9Co(\eta^5-C_5H_5)$;[41,42] $C_2B_7H_7Me_2Fe(CO)_3$;[50] $C_2B_6H_8Co_2(\eta^5-C_5H_5)_2$;[41,51] $C_2B_5H_7Co_3(\eta^5-C_5H_5)_3$ [30]
11	Octadecahedron	$[CB_9H_{10}Co(\eta^5-C_5H_5)]^-$;[52] $[(C_2B_8H_{10})_2Co]^-$;[42] $C_2B_8H_{10}Co(\eta^5-C_5H_5)$;[41,42] $C_2B_8H_{10}IrH(PPh_3)_2$;[53] $C_2B_7H_9Co_2(\eta^5-C_5H_5)_2$ [54]
12	Icosahedron	$[B_{11}H_{11}Ni(\eta^5-C_5H_5)]^-$;[47] $[B_{10}H_{10}CoNi(\eta^5-C_5H_5)_2]^-$;[47] $C_2B_9H_{11}ML_n$ [$ML_n = Co(\eta^5-C_5H_5)$,[55,56] $Pt(PR_3)_2$,[57] $Ni(PR_3)_2$,[58] Ge, Sn, Pb,[59] Tl^-,[60] AlEt,[61] $BeNMe_3$,[62] $Ru(CO)_3$,[63] $RhH(PEt_3)_2$];[64] $[(C_2B_9H_{11})_2M]^{x-}$ [M = Fe(II), Co(III), Ni(IV)];[65–68] $[(CB_{10}H_{11})_2M]^{y-}$ [M = Co(III), Ni(IV)];[69] $[(B_{10}H_{10}S)_2M]^{z-}$ [M = Fe(II), Co(III)];[70] $B_{10}H_{10}SCo(\eta^5-C_5H_5)$;[70] $C_2B_8H_{10}Co_2(\eta^5-C_5H_5)_2$;[42,54,71] $C_2B_7H_9Co_3(\eta^5-C_5H_5)_3$;[54] $CB_9H_{10}AsCo(\eta^5-C_5H_5)$ [42,71]

(continued overleaf)

Table 1 (*cont.*)

Number of skeletal atoms	Shape	Examples
13	1,5,6,1 polyhedron	$C_2B_{10}H_{12}Co(\eta^5-C_5H_5)$;[41,72] $C_2B_9H_{11}Co_2(\eta^5-C_5H_5)_2$[54]
14	Bicapped hexagonal antiprism	$C_2B_{10}H_{12}Co_2(\eta^5-C_5H_5)_2$;[73] $C_4B_8H_8Me_4Fe_2(\eta^5-C_5H_5)_2$[74]

of five skeletal electrons, it is more likely to act as a monovalent substituent on the periphery of a cluster than to be incorporated into the cluster skeleton, unless some of the carbonyl groups are acting as bridges between the cobalt atom and other skeletal atoms.

Table 2. Some Metallaboranes and Metallacarboranes with *Nido* Structures

Number of skeletal atoms	Shape	Examples
4	Butterfly	$B_3H_8Mn(CO)_3$[75]
5	Square pyramid	$B_4H_8Fe(CO)_3$;[21] $B_4H_8Co(\eta^5-C_5H_5)$;[22,27] $B_3H_7Fe_2(CO)_6$[76]
6	Pentagonal pyramid	$B_5H_{10}Fe(\eta^5-C_5H_5)$;[77] $B_5H_9Co(\eta^5-C_5H_5)$;[78] $C_2B_3H_7Fe(CO)_3$;[79] $C_2B_3H_7Co(\eta^5-C_5H_5)$;[80] $C_2B_2R_4SFe(CO)_3$;[81] $C_3B_2R_5Ni(\eta^5-C_5H_5)$;[82] $C_4BH_5Fe(CO)_3$;[83] $C_2B_3H_5Me_2Co(C_2B_4H_3Me_2R)$[36]
7	Hexagonal pyramid	$B_3N_3Et_6Cr(CO)_3$;[84] $B_2Me_2N_2Me_2C_2Et_2Cr(CO)_3$;[85] $C_5H_5BPhMn(CO)_3$[86]
9	Capped square antiprism	$B_5H_5Ni_4(\eta^5-C_5H_5)_4$;[199] $C_2B_6H_6R_2Pt(PR_3)_2$[88]
10	$B_{10}H_{14}$-type	$B_9H_{13}Co(\eta^5-C_5H_5)$;[89] $B_9H_{11}PEt_3Pt(PEt_3)_2$;[90] $C_2B_7H_{11}Co(\eta^5-C_5H_5)$;[91] $C_2B_7H_9Me_2Ni(PR_3)_2$[92]
11	Icosahedral fragment	$[(B_{10}H_{12})_2Zn]^{2-}$;[93] $[B_{10}H_{12}Ni(\eta^5-C_5H_5)]^-$;[94] $B_9H_{10}SPtH(PEt_3)_2$[90]

Table 3. Skeletal Electron Contributions ($v + x - 2$) that Main Group Cluster Units May Make[a]

Group number ($= v$)	Element	E ($x = 0$)	EH, EX ($x = 1$)	EH$_2$, EL ($x = 2$)
1	Li, Na		0	1
2	Be, Mg, Zn, Cd	0	1	2
3	B, Al, Ga, In, Tl	1	2	3
4	C, Si, Ge, Sn, Pb	2	3	4
5	N, P, As, Sb, Bi	3	4	
6	O, S, Se, Te	4	5	
7	F, Cl, Br, I	5		

[a] v = number of valence shell electrons on E; x = number of electrons donated by ligands; X = a one-electron ligand, L = a two-electron ligand.

Table 4. Skeletal Electron Contributions ($v + x - 12$) that Transition Metal Cluster Units May Make[a]

Valence shell electrons ($= v$)	Transition metal	M(CO)$_2$ ($x = 4$)	M(η^5–C$_5$H$_5$) ($x = 5$)	M(CO)$_3$ ($x = 6$)	M(CO)$_4$ ($x = 8$)
6	Cr, Mo, W		−1	0	2
7	Mn, Tc, Re	−1	0	1	3
8	Fe, Ru, Os	0	1	2	4
9	Co, Rh, Ir	1	2	3	5
10	Ni, Pd[b], Pt[b]	2	3	4	

[a] v = number of valence shell electrons on M; x = number of electrons donated by ligands.

[b] The tendency of these elements to form 16-electron complexes often boosts their skeletal electron contribution by 2.

2.3. Isolobal Relationships between Cluster Units

The classification of units like Fe(CO)$_3$, Co(η^5–C$_5$H$_5$), Ni(PR$_3$)$_2$, and isoelectronic species as formally analogous to a BH unit, capable of functioning as sources of two electrons and three AO's for skeletal bonding, was originally suggested as an electronic bookkeeping device to facilitate electron-counting, and has occasionally been misinterpreted as implying that these transition metal units use only their s and p AO's for skeletal bonding. That this is not so has been shown by an analysis[95,96] of the types of orbital that pyramidal units such as $M(CO)_3$, $M(\eta^5-C_5H_5)$, and $M(CO)_4$ can make available for bonding to neighboring atoms, whether in cluster systems or other types of compound. The orbitals that a pyramidal $M(CO)_3$ unit can make use of in cluster bonding may be viewed in two ways, illustrated in Figure 5. On the one hand, the pyramidal $M(CO)_3$ unit can be regarded as a fragment of an octahedral $M(CO)_6$ complex from which three adjacent carbonyl ligands have been removed. As such, it has a set of three equivalent (d^2sp^3) AO's available for bonding in the directions of the missing ligands, much as a BH unit, with sp^3-hybridized boron, would have a set of three equivalent AO's available for skeletal bonding. Conversely, the $M(CO)_3$ unit can be regarded as providing one unique ($sp_zd_{z^2}$ hybrid) AO pointing along the three-fold axis away from the carbonyl ligands, and a pair of pd hybrid AO's (p_xd_{xz} and p_yd_{yz}) as the counterparts of the radially orientated sp_z, and tangential p_x and p_y AO's of a BH unit.

Despite the d-orbital component of the transition metal cluster units, the lobal characteristics of the available orbitals as viewed from the cluster center are like those of a BH unit, and the term "isolobal" has been coined to describe the relationship between $Fe(CO)_3$ and BH.[95,96] Units are said to be isolobal if they have equal numbers of frontier orbitals of similar symmetry properties, energies, and extent in space. A pyramidal η^5-cyclopentadienyl–metal unit like $Co(\eta^5-C_5H_5)$ is a member of the same isolobal family as $Fe(CO)_3$ and BH. However, a butterfly-shaped metal–tetracarbonyl unit $M(CO)_4$ (where M = Cr, Mo, or W), which in electron counting terms resembles the units $Fe(CO)_3$ and $Co(\eta^5-C_5H_5)$ in being capable of contributing a pair of skeletal bonding electrons, nevertheless differs from the latter two units in not being isolobal with them. Regarded as a fragment of an octahedral parent species $M(CO)_6$, it has two (empty) d^2sp^3 equivalent AO's available through which it can bond to other units. In this respect it resembles a V-shaped BH_2^+ unit more closely than it does a BH unit; indeed, such a butterfly-shaped $M(CO)_4$ unit and V-shaped BH_2 or CH_2 units are isolobal species with 0, 1, and 2 electrons, respectively, available for use in the two AO's that each unit can use for bonding to other units. Similarly, units like pyramidal BH_3 or CH_3, with one AO available to bond to other atoms, may be regarded as isolobal with a square pyramidal $M(CO)_5$ unit (M = Cr, Mo, W, Mn, Tc, or Re, etc.)

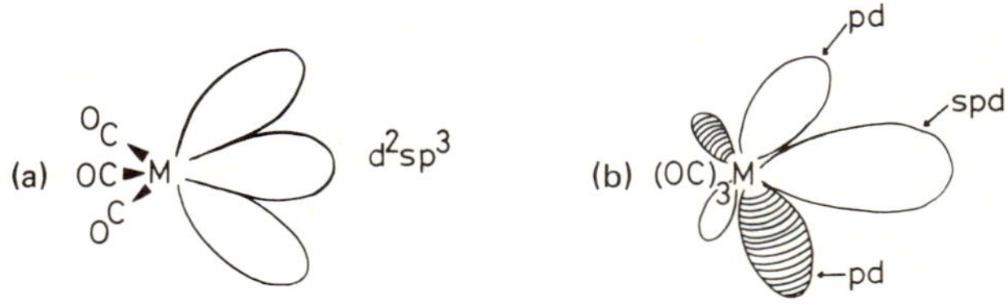

Figure 5. Orbitals made available for cluster bonding by a pyramidal $M(CO)_3$ unit: (a) three equivalent d^2sp^3 hybrid AO's; (b) A (radial) $p_zd_{z^2}$ hybrid AO and two (tangential) pd hybrid AO's.

Such isolobal relationships are helpful in showing where, in a relatively complicated molecule, one unit may be replaceable by another. For example, they not only allow one to rationalize the substitution of a BH unit of a *closo*-carborane by a metal unit like $Fe(CO)_3$ or $Co(\eta^5-C_5H_5)$, but also allow one to see such metallaboranes as $B_3H_8Mn(CO)_4$ or $B_3H_8Cu(PR_3)_2$ as derived from B_4H_{10} by replacement of a BH_2 unit by the isolobal $Mn(CO)_4$ or $Cu(PR_3)_2$. They do not invalidate the electron counting arguments already presented, and summarized in Tables 3 and 4. A BH_2 unit can be regarded either as a source of one electron and two AO's, or as a source of three electrons and three AO's, for cluster bonding, provided that, in the latter case, one of the two B—H bonds lies in the pseudospherical surface of the cluster.

3. RELATED CLUSTER SYSTEMS

We have already noted that metallaboranes and metallacarboranes bridged the gap between boron clusters on the one hand and various other cluster and ring systems on the other, by demonstrating the wide range of units that, like the BH and CH groups of boranes and carboranes, could form clusters effectively by use of three AO's apiece. This allowed such apparently disparate systems as boranes and metal–carbonyl clusters, metal–hydrocarbon π-complexes, nonclassical carbo-cations, and anionic tin clusters to be seen as members of the same general family, allowing developments in one area to be used to make predictions or stimulate new work in the other areas. In the present section we consider, in turn, some metal–carbonyl clusters, metal–hydrocarbon π-complexes, and main group clusters to illustrate their relationship to boranes and metallaboranes, noting both similarities and differences.

3.1. Metal–Carbonyl Clusters

Metal-carbonyl cluster chemistry is a rapidly expanding field too complex to review in detail here. A useful lead into the literature is provided by Ref. 97. Known systems range from simple triangular or tetrahedral clusters for which localized 2-center bond descriptions are quite adequate, through to high nuclearity clusters with very complicated structures for which no simple bonding schemes are as yet available. Examples of the former type are provided by the triangular dodecacarbonyls $M_3(CO)_{12}$ formed by iron, ruthenium, and osmium, and the tetrahedral dodecacarbonyls $M_4(CO)_{12}$ of cobalt, rhodium, and iridium.[97] The anionic clusters $[Rh_{13}(CO)_{24}H_{5-n}]^{n-}$,[98] $[Rh_{15}(CO)_{28}C_2]^-$,[99] and $[Pt_{15}(CO)_{30}]^{2-}$[100] fall into the latter category. It is between these extremes that metal analogs of boron clusters are to be found, and these intermediate-sized clusters will be our concern here. The formulas of some representative examples, classified by the number of skeletal electron pairs they formally contain, are given in Tables 5 and 6.[97,101-124] Their skeletal structures, illustrated in Figure 6, not only show the dependence on electron numbers expected of

Table 5. Some Metal–Carbonyl Clusters that Formally Contain Six Skeletal Bond Pairs[a]

A[b]	V[c]	Type	Shape	Examples
6	84	Capped *closo*	Capped trigonal bipyramid	$Os_6(CO)_{18}$ [101]
5	72	*Closo*	Trigonal bipyramid	$Os_5(CO)_{16}$; [102] $[HOs_5(CO)_{15}]^-$ [103]
4	60	*Nido*	Tetrahedron	$M'_4(CO)_{12}$; [97][d] $[M'_4(CO)_{10}H_2]^{2-}$ [104] $M_4(CO)_{12}H_4$; [105][e] $(\eta^5-C_5H_5)M'_2M_2(CO)_8$ [97] $[Re_4(CO)_{12}H_6]^{2-}$ [106]
3	48	*Arachno*	Triangle	$M'_3(CO)_{12}$ [97]

[a] Species for which $V = 12A + 12$.
[b] A = number of skeletal atoms.
[c] V = total number of valence shell electrons.
[d] M' = Co, Rh, Ir.
[e] M = Fe, Ru, Os.

Table 6. Some Metal–Carbonyl Clusters that Formally Contain Seven Skeletal Bond Pairs[a]

A[b]	V[c]	Type	Shape	Examples
10	134	Tetracapped *closo*	Tetracapped octahedron	$[Os_{10}(CO)_{24}C]^{2-}$ [107]
7	98	Capped *closo*	Capped octahedron	$Os_7(CO)_{21}$; [108] $[Rh_7(CO)_{16}]^{3-}$; [109] $[Rh_7(CO)_{16}I]^{2-}$ [110]
6	86	*Closo*	Octahedron	$Rh_6(CO)_{16}$; [111] $H_2Ru_6(CO)_{18}$; [112] $[HM_6(CO)_{18}]^-$; [113,114][d] $[Os_6(CO)_{18}]^{2-}$; [114] $[Co_6(CO)_{14}]^{4-}$; [115] $[Co_4Ni_2(CO)_{14}]^{2-}$; [116] $[Ni_6(CO)_{12}]^{2-}$; [117] $Ru_6(CO)_{17}C$; [118] $[Fe_6(CO)_{16}C]^{2-}$ [119]
6	86	Capped *nido*	Capped square pyramid	$H_2Os_6(CO)_{18}$; [114] $Os_6(CO)_{16}(CMe)_2$ [120]
5	74	*Nido*	Square pyramid	$M_5(CO)_{15}C$ [121,122]
4	62	*Arachno*	Butterfly	$Os_4(CO)_{12}H_3I$; [123] $[Fe_4(CO)_{13}H]^-$ [124]

[a] Species for which $V = 12A + 14$.
[b] A = number of skeletal atoms.
[c] V = total number of valence shell electrons.
[d] M = Fe, Ru, Os.

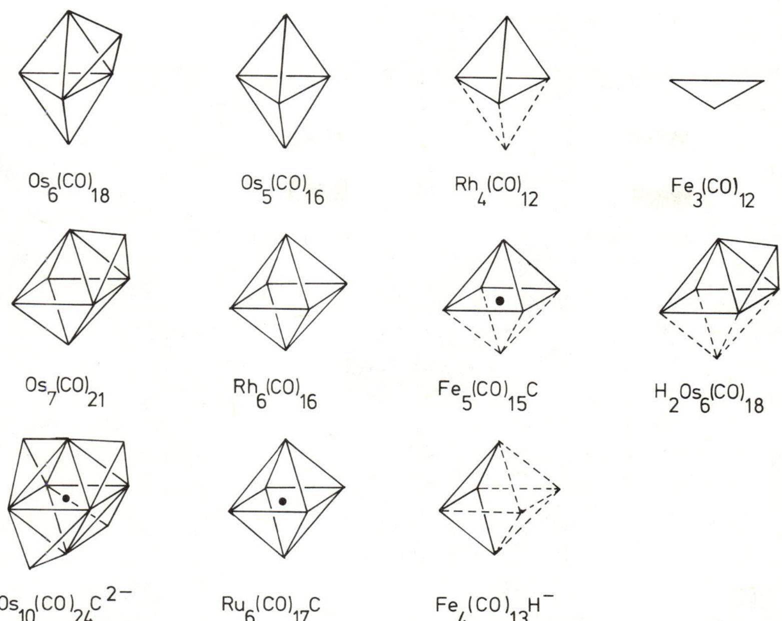

Figure 6. Skeletal structures of metal–carbonyls formally related to boranes with six or seven skeletal bond pairs.

borane-type clusters, but add significantly to the range of structural types.

Before considering these types, and noting their implications for metallaborane chemistry, it should be stressed that the assignment of a skeletal bonding role to a specified number of electrons in metal-carbonyl clusters is necessarily arbitrary. Unlike boranes and carboranes, which contain relatively few valence shell electrons, readily assigned to either ligand bonding or skeletal bonding, metal-carbonyl clusters contain so many valence shell electrons as to fill not only the skeletal bonding MO's, but also some that are non- or antibonding with respect to skeletal interactions. The classification of the systems in Tables 5 and 6, as six and seven skeletal bond pair systems, respectively, has been made using the electron counting approach indicated in Tables 3 and 4, i.e., assuming each metal atom uses three AO's for skeletal bonding, which appears justified in view of the isolobal relationships outlined above. However, the electrons dismissed as nonbonding in this approach may well make a small net bonding contribution to the skeleton. For example, for such octahedral clusters as $Rh_6(CO)_{16}$,[111] $H_2Ru_6(CO)_{18}$,[112] and $[Os_6(CO)_{18}]^{2-}$,[14] the skeletal bonding may well be rather stronger than the effect of seven bond pairs spread over twelve 2-center links (the octahedral edges) may seem to imply, although a localized bond description, allowing eleven 2-center electron pair bonds to resonate among the twelve octahedral edges, probably overestimates the skeletal bond orders.[7,125] It has been suggested that these octahedral metal clusters resemble $B_6H_6^{2-}$ less in the number of filled bonding MO's than in the number of empty antibonding MO's.[115,126]

The examples in Figure 6 and Tables 5 and 6 nevertheless clearly reveal a link between the structures of metal–carbonyl clusters and the total number of valence shell electrons(V) they contain. Designating the number of skeletal atoms by S, one can see that systems for which $V = 12S + 12$ have structures that can be related to a trigonal bipyramid, while systems for which $V = 12S + 14$ have structures that can be related to an octahedron, the *closo* systems in the two series having $V = 72$ and $V = 86$, respectively.

One important difference between metal–carbonyl clusters and their boron counterparts is that the former may contain core atoms whereas the latter do not. For example, an octahedrally coordinated carbon atom is located at the center of the *closo* clusters $Ru_6(CO)_{17}C^{[118]}$ and $[Fe_6(CO)_{16}C]^{2-}$,[119] where its s and p AO's can stabilize the A_{1g} and T_{1u} skeletal bonding MO's, while a pentacoordinate carbon atom is located just below the basal plane of the square pyramidal *nido* cluster $Fe_5(CO)_{15}C$.[121,122] In both cases the carbon atom clearly uses all its valence shell electrons and AO's for cluster bonding. Similar core atoms have not been found at the centers of boranes or carboranes, or of their mono- or dimetallo derivatives, nor are they expected, since the cavity for the core atom would either be too small (in the case of octahedral or similarly sized clusters) or else (in the case of icosahedral or similarly large clusters) offer such a highly coordinated environment as to allow the carbon valence shell AO's to stabilize too few of the skeletal bonding MO's. Nevertheless, core atoms might be accommodated in metallaboranes or metallacarboranes rich enough in metal atoms to generate suitably large cavities. For example, square pyramidal species $Fe_3(CO)_9(BH)_2C$ or $Fe_4(CO)_{12}(BH)C$ might be envisaged as possible members of the family of metallaboranes related to B_5H_9 and $Fe_5(CO)_{15}C$, a family that has incidentally been the subject of detailed structural, spectroscopic, and theoretical studies[10] that have done much to illustrate both the validity and limitations of treating BH and $Fe(CO)_3$ units as interchangeable cluster components.

Another distinctive feature of the series of metal–carbonyl clusters in Figure 6 and Tables 5 and 6 is the presence of species with capped structures, in which one or more metal atoms occupy sites capping faces of the fundamental polyhedron. The cluster $Os_6(CO)_{18}$[101] for example has a capped trigonal bipyramidal (or bicapped tetrahedral) structure. The trigonal bipyramid in its structure is intelligible in that this cluster formally contains six skeletal bond pairs, the number appropriate for the five-vertex polyhedron. The position of the "surplus" osmium atom, capping a face, where its three skeletal AO's can be used effectively in bonding to the three adjacent metal atoms, is also readily intelligible, as it can use what would otherwise be skeletally nonbonding electrons. Similar arguments apply to the capped octahedral structures of the seven-atom seven-bond pair systems $Os_7(CO)_{21}$,[108] $[Rh_7(CO)_{16}]^{3-}$,[109] and $[Rh_7(CO_{16}I]^{2-}$,[110] and even to the tetracapped carbonyl carbide $[Os_{10}(CO)_{24}C]^{2-}$.[107] The capped *nido* structure of $H_2Os_6(CO)_{18}$,[114] which may be contrasted with the *closo* structures of $[Os_6(CO)_{18}]^{2-}$ [114] and

$H_2Ru_6(CO)_{18}$,[112] illustrates the subtle ramifications of metal cluster structures, and interestingly provides a skeleton with the eleven edges for which there are bond pairs available on a localized bond treatment.

Of crucial importance in the formation of such capped structures is the availability of electrons and AO's on the capped atoms for the capping atom to exploit. Similar structures are, therefore, unlikely in the case of boranes and carboranes, although possible (and indeed known) for metallaboranes and metallacarboranes, when it significantly is a metal-containing face that is capped.[26,28]

3.2. Metal–Hydrocarbon π-Complexes

Callahan and Hawthorne's earliest studies on metallacarboranes,[24] by showing how the anion $C_2B_9H_{11}{}^{2-}$ resembles $C_5H_5{}^-$ in its capacity to bond pentahapto to metals, indicated that a close relationship exists between metallacarboranes and metal–hydrocarbon π-complexes. This relationship has subsequently been touched on in various articles.[1,5,7,9,128-130] It can be illustrated by considering metal–hydrocarbon π-complexes as mixed metal-carbon clusters whose structures (Table 7 and Figure 7)[131-155] can be classified as *closo, nido, arachno,* etc. The skeletal electron counting follows the usual procedure, for example associating eight skeletal bond pairs with the six skeletal atoms of a pentahapto cyclopentadienyl–metal complex such as $Mn(CO)_3(\eta^5-C_5H_5)$, regarded as a *nido* cluster with a structure based on a pentagonal bipyramid. The skeletal electron count thus includes the five carbon–carbon sigma bonding pairs of the cyclopentadienyl ring as well as the three π-bonding pairs. The latter are those primarily responsible for the metal–ligand bonding, and, therefore, those counted as donated to the metal valence shell on the usual 18-electron rule treatment. (In conventional organometallic/coordination chemistry terminology, the neutral pentahapto-cyclopentadienyl ligand is treated as a source of five electrons that are contributed to the valence shell of the metal, which is equivalent to describing an anionic ligand $(\eta^5-C_5H_5)^-$ as a six-electron ligand). Thus, since all nine metal valence shell AO's are expected to be filled using the skeletal electron counting approach outlined previously (Table 4), organometallic complexes that obey the 18-electron rule[156] will have structures (*closo, nido, arachno,* etc.) that are predictable using the borane analogy. However, complexes in which the metal effectively makes use of fewer AO's, such as the 16-electron complexes that are not uncommon for metals of the cobalt, or more especially the nickel subgroup, will have more open skeletal structures than Table 4 might appear to imply. We shall return to this point later.

Treating metal–hydrocarbon complexes as mixed metal-carbon clusters of the metallaborane type draws attention to some important features, which include:

(i) the coordination-site preferences of different cluster-forming atoms or groups (the carbon atoms tend to be found in low-coordinate sites, while the metal atoms can generally, although not invariably, tolerate the more highly coordinated sites);

(ii) the relationship to aromatic ring systems (charged or uncharged aromatic C_nH_n ring systems like $C_4H_4^{2-}$, $C_5H_5^-$, C_6H_6, and $C_7H_7^+$ are classifiable as *arachno* species); and

(iii) the numbers of electrons and AO's involved in bonding (the division of the eight skeletal bond pairs associated with the six skeletal atoms of $(\eta^5-C_5H_5)Mn(CO)_3$ into five pairs that are C–C σ-bonding, and three pairs that are metal–ligand bonding, underlines the need for the metal to make available three AO's for skeletal bonding.).

The relationship between aromatic ring systems and borane-type clusters can profitably be taken further. Aromatic C_nH_n rings demonstrate one effective way that units like CH that are potentially tervalent can use their three AO's to form only two bonds, each of which has some multiple character. Such rings are not expected to be the most stable structure for related systems if the π-bonding capacity of the ring atoms is markedly less than the σ-bonding capacity, as is generally true for elements lower in the periodic table than carbon, when three-dimensional aggregates of the cluster forming units, therefore, are, more likely. The $Fe(CO)_3$ unit of the 1,3-butadiene complex $(\eta^4-CH_2CHCHCH_2)Fe(CO)_3$ (formally an *arachno* complex for which a planar FeC_4 five-membered ring structure, with two Fe—H—C bonds, would be an alternative possible geometry) prefers to bond moderately strongly to all four carbon atoms of the ligand, rather than very strongly to only two of them. Similar factors are seen to operate in metallaborane or metallacarborane shapes.

Comparison with aromatic ring systems also draws attention to the alternative shapes possible for clusters with nine or more skeletal bond pairs. The shapes of benzene and the cycloheptatrienyl cation $C_7H_7^+$, regarded as the *arachno* species with nine and ten skeletal bond pairs, respectively, are clearly related to bipyramidal parent polyhedra (Figure 7), not the D_{2d} dodecahedron and D_{3h} tricapped trigonal prism of $B_8H_8^{2-}$ and $B_9H_9^{2-}$. The borabenzene complex $(\eta^6-C_5H_5BPh)Mn(CO)_3$[151] and related species[6] are examples of metallacarboranes with nine skeletal bond pairs and structures related to the hexagonal bipyramid rather than to the D_{2d} dodecahedron.

The treatment of pentahapto-cyclopentadienyl complexes as *nido* clusters with structures based on a pentagonal *bi*pyramid as the parent polyhedron prompted the suggestion[128] that aromatic ring systems, in general, might be cappable on both sides, coordinating simultaneously to metal atoms above and below the ring plane. Although examples of such species are still scarce, impressive progress has been made with the building of multidecker sandwich compounds in which five-membered C_2B_3-, C_4B-, and C_2B_2S-rings have been sandwiched between pairs of metal atoms.[6]

As a further corollary of the relationship between aromatic ring systems and borane-type clusters, one can regard the bicapped antiprismatic polyhedra of $B_{10}H_{10}^{2-}$ and $B_{12}H_{12}^{2-}$ and their carba- and metalla-derivatives as derivable

Table 7. Metal–Hydrocarbon π-Complexes and Related Hydrocarbon Systems Classified as Borane-Type Clusters

Fundamental polyhedron	S^a	*Closo* species	*Nido* species	*Arachno* species
Trigonal bipyramid	6	$Fe_3(CO)_9C_2Ph_2$ [131]	$Co_2(CO)_6C_2Ph_2$; [137] $Ni_2Cp_2C_2Ph_2$; [138][b] $W_2Cp_2(CO)_4C_2R_2$; [139] $C_4Bu^t_4$; [140] $Co(CO)_3C_3Ph_3$; [154] $Co_3(CO)_9CR$ [155]	$Pt(PPh_3)_2C_2H_4$; [138] $C_3H_7^+$
Octahedron	7	$Co_4(CO)_{10}C_2Et_2$; [132] $Ru_4(CO)_{12}C_2Ph_2$; [133] $Ir_4(CO)_5(C_8H_{12})$ (C_8H_{10}) [134]	$CoCp(\eta^4-C_4H_4)$; [138][b] $Fe(CO)_3(\eta^4-C_4H_4)$; [141] $Os_3(CO)_{10}C_2Ph_2$; [142] $C_5H_5^+$ [12,143-147]	$Co(CO)_3(\eta^3$-allyl); bicyclobutane; [150] (cyclobutadiene)$^{2-}$
Pentagonal bipyramid	8	$Fe_3(CO)_8C_4R_4$ [135]	$Mn(CO)_3(\eta^5-C_5H_5)$; [138] $Fe_2(CO)_6(C_4R_2X_2)$; [148] $C_6Me_6^{2+}$ [149]	$Fe(CO)_3$ $(\eta^4-CH_2CHCHCH_2)$ [138]; cyclopentadienyl anion
Hexagonal bipyramid	9		$M(\eta^6-C_6H_6)_2$; $M(CO)_3(\eta^6-C_6H_6)$ [138] $(M=Cr,Mo,W)$; $Mn(CO)_3$ $(\eta^6-C_5H_5BPh)$ [151]	Benzene
Dodecahedron	9	$Fe_4(CO)_{11}(HCCEt)_2$ [136]		$AlMe_2(\eta^3-C_5H_5)$; [152] benzvalene; $C_6Me_6H^+$ [153]
Heptagonal bipyramid	10		$V(CO)_3(\eta^7-C_7H_7)$ [138]	Cycloheptatrienyl cation

^{a} S = number of skeletal electron pairs.
^{b} $Cp = \eta^5-C_5H_5$.

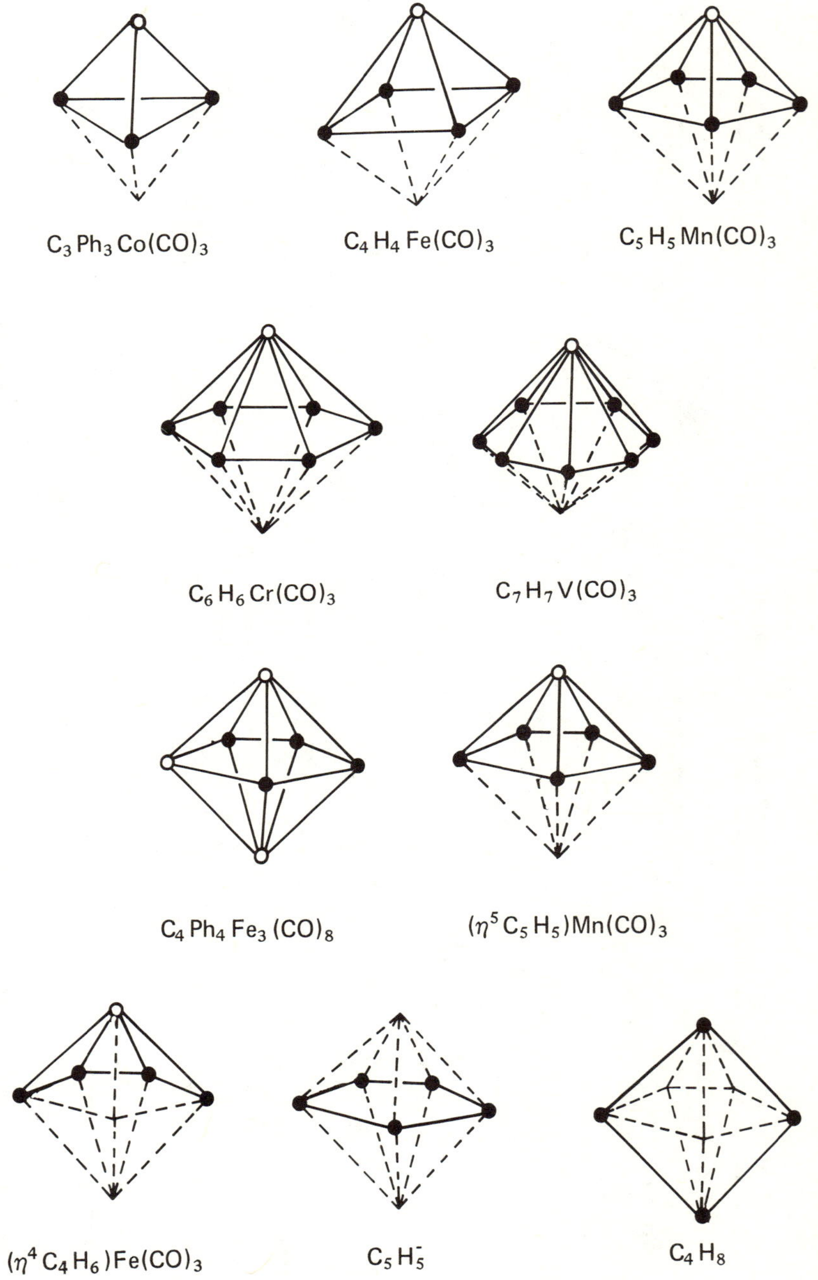

Figure 7. Skeletal structures of metal-hydrocarbon π-complexes and hydrocarbon systems formally related to boranes.

from pairs of pyramidal units, brought together base to base in a staggered configuration (Figure 8). This would correspond to the hypothetical reactions:

$$B_5H_5{}^{4-} + B_5H_5{}^{4-} \longrightarrow B_{10}H_{10}{}^{2-} + 6e^-$$

and

$$B_6H_6{}^{4-} + B_6H_6{}^{4-} \longrightarrow B_{12}H_{12}{}^{2-} + 6e^-$$

The six spare electrons produced in each case become redundant because the π-systems of the basal planes of the *nido* fragments effectively fuse into one system, requiring only three electron pairs altogether, as these fragments are brought together in a staggered configuration. Although these reactions are hypothetical, metal-assisted reactions related to these, between two *nido* fragments brought together by a metal atom initially sandwiched between them, have been used in metallacarborane chemistry to generate a variety of large tetra-carbon species such as $(\eta^5-C_5H_5)CoC_4B_7H_{11}$, $(\eta^5-C_5H_5)_2Co_2C_4B_6H_{10}$,[157] $(\eta^5-C_5H_5)CoC_4Me_4B_7H_6OEt$,[158] and $C_4Me_4B_8H_8$[37] from small dicarba precursors.

In concluding this brief discussion of the relationship between metal–hydrocarbon π-complexes and metallaboranes and metallacarboranes, we should acknowledge that our emphasis has been on the contribution, in terms of AO's and skeletal electrons, that CH, BH, and metal-containing units can make to the cluster as a whole. In this way, noting the preference of carbon atoms for sites

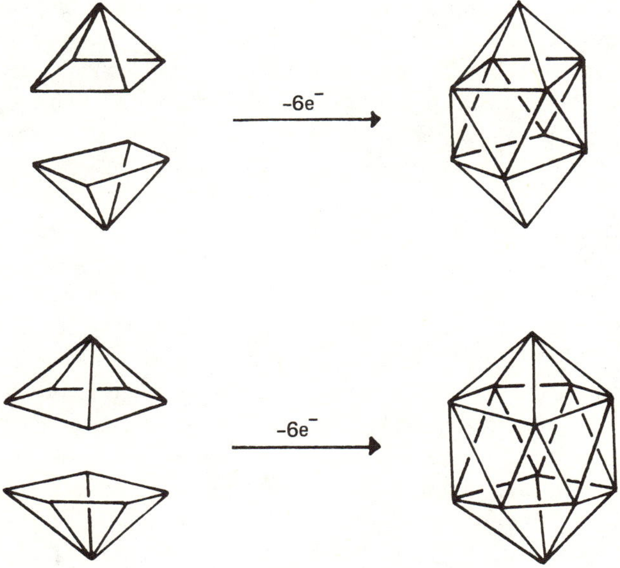

Figure 8. Hypothetical formation of 10- and 12-vertex polyhedra from pairs of smaller *nido* fragments.

of low-coordination number, one can rationalize the shapes defined by the skeletal atoms of mixed metal-carbon clusters. Detailed consideration of the metal-carbon bonding in metal-hydrocarbon π-complexes, or of the metal-boron and metal-carbon bonding in metallaboranes and metallacarboranes, requires a consideration not only of what electrons and AO's the metal units can contribute to the cluster, but what electrons and MO's the rest of the cluster can contribute to the metal. Just as organic ligands are conventionally classified according to their hapticities and the number of electrons they formally contribute to the metal valence shell, so can the borane or carborane residues of metallaboranes and metallacarboranes be usefully classified in a similar manner[159] as a prelude to the detailed treatment of the metal-ligand bonding, as other chapters in the present work illustrate.

3.3. Other Related Main Group Cluster Systems

In addition to the metal carbonyl clusters, metal-hydrocarbon π-complexes, and aromatic systems already considered, there are several other categories of compounds that provide further examples of, and variations on, the borane/metallaborane structural and bonding pattern. They include nonclassical carbocations, anionic clusters of group IV metals, and cationic clusters of group V metals. Their relationship to boranes is considered briefly in this section.

Most carbocations can formally be classified as derivatives of the methyl cation CH_3^+ and are referred to as carbenium ions. Like their short-lived parent, they contain a trigonally coordinated carbon atom as the cationic center, and have no unusual structural features; their bonding can generally be described quite satisfactorily in terms of localized 2-center bonds, although a small number of examples is known with nonclassical, bridged structures in which carbon or hydrogen atoms bridge two potential centers of the cationic charge. It is these systems, and the rarer carbonium ions (formally derivatives of the methanonium cation CH_5^+) that require 3-center bonded descriptions, that are formally analogous to borane clusters.

Two examples have already been noted by inclusion in Table 7, viz. the species $C_5H_5^+$ (theoretical treatments of which[12,143-147] support a square pyramidal structure like that of the isoelectronic B_5H_9), and the dication $C_6Me_6^{2+}$, which is preparable in solution in highly acidic media, for which the ^1H and ^{13}C nmr spectra[149] indicate a pentagonal pyramidal structure like that of B_6H_{10}. Although no definitive structural information, such as might be provided by X-ray crystallographic studies, is as yet available on nonclassical carbocations, there is considerable spectroscopic evidence (e.g., from ^{13}C chemical shifts) for nonclassical structures for several other carbocations also, including those shown in Figure 9. This illustrates their relationship to analogous boranes by showing the 2- and 3-center bond networks that can be used to describe these structures. Detailed discussion of these and other carbocations will be found in Refs. 160-162.

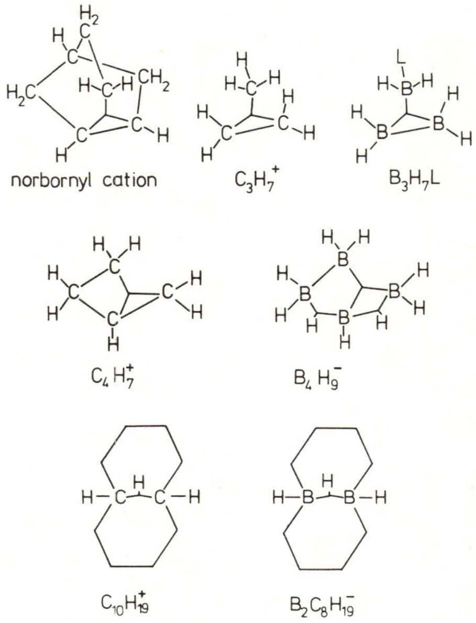

Figure 9. Bond networks for some non-classical carbocations and related boranes.

In Table 3, bare atoms of the group IV elements C, Si, Ge, Sn, and Pb were treated as sources of two electrons for skeletal bonding, and so capable of replacing the BH units of boranes and carboranes. This point is illustrated by the formation of *closo*-metallacarboranes of formulas $SnC_2B_{n-2}H_n$ from $SnCl_2$ and *nido* anions $[C_2B_{n-2}H_n]^{2-}$.[59] Further illustrations are provided by the anionic clusters, of the general formulas M_n^{2-} or M_n^{4-}, formed by germanium, tin, and lead. Although few such systems have yet been structurally characterized, those that have (been structurally characterized) adopt the expected *closo* structures when they are of formulas M_n^{2-} (cf. $B_nH_n^{2-}$), while *nido* structures appear normal for the anions M_n^{4-}.[163–7]

Progressing from group IV to group V, where each atom can supply an extra electron for cluster bonding, one finds that borane-type clusters based on these elements are necessarily cationic. Examples include the *closo*-trigonal bipyramidal Bi_5^{3+},[168] the *arachno*-square antiprismatic Bi_8^{2+},[168] and the *nido*-capped square antiprismatic Bi_9^{5+}.[169,170] Most of the work on such systems has been carried out by J. D. Corbett, who was the first[168] to draw attention to the similar bonding characteristics of a Bi^+ cation and a BH unit.

Yet more distant relatives of boranes and metallaboranes are to be found among the ring and cage systems formed by group VI elements, such as the tetranuclear cationic species S_4^{2+}, Se_4^{2+}, Te_4^{2+}, $Te_2Se_2^{2+}$ etc. (cf. $C_4H_4^{2-}$), which may be regarded as *arachno* species containing seven skeletal bond pairs, and hexanuclear cationic species like Te_6^{4+}, the trigonal prismatic structure of which may be regarded as being appropriate for such a 6-atom, 10-bond pair system. However, since the connectivities of the atoms in these systems are generally significantly lower than those in boranes, localized bond schemes are often adequate to rationalize their shapes. A rationalization based on the close-packing of electron pairs is suggested in Ref. 171.

4. SOME SPECIFIC METALLACARBORANE STRUCTURAL TYPES

The earlier sections of this survey have illustrated the wide range of systems whose structures are classifiable as *closo*, *nido*, *arachno*, etc., like boranes. In the present section we consider some specific systems that have been the subject of more detailed theoretical study, or that reflect more subtle aspects of the structural and bonding pattern.

4.1. Commo Complexes $[M(C_2B_9H_{11})_2]^{n-4}$ with Slipped Structures, and Related Systems

Commo complexes $[M(C_2B_9H_{11})_2]^{n-4}$, in which a metal ion M^{n+} is sandwiched between two *nido* $[C_2B_9H_{11}]^{2-}$ ions, were among the first metallacarboranes to be prepared and structurally characterized.[24] Their structures reflect the number of electrons on the metal ion M^{n+}. When this has a d^6 configuration, as in the case of Fe^{2+}, Co^{3+}, or Ni^{4+}, a ferrocene-like structure is adopted, with the metal on the pseudo-five-fold axis common to the two $[C_2B_9H_{11}]^{2-}$ ligands [Figure 10(a)]. Slipped structures [Figure 10(b)] are

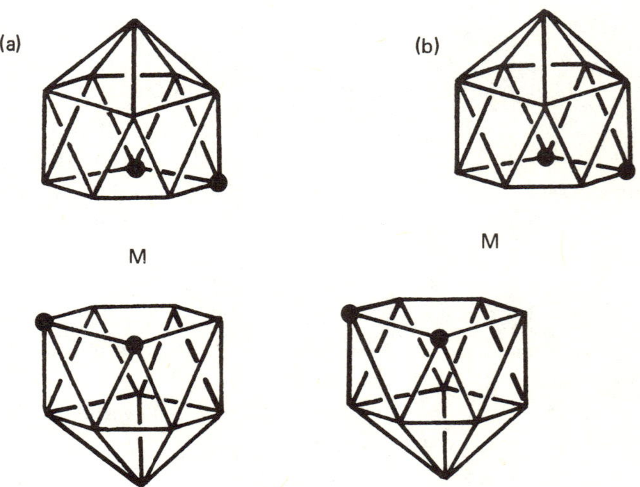

Figure 10. Structures of *commo* species $[M(C_2B_9H_{11})_2]^{n-4}$. (a) Symmetrical *closo* structure (M = Fe^{2+}, Co^{3+}, Ni^{4+}). (b) Slipped *nido* structure (M = Ni^{2+}, Cu^+, Au^{3+}).

found in the case of complexes in which the metal ion has more than six electrons.[172]

These structures can be rationalized at a simplistic level by a skeletal electron count. A d^6 electronic configuration for the sandwiched metal ion allows it to make three empty AO's available for bonding to each icosahedral residue by occupying the vacant icosahedral vertex. Each of the icosahedra linked by the metal on a common vertex thus has the appropriate 13 skeletal bond pairs. Any extra electrons on the metal ion formally increase the number of electrons available for skeletal bonding, thus increasing the number of vertices on the expected fundamental polyhedron. A slipped structure is thus seen as that which places the 12 skeletal atoms (MC_2B_9) on all but one of the 13 vertices of the larger polyhedron that becomes appropriate.

An alternative metal-oriented (as opposed to ligand-oriented) view of the bonding invokes the 18-electron rule[156] and focuses attention on the number of electrons formally donated to the metal ion by the $[C_2B_9H_{11}]^{2-}$ ligands. When bonding pentahapto to a metal ion on the remaining icosahedral vertex, these ligands function as donors of six electrons apiece (the six "π-bonding" electrons associated with the open pentagonal face of the ligand—cf. $[\eta^5-C_5H_5]^-$). Two such ligands can, therefore, be accommodated by a single metal ion only if that ion has six or fewer valence shell electrons to start with. A d^8 metal ion like Ni^{2+}, for example, would acquire a total of 20 valence shell electrons if coordinating simultaneously two $[\eta^5-C_2B_9H_{11}]^{2-}$ ligands, so the slipped structure of $[Ni(C_2B_9H_{11})_2]^{2-}$ is seen as that which, by reducing the hapticities of the ligands, reduces the number of electrons associated with the valence shell of the metal. The bonding problem posed by this nickel complex is thus analogous to that posed by nickelocene $Ni(C_5H_5)_2$.

In the above treatment, it is worth stressing that the neutral $C_2B_9H_{11}$ icosahedral residue functions as a four-electron ligand to metals when it coordinates pentahapto (contrast a neutral pentahapto cyclopentadienyl ligand, which acts as a source of five electrons).

Although the treatments already outlined allow a superficial rationalization of the structures of sandwich complexes $[M(C_2B_9H_{11})_2]^{n-4}$ and related metal-lacarboranes $MX_nC_2B_9H_{11}$, closer examination reveals a more complicated picture. Figure 11 illustrates the different positions adopted by the metal with respect to the open face of the icosahedral-fragment ligand $[C_2B_9H_{11}]^{2-}$ in the series of complexes $[Re(CO)_3C_2B_9H_{11}]^-$,[173] $Au(S_2CNEt_2)C_2B_9H_{11}$,[174] $Hg(PPh_3)C_2B_9H_{11}$,[174] and $[TlC_2B_9H_{11}]^-$.[174] The rhenium(I) complex has the expected icosahedral structure,[173] with the metal above the centroid of the pentagonal face of the ligand, as appropriate for a complex of a d^6 metal ion that formally contains 13 electron pairs available for cluster bonding. From a skeletal electron count, a similar structure might have been expected for the gold, mercury, and thallium complexes, since these too formally contain the 13 skeletal bond pairs required for an icosahedral structure. For example, the dithiocarbamato–gold unit AuS_2CNEt_2, in which the bidentate dithiocarbamate residue acts as a three-electron ligand to the metal, in principle might function as a source of $11 + 3 - 12$, i.e., 2 electrons, for cluster bonding, and so occupy the same type of site as a BH unit. However, it is displaced away from the pseudo-five-fold axis of the $C_2B_9H_{11}$ residue, away from the carbon atoms (Figure 11), while a greater displacement in the same general direction is shown by the $HgPPh_3$ residue in the mercury complex. Only in the case of the anionic thallium(I) complex $[T1C_2B_9H_{11}]^-$ is the "expected" structure adopted, and it has metal–ligand bonds of such length as to suggest that the bonding is essentially ionic rather than covalent.

Rationalization of these structures can take various forms. On the one hand, it is well known that the heavier elements of the later transition metal subgroups tend not to use all of their valence shell AO's in their typical complexes. For example, 16-electron rather than 18-electron configurations are common in the coordination chemistry of elements like platinum and gold; the remaining metal orbital in square planar Pt^{II} complexes is of too high energy to interact very strongly with a fifth ligand. Since the electron counting procedures of PSEPT assume that all nine valence shell AO's are used, a 16-electron configuration necessarily implies a more open metallaborane or metallacarborane cluster shape, such as a *nido* geometry where a *closo* shape was expected. Several examples of platina-carboranes with unexpectedly open structures explicable in this way are discussed in Ref. 175, including $Pt(PEt_3)_2C_2Me_2B_6H_6$, $Pt(PEt_3)_2C_2Me_2B_7H_7$, and $Pt(PMe_3)_2C_2B_8H_{10}$.

A more detailed treatment of such "anomalous" structures is possible if one considers the frontier orbitals of the metal and carborane units, and moreover takes note of the slight distortions shown by the open pentagonal face (not illustrated in Figure 11) that tend to accompany the metal slippage. For example, with a d^6 metal such as Re(I) in position 1 (Figure 11), both the empty d_{xz} and d_{yz} metal AO's can accept electronic charge from filled e-symmetry π-bonding MO's of the $C_2B_9H_{11}{}^{2-}$ ligand, whereas only one such metal AO remains available for metal–ligand bonding in the case of a d^8 complex like $Au(PPh_3)C_2B_9H_{11}$, when the metal accordingly shifts to position 2, and the pentagonal face folds.

Figure 11. Positions of the metal atoms over the open pentagonal carborane face in metallacarboranes $MX_nC_2B_9H_{11}$. (1) $MX_n =$ $[Re(CO)_3]^-$;[173] (2) $MX_n = Au(S_2CNEt_2)$;[174] (3) $MX_n =$ $HgPPh_3$;[174] (4) $MX_n = T1^-$.[174]

For a d^{10} configuration [as in $Hg(PPh_3)C_2B_9H_{11}$] both the d_{xz} and d_{yz} AO's are filled, leaving a metal sp hybrid AO to bond effectively to only one boron atom of the pentagonal face (position 3). The $d^{10}s^2$ configuration of the thallous ion in $[Tl(C_2B_9H_{\overline{11}})]^-$ allows only ionic metal–ligand bonding (position 4).

Another factor that should not be overlooked in considerations of the detailed shapes of metallacarboranes is that the number and positions of the carbon atoms in the open face of the carborane residue can markedly influence its ligand properties. This is strikingly illustrated by the three palladium-group complexes $Pd(CNBu^t)_2(1-Me_3NCB_{10}H_{10})$, [176] $Pt(PMe_2Ph)_2(2,5-Me_2C_2B_9H_9)$, [177] and $Pt(PEt_3)_2(3,4-C_2B_9H_{11})$. [178] Each of these three formally isoelectronic metallacarboranes has a unit ML_2 (M = Pd or Pt; L = a two-electron ligand, an isonitrile, or a phosphine) coordinated to the *nido*-shaped icosahedral-fragment carborane residue in a manner that allows the whole to be classed as a *closo* cluster—the metal atom is located near enough to the pseudo-five-fold axis of the carborane residue to be regarded as occupying the twelfth icosahedral vertex. However, the metal atoms do show small yet significant displacements from this axis, and the orientation of the ML_2 ligand with respect to the open face of the ligand (Figure 12) shows a marked dependence on where the carbon atoms are. The ML_2 unit lies in the mirror plane in the case of the compound $Pt(PMe_2Ph)_2(2,5-Me_2C_2B_9H_9)$, but perpendicular to this plane in the other two cases. Since in each case the ML_2 orientation gets the ligands L well away from the carbon substituents, a steric contribution to these differences should not be overlooked. However, detailed analysis of the differences, which include a slight folding of the open carborane face that in each case shifts the carbon atoms slightly away from the metal atom, suggests an electronic origin as follows (Figure 13). [178]

A V-shaped ML_2 residue like that present in these species has, as HOMO, a $d_{xz} - p_x$ hybrid AO lying in the ML_2 plane, while the LUMO is an sp_z hybrid AO. To maximize the metal–carborane bonding, these orbitals need to overlap as effectively as possible with the LUMO and HOMO, respectively, of the carborane residue, which for the neutral *nido*-shaped carborane residues will be the pair of face "π-bonding" MO's that in the parent borane, a hypothetical *nido*-shaped $B_{11}H_{11}{}^{2-}$ residue, would be a degenerate pair of e_1 symmetry. The degeneracy is removed by the presence of the carbon atoms in the face. When there is only one carbon atom or when there are two adjacent carbon atoms in the

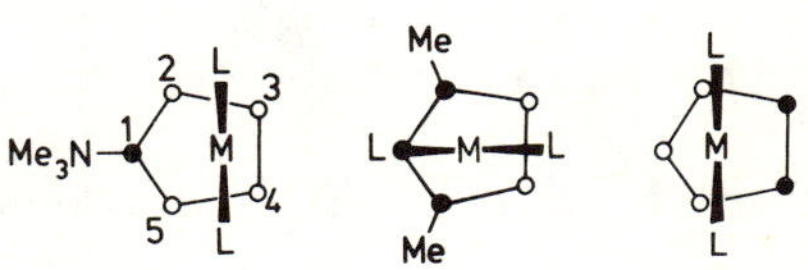

Figure 12. Orientations possible for the ML_2 unit (M = Pd or Pt; L = CNR or PR_3) with respect to an eleven-atom *nido*-carborane residue as a function of the positions of the carbon atoms in the open pentagonal face. (a) $Pd(CNBu^t)_2(1-Me_3NCB_{10}H_{10})$. [176] (b) $Pt(PMe_2Ph)_2(2,5-Me_2C_2B_9H_9)$. [177] (c) $Pt(PEt_3)_2(3,4-C_2B_9H_{11})$. [178]

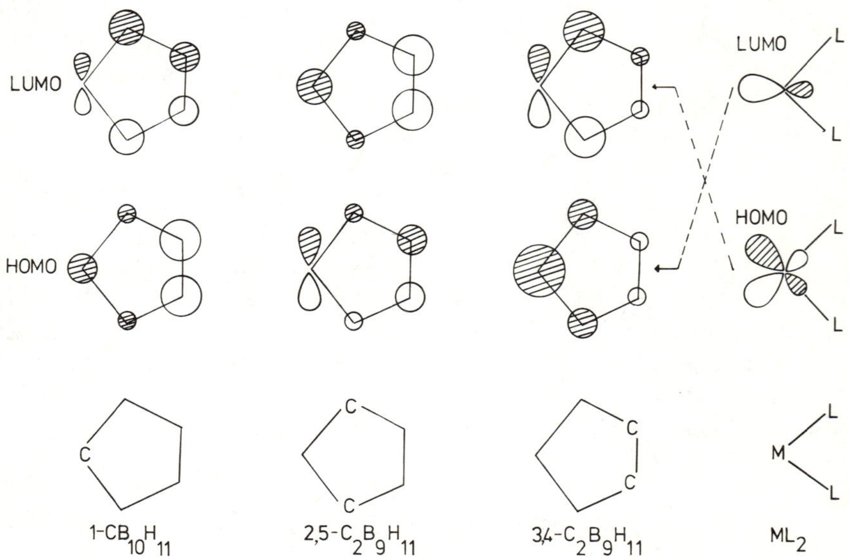

Figure 13. Frontier orbitals of the ML_2 and carborane units illustrated in Figure 12.

face, then the carborane LUMO is that orbital with a nodal plane in the face mirror plane, and so the ML_2 residue can release charge into this orbital from its HOMO most effectively if it lies perpendicular to this plane, which is the orientation observed in the case of $Pd(CNBu^t)_2(1-Me_3NCB_{10}H_{10})$ and $Pt(PEt_3)_2(3,4-C_2B_9H_{11})$. The different ML_2 orientation in $Pt(PMe_2Ph)_2-(2,5-Me_2C_2B_9H_9)$ arises because of a reversal of the relative energies of the carborane frontier orbitals; when the carbon atoms are nonadjacent in the face, the MO with a nodal plane in the face mirror plane is the HOMO (Figure 13). Close analysis of the contributions each boron and carbon AO makes to these MO's allows the slip and folding distortions to be rationalized as well.[178,179]

The pallada-carboranes $Pd(Me_2NCH_2CH_2NMe_2)(3,4-C_2B_9H_{11})$ and $Pd(PMe_3)_2(3,4-C_2B_9H_{11})$[180] have structures that reveal yet further nuances of metal–carborane bonding interactions. Both have the PdL_2 unit lying in a plane perpendicular to the carborane face mirror plane, with pronounced slip distortion in the former but not in the latter. The difference is, in part, attributable to the π-acidity of phosphine but not amine ligands, and underlines the range of factors that influence the structures of these heavy metal derivatives, all of which are predicted by a skeletal electron count to have *closo* structures.

4.2. Multidecker Sandwich Metallacarboranes

Comparison of metal–hydrocarbon π-complexes with borane-type clusters, and the description of metal–C_nH_n ring complexes as *nido* MC_n clusters formally related to parent *bi*pyramidal polyhedra prompted the suggestion[128] that aromatic ring systems such as $C_4H_4{}^{2-}$, $C_5H_5{}^{-}$, C_6H_6, or $C_7H_7{}^{+}$ might coordinate simultaneously to two metal atoms, one above and the other below the ring plane, thus generating M_2C_n *closo* clusters. Although there is as yet little evidence that C_nH_n rings can form multidecker sandwich compounds in this way, there is now considerable evidence that aromatic ring systems incorporating boron atoms can perform such a function, possibly because of the charge alternation between the ring and the metal atom generated by the replacement of CH units by BH units.

Detailed discussion of such multidecker sandwich systems is to be found in two recent reviews.[6,181] Here, there is room merely to note their bonding implications.

Carborane ring systems that have been found to be capable of forming multidecker sandwich compounds include the C_2B_2S-thiadiborolene ring, and the C_3B_2- and C_2B_3-di- and tri-boracyclopentadienyl rings, which would need to bear anionic charges of 2, 3, and 4, respectively, to acquire full complements of six π-electrons, and so become classifiable as 5-atom, eight-skeletal bond pair *arachno* systems:

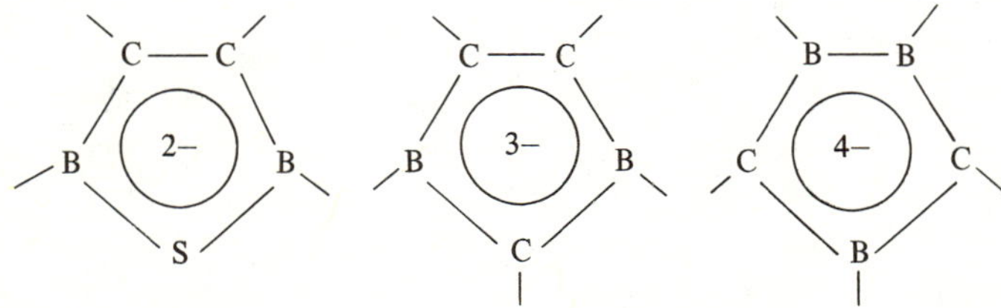

Examples of triple-decker sandwich compounds incorporating such rings are shown in Figure 14. In the iron compound $(C_5H_5)Fe(C_2Et_2B_2Me_2S)Fe(C_5H_5)$, for example, one way of viewing the bonding would be to regard it as containing the aromatic anionic ring system $(C_2Et_2B_2Me_2S)^{2-}$ sandwiched between two $[(C_5H_5)Fe]^{+}$ cations, or more generally as two Fe^{2+} cations sandwiched between the aromatic anions $(C_5H_5)^{-}$, $(C_2Et_2B_2Me_2S)^{2-}$, and $(C_5H_5)^{-}$. It can, thus, be regarded as a $\pi^6 d^6 \pi^6 d^6 \pi^6$ system. In cluster terminology, it consists of a central pentagonal bipyramidal $Fe_2C_2B_2S$ *closo* cluster sharing its axial vertices (the iron atoms) with two pentagonal pyramidal FeC_5 *nido* clusters. Both approaches indicate that it has a closed shell electronic configuration.

Similar arguments can be applied to the other multidecker species shown in Figure 14. The cobalt thiadiborolene complex,[183] for example, can be regarded as another $\pi^6 d^6 \pi^6 d^6 \pi^6$ system if written as $(C_2Et_2B_2Me_2S)^{2-}Co^{3+}(C_2Et_2B_2-Me_2S)^{2-}Co^{3+}(C_2Et_2B_2Me_2S)^{2-}$, although it should be stressed that this way of representing the complex is to illustrate the way the electrons can easily be counted, rather than to suggest that it has an ionic constitution. The cobalt triboracyclopentadienyl complex in Figure 14(d)[186] can, in turn, be repre-

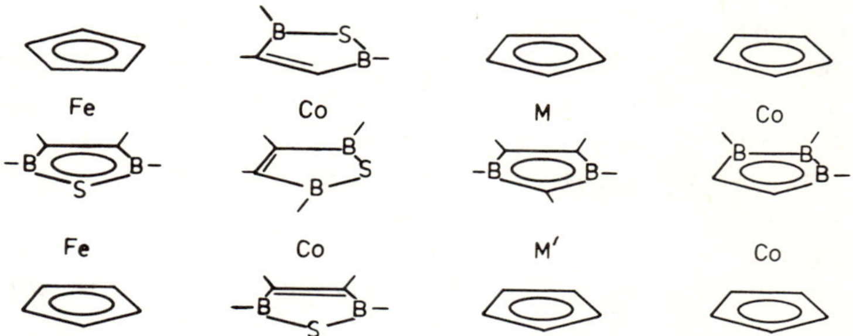

Figure 14. Some triple-decker sandwich metallacarboranes. (a) $(C_5H_5)Fe(\eta\text{-}C_2Et_2B_2Me_2S)$ $Fe(C_5H_5)$.[182] (b) $LCo(\mu\text{-}L)CoL$ $(L = C_2Et_2B_2Me_2S)$.[183] (c) $(C_5H_5)M(C_3R_3B_2R'_2)$ $M'(C_5H_5)$:[184,185]

M = Fe	Co	Co	Ni
M' = Co	Co	Ni	Ni

(d) $(C_5H_5)Co(C_2R_2B_3R'_3)Co(C_5H_5)$.[186]

sented as $(C_5H_5)^-Co^{3+}(C_2R_2B_3R'_3)^{4-}Co^{3+}(C_5H_5)^-$. Treated in this way, it is apparent that the sandwiched metal atoms in these complexes need to acquire formal oxidation states corresponding to d^6 electronic configurations.

For triple-decker sandwich compounds in general, containing two metal atoms sandwiched between three rings, $(ring)M(\mu\text{-}ring')M'(ring'')$, a closed shell electronic configuration will result if $(V_M + V_{M'} - x) = 12$, where V_M and $V_{M'}$ are the number of valence shell electrons on the atoms M and M', respectively, and x is the total number of electrons formally required by the three ring systems to give each an aromatic sextet. Thus, $x = 3$ if all three rings are cyclopentadienyl rings, $x = 6$ if all three are thiadiborolene rings, and so on. Similarly, for tetra-decker sandwich compounds like LCoLFeLCoL ($L = C_2Et_2B_2Me_2S$)[187] a closed shell electronic configuration requires $(V_M + V_{M'} + V_{M''} - x) = 18$.

An alternative way of expressing the electron count for these species is as follows. If one counts the total number of valence shell electrons on the metal atoms in the usual way (a neutral cyclopentadienyl ligand, whether terminal or bridging, acts as a source of five electrons, a thiadiborolene ring as a source of four, the diboracyclopentadienyl ring as a source of three, and the triboracyclopentadienyl ring as a source of two electrons), then the total number of electrons needed for a closed shell triple-decker complex will be 30 electrons. For example, for the complex $(C_5H_5)Fe(C_2Et_2B_2Me_2S)Fe(C_5H_5)$, the count would be $5 + 8 + 4 + 8 + 5 = 30$ electrons. For a tetra-decker complex, the total number of electrons expected for a closed shell configuration would be 42.

These generalizations do not preclude the formation of multidecker complexes with other than closed shell configurations, although they identify the systems most likely to be preparable. Although the first X-ray structural characterization of a triple-decker complex was that of neutral $(C_5H_5)Co(MeC_2B_3H_4)$-$Co(C_5H_5)$,[186] the initial report of a triple-decker concerned the cation $[(C_5H_5)Ni(\mu-C_5H_5)Ni(C_5H_5)]^+$,[188] which is actually a 34-electron species. Extended Hückel calculations[189] show this to be a closed shell configuration—the extra four electrons are accommodated in a degenerate pair of essentially non-bonding MO's. The species $(C_5H_5)Fe(\mu-C_3R_3B_2R'_2)Co(C_5H_5)$ and $[(C_5H_5)$-$Ni(\mu-C_5H_5)Ni(C_5H_5)]^+$ thus represent limiting members of a series of triple-decker complexes with 30 to 34 valence shell electrons. Other members are now known, and include the diboracyclopentadienyl complexes $(C_5H_5)M$-$(\mu-C_3R_3B_2R'_2)M'(C_5H_5)$ where $M = M' = Co$ (31 electrons), $M = Co$, $M' = Ni$ (32 electrons), $M = M' = Ni$ (33 electrons), and a singly charged anionic variant on this last example (34 electrons), the species with 31, 32, or 33 electrons being paramagnetic.[185]

These last examples would, therefore, be regarded as anomalous if treated as borane-type clusters, in that they adopt a *closo*-pentagonal bipyramidal geometry even though an electron count would suggest that a more open structure was likely. Most of the known multidecker systems, however, contain the numbers of electrons expected from the borane analogy.

4.3. Further Systems with Apparently Anomalous Electron Numbers

A few further systems that have structures or electron numbers different from those expected from the conventional borane cluster rules [*closo* for $(n+1)$ bond pairs, *nido* for $(n+2)$ pairs, etc.] are worth noting briefly.

The electron counting procedure used as the basis for these rules, and outlined in Section 2.2., assumes that each skeletal atom uses all its valence shell AO's, three of them for skeletal bonding, the remainder to bond ligands or accommodate lone pairs. Clearly, structural predictions made on this basis will fail if these assumptions are unjustified. We have already seen that unexpectedly open *nido* instead of *closo* structures may be adopted by metalla-carboranes of the heavier later transition metals because one or more of their valence shell AO's may remain unused. At the other extreme, among derivatives of the earlier transition metals, it is not surprising that incomplete occupancy of the valence shell AO's leads to apparent anomalies, such as *closo* structures (rather than capped structures) for systems that appear to contain fewer than $(n+1)$ skeletal bond pairs. As examples, one could cite the regular *closo commo* structures, of the type illustrated in Figure 10(a), adopted by the chromium(III) species $[Cr(C_2Me_2B_9H_9)_2]^{-}$[190] and the titanium(II) species $[Ti(C_2Me_2B_{10}H_{10})_2]^{2-}$.[191] These contain 24 and 28 skeletal bond pairs respectively, shared equally between the linked polyhedra even though they appear to be electron deficient using the electron counting rules outlined previously. The apparent deficiency arises because the three metal AO's not required for skeletal bonding are assumed to be filled, whereas, as nonbonding AO's, they will not be filled unless there are enough electrons to fill all the skeletal bonding MO's too. Instead of formally sandwiching a metal ion with a d^6 configuration, the $[C_2Me_2B_nH_n]^{2-}$ ligands in such species effectively sandwich a d^0–d^5 metal ion.[190–192]

Dimetallacarboranes that are formally electron deficient in this sense, may be paramagnetic as a result of the unpaired electrons on the metal atoms,[194] although spin-pairing may occur, particularly when the metal atoms occupy adjacent polyhedral sites, allowing direct metal–metal interactions additional to those normal for such clusters.[193-197] For example, the 10-atom *closo* cluster $(C_5H_5)_2Fe_2C_2B_6H_8$,[195] which formally contains only ten skeletal bond pairs, exists in the form of both paramagnetic and diamagnetic isomers, the metal-metal bonding in the latter leading to its adopting a *closo* structure that is a variant on the usual bicapped square antiprismatic 10-vertex polyhedron. The deficiency of electrons in the mixed metal species $Me_4C_4B_8H_8FeCo(C_5H_5)$ leads to the presence of a direct metal–metal bond, and a capping BH unit.[196] A capped structure (with a BH unit capping the Co_3 face) is also found in the case of the seven-atom, seven-bond pair cobaltaborane $(C_5H_5)_3Co_3B_4H_4$.[28]

Other anomalous systems worthy of brief consideration are the tetrametalla-boranes $(C_5H_5)_4Co_4B_4H_4$[198] and $(C_5H_5)_4Ni_4B_4H_4$,[199] both of which have D_{2d} dodecahedral structures (Figure 15) even though they contain eight and ten skeletal bond pairs, respectively, instead of the nine bond pairs normally associated with a *closo* eight-atom cluster like $B_8H_8{}^{2-}$ or $C_2B_6H_8$. They differ in the positions occupied by their metal atoms, which are in the sites of the higher coordination number in the cobalt compound, but in the sites of the lower coordination number in the nickel compound. In this respect, a $Co(C_5H_5)$ unit appears to resemble a BH unit, and a $Ni(C_5H_5)$ unit appears to resemble a CH unit, in their site preferences, the latter units being found in the sites where the skeletal electron density tends to be greater.

In considering their structures, it is helpful to consider first why the eight-vertex D_{2d} deltahedron apparently differs from other *closo* n-vertex polyhedra in being able to accommodate n, $(n+1)$, or $(n+2)$ skeletal bond pairs, as represented by $(C_5H_5)_4Co_4B_4H_4$, $B_8H_8{}^{2-}$, and $(C_5H_5)_4Ni_4B_4H_4$, respectively,

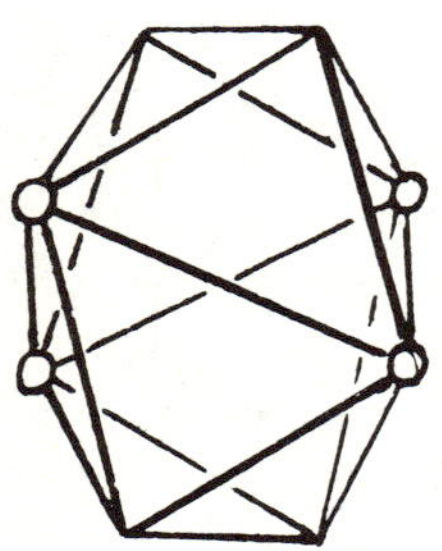
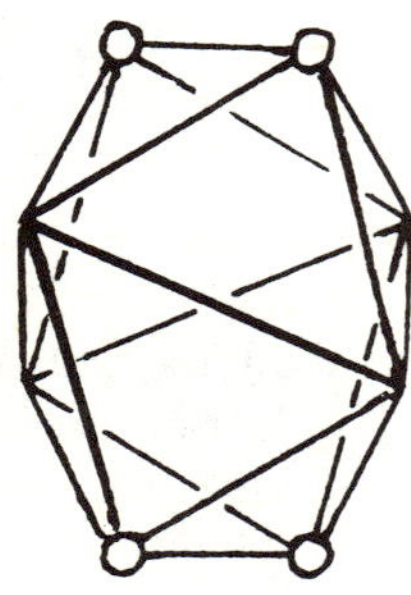

Figure 15. The D_{2d} dodecahedral skeletons of the metallaboranes $(C_5H_5)_4Co_4B_4H_4$[198] and $(C_5H_5)_4Ni_4B_4H_4$.[199]

without a change of symmetry. This can probably be attributed to the nondegeneracy of the HOMO (which has b_2 symmetry) and the LUMO (a_2 symmetry) of $B_8H_8^{2-}$.[200] Whether these MO's are both empty, as in the case of an analog of the hypothetical B_8H_8 [like $(C_5H_5)_4Co_4B_4H_4$], or both filled as in the case of an analog of the hypothetical $B_8H_8^{4-}$ [like $(C_5H_5)_4Ni_4B_4H_4$], the result is a closed-shell electronic configuration with no unpaired electrons (although with rather weaker skeletal bonding than when nine bond pairs are present).[207,208]

By contrast, the shapes of most of the other *closo*-borane polyhedra are such as to cause the HOMO's and LUMO's of the anions $B_nH_n^{2-}$ to be degenerate.[11,15,23,201] Adding two electrons to most *closo*-boranes, or removing two electrons from them, without changing their shapes would generate paramagnetic species prone to Jahn–Teller distortion.[3,202] Only the nine-vertex tricapped trigonal prism of $B_9H_9^{2-}$ and the eleven-vertex octadecahedron of $B_{11}H_{11}^{2-}$ among the other *closo*-borane anions have nondegenerate HOMO's and LUMO's, so these too are, in principle, capable of accommodating n, $(n+1)$, or $(n+2)$ bond pairs (interestingly, the cationic nine-atom eleven-bond pair bismuth cluster Bi_9^{5+} has a *closo*-tricapped trigonal prismatic shape.[203]

These considerations of the degeneracies of the HOMO's and LUMO's do not explain why $(C_5H_5)_4Co_4B_4H_4$ and $(C_5H_5)_4Ni_4B_4H_4$ have D_{2d} dodecahedral shapes, but at least show why these shapes are compatible with the numbers of electrons present. Theoretical studies[11,15,23,204,205] of the alternative shapes possible for *closo* clusters like $B_8H_8^{2-}$ suggest that there is little energy difference between various possible shapes, and whether an eight-atom ten-bond pair cluster adopts an expanded *closo* structure like that of $(C_5H_5)_4Ni_4B_4H_4$ or a *nido* structure like that of B_8H_{12}[206] probably depends crucially on the bonding requirements of the ligands (especially the "extra" four hydrogen atoms of B_8H_{12}) as well as the identities of the skeletal atoms.

5. CONCLUSION

It was the intention in this chapter to illustrate the key role played by metallaboranes and metallacarboranes in relating borane clusters to other cluster and ring systems. They illustrate the range of units that can use three AO's apiece for bonding, whether to bond to two neighboring atoms as in the aromatic rings, to three neighbors as in the conventionally two-center bonded systems, or to more neighbors as in clusters, and stimulated development of the isolobal concept. Skeletal electron counting approaches developed to rationalize their shapes have drawn attention to the relative merits of localized bond and molecular orbital treatments and allowed useful predictions to be made. Where predictions proved incorrect or anomalies were discovered, useful conclusions have been drawn about previously overlooked factors, variables that had been treated as invariant, or about the need for more detailed treatments. Metallaboranes and metallacarboranes have done much to advance our knowledge of structural chemistry and bonding theory. It is to be hoped that this short survey has given some indication of this.

REFERENCES

1. K. Wade, *Adv. Inorg. Chem. Radiochem.,* **18**, 1 (1976).
2. D. M. P. Mingos, *Nat. Phys. Sci.,* **236**, 99 (1972).
3. R. W. Rudolph, *Accts. Chem. Res.,* **9**, 446 (1976).
4. R. N. Grimes, *Ann. N.Y. Acad. Sci.,* **239**, 180 (1974).
5. R. Mason and D. M. P. Mingos, *M.T.P. Int. Rev. Sci., Phys. Chem. Ser. 2,* **11**, 121 (1975).
6. R. N. Grimes, *Coord. Chem. Rev.,* **28**, 47 (1979).
7. C. E. Housecroft and K. Wade, *Gaz. Chim. Ital.,* **110**, 87 (1980).
8. R. W. Rudolph, in: *Boron Chemistry–4* (R. W. Parry and G. Kodama, eds.), pp. 11-22, Pergamon, Oxford (1980).
9. K. Wade, in: *Boron Chemistry–4* (R. W. Parry and G. Kodama, eds.), pp. 23-32, Pergamon, Oxford (1980).
10. T. P. Fehlner, in: *Boron Chemistry–4* (R. W. Parry and G. Kodama, eds.), pp. 95-107, Pergamon, Oxford (1980).
11. W. N. Lipscomb, *Boron Hydrides,* Benjamin, New York (1963).
12. R. E. Williams, *Inorg. Chem.* **10**, 210 (1971).
13. R. E. Williams, *Adv. Inorg. Chem. Radiochem.,* **18**, 66 (1976).
14. T. Onak, in: *Boron Hydride Chemstry* (E. L. Muetterties, ed.), pp. 349-382, Academic Press, New York (1975).
15. W. N. Lipscomb, in: *Boron Hydride Chemistry* (E. L. Muetterties, ed.), pp. 39-78, Academic Press, New York (1975).
16. W. N. Lipscomb, *Science,* **196**, 1047 (1977).
17. W. N. Lipscomb, in: *Boron Chemistry–4* (R. W. Parry and G. Kodama, eds.) pp. 1-10, Pergamon, Oxford (1980).
18. J. C. Calabrese, and L. F. Dahl, *J. Am. Chem. Soc.,* **93**, 6042 (1971).
19. S. J. Lippard and K. M. Melmed, *Inorg. Chem.,* **8**, 2755 (1969).
20. L. J. Guggenberger, *Inorg. Chem.,* **9**, 367 (1970).
21. N. N. Greenwood, C. G. Savory, R. N. Grimes, L. G. Sneddon, A. Davison, and S. S. Wreford, *Chem. Comm.,* 718 (1974).
22. V. R. Miller and R. N. Grimes, *J. Am. Chem. Soc.,* **95**, 5078 (1973).
23. H. C. Longuet-Higgins, *Quart. Rev.,* **11**, 121 (1957).
24. K. P. Callahan and M. F. Hawthorne, *Adv. Organometal. Chem.,* **14**, 145 (1976).
25. J. A. Ulman and T. P. Fehlner, *J.C.S. Chem. Comm.,* 632 (1976).
26. J. R. Pipal and R. N. Grimes, *Inorg. Chem.,* **18**, 252 (1979).
27. V. R. Miller, R. Weiss, and R. N. Grimes, *J. Am. Chem. Soc.,* **99**, 5646 (1977).
28. J. R. Pipal and R. N. Grimes, *Inorg. Chem.,* **16**, 3255 (1977).
29. V. R. Miller and R. N. Grimes, *J. Am. Chem. Soc.,* **98**, 1600 (1976).
30. V. R. Miller, L. G. Sneddon, D. G. Beer, and R. N. Grimes, *J. Am. Chem. Soc.,* **96**, 3090 (1974).
31. R. N. Grimes, *J. Am. Chem. Soc.,* **93**, 261 (1971).
32. R. N. Grimes, W. J. Rademaker, M. L. Denniston, R. F. Bryan, and P. T. Greene, *J. Am. Chem. Soc.,* **94**, 1865 (1972).
33. D. C. Beer, V. R. Miller, L. G. Sneddon, R. N. Grimes, M. Matthew, and G. J. Palenik, *J. Am. Chem. Soc.,* **95**, 3046 (1973).

34. G. K. Barker, M. Green, F. G. A. Stone, and A. J. Welch, *J.C.S. Dalton,* 1186 (1980).
35. R. N. Grimes, E. Sinn, and R. B. Maynard, *Inorg. Chem.,* **19,** 2384 (1980).
36. J. R. Pipal, W. M. Maxwell, and R. N. Grimes, *Inorg. Chem.,* **17,** 1447 (1978).
37. W. M. Maxwell, V. R. Miller, and R. N. Grimes, *Inorg. Chem.,* **15,** 1343 (1976).
38. J. W. Howard and R. N. Grimes, *Inorg. Chem.,* **11,** 263 (1972).
39. G. E. Herberich, J. Hengesbach, U. Kolle, G. Huttner, and A. Frank, *Angew. Chem. Int. Ed. Eng.,* **15,** 433 (1976).
40. K. S. Wong and R. N. Grimes, *Inorg. Chem.,* **16,** 2053 (1977).
41. D. F. Dustin, W. J. Evans, C. J. Jones, R. J. Wiersema, H. Gong, S. Chan, and M. F. Hawthorne, *J. Am. Chem. Soc.,* **96,** 3085 (1974).
42. W. J. Evans, G. B. Dunks, M. F. Hawthorne, *J. Am. Chem. Soc.,* **95,** 4565 (1973).
43. A. D. George and M. F. Hawthorne, *Inorg. Chem.,* **8,** 1801 (1969).
44. A. J. Welch, *J.C.S. Dalton,* 225 (1976).
45. R. N. Grimes, A. Zalkin, and W. T. Robinson, *Inorg. Chem.,* **15,** 2274 (1976).
46. D. F. Dustin and M. F. Hawthorne, *Chem. Comm.,* 1329 (1972).
47. R. N. Leyden, B. P. Sullivan, R. T. Baker and M. F. Hawthorne, *J. Am. Chem. Soc.,* **100,** 3758 (1978).
48. G. E. Hardy, K. P. Callahan and M. F. Hawthorne, *Inorg. Chem.,* **17,** 1662 (1978).
49. D. St. Clair, A. Zalkin and D. H. Templeton, *Inorg. Chem.,* **11,** 377 (1972).
50. W. M. Maxwell, K. S. Wong and R. N. Grimes, *Inorg. Chem.,* **16,** 3094 (1977).
51. E. L. Hoel, C. E. Strouse and M. F. Hawthorne, *Inorg. Chem.,* **13,** 1388 (1974).
52. J. Dolansky, K. Base and B. Stibr, *Chem. Ind.,* 853 (1976).
53. C. W. Jung and M. F. Hawthorne, *Chem. Comm.,* 499 (1976).
54. W. J. Evans and M. F. Hawthorne, *Inorg. Chem.,* **13,** 869 (1974).
55. M. F. Hawthorne and T. D. Andrews, *Chem. Comm.,* 443 (1965).
56. M. F. Hawthorne, D. C. Young, T. D. Andrews, D. V. Howe, R. L. Pilling, A. D. Pitts, M. Reintjes, L. F. Warren and P. A. Wegner, *J. Am. Chem. Soc.,* **90,** 879 (1968).
57. M. Green, J. L. Spencer, F. G. A. Stone and A. J. Welch, *Chem. Comm.,* 786 (1976).
58. S. B. Miller and M. F. Hawthorne, *Chem. Comm.,* 786 (1976).
59. R. W. Rudolph, R. L. Voorhees and R. E. Cochoy, *J. Am. Chem. Soc.,* **92,** 3351 (1970).
60. H. M. Colquhoun, T. J. Greenhough and M. G. H. Wallbridge, *Acta Cryst.* **B34,** 2373 (1978).
61. M. R. Churchill and A. H. Reis, *J.C.S. Dalton,* 1317 (1972).
62. G. Popp and M. F. Hawthorne, *J. Am. Chem. Soc.,* **90,** 6553 (1968).
63. A. R. Siedle, *J. Organomet. Chem.,* **90,** 249 (1975).
64. W. C. Kalb, C. W. Kreimendahl, D. C. Busby and M. F. Hawthorne, *Inorg. Chem.,* **19,** 1590 (1980).
65. A. Zalkin, T. E. Hopkins and D. H. Templeton, *Inorg. Chem.* **5,** 1189 (1966).
66. B. G. DeBoer, A. Zalkin and D. H. Templeton, *Inorg. Chem.* **7,** 2288 (1968).
67. M. R. Churchill, K. Gold, J. N. Francis and M. F. Hawthorne, *J. Am. Chem. Soc.,* **91,** 1222 (1969).
68. D. St. Clair, A. Zalkin and D. H. Templeton, *J. Am. Chem. Soc.,* **92,** 1173 (1970).
69. W. H. Knoth, *Inorg. Chem.,* **10,** 598 (1971).
70. W. R. Hertler, F. Klanberg and E. L. Muetterties, *Inorg. Chem.,* **6,** 1696 (1967).
71. W. J. Evans and M. F. Hawthorne, *J. Am. Chem. Soc.,* **96,** 301 (1974).
72. D. F. Dustin, G. B. Dunks and M. F. Hawthorne, *J. Am. Chem. Soc.,* **95,** 1109 (1973).
73. M. R. Churchill and B. G. DeBoer, *Chem. Comm.,* 1326 (1972).
74. J. R. Pipal and R. N. Grimes, *Inorg. Chem.,* **17,** 6 (1978).
75. S. J. Hildebrandt, D. F. Gaines and J. C. Calabrese, *Inorg. Chem.,* **17,** 790 (1978).
76. E. L. Anderson, K. J. Haller and T. P. Fehlner, *J. Am. Chem. Soc.,* **101,** 4390 (1979).
77. R. Weiss and R. N. Grimes, *J. Am. Chem. Soc.,* **99,** 8087 (1977).
78. R. Wilczynski and L. G. Sneddon, *Inorg. Chem.,* **18,** 864 (1979).

79. J. P. Brennan, R. N. Grimes, R. Schaeffer and L. G. Sneddon, *Inorg. Chem.*, **12**, 2266 (1973).

80. R. N. Grimes, D. C. Beer, L. G. Sneddon, V. R. Miller and R. Weiss, *Inorg. Chem.*, **13**, 1138 (1974).

81. W. Siebert, R. Full, J. Edwin, K. Kinberger and C. Kruger, *J. Organomet. Chem.*, **131**, 1 (1977).

82. W. Siebert, J. Edwin and M. Bochmann, *Angew. Chem. Internat. Edn.*, **17**, 868 (1978).

83. T. P. Fehlner, *J. Am. Chem. Soc.*, **100**, 3250 (1978).

84. G. Huttner and B. Krieg, *Chem. Ber.*, **105**, 3437 (1972).

85. W. Siebert and R. Full, *Angew. Chem. Internat. Edn.*, **15**, 45 (1976).

86. G. Huttner and W. Gartzke, *Chem. Ber.*, **107**, 3786 (1974).

87. J. R. Bowser and R. N. Grimes, *J. Am. Chem. Soc.*, **100**, 4623 (1978).

88. M. Green, J. L. Spencer, F. G. A. Stone and A. J. Welch, *Chem. Comm.*, 794 (1974).

89. J. R. Pipal and R. N. Grimes, *Inorg. Chem.*, **16**, 3251 (1977).

90. A. R. Kane, L. J. Guggenberger and E. L. Muetterties, *J. Am. Chem. Soc.*, **92**, 2571 (1970).

91. K. P. Callahan, F. Y. Lo, C. E. Strouse, A. L. Sims and M. F. Hawthorne, *Inorg. Chem.*, **13**, 2842 (1974).

92. M. Green, J. Howard, J. L. Spencer and F. G. A. Stone, *Chem. Comm.*, 153 (1974).

93. N. N. Greenwood, J. A. McGinnety and J. D. Owen, *J. Am. Chem. Soc. (A)*, 809 (1971).

94. R. N. Leyden, B. P. Sullivan, R. T. Baker and M. F. Hawthorne, *J. Am. Chem. Soc.*, **100**, 3758 (1978).

95. M. Elian and R. Hoffman, *Inorg. Chem.*, **14**, 1058 (1975).

96. M. Elian, M. M. L. Chen, D. M. P. Mingos and R. Hoffmann, *Inorg. Chem.*, **15**, 1148 (1976).

97. B. F. G. Johnson, *Transition Metal Clusters*, Wiley, London (1980).

98. V. G. Albano, G. Ciani, S. Martinengo and A. Sironi, *J. C. S. Dalton*, 978 (1979).

99. V. G. Albano, P. Chini, S. Martinengo, M. Sansoni and D. Strumolo, *J. C. S. Chem. Comm.*, 299 (1974).

100. J. C. Calabrese, L. F. Dahl, P. Chini, G. Longoni and S. Martinengo, *J. Am. Chem. Soc.*, **96**, 2614 (1974).

101. R. Mason, K. M. Thomas and D. M. P. Mingos, *J. Am. Chem. Soc.*, **95**, 3802 (1973).

102. B. F. G. Johnson, C. R. Eady, J. Lewis. B. E. Reichert and G. M. Sheldrick, *J. C. S. Chem. Comm.*, 271 (1976).

103. J. J. Guy and G. M. Sheldrick, *Acta Cryst.*, **B34**, 1722 (1978).

104. V. G. Albano, G. Ciani, M. Manassero, F. Canziani, G. Giordano, S. Martinengo and P. Chini, *J. Organometal. Chem.*, **150**, C17 (1978).

105. R. D. Wilson, S. M. Wu, R. A. Love and R. Bau, *Inorg. Chem.*, **17**, 1271 (1978).

106. H. D. Kaesz, B. Fontal, R. Bau, S. W. Kirtley and M. R. Churchill, *J. Am. Chem. Soc.*, **91**, 1021 (1969).

107. P. F. Jackson, B. F. G. Johnson, J. Lewis, M. McPartlin and W. J. H. Nelson, *J. C. S. Chem. Comm.*, 224 (1980).

108. C. R. Eady, B. F. G. Johnson, J. Lewis, R. Mason, P. B. Hitchcock and K. M. Thomas, *J. C. S. Chem. Comm.*, 385, (1977).

109. V. G. Albano, P. L. Bellon and G. F. Ciani, *Chem. Comm.*, 1024 (1969).

110. V. G. Albano, G. Ciani, S. Martinengo, P. Chini and G. Giordano, *J. Orangometal. Chem.*, **88**, 381 (1975).

111. E. R. Corey, L. F. Dahl and W. Beck, *J. Am. Chem. Soc.*, **85**, 1202 (1963).

112. M. R. Churchill and J. Wormald, *J. Am. Chem. Soc.*, **93**, 5670 (1971).

113. C. R. Eady, B. F. G. Johnson, J. Lewis, M. C. Malatesta, P. Machin and M. McPartlin, *J. C. S. Chem. Comm.*, 945 (1976).

114. M. McPartlin, C. R. Eady, B. F. G. Johnson and J. Lewis, *J. C. S. Chem. Comm.*, 883 (1976).

115. D. M. P. Mingos, *J. C. S. Dalton*, 113 (1974).

116. V. G. Albano, G. Ciani and P. Chini, *J. C. S. Dalton*, 432 (1974).

117. J. C. Calabrese, L. F. Dahl, A. Cavalieri, P. Chini, G. Longoni and S. Martinengo, *J. Am. Chem. Soc.*, **96**, 2616 (1974).

118. A. Sirigu, M. Bianchi and E. Benedetti, *Chem. Comm.*, 596 (1969).

119. M. R. Churchill and J. Wormald, *J. C. S. Dalton*, 2410 (1974).

120. C. R. Eady, J. M. Fernandez, B. F. G. Johnson, J. Lewis, P. R. Raithby and G. M. Sheldrick, *J. C. S. Chem. Comm.*, 421 (1978).

121. E. H. Braye, L. F. Dahl, W. Hübel and D. L. Wampler, *J. Am. Chem. Soc.*, **84**, 4633 (1962).

122. C. R. Eady, B. F. G. Johnson, J. Lewis and T. Matheson, *J. Organometal. Chem.*, **57**, C82 (1973).

123. B. F. G. Johnson, J. Lewis, P. R. Raithby, G. M. Sheldrick, K. Wong and M. McPartlin, *J. C. S. Dalton*, 673 (1978).

124. C. J. Commons and B. F. Hoskins, *Austral. J. Chem.*, **28**, 1663 (1975).

125. C. E. Housecroft, K. Wade and B. C. Smith, *J. C. S. Chem. Comm.*, 765 (1978).

126. J. W. Lauher, *J. Am. Chem. Soc.*, **100**, 5305 (1978).

127. G. Ciani and A. Sironi, *J. Organometal. Chem.*, **197**, 233 (1980).

128. K. Wade, *Inorg. Nucl. Chem. Lett.*, **8**, 563 (1972).

129. K. Wade, *Chem. in Brit.*, **11**, 177 (1975).

130. D. M. P. Mingos, *Advan. Organometal. Chem.*, **15**, 1 (1977).

131. J. F. Blount, L. F. Dahl, C. Hoogzand and W. Hübel, *J. Am. Chem. Soc.*, **88**, 292 (1966).

132. L. F. Dahl and D. L. Smith, *J. Am. Chem. Soc.*, **84**, 2450 (1962).

133. B. F. G. Johnson, J. Lewis, B. E. Reichert, K. T. Schorpp and G. M. Sheldrick, *J. C. S. Dalton*, 1417 (1977).

134. G. F. Stuntz, J. R. Shapley and C. G. Pierpoint, *Inorg. Chem.*, **17**, 2596 (1978).

135. R. P. Dodge and V. Schomaker, *J. Organometal. Chem.*, **3**, 274 (1965).

136. E. Sappa, A. Tiripicchio and M. T. Camellini, *J. C. S. Dalton*, 419 (1978).

137. W. G. Sly, *J. Am. Chem. Soc.*, **81**, 18 (1959).

138. G. E. Coates, M. L. H. Green and K. Wade, *Organometallic Compounds; Vol. 2 The Transition Elements*, 3rd edn., Methuen, London (1968).

139. D. S. Ginley, C. R. Bock, M. S. Wrighton, B. Fischer, D. L. Tipton and R. Bau, *J. Organometal. Chem.*, **157**, 41 (1978).

140. G. Maier, S. Pfriem, U. Schäfer and R. Matusch, *Angew. Chem. Internat. Edn.*, **17**, 520 (1978).

141. R. P. Dodge and V. Schomaker, *Acta. Cryst.*, **18**, 614 (1965).

142. C. G. Pierpoint, *Inorg. Chem.*, **16**, 636 (1977).

143. W.-D. Stohrer and R. Hoffmann, *J. Am. Chem. Soc.*, **94**, 1660 (1972).

144. S. Masamune, M. Sakai and H. Ona, *J. Am. Chem. Soc.*, **94**, 8955 (1972).

145. S. Masamune, M. Sakai, H. Ona and A. J. Jones, *J. Am. Chem. Soc.*, **94**, 8956 (1972).

146. M. J. S. Dewar and R. C. Haddon, *J. Am. Chem. Soc.*, **95**, 5836 (1973).

147. W. J. Hehre and P. von R. Schleyer, *J. Am. Chem. Soc.*, **95**, 5837 (1973).

148. A. A. Hock and O. S. Mills, *Acta. Cryst.*, **14**, 139 (1961).

149. H. Hogeveen and P. W. Kwant, *J. Am. Chem. Soc.*, **96**, 2208 (1974).

150. K. W. Cox and M. D. Harmony, *J. Chem. Phys.*, **50**, 1976 (1969).

151. G. Huttner and W. Gartzke, *Chem. Ber.,* **107,** 3786 (1974).
152. D. A. Drew and A. Haaland, *Acta. Chem. Scand.,* **27,** 3735 (1973).
153. H. Hogeveen and P. W. Kwant, *J. Am. Chem. Soc.,* **96,** 2208 (1974).
154. T. Chiang, R. C. Kerber, S. D. Kimball and J. W. Lauher, *Inorg. Chem.,* **18,** 1687 (1979).
155. B. R. Penfold and B. H. Robinson, *Accounts Chem. Res.,* **6,** 73 (1973).
156. C. A. Tolman, *Chem. Soc. Rev.,* **1,** 337 (1972).
157. K. S. Wong, J. R. Bowser, J. R. Pipal and R. N. Grimes, *J. Am. Chem. Soc.,* **100,** 5045 (1978).
158. J. R. Pipal and R. N. Grimes, *J. Am. Chem. Soc.,* **100,** 3083 (1978).
159. N. N. Greenwood and I. M. Ward, *Chem. Soc. Rev.,* **3,** 231 (1974).
160. H. C. Brown, *The Non-Classical Ion Problem,* Plenum Press, New York (1977).
161. G. A. Olah, G. K. S. Prakash and J. Sommer, *Science,* **206,** 13 (1979).
162. R. E. Williams, L. D. Field, in: *Boron Chemistry–4,* (R. W. Parry and G. Kodama, eds.), pp. 131–150, Pergamon, Oxford (1980).
163. J. D. Corbett and P. A. Edwards, *J. C. S. Chem. Comm.,* 984 (1975).
164. C. H. E. Belin, J. D. Corbett and A. Cisar, *J. Am. Chem. Soc.,* **99,** 7163 (1977).
165. J. D. Corbett and P. A. Edwards, *Inorg. Chem.,* **16,** 903 (1977).
166. J. D. Corbett and P. A. Edwards, *J. Am. Chem. Soc.,* **99,** 3313 (1977).
167. R. W. Rudolph, W. L. Wilson, F. Parker, R. C. Taylor and D. C. Young, *J. Am. Chem. Soc.,* **100,** 4629 (1978).
168. J. D. Corbett, *Inorg. Chem.,* **7,** 198 (1968).
169. A. Hershaft and J. D. Corbett, *Inorg. Chem.,* **2,** 979 (1963).
170. R. M. Friedman and J. D. Corbett, *Inorg. Chem.,* **12,** 1134 (1973).
171. R. J. Gillespie, *Chem. Soc. Rev.,* **8,** 315 (1979).
172. R. M. Wing, *J. Am. Chem. Soc.,* **89,** 5599 (1967); **90,** 4828 (1968).
173. A. Zalkin, T. E. Hopkins and D. H. Templeton, *Inorg. Chem.,* **6,** 1911 (1967).
174. H. M. Colquhoun, T. J. Greenhough and M. G. H. Wallbridge, *J. C. S. Chem. Comm.,* 1019 (1976); 737 (1977).
175. M. Green, J. L. Spencer and F. G. A. Stone, *J. C. S. Dalton,* 1679 (1979).
176. W. E. Carroll, M. Green, F. G. A. Stone and A. J. Welch, *J. C. S. Dalton,* 2263 (1975).
177. A. J. Welch, *J. C. S. Dalton,* 1473 (1975).
178. D. M. P. Mingos, M. I. Forsyth and A. J. Welch, *J. C. S. Chem. Comm.,* 605 (1977).
179. D. M. P. Mingos, *J. C. S. Dalton,* 602 (1977).
180. H. M. Colquhoun, T. J. Greenhough and M. G. H. Wallbridge, *J. C. S. Chem. Comm.,* 322 (1978).
181. W. Siebert, in: *Boron Chemistry–4,* (R. W. Parry and G. Kodama, eds.), pp. 81–94, Pergamon, Oxford (1980).
182. W. Siebert, T. Renk, K. Kinberger, M. Bochmann and C. Krüger, *Angew. Chem. Internat. Edn.,* **15,** 779 (1976).
183. W. Siebert and W. Rothermel, *Angew. Chem. Internat. Edn.,* **16,** 333 (1977).
184. W. Siebert and M. Bochmann, *Angew. Chem. Internat. Edn.,* **16,** 857 (1977).
185. W. Siebert, J. Edwin and M. Bochmann, *Angew. Chem. Internat. Edn.,* **17,** 869 (1978).
186. D. C. Beer, V. R. Miller, L. G. Sneddon, R. N. Grimes, M. Mathew and G. J. Palenik, *J. Am. Chem. Soc.,* **95,** 3046 (1973).
187. W. Siebert, W. Rothermel, C. Böhle, C. Krüger and D. J. Brauer, *Angew. Chem. Internat. Edn.,* **18,** (1979).
188. A. Salzer and H. Werner, *Angew. Chem. Internat. Edn.,* **11,** 930 (1972).
189. J. W. Lauher, M. Elian, R. H. Summerville and R. Hoffmann, *J. Am. Chem. Soc.,* **98,** 3219 (1976).

190. D. St. Clair, A. Zalkin and D. H. Templeton, *Inorg. Chem.*, **10**, 2587 (1971).
191. F. Y. Lo, C. E. Strouse, K. P. Callahan, C. B. Knobler and M. F. Hawthorne, *J. Am. Chem. Soc.*, **97**, 428 (1975).
192. C. G. Salentine and M. F. Hawthorne, *Inorg. Chem.*, **15**, 2872 (1976).
193. E. K. Nishimura, *J. C. S. Chem. Comm.*, 858 (1978).
194. W. J. Wiersema and M. F. Hawthorne, *J. Am. Chem. Soc.*, **96**, 761 (1974).
195. K. P. Callahan, W. J. Evans, F. Y. Lo, C. E. Strouse and M. F. Hawthorne, *J. Am. Chem. Soc.*, **97**, 296 (1975).
196. W. M. Maxwell, E. Sinn and R. N. Grimes, *J. Am. Chem. Soc.*, **98**, 3490 (1976).
197. C. G. Salentine and M. F. Hawthorne, *Inorg. Chem.*, **17**, 1498 (1978).
198. J. R. Pipal and R. N. Grimes, *Inorg. Chem.*, **18**, 257 (1979).
199. J. R. Bowser, A. Bonny, J. R. Pipal and R. N. Grimes, *J. Am. Chem. Soc.*, **101**, 6229 (1979).
200. D. A. Kleir and W. N. Lipscomb, *Inorg. Chem.*, **18**, 1312 (1979).
201. L. J. Guggenberger, *Inorg. Chem.*, **7**, 2260 (1968).
202. C. Glidewell, *J. Organometal. Chem.*, **128**, 13 (1977).
203. J. D. Corbett, *Progr. Inorg. Chem.*, **21**, 140 (1976).
204. F. Klanberg, D. R. Eaton, L. J. Guggenberger and E. L. Muetterties, *Inorg. Chem.*, **6**, 1271 (1967).
205. E. L. Muetterties and B. F. Beier, *Bull. Soc. Chim. Belg.*, **84**, 397 (1975).
206. R. E. Enrione, F. P. Boer and W. N. Lipscomb, *J. Am. Chem. Soc.*, **86**, 1451 (1964); *Inorg. Chem.*, **3**, 1659 (1964).
207. D. N. Cox, D. M. P. Mingos, and R. Hoffmann, *J. Chem. Soc. Dalton*, 1788 (1981).
208. M. E. O'Neill and K. Wade, *Inorg. Chem.*, **21**, 461 (1982).

2

Transition-Metal Derivatives of Nido-Boranes and Some Related Species

Norman N. Greenwood and John D. Kennedy

1. INTRODUCTION

The preparation of stable metallaborane clusters, and the recognition that boranes and their anions can function as excellent *polyhapto* ligands have, together, constituted one of the most exciting and significant developments in inorganic chemistry during the past 15 years. This chapter reviews the wide variety of metallaboranes that have been made by the reactions of transition-metal compounds with *nido*-boranes or their anions. *Nido*-boranes have the formula B_nH_{n+4} and are represented by the stable neutral boranes B_2H_6, B_5H_9, B_6H_{10}, $B_{10}H_{14}$, $n-B_{18}H_{22}$, and $i-B_{18}H_{22}$, and the unstable species $\{B_3H_7\}$, $\{B_4H_8\}$, $\{B_8H_{12}\}$, and $\{B_9H_{13}\}$.[1] Related *nido*-borane anions are BH_4^-, $B_5H_8^-$, $B_6H_9^-$, $B_9H_{12}^-$, $B_{10}H_{13}^-$, $B_{10}H_{12}^{2-}$, $B_{11}H_{14}^-$, $B_{11}H_{13}^{2-}$, etc. In addition there are the various stable geometrical isomers of the B—B bonded conjuncto-nido-boranes $(B_5H_8)_2$ and $(B_{10}H_{13})_2$. In this review we concentrate mainly on preparative and structural aspects although, where appropriate, we discuss presumed reaction mechanisms or postulated modes of bonding. The literature has been surveyed to the end of 1979 for the main body of the text but the Appendix summarizes some important further results appearing up to early 1982.

In reviewing transition-metal derivatives of *nido*-boranes it is possible to classify the compounds considered *either* on the basis of the number of boron atoms in the starting borane *or* according to the number of boron atoms in the product metallaborane. Rigid application of either scheme has several disadvantages and we have chosen to use both systems simultaneously, with appro-

Norman N. Greenwood and John D. Kennedy ● Department of Inorganic and Structural Chemistry, University of Leeds, Leeds LS2 9JT, England.

priate cross references, so as to obtain the advantage of a flexible narrative style within the framework of a comprehensive overview of the subject. The review is principally concerned with transition metal-*nido*-borane clusters which involve direct M—B contacts, although it is convenient occasionally to refer to compounds with M—H—B bonds or to clusters having *closo*- or *arachno*-borane electron counts. Such compounds are dealt with more fully in other chapters, as are main-group element metallaborane clusters and other heteroatom clusters such as metallacarborane compounds. Various aspects of the subject have previously been reviewed.[2-7]

2. MONOBORON AND DIBORON COMPOUNDS

One of the first metallaborane complexes to be characterized was $[Mn_3(B_2H_6)(CO)_{10}H]$ formed as a by-product (15% yield) in the synthesis of $[Mn(CO)_4H]_3$ from $[Mn_2(CO)_{10}]$ and an excess of $NaBH_4^-$. The compound was obtained as dark-red diamagnetic needles which were moderately stable in air.[8] X-ray analysis revealed the structure shown in Figure 1. Notable features are the direct B—B bond (176 pm) in the central $\{B_2H_6\}$ unit and the six B—H—Mn bonds, two to each Mn atom. All three Mn atoms are six-coordinate but, whereas the unique Mn atom carries four terminal CO groups, the other two have three terminal CO groups and a bridging H atom. The presence of the B—B bond suggests that the borane ligand can be considered as the *arachno*-anion $\{B_2H_6\}^{2-}$, isoelectronic with C_2H_6. If the bridging hydrido ligand is taken as H_μ^- for the sake of electron counting, then each Mn atom is in the +1 oxidation state (d^6) and the six ligands around each bring its electron count to 18. Another notable feature of the structure is that the Mn⋯Mn distance in the three-center Mn—H—Mn bond is only 284.5 pm which is significantly shorter than the directly-bonded Mn—Mn distance of 292.3 pm in $[(CO)_5MnMn(CO)_5]$.

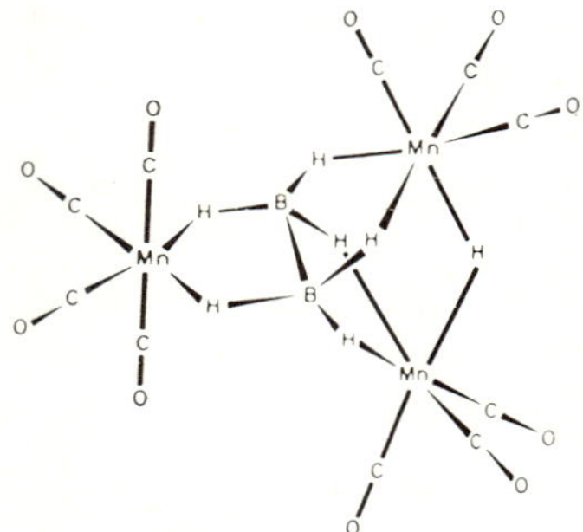

Figure 1. Molecular configuration of $[Mn_3(\eta^6\text{-}B_2H_6)(CO)_{10}(\mu\text{-}H)]$ showing the direct two-center B–B bond and the three-center Mn–Hμ–Mn interaction. The borane moiety can be thought of as an anionic *tris*-bidentate $\{\eta^6\text{-}B_2H_6\}^{2-}$ *arachno*-borane ligand. The four independent Mn–B diatances are equivalent with a mean value of 230 pm (individual esd 2pm).

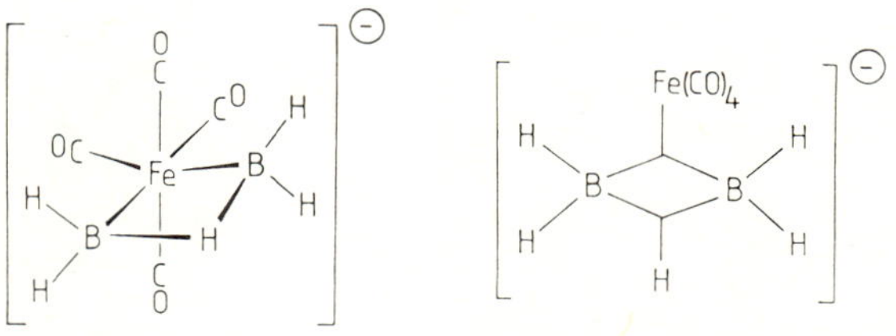

Figure 2. Proposed structure and bonding for the anion $[Fe(\eta^2\text{-}B_2H_5)(CO)_4]^-$.

More recently metallaboranes have been prepared in which a bridging H atom in B_2H_6 has been formally replaced by a coordinated Fe atom:[9,10]

$$K_2^+[Fe(CO)_4]^{2-} + 3[(C_4H_8O)BH_3] \longrightarrow K^+[\mu\text{-}\{Fe(CO)_4\}(B_2H_5)]^- + KBH_4 + 3C_4H_8O$$

$$K^+[Fe(\eta^5\text{-}C_5H_5)(CO)_2] + 3/2B_2H_6 \xrightarrow{Me_2O/-78^\circ} [Fe(\eta^5\text{-}C_5H_5)(CO)_2(\eta^2\text{-}B_2H_5)] + KBH_4$$

In the first reaction the product is obtained in a 98% yield as a dark-brown solid which was stable at room temperature for several hours when sealed under vacuum.[9] The structure (Figure 2), proposed on the basis of ^1H and ^{11}B nmr spectra, has a bridging $\{Fe(CO)_4\}$ unit replacing a bridging proton H_μ^+, both species contributing one vacant orbital and no electrons to the bonding. Expressed in other words $\{\eta^2\text{-}B_2H_5\}^-$ is acting as a *bihapto* two-electron donor to form a three-center BFeB bond in which the $Fe^{\underline{0}}$ atom is formally dsp^3 hybridized as in the direct isoelectronic analog $[Fe(\eta^2\text{-}C_2H_4)(CO)_4]$. In the product of the second reaction the additional bonding electron in the bridging $\{Fe(\eta^5\text{-}C_5H_5)(CO)_2\}$ group enables it to formally replace an H atom (rather than a proton) of B_2H_6 to give the neutral metalladiborane $[Fe(\eta^5\text{-}C_5H_5)(CO)_2(\eta^2\text{-}B_2H_5)]$ which is a volatile yellow crystalline solid (Figure 3).

In this section should also be mentioned the yellow-brown air-sensitive liquid $[Fe_2(CO)_6(B_2H_6)]$ which is formed in 1–10% yield by an obscure reaction between $[Fe(CO)_5]$ and B_5H_9 in the presence of $LiAlH_4$ in ether solution at room temperature; subsequent treatment with HCl gives $[Fe(CO)_3(B_4H_8)]$ (see p. 50) together with the less volatile diferricdiborane.[11] Spectroscopic properties are consistent with the structure shown in Figure 4; the compound can be viewed *either* as a complex of the eight-electron donor $\{\eta^4\text{-}B_2H_6\}^{2-}$ and the $\{Fe_2(CO)_6\}^{2+}$ dimer, thereby satisfying the 18-electron rule, *or* as a

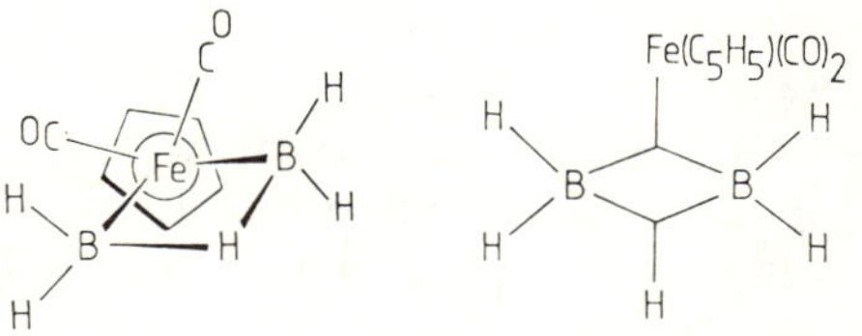

Figure 3. Proposed structure and bonding for the neutral species $[Fe(\eta^2\text{-}B_2H_5)(\eta^5\text{-}C_5H_5)(CO)_2]$.

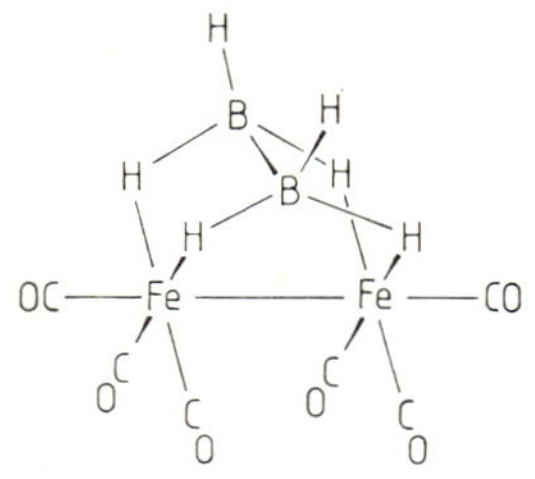

Figure 4. Proposed structure of $[Fe_2(B_2H_6)(CO)_6]$ showing the presumed presence of a B—B and an Fe—Fe bond, and four three-center $B-H_\mu-Fe$ bridging bonds.

four-atom *nido*-cluster having 12 skeletal electrons (two from each $\{Fe(CO)_3\}$ unit, two from each $\{BH\}$ unit, and one from each $\{BHFe\}$ unit). The former description emphasizes the analogy with the $\{\eta^6-B_2H_6\}^{2-}$ trimanganese complex in Figure 1.

3. TRIBORON COMPOUNDS

Nido-triborane, B_3H_7, is an unstable transient species which has not been isolated or characterized, although reactions involving the more stable adducts B_3H_7L may well prove to be of synthetic utility in generating $\{B_3H_7\}$ for the preparation of metal triboranes. As a special case of B_3H_7L (where $L = H^-$), the *arachno*-borane anion $B_3H_8^-$ is known to form *di*- and *trihapto*-complexes via three-center B—H—M bonds (see other chapters). Complexes involving direct B—M contacts that have so far been prepared are of three structural types, as illustrated schematically in Figure 5, and these will be considered in turn.

A general route to complexes of the type $[M^{(II)}(\eta^3-B_3H_7)L_2]$ for M = Ni, Pd, Pt is the reaction of a *cis*-bis (phosphine) complex with $B_3H_8^-$ in the presence of a base such as NEt_3:[12,13]

$$cis-[PtCl_2(PR_3)_2] + CsB_3H_8 + NEt_3 \longrightarrow [Pt(\eta^3-B_3H_7)(PR_3)_2] + CsCl + Et_3NHCl$$

For the platinum complexes, which are the most stable, yields are in the range of 10-55%; the compounds are white crystalline solids and typical melting points (mp's) are: $L = PEt_3$ 93°, PMe_2Ph 112°, PPh_3 196°(d), and $P(p$-tolyl$)_3$ 180°(d). The off-white palladium complex $[Pd(\eta^3-B_3H_7)(Ph_2PCH_2CH_2PPh_2)]$ melts with decomposition at 176°. The nickel analogs are much less stable. A key to the unusual structure of these complexes derives from their ^1H nmr spectra which, in the BH region, show four resonances with an intensity ratio 2:1:2:2. The positions of these resonances and the stereochemical rigidity of the $\{B_3H_7\}^{2-}$ ligand contrasts markedly with the facile fluxionality of $B_3H_8^-$. The crystal structure of $[Pt(\eta^3-B_3H_7)(PMe_2Ph)_2]$[13] shows a σ,μ-borallyl disposition of the ligand which is very reminiscent of some complexes of the isoelectronic allyl ligand $\{C_3H_5\}^-$ (see Figure 6). A more detailed view of the coordination geometry about the Pt atom is shown in Figure 7 which illustrates a second orientation of the $\{\eta^3-B_3H_7\}^{2-}$ ligand with respect to the metal atom in the dis-

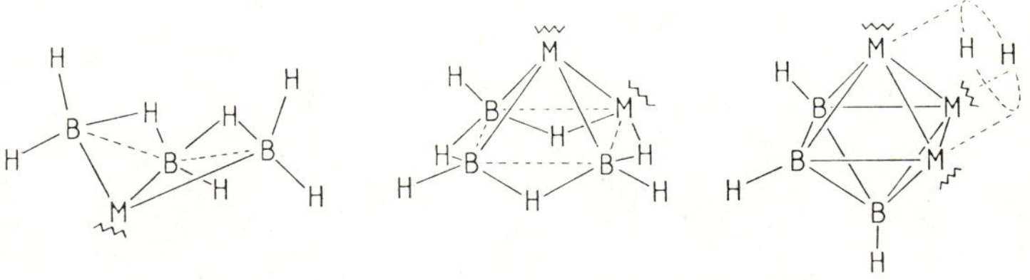

Figure 5. Schematic representation of *trihapto* triborane complexes involving B—M bonds to one- two-, and three-metal centers.

ordered crystal. The complex is best described as a dsp^2 bonded square-planar complex of Pt(II) and the presence of Pt(II) is also consistent with the Pt $4f_{7/2}$ ESCA spectra of the complexes.

More recently, a second, related route to $\{\eta^3\text{-}B_3H_7\}$ complexes has been reported which involves ligand replacement of Cl^- by $B_3H_8^-$ in an Ir(I) complex followed by internal oxidative addition to give the product in a 40% yield:[14]

$$\textit{trans-}[Ir^{(I)}(CO)Cl(PPh_3)_2] + TlB_3H_8 \xrightarrow{C_6H_6/20°} TlCl + [Ir^{(III)}(\eta^3\text{-}B_3H_7)(CO)(PPh_3)_2]$$

The compound is a white, air-stable solid, mp 150°(d), and the structure proposed on the basis of ^1H, ^{11}B, and ^{31}P nmr spectra is shown in Figure 8 (cf. Figure 5). The compound was interpreted as being an iridatetraborane(8) in which one "hinge" {BH} group in *nido*-B_4H_8 (2112 topology with one vacant orbital) has been replaced by the {Ir(CO)H(PPh$_3$)$_2$} group, although its alternative descrip-

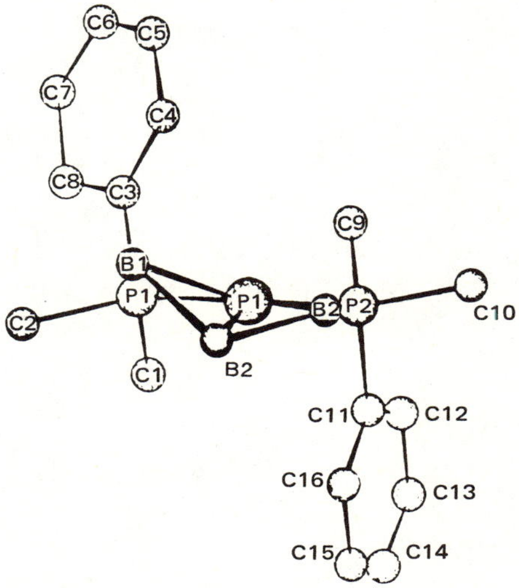

Figure 6. Molecular structure of $[Pt(\eta^3\text{-}B_3H_7)(PMe_2Ph)_2]$ with H atoms omitted, viewed approximately in the P(1)—Pt—P(2) plane. The "butterfly" configuration of the four-vertex *nido*-metallatetraborane unit is apparent.

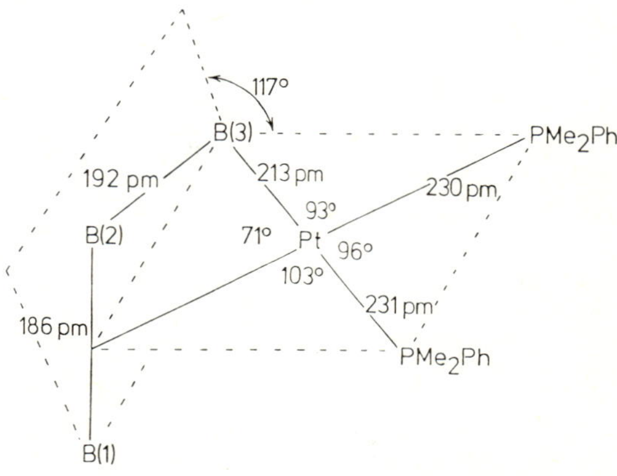

Figure 7. Detailed coordination geometry about the Pt(II) atom and one orientation of the disordered $\{B_3H_7\}^{2-}$ ligand in $[Pt(\eta^3\text{-}B_3H_7)(PMe_2Ph)_2]$. The Pt atom is coplanar with P(1), P(2), B(3) and the midpoint of B(1)–B(2) [compare with Figure 46]. The B(1)–B(2)–B(3) plane makes an angle of ∿117° with this plane. The distances Pt–B(1) 236 pm and Pt–B(2) 217 pm are longer than Pt–B(3) 213 pm and the distance B(1) ⋯ B(3) 315 pm, is nonbonding (compare 178 pm in $B_3H_8^-$). H atoms on the $\{B_3H_7\}^{2-}$ ligand could not be refined.

tion as an analog of *arachno*-$B_4H_9^-$ (no vacant orbitals) cannot be ruled out on present evidence (see Section 14.1). By contrast when $B_3H_8^-$ is allowed to react with *trans*-$[Rh^{(I)}(CO)Cl(PPh_3)_2]$ or with various complexes of Co(I) or Co(II) under similar conditions, the main boron-containing products are B_3H_7L together with $B_2H_4L_2$ and BH_3L (where L = pyridine or tertiary phosphine).

The second class of metallaborane involving the $\{B_3H_7\}$ ligand is exemplified, at present, by the sole example of the *nido*-diferrapentaborane $[Fe_2(B_3H_7)\text{-}(CO)_6]$.[15] This is formed in a 1% yield as volatile, air-sensitive, orange crystals, mp 51°, by a variant of the reaction mentioned on p. 71:

$$B_5H_9 + Fe(CO)_5 + LiAlH_4 \xrightarrow[60-70°]{(MeOCH_2-)_2} [Fe_2(B_3H_7)(CO)_6] + \cdots$$

The compound is thermally stable (45° for 2 months) and has the structure shown in Figure 9. The relationship to B_5H_9 is clear, the apex $\{BH\}$ unit and one basal $\{BH\}$ unit having been replaced by the isolobal $\{Fe(CO)_3\}$ groups. Ac-

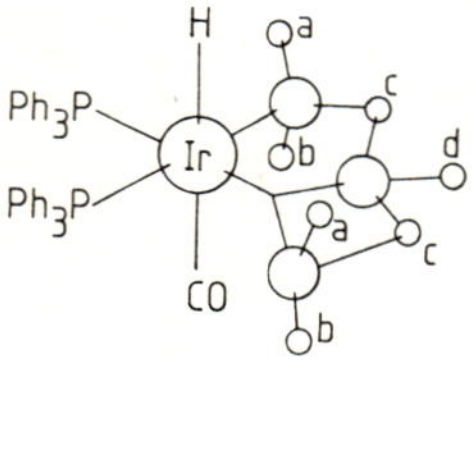

Figure 8. Proposed structure and bonding of $[Ir(\eta^3\text{-}B_3H_7)(CO)H(PPh_3)_2]$; the two "terminal" B atoms are equivalent in solution (one hybrid shown) and the ^1H–$\{^{11}B\}$ nmr spectrum shows the expected intensity rations 1:2:2:2:1, the last resonance being a high-field triplet due to Ir–H coupled to the two *cis* ^{31}P nuclei. (See Figure 70).

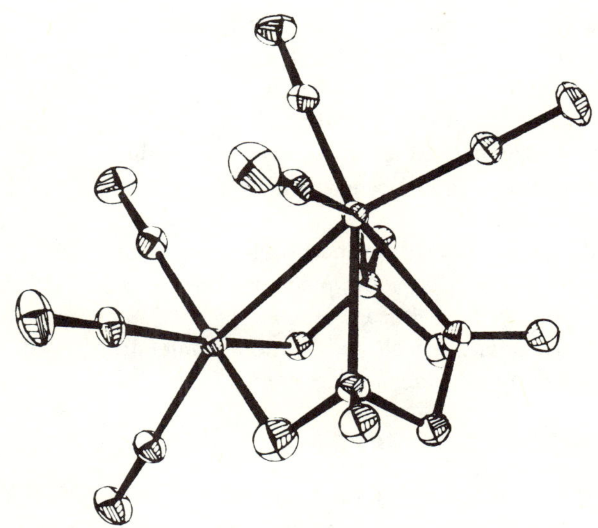

Figure 9. Structure of $[Fe_2(B_3H_7)(CO)_6]$. The Fe—Fe distance is 256 pm (single bond) and the three Fe_{apex}—B distances are 206 ± 1 pm. The two Fe_{base}—H—B distances are rather longer (225, 227 pm) and the B—B, B—H_t and B—H_μ distances fall within normal ranges.

cordingly, the bonding can be described as indicated in Figure 10(a). Alternatively, the compound can be thought of as a complex of the ligand $\{B_3H_7\}^{2-}$ which acts as a donor to both Fe atoms in $\{Fe_2(CO)_6\}^{2+}$ as in Figure 10(b), and as such has similarities to the complex between $\{B_2H_6\}^{2-}$ and $\{Fe_2(CO)_6\}^{2+}$ in Figure 4. The $\{B_3H_7\}^{2-}$ moiety is stereochemically rigid on the nmr time scale, but differs from that in the Pt(II) and Ir(III) complexes just considered in its mode of attachment; the ^1H-$\{^{11}B\}$ decoupled nmr spectrum shows three

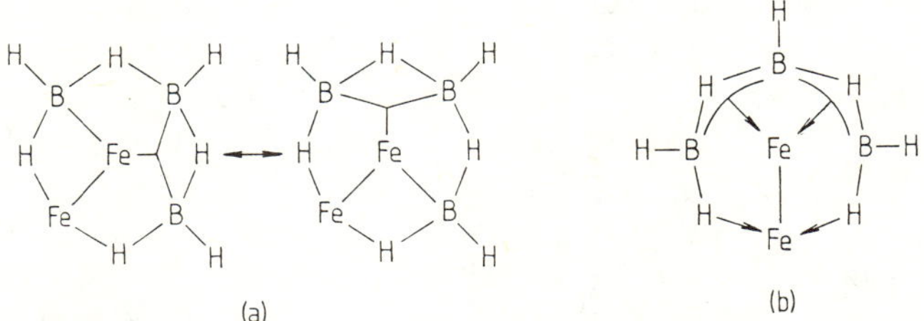

(a)   (b)

Figure 10. Alternative descriptions of the bonding mode in $[Fe_2(B_3H_7)(CO)_6]$: (a) two valence bond structures showing the analogy to *nido*-pentaborane, B_5H_9; note also the similarities to the bonding in $[(IrB_3H_7)(CO)H(PPh_3)_2]$ (Figure 8); (b) a representation of the molecule as resulting from the donation of four electrons from the "π-system" of the pseudoallyl anion to one Fe center, and the donation of four electrons in two B—H bonds to the second Fe atom; with a single bond between the Fe atoms the 18–electron rule is satisfied.

resonances with the intensity ratio 2:3:2, presumably arising from the two $B-H_\mu-B$ protons, the three (almost equivalent) $B-H_t$ protons, and the two $B-H_\mu-Fe$ protons.

The final structure type alluded to in Figure 5 is the octahedral cluster $\{M_3B_3\}$. An example is the brown crystalline solid $[\{Co(\eta^5-C_5H_5)\}_3(BH)_3-(\mu-H)_2]$ which has a *fac* arrangement of the $\{Co_3B_3\}$ skeleton.[16] Each Co atom is coordinated by an $\{\eta^5-C_5H_5\}$ ligand and each B atom carries a terminal H atom; the final two H atoms occupy bridging sites on the Co_3 face, but disorder and fluxionality have so far precluded their location. The compound is one of the many products formed from the very fruitful reaction between $CoCl_2$, $C_5H_5^-$, and $B_5H_8^-$ and is dealt with in more detail in the pentaboron section.

Further triboron metallaboranes are mentioned in Section 14.1.

4. TETRABORON COMPOUNDS

As with *nido*-triborane(7), the *nido*-tetraborane(8) B_4H_8 has been postulated only as a transient reactive intermediate and its direct interaction with metal compounds has not yet been investigated. Again, it appears that dissociation of the more stable *arachno* base-adducts such as $B_4H_8 \cdot CO$ may have a potential synthetic utility that remains to be exploited. Several metal tetraboranes have, however, been prepared from various other tetraborane or pentaborane species.

A key compound in which a metal center has clearly replaced a $\{BH\}$ group in a borane cluster is the *nido*-ferrapentaborane $[Fe(\eta^4-B_4H_8)(CO)_3]$.[17] The compound was first prepared (10–20% yield) by the reaction of B_5H_9 with $[Fe(CO)_5]$ in a hot/cold reactor at $220°/20°$; the yield depends critically upon conditions and falls almost to zero above $240°$ or below $200°$ (see also p. 58). An alternative, although less effective route (2–3% yield), is from B_4H_{10} and $[Fe_2(CO)_9]$ at low temperatures. $[Fe(\eta^4-B_4H_8)(CO)_3]$ is an orange liquid, freezing point (fp) $\sim 5°$, which bears an obvious structural resemblance to B_5H_9, the apical $\{BH\}$ being formally replaced by the isoelectronic and isolobal group $\{Fe(CO)_3\}$ as shown in Figure 11. The relationships to the isoelectronic cyclobutadiene complex $[Fe(\eta^4-C_4H_4)(CO)_3]$ and to the carbido cluster-compound $[\{Fe(CO)_3\}_5C]$ are readily apparent. See also Figure 9 in which two $\{BH\}$ groups have been replaced by $\{Fe(CO)_3\}$ units. The ultraviolet photoelectron spectrum of $[Fe(\eta^4-B_4H_8)(CO)_3]$ and several MO calculations on the molecule have been reported.[18,19]

The closely related *nido*-cobaltapentaboranes $[1-\{Co(\eta^5-C_5H_5)\}(\eta^4-B_4H_8)]$ and $[2-\{Co(\eta^5-C_5H_5)\}(\eta^4-B_4H_8)]$ are also known.[20] The latter species is among the many products of the previously mentioned reaction between anhydrous $CoCl_2$, NaB_5H_8, and NaC_5H_5, in THF below $-20°$, and the red basal (or 2) isomer can be converted to the yellow-apical (or 1) isomer by heating at $200°$ for 30 min.[21] Structures based on spectroscopic data are in Figure 12 and a detailed X-ray single crystal structure on the red isomer con-

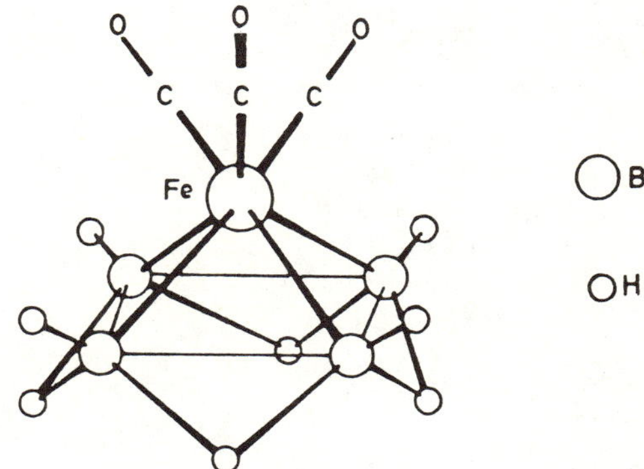

Figure 11. Proposed structure of [Fe-$(B_4H_8)(CO)_3$]. The structural intermediacy between B_5H_9 and [Fe_2-$(B_3H_7)(CO)_6$] (Figure 9) is apparent.

firmed this:[22] selected interatomic distances are $Co-B_{apex}$ 201 pm, $Co-B_{base}$ 214 pm, $B_{apex}-B_{base}$ 167 pm, $B_{base}-B_{base}$ 176 pm, and $Co-H_\mu$ 143 pm.

A summary of the metallaborane products obtained from the general reaction is given in Table 1.[21] In a convenient variant of the reaction, which leaves the product distribution essentially unchanged, $C_5H_5^-$ is used both to deprotonate B_5H_9 and to coordinate to Co: the preferred technique is to prepare NaC_5H_5 by deprotonating C_5H_6 with NaH in THF and then to add B_5H_9 and anhydrous $CoCl_2$ at $-196°$; dissolution is effected at $-20°$ and the reaction then allowed to proceed for several hours at $-78°$ before removal of the solvent and work up in air. All the tabulated products can be regarded as diamagnetic Co(III) species. In addition, the reaction yields the orange organometallic sandwich species 1,3-cyclohexadienyl cobalt(II) cyclopentadienyl, [$Co(C_6H_8)(C_5H_5)$], as an unexplained major product and there is a trace of the novel red carbocyclic *closo*-metallacarbaborane complex [$1,7-\{Co(\eta^5-C_5H_5)\}_2-\mu(2,3)-C_3H_4-2,3-C_2-$

(a)　　　　　　　　　　　　　(b)

Figure 12. Structures of [$2-\{Co(\eta^5-C_5H_5)\}(B_4H_8)$] (a) and [$1-\{Co(\eta^5-C_5H_5)\}B_4H_8$] (b); that for the 2–isomer has been confirmed by single-crystal X-ray diffraction analysis (see text).

Table 1. Metallaborane Products from the Reaction of $CoCl_2$ with
$B_5H_8{}^-$ and $C_5H_5{}^-$

Compound	Color	Yield	Structure
Nido $(2n + 4)$–electron systems			
$[1\text{-}\{Co(\eta^5\text{-}C_5H_5)\}(B_4H_8)]$	Yellow	[a]	Figure 12
$[2\text{-}\{Co(\eta^5\text{-}C_5H_5)\}(B_4H_8)]$	Red	4.7%	Figure 12
$[5\text{-}\{Co(\eta^5\text{-}C_5H_5)\}(B_9H_{13})]$	Red	[b]	Figure 54
Closo $(2n + 2)$–electron systems			
$[1,2\text{-}\{Co(\eta^5\text{-}C_5H_5)\}_2(B_4H_6)]$	Violet	1%[c]	Figure 13
$[1,2\text{-}\{Co(\eta^5\text{-}C_5H_5)\}_2\text{-}3\text{-}\sigma\text{-}C_5H_9\text{—}(B_4H_5)]$	Violet	[b]	Figure 13
$[1,2\text{-}\{Co(\eta^5\text{-}C_5H_5)\}_2\text{-}4\text{-}\sigma\text{-}C_5H_9\text{—}(B_4H_5)]$	Violet	[b]	Figure 13
$[1,2,3\text{-}\{Co(\eta^5\text{-}C_5H_5)\}_3(B_3H_5)]$	Brown	[c]	Figure 5
$[1,2,3\text{-}\{Co(\eta^5\text{-}C_5H_5)\}_3C_5H_7(B_3H_4)]$	Brown	[b]	–
$2n$–electron systems			
$[1,2,3\text{-}\{Co(\eta^5\text{-}C_5H_5)\}_3(B_4H_4)]$	Yellow	[c]	Figure 14
$[3,4,5,6\text{-}\{Co(\eta^5\text{-}C_5H_5)\}_4(B_4H_4)]$	Green	[c]	Figure 14

[a] Not obtained in original reaction; formed in a 30% yield by isomerization of 2–isomer at 220° (Ref. 21; T. L. Venable and R. N. Grimes, unpublished results).

[b] Except where otherwise stated, yields were in the range 0.01–0.1% based on B_5H_9 used.

[c] Obtained in a larger yield in the reaction of the $(C_5H_5)CoB_4H_7{}^-$ ion with $CoCl_2$ and $C_5H_5{}^-$; see text, pp. 53–54.

$(B_3H_3)]$. Of the tetraboron products, crystal structures and detailed interatomic distances are available for violet $[1,2\text{-}\{Co(\eta^5\text{-}C_5H_5)\}_2(B_4H_6)]$,[23] yellow $[1,2,3\text{-}\{Co(\eta^5\text{-}C_5H_5)\}_3(B_4H_4)]$,[16] and green $[3,4,5,6\text{-}\{Co(\eta^5\text{-}C_5H_5)\}_4\text{-}(B_4H_4)]$[24] in addition to that for red $[2\text{-}\{Co(\eta^5\text{-}C_5H_5)\}(B_4H_8)]$ mentioned previously,[22] but the geometrical features are more readily discerned from the diagrams in Figures 12-14. Notable features include the two face-bridging H atoms on the Co_2B faces of violet $[1,2\text{-}\{Co(\eta^5\text{-}C_5H_5)\}_2(B_4H_6)]$, and the *capped-closo*-structure of yellow $[1,2,3\text{-}\{Co(\eta^5\text{-}C_5H_5)\}_3(B_4H_4)]$ in which a $\{BH\}$ group lies centrally above the $\{Co_3\}$ face. The $\{Co_4B_4\}$ *closo*-polyhedron in green $[3,4,5,6\text{-}\{Co(\eta^5\text{-}C_5H_5)\}_4(B_4H_4)]$ is very close to an idealized D_{2d} dodecahedron analogous to the $B_8H_8{}^{2-}$ anion, although (like B_8Cl_8) it has only 16 framework valence electrons rather than the 18 expected (as in $B_8H_8{}^{2-}$) on the basis of Wade's rules [25] (See also p. 102).

The variety of products from this reaction is intriguing, but the mechanism is not understood. It is remarkable that no pentaboron species are formed. The product distribution has been taken to suggest that the major process occurring is the replacement of a B atom in $B_5H_8{}^-$ by cobalt to give $[2\text{-}\{Co(\eta^5\text{-}C_5H_5)\}B_4H_8]$ with subsequent addition of more cobalt to produce a metal-rich species.[21] That $C_5H_5{}^-$ plays a major role is apparent from the formation of the

Figure 13. Structure of $[1,2\text{-}\{Co(\eta^5\text{-}C_5H_5)\}_2(B_4H_6)]$; in addition to the four terminal H atoms bound to B(3)–B(6) there are two bridging H atoms which are associated with the two deltahedral faces containing the Co(1)–Co(2) edge.

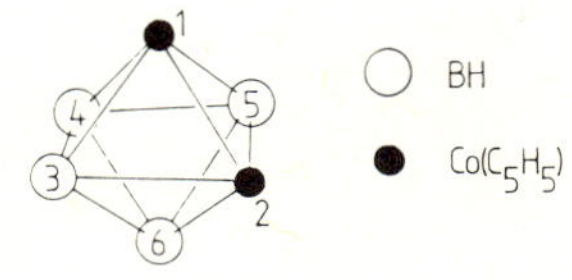

hydrocarbon derivatives $[1,2\text{-}\{Co(\eta^5\text{-}C_5H_5)\}_2\text{-}3\text{-}\sigma\text{-}C_5H_9(B_4H_5)]$, $[1,2\text{-}\{Co(\eta^5\text{-}C_5H_5)\}_2\text{-}4\text{-}\sigma\text{-}C_5H_9(B_4H_5)]$, and, in particular $[1,2,3\text{-}\{Co(\eta^5\text{-}C_5H_5)\}_3C_5H_7(B_3H_4)]$, which suggests a direct attack of the Lewis base $C_5H_5^-$ on one or more of the borane substrates during the course of the reaction. No neutral metallaboranes have, thus far, been isolated in the absence of cyclopentadiene and it is suspected that $C_5H_5^-$ probably facilitates the abstraction of boron from, and the introduction of Co atoms into, the monocobaltaborane species initially formed. Evidence for $[2\text{-}\{Co(\eta^5\text{-}C_5H_5)\}B_4H_8]$ and its anion being key intermediates in the process is provided by its reaction, in turn, with anhydrous $CoCl_2$ and NaC_5H_5 as discussed below.

Although most of the products tabulated above are obtained in excruciatingly small yields (0.01–0.1%) further work on the major *nido*-cobaltapentaborane product $[2\text{-}\{Co(\eta^5\text{-}C_5H_5)\}(B_4H_8)]$ has shown it to be a useful source of several of the other cobaltapolyboranes.[26] Thus, bridge deprotonation of red $[2\text{-}\{Co(\eta^5\text{-}C_5H_5)\}(B_4H_8)]$ with NaH in THF below room temperature yields $[2\text{-}\{Co(\eta^5\text{-}C_5H_5)\}(B_4H_7)]^-$ which can be used as the starting point of several further reactions. Reprotonation of the anion with dry HCl regenerates the original metallaborane in 84% yield and it has been shown that the proton abstracted is from a B–H–Co rather than a B–H–B location. Further reaction of $[2\text{-}\{Co(\eta^5\text{-}C_5H_5)\}(B_4H_7)]^-$ with anhydrous $CoCl_2$ and NaC_5H_5 in THF at room temperature gives violet $[1,2\text{-}\{Co(\eta^5\text{-}C_5H_5)\}_2(B_4H_6)]$ (14.9% yield based on the amount of red isomer consumed), yellow $[1,2,3\text{-}\{Co(\eta^5\text{-}C_5H_5)\}_3(B_4H_4)]$

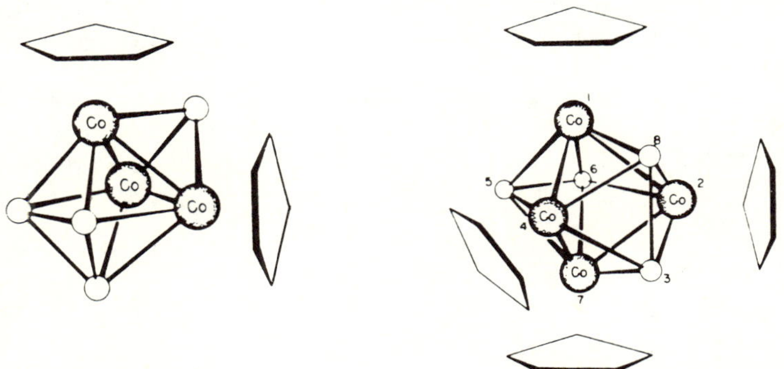

Figure 14. Structures of the $2n$-electron cobaltaboranes $[1,2,3\text{-}\{Co(\eta^5\text{-}C_5H_5)\}_3(B_4H_4)]$ (left, with one C_5H_5 ring omitted for clarity) and $[3,4,5,6\text{-}\{Co(\eta^5\text{-}C_5H_5)\}_4(B_4H_4)]$ (right).

(2% yield), brown $[1,2,3-\{Co(\eta^5-C_5H_5)\}_3(B_3H_5)]$ (1.75% yield), and green $[3,4,5,6-\{Co(\eta^5-C_5H_5)\}_4(B_4H_4)]$ (0.72% yield).

Mixed-metal metallaborane clusters can also be prepared;[26] for example, the room temperature reaction of red $[2-\{Co(\eta^5-C_5H_5)\}(B_4H_8)]$ with $[Fe(CO)_5]$ in heptane under ultraviolet light gives a 1-2% yield of the novel brown cluster complex $[1,2,3-\{Co(\eta^5-C_5H_5)\}_2\{Fe(CO)_4\}(B_3H_3)]$. Nmr studies lead to the structure proposed in Figure 15 which has a *facial* disposition of the three metal atoms with two terminal and two bridging CO groups. Other products of this reaction are the expected $[Fe_2(CO)_9]$ and $[Fe_3(CO)_{12}]$ and the previously known cluster complex $[Fe_2(CO)_9Co(\eta^5-C_5H_5)]$, of which each is obtained in a 1-2% yield. Additionally, reaction of red $[2-\{Co(\eta^5-C_5H_5)\}-(B_4H_8)]$ with C_2H_2 produces the known metallacarbaborane $[1,2,3-\{Co(\eta^5-C_5H_5)\}C_2B_3H_7]$ (see other chapters) in a reaction that is clearly analogous to the insertion of C_2H_2 into the parent B_5H_9 to yield $C_2B_4H_8$.

So far in this section we have considered tetraboron clusters in which Fe and/or Co atoms have been the heteroatoms. Nickel and copper have also been incorporated to form metal-tetraborane clusters. In a reaction inspired by the fertile juxtaposition of $CoCl_2/NaB_5H_8/NaC_5H_5$ discussed at length above, $NiBr_2$ was similarly treated, but only the larger clusters $[1,6-\{Ni(\eta^5-C_5H_5)\}_2(B_8H_8)]$ (p. 81) and $[1,7-\{Ni(\eta^5-C_5H_5)\}_2(B_{10}H_{10})]$ (p. 94) have so far been isolated.[27] By contrast, when nickelocene $[Ni(\eta^5-C_5H_5)_2]$ itself is allowed to react in THF with NaB_5H_8 and stoichiometric amounts of sodium amalgam, variable yields of tetranickel clusters with four or five boron atoms are produced.[27, 28] Under optimum conditions a 17% yield of brown *closo*-$[1,2,7,8-\{Ni(\eta^5-C_5-H_5)\}_4(B_4H_4)]$ and 10% of green *nido*-$[\{Ni(\eta^5-C_5H_5)\}_4(B_5H_5)]$ (p. 71) can be obtained. An X-ray crystal study of the brown tetraboron cluster confirmed the D_{2d} *closo*-structure in Figure 16 despite the 20 skeletal valence electrons, i.e. $(2n + 4)$, which might lead one to expect a *nido*-cluster.[27] It will be noticed also that the Ni atoms (interestingly) occupy the low-coordinate vertices in the dodecahedron whereas in the cobalt D_{2d} analog (Figure 14) the metal atoms occupy the high-coordinate vertices and form a contiguous belt around the

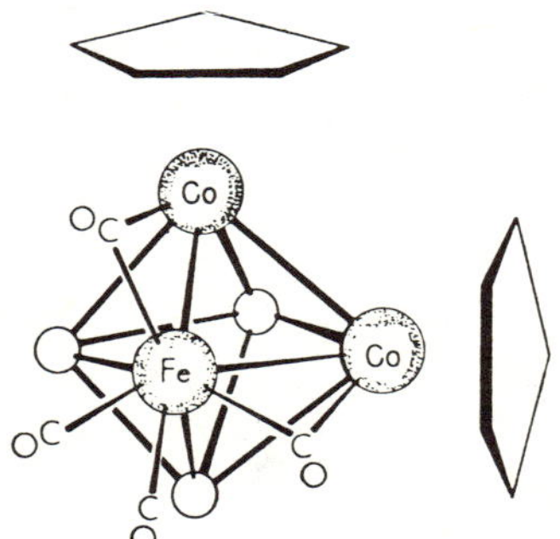

Figure 15. Proposed structure of the *closo*-ferradicobaltahexaborane $[1,2-\{Co(\eta^5-C_5H_5)\}_2-3-\{Fe(CO)_4\}(B_3H_3)]$.

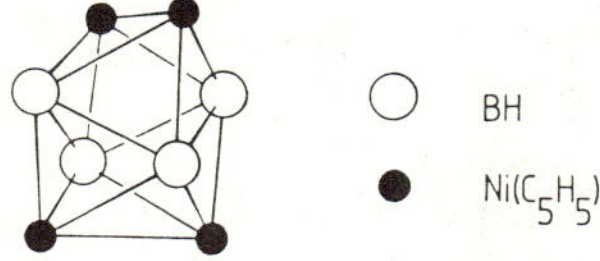

Figure 16. Representation of the D_{2d} structure of the central tetrametallaborane cluster in *closo*-[1,2,7,8-$\{Ni(\eta^5\text{-}C_5H_5)\}_4(B_4H_4)]$; contrast with the cobalt analog in Figure 14.

middle of the cluster; moreover the Ni—Ni distance of 235 pm is the shortest M—M link of any type yet found in metallaboron polyhedra and may indicate localized Ni—Ni bonding. (See also Section 14.2).

Copper-tetraboron clusters are at present represented by the sole example of $[Cu(\eta^3\text{-}B_4H_9)(PPh_3)_2]$. This is a cream white solid which is remarkably stable in air when compared with free salts of the *arachno*-tetraborane anion $B_4H_9^-$ which may be made by the deprotonation of B_4H_{10} with KH. The copper compound can be made by the reaction of $[CuCl(PPh_3)_3]$ with either KB_4H_9 or KB_5H_{12} in THF/CH_2Cl_2 solution at $-45°$:[29]

$$B_4H_9^- + [CuCl(PPh_3)_3] \longrightarrow [Cu(B_4H_9)(PPh_3)_2] + Cl^- + PPh_3$$

$$B_5H_{12}^- + [CuCl(PPh_3)_3] \longrightarrow [Cu(B_4H_9)(PPh_3)_2] + Cl^- + Ph_3PBH_3$$

Spectroscopic data suggest a square pyramidal structure with (uniquely for a copper-*arachno*-borane) no evidence for Cu—H—B bridge bonding (see Figure 17.) The ^1H nmr spectrum is independent of temperature between $-66°$ and $0°$ and gives no evidence of dynamic character, in contrast to the highly fluxional nature of the uncoordinated $B_4H_9^-$ ion. This stereochemical rigidity in $\{\eta^3\text{-}B_4H_9\}^-$ is reminiscent of the coordinated $\{\eta^3\text{-}B_3H_7\}^{2-}$ (p. 47). A similar reaction of KB_4H_9 with the nickel complex $[NiBr_2(Ph_2PCH_2CH_2PPh_2)]$ gives the nickelaborane $[NiBr(B_4H_9)\,(dppe)]$ which may also have a square-based pyramidal structure with the Ni atom occupying one of the basal sites.[30]

5. PENTABORON COMPOUNDS

Nido-pentaborane, B_5H_9, is thermally the most stable and also one of the most readily available of the lower boranes and, accordingly, its metal-derivative chemistry has been extensively studied. Formal replacement of terminal or bridging H atoms by metal centers leads to the clusters shown schematically in Figure 18(a) and (b) and more complete incorporation of the metal center can

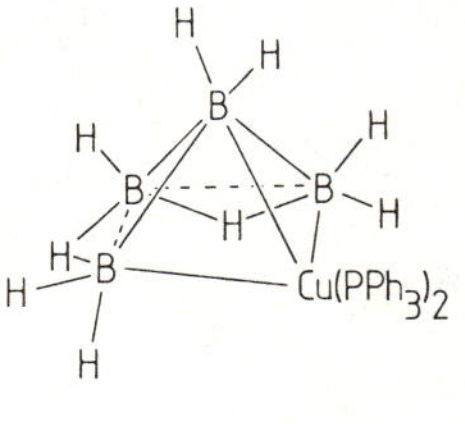

Figure 17. Proposed structure for $[Cu(\eta^3\text{-}B_4H_9)(PPh_3)_2]$.

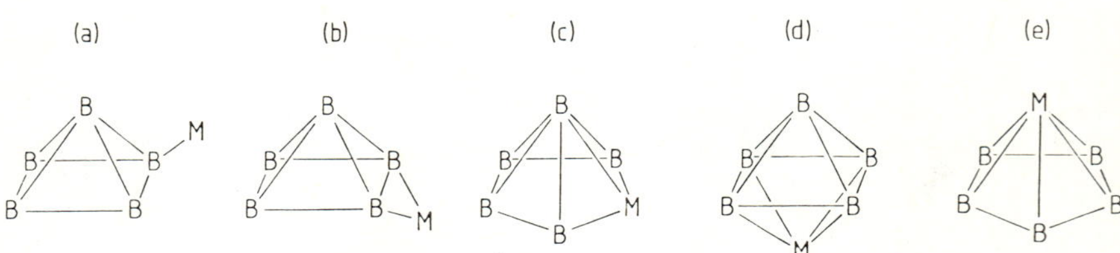

Figure 18. Schematic representation of the geometry of some known metal–pentaboron clusters; terminal and bridging H atoms have been omitted so as to emphasize the cluster geometry and the various ligands attached to the metal atoms M have also been omitted for generality.

yield *nido*- or *closo*-metallahexaboranes as in Figure 18(c) and (d). Even greater modification of the pentaborane cluster occurs in the isomeric *nido*-metalla-hexaborane [Figure 18(e)] and, as is to be seen later, incorporation of two, three, or four metal atoms in a pentaboron cluster can lead to geometries far removed from the original square pyramid of B_5H_9. Because of the large number of structure types involved and the diversity of the chemistry, it is convenient to subdivide this section according to the various preparative routes which have been used. The sequence of presentation is:

1. direct reaction of B_5H_9 with metal atoms;
2. direct reaction of B_5H_9 with metal compounds;
3. reaction of $B_5H_8^-$ with complex metal halides;
4. reaction of halopentaboranes B_5H_8X with anionic metal complexes;
5. insertion by oxidative addition;
6. reaction of B_5H_9 or $B_5H_8^-$ with transition-metal compounds in the presence of reductive bases;
7. insertion of further metal centers into a preformed metalla-*nido*-pentaborane.

Using these various routes some 20 different metals have now been inserted into pentaborane species. The most extensively studied systems involve metals in the second half of the transition element series (Mn, Re; Fe; Co, Rh, Ir; Ni, Pd, Pt; Cu, Ag, Au; Cd, Hg) but a Ti(III) complex has also been reported as well as derivatives of the main-group elements, notably Be, Si, Ge, Sn and Pb (p. 64).

5.1. Direct Reaction of B_5H_9 with Metal Atoms

Cocondensation of Co atoms with B_5H_9 and cyclopentadiene at $-196°$ in an approximate mole ratio 1:50:50 followed by warming to room temperature yields several products including the new dark-green air-stable compound $[\{Co(\eta^5-C_5H_5)\}_3(B_5H_5)]$ and the yellow–green cyclopentyl substituted derivative $[\{Co(\eta^5-C_5H_5)\}_3\{(\sigma\text{-cyclo-}C_5H_7)B_5H_4\}]$.[31] The compounds are $2n$-electron systems (8 framework atoms and 16 electrons) and so might adopt the capped pentagonal bipyramidal structure in Figure 19(a), but the fact that the isoelectronic cluster $[\{Co(\eta^5-C_5H_5)\}_4(B_4H_4)]$ adopts the D_{2d} dodecahedral

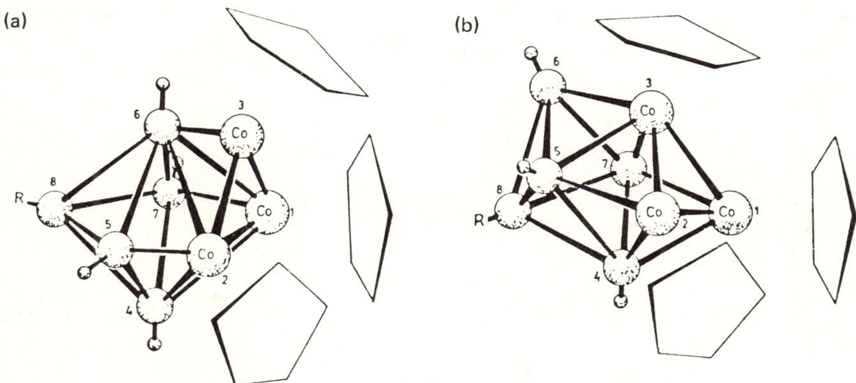

Figure 19. Possible structures for $[\{Co(C_5H_5)\}_3(B_5H_5)]$ and its derivatives: (a) capped pentagonal bipyramidal and (b) dodecahedral.

configuration (Figure 14) suggests that a structure like that in Figure 19(b) might also be possible. Present nmr data do not allow a choice between these two geometries and an X-ray structural determination is awaited with interest. It will be noted that, whereas the reaction $CoCl_2/B_5H_8{}^-/C_5H_5{}^-$ yields mainly B_3 and B_4 products and no B_5-containing clusters, the metal–atom reaction yields B_5 products as well as several of the previously obtained cobaltaborane clusters. The method clearly has considerable synthetic potential. (See also Section 14.3).

5.2. Direct Reaction of B_5H_9 with Metal Compounds

We have already referred to the reaction of B_5H_9 with $[Fe(CO)_5]$ in a hot/cold reactor to yield $[Fe(\eta^4\text{-}B_4H_8)(CO)_3]$ (p. 50). The reaction also produces species such as $[Fe(B_5H_5CO)(CO)_3]$ and $[Fe\{B_6H_4(CO)_2\}(CO)_3]$ of unknown structure and $[Fe\{B_5H_3(CO)_2\}(CO)_3]$ which is believed to have the closed polyhedral structure shown in Figure 20 (isoelectronic with $B_6H_6{}^{2-}$).[32] The compound is a red liquid, less volatile than its coproduct $[Fe(\eta^4\text{-}B_4H_8)\text{-}(CO)_3]$ and it freezes to an orange solid at $\sim5°$. If the structure is confirmed it will be the first example of a carbonyl derivative of a *closo*-ferraborane. Yet another product is the unstable red liquid $[Fe(B_5H_9)(CO)_3]$ which is sensitive to both heat and light and appears to be the precursor of the more stable $[Fe\text{-}(B_4H_8)(CO)_3]$. Nmr evidence indicates insertion of the $\{Fe(CO)_3\}$ moiety between two basal B atoms and rapid fluxionality of the Fe–H–B hydrogen atom between the two equivalent positions (Figure 21).[33,34] A kinetic study of the cothermolysis of B_5H_9 and $[Fe(CO)_5]$ at $217°$ indicated that the reaction is initiated by the rate-determining first-order decomposition of $[Fe(CO)_5]$, followed by the production of $[Fe(B_5H_9)(CO)_3]$; if the products are chilled at this stage then this is the major product of the reaction, but further thermolytic decomposition of this unstable species leads to the more stable $[Fe(B_4H_8)(CO)_3]$

and the presumed fugitive species $\{Fe(B_6H_{10})(CO)_3\}$ which then decomposes to intractable materials of lower volatility:[34]

$$[Fe(CO)_5] \xrightarrow{220°} \{Fe(CO)_4\} + CO$$

$$B_5H_9 + \{Fe(CO)_4\} \xrightarrow{220°} [Fe(B_5H_9)(CO)_3] + CO$$

$$2[Fe(B_5H_9)(CO)_3] \xrightarrow{220°} [Fe(B_4H_8)(CO)_3] + \{Fe(B_6H_{10})(CO)_3\}$$

$$\{Fe(B_6H_{10})(CO)_3\} \xrightarrow{220°} \text{intractable materials}$$

$[Fe(B_5H_9)(CO)_3]$ is clearly an analog of the isoelectronic and isostructural *nido*-hexaborane B_6H_{10} in which one basal $\{BH\}$ unit has been replaced by a $\{Fe(CO)_3\}$ group. Like B_6H_{10} it can be deprotonated by a strong base to give the corresponding anion:

$$[Fe(B_5H_9)(CO)_3] + KH \xrightarrow[-78°]{Me_2O} K^+[Fe(B_5H_8)(CO)_3]^- + H_2$$

Metathesis with Bu_4^nNBr in CH_2Cl_2 at $-23°$ gives the corresponding tetra-butylammonium salt of which crystals suitable for X-ray analysis have been obtained.[33] The structure of the anion $[Fe(B_5H_8)(CO)_3]^-$ is shown in Figure 22 which emphasizes the strucural relationship with $B_6H_9^-$; the proton has been abstracted from the $B(3)$—H—$B(4)$ bridge in the neutral ferraborane rather than from the $B(3)$—H—Fe bridge bond. Reaction of the anion to form a mixed cupra-ferraborane is discussed later on p. 72.

The neutral compound can also be protonated to give the cation $[Fe(B_5H_{10})(CO)_3]^+$, isolated below $-30°$ as its yellow salt with BCl_4^-:

$$[Fe(B_5H_9)(CO)_3] \xrightarrow{HCl/BCl_3} [Fe(B_5H_{10})(CO)_3]^+[BCl_4]^-$$

The properties are reported to be very similar to those of $[B_6H_{11}]^+Br^-$ and suggest that protonation has occurred at the remaining Fe—$B(basal)$ site.

The reaction between B_5H_9 and $[Co_2(CO)_8]$ has also been reported to give a *nido*-2-metallahexaborane $[Co(B_5H_8)(CO)_3]$ although in very small yields.[35] The neutral compound was thought to have no B—H—Co bridge bonds, although it could be protonated to $[Co(B_5H_9)(CO)_3]^+$ which would have such a bond. It

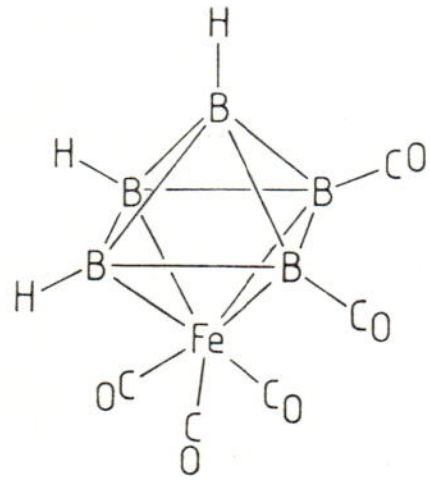

Figure 20. Proposed structure of $[Fe\{B_5H_3(CO)_2\}(CO)_3]$.

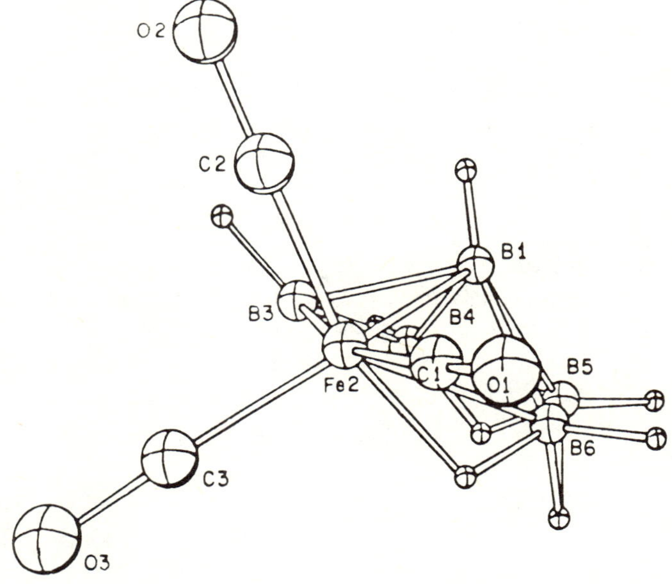

Figure 21. Structures illustrating the proposed nature of the fluxionality of [Fe-$(B_5H_9)(CO)_3$].

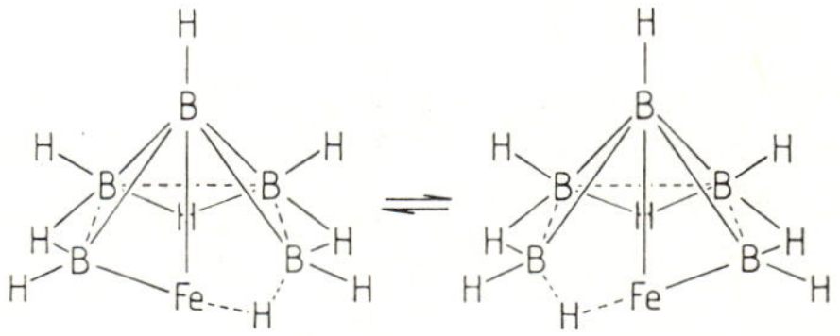

was also reported that treatment with KH failed to deprotonate the compound but resulted, rather, in the addition of H^- to give the *arachno* anion $[Co(B_5H_9)-(CO)_3]^-$. Further details of work on this system are awaited with interest. In a similar system, reaction between B_5H_9 and $[Co(\eta^5\text{-}C_5H_5)(CO)_2]$ in a hot/cold reactor at $225°/75°$ gives a 2% yield of a moderately air-sensitive yellow crystalline compound $[Co(B_5H_9)(\eta^5\text{-}C_5H_5)]$.[36] Other products are the known tetraboron clusters $[2\text{-}\{Co(\eta^5\text{-}C_5H_5)\}(B_4H_8)]$ and $[1,2\text{-}\{Co(\eta^5\text{-}C_5H_5)\}_2\text{-}(B_4H_6)]$ (pp. 51–53) together with the 2-isomer of the known nonaboron cluster $[5\text{-}\{Co(\eta^5\text{-}C_5H_5)\}(B_9H_{13})]$ (pp. 85–87). The yield is mysteriously doubled when a substoichiometric amount of 2-butyne is added to the reactants. Spectroscopic data encourage the belief that the product is the apex-metallated isomer $[1\text{-}\{Co(\eta^5\text{-}C_5H_5)\}(B_5H_9)]$ shown in Figure 23; again the relation to B_6H_{10} is

Figure 22. Structure of $[Fe(B_5H_8)(CO)_3]^-$ showing that deprotonation has occurred by removal of the bridging H^+ between B(3) and B(4). The Fe atom is almost coplanar with the basal B atoms and the three B—Fe distances are 216 pm [to B(1)], 208 pm [to B(3)], and 213 pm [to B(6)]; these are somewhat shorter than the B—Fe distances in $[Fe(B_7H_{12})-(CO)_4]^-$ (Figure 44) in which the Fe atom occupies a bridging position below the basal plane.

clear. The compound is also isostructural with the isoelectronic cobalticenium cation $[Co(\eta^5\text{-}C_5H_5)_2]^+$.

Finally, the reaction between B_5H_9 and metal carbonyls has been extended to manganese.[37] Thermolysis of B_5H_9 in the presence of $[Mn_2(CO)_{10}]$ and H_2 at 140° for 6 days gives a 9.3% yield of a pale-yellow, air-stable (weeks) liquid (fp −2.5°), which has been shown to be $[2\text{-}\{Mn(CO)_3\}(B_5H_{10})]$:

$$B_5H_9 + \tfrac{1}{2}[Mn_2(CO)_{10}] + \tfrac{1}{2}H_2 \xrightarrow{140°} [Mn(B_5H_{10})(CO)_3] + 2CO$$

This is a direct analog of the iron compound $[2\text{-}\{Fe(CO)_3\}(B_5H_9)]$, the extra bridge H atom supplying the additional electron to make the Mn species isoelectronic with the Fe cluster. Rather more modest yields are obtained from the direct hot/cold thermolysis of B_5H_9 and $[Mn(CO)_5H]$ at 220°/22°, and ultraviolet photolysis of $B_5H_9/[Mn_2(CO)_{10}]/H_2$ in isopentane at room temperature likewise offers no advantage. Direct bromination of the product with Br_2/CH_2Cl_2 at subambient temperatures gives a 27% yield of $[1\text{-}Br\text{-}2\text{-}\{Mn(CO)_3\}(B_5H_9)]$, an orange, air-stable (months) solid (mp 112°) (see Figure 24). The apical position of bromination contrasts with the basal bromination of B_6H_{10} itself. NaH deprotonates $[2\text{-}\{Mn(CO)_3\}(B_5H_{10})]$ by removal of the bridge hydrogen on the symmetry plane, i.e., between $B(4)$ and $B(5)$:

$$[2\text{-}\{Mn(CO)_3\}(B_5H_{10})] + NaH \xrightarrow[Et_2O]{-196° \text{ to } -22°} Na[2\text{-}\{Mn(CO)_3\}(B_5H_9)] + H_2$$

The anion is moderately stable in solution at room temperature, reprotonating over a period of weeks to regenerate the parent cluster.

5.3. Reaction of $B_5H_8^-$ with Complex Metal Halides

This route to metallaboranes is unusual in being one of the few that can at present be mechanistically rationalized. Deprotonation of B_5H_9 by a base such as KH results in a negative charge associated with the basal B—B bond formed by removal of a bridging proton; this can now act as the donor site to a metal center, thus enabling the $B_5H_8^-$ ligand to displace a halide ion from a suitable complex:

$$B_5H_9 + KH \xrightarrow[\text{temp}]{THF/low} KB_5H_8 + H_2$$

$$[L_nMX] + B_5H_8^- \xrightarrow{THF/CH_2Cl_2} [L_nM(B_5H_8)] + X^-$$

The first example of this reaction type was the preparation of the white Cu(I) complexes $[Cu(B_5H_8)(PPh_3)_2]$ and $[Cu(B_6H_9)(PPh_3)_2]$.[38] These compounds were also the first air-stable derivatives of pentaborane(9) and hexa-

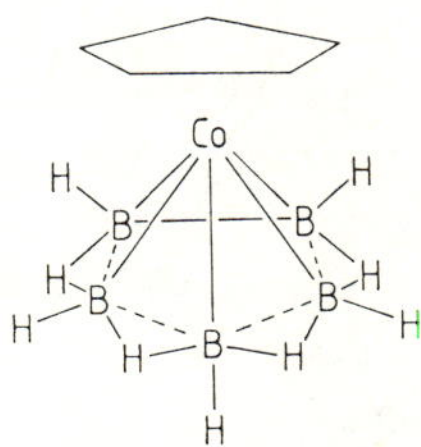

Figure 23. Proposed structure of $[1-\{Co(\eta^5-C_5H_5)\}(B_5H_9)]$.

borane(10) to be made. Reprotonation with dry HCl regenerates B_5H_9 (and B_6H_{10}) in high yield ($\sim80\%$):[38,39]

$$[CuCl(PPh_3)_3] + B_5H_8^- \xrightarrow[\text{low temp.}]{\text{THF/CH}_2\text{Cl}_2} [Cu(B_5H_8)(PPh_3)_2] + Cl^- + PPh_3$$

$$[Cu(B_5H_8)(PPh_3)_2] + HCl \xrightarrow{\text{Me}_2\text{O}/-35^\circ} B_5H_9 + \cdots$$

The apex and base substituted anions $1\text{-MeB}_5H_7^-$, $1\text{-BrB}_5H_7^-$, and $4\text{-MeB}_5H_7^-$ react similarly to give the analogous complexes in 53, 45, and 65% yields, respectively.[39,40] The detailed infrared and nmr spectra of $[Cu(B_5H_8)(PPh_3)_2]$[39,41] are fully consistent with the basal-bridged *dihapto* structure of the complex established by single-crystal X-ray diffractometry.[42] The structure is in Figure 25 which shows clearly that a three-center B—H—B bond in B_5H_9 has been replaced by a three-center B—Cu—B bond in $[Cu(\eta^2-B_5H_8)(PPh_3)_2]$. The Cu atom is essentially coplanar with a triangular face of the borane cluster (dihedral angle 178.4°) and is well below the plane of the four basal boron atoms. Close contacts are Cu—B(2) 221 pm and Cu—B(3) 224 pm. The Cu(I) atom is formally a 16- electron species and the bonding about the metal is essentially trigonal planar with P—Cu—P 123.4° with the angles subtended at the Cu atom by the midpoint of B(2)—B(3) and the two P atoms being 122.3° and 114.0°. The B(2)—B(3) vector is approximately perpendicular to the Cu-coordination plane, deviating from the normal by 16°.

Many other complexes of $\{\eta^2-B_5H_8\}^-$ or $\{\eta^2-B_5H_7X\}^-$ have been prepared by halide-ion displacement from a transition-metal complex and examples involving Ti(III), Ni, Pd, Pt, Cu, Ag, (Au), Cd, and Hg are listed in Table 2. Yields are generally in the range 50–75%. The reaction also works for main-group elements such as Be, B, Si, Ge, Sn, Pb, and P, and examples of these are also included

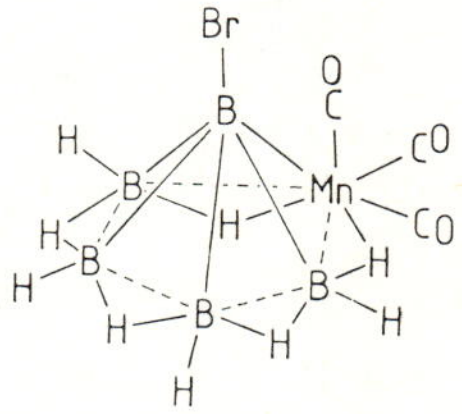

Figure 24. Proposed structure of $[1\text{-Br-}2\text{-}\{Mn(CO)_3\}(B_5H_9)]$.

in Table 2 for completeness, although they fall outside the main area of this particular chapter (see, however, Chapter 4). Stabilities of $(\eta^2\text{-}B_5H_8)$ complexes apparently decrease in the sequences $Ni>Pt>Pd$ and $Cu>Ag\gg Au$. For the former group the thermal stability has also been assessed as a function of group V ligand and the qualitative order is $PPh_3\approx AsPh_3\approx dppe>PMe_2Ph>PMe_3$. Detailed nmr studies have shown that the complexes have a static structure in solution[44] despite earlier suggestions to the contrary for the Ni complexes.[43] In the square planar Pt(II) complexes (as in the Ni and Pd complexes) the two phosphorus ligands differ since one is *trans* to the $\{\eta^2\text{-}B_5H_8\}$ ligand and the other is *trans* to the other substituent. Comparison of the two platinum–phosphorus coupling constants $^1J(^{195}Pt-^{31}P)$ provides a measure of the σ-electron demand of the $\{\eta^2\text{-}B_5H_8\}$ ligand which can thereby be placed in the sequence $\sigma\text{-alkyl}>\{\eta^2\text{-}B_5H_8\}\gg I>Br>Cl$.[45] This sequence can be compared with the results obtained from interatomic distances by single-crystal X-ray diffraction analysis of the compounds $[Ir(\eta^1\text{-}B_5H_8)Br_2(CO)(PMe_3)_2]$ and $[(IrB_5H_8)(CO)(PPh_3)_2]$ discussed below.

The mercury compound $[Hg(\eta^2\text{-}B_5H_8)_2]$[47] is unique in having two digonally oriented three-center B–Hg–B bonds and no other ligands about the metal center. It is readily prepared in almost quantitative yield by addition of $HgCl_2$ to a solution of NaB_5H_8 in THF at $-30°$. The proposed structure shown in Figure 26 is fully consistent with the ^1H and ^{11}B nmr spectra and with the presence of "strong" coupling $^2J(^{199}Hg-B-^1H_t)$ of 141 Hz.

All the compounds mentioned so far in this section are diamagnetic and contain a *dihapto* $\{\eta^2\text{-}B_5H_8\}$ group. However, neither feature is universal and the reaction of $B_5H_8^-$ with metal halide complexes can lead to paramagnetic products, as in the Ti(III) compound $[Ti(\eta^2\text{-}B_5H_8)(\eta^5\text{-}C_5H_5)]$,[30] or to compounds with other structural features. In this latter regard,[6,53] reaction of $B_5H_8^-$ with $[Fe(\eta^5\text{-}C_5H_5)(CO)_2I]$ affords the σ-bonded product $[Fe(2\text{-}B_5H_8)\text{-}(\eta^5\text{-}C_5H_5)(CO)_2]$ in high yield (86%) rather than the μ-bonded derivative $[Fe(\eta^2\text{-}B_5H_8)(\eta^5\text{-}C_5H_5)(CO)_2]$ although this may be an intermediate product which rearranges.[54] $[Fe(2\text{-}B_5H_8)(\eta^5\text{-}C_5H_5)(CO)_2]$ is an air-stable orange-yellow solid with mp of $21°$. The structure is shown in Figure 27. It is not clear why the terminal position should be preferred by the metal although it is noted that the Fe center is an 18-electron rather than a 16-electron species in this complex. Certainly there seems to be no steric bar to the bridging position since $[Fe(\eta^2\text{-}B_2H_5)(\eta^5\text{-}C_5H_5)(CO)_2]$ (Figure 3) and $[Fe(\eta^2\text{-}C_2B_4H_7)(\eta^5\text{-}C_5H_5)\text{-}(CO)_2]$ are known.[55] Deprotonation of $[Fe(2\text{-}B_5H_8)(\eta^5\text{-}C_5H_5)(CO)_2]$ with KH in monoglyme yields the anion $[Fe(2\text{-}B_5H_7)(\eta^5\text{-}C_5H_5)(CO)_2]^-$ which can be isolated in high yield as the orange–yellow $[PPh_4]^+$ or $[NBu_4^n]^+$ salts. Alternatively, further reaction of the anion with a second mole of $[Fe(\eta^5\text{-}C_5H_5)\text{-}(CO)_2I]$ gives $[\{Fe(\eta^5\text{-}C_5H_5)(CO)_2\}_2(2,4\text{-}B_5H_7)]$ as air-stable yellow needles, mp $110°$, in 11% yield. This was the first example of a compound with two transition-metal atoms bound to the pentaborane skeleton and the *trans*-basal distribution of the two Fe atoms was established spectroscopically.[54]

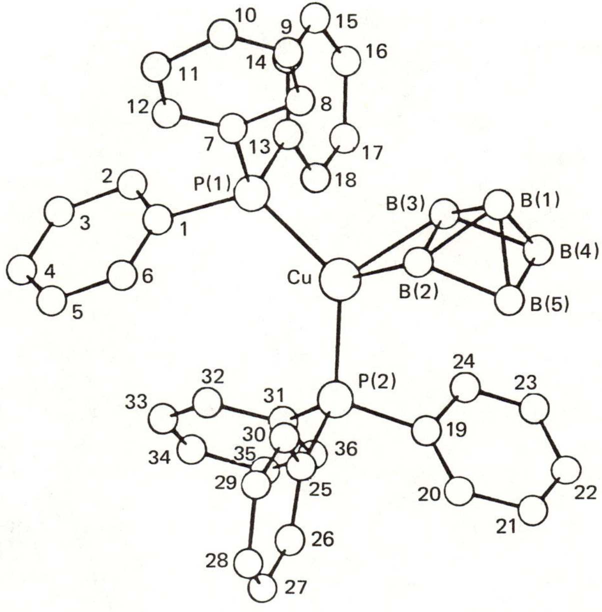

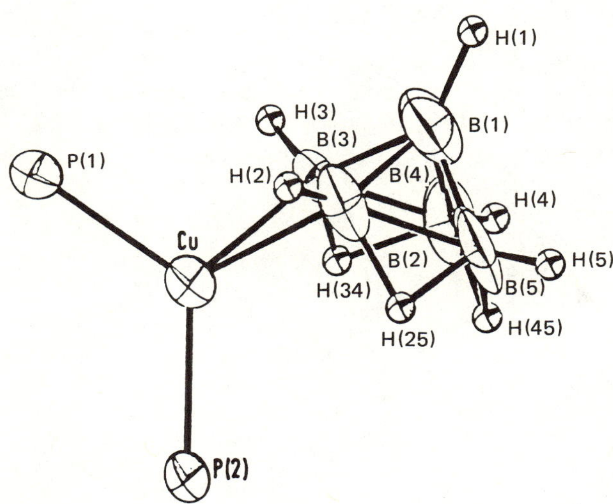

Figure 25. The molecular structure of $[Cu(\eta^2\text{-}B_5H_8)(PPh_3)_2]$ showing (upper diagram) the positions of the heavier atoms and the disposition of the $\{B_5H_8\}$ unit with respect to the two phosphine ligands, and (lower diagram) the detailed structure of the central $\{Cu\text{-}(B_5H_8)P_2\}$ unit. Note the considerable anistropy in the thermal motion of the borane cluster.

Table 2. 2,3-*Dihapto* Complexes of $(B_5H_8)^-$ and $(B_5H_7X)^-$

Complex	Color	mp	Reference
$[Ti(\eta^2-B_5H_8)(\eta^5-C_5H_5)_2]$ (paramagnetic)	Blue		(30)
$[Ni(\eta^2-B_5H_8)Cl(dppe)]$[a]	Orange	190°	(43, 44)
$[Ni(\eta^2-B_5H_7-1-Br)Cl(dppe)]$[a]	Orange	210°	(43)
$[Ni(\eta^2-B_5H_8)Br(dppe)]$[a]	Red	220°	(43)
$[Ni(\eta^2-B_5H_7-1-Br)Br(dppe)]$[a]	Red	241°	(43)
$[Ni(\eta^2-B_5H_8)I(dppe)]$[a]	Brick-red	209°	(43)
$[Ni(\eta^2-B_5H_7-1-Br)I(dppe)]$[a]	Brick-red	220°	(43)
$[Ni(\eta^2-B_5H_8)Br(dppp)]$[b]	Orange	190°	(43)
$[Pd(\eta^2-B_5H_8)Cl(dppe)]$[a]	Beige or white	$d>170°$	(44,45)
cis-$[Pd(\eta^2-B_5H_8)Cl(PPh_3)_2]$	Beige or white	$d>170°$	(45)
cis-$[Pd(\eta^2-B_5H_8)Cl(PMe_2Ph)_2]$	Beige or white	$d>170°$	(45)
$[Pt(\eta^2-B_5H_8)Cl(dppe)]$[a]	Beige or white	$d>170°$	(44,45)
cis-$[Pt(\eta^2-B_5H_8)Cl(PPh_3)_2]$	Beige or white	$d>170°$	(45)
cis-$[Pt(\eta^2-B_5H_8)Cl(PMe_2Ph)_2]$	Beige or white	$d>170°$	(45)
cis-$[Pt(\eta^2-B_5H_8)Cl(PMe_3)_2]$	Beige or white	$d>170°$	(45)
$[Pt(\eta^2-B_5H_8)Br(dppe)]$[a]	Beige or white	$d>170°$	(44,45)
cis-$[Pt(\eta^2-B_5H_8)Br(PPh_3)_2]$	Beige or white	$d>170°$	(45)
cis-$[Pt(\eta^2-B_5H_8)Br(PMe_2Ph)_2]$	Beige or white	$d>170°$	(45)
$[Pt(\eta^2-B_5H_8)I(dppe)]$[a]	Beige or white	$d>170°$	(44,45)
cis-$[Pt(\eta^2-B_5H_8)I(PPh_3)_2]$	Beige or white	$d>170°$	(45)
cis-$[Pt(\eta^2-B_5H_8)I(PMe_2Ph)_2]$	Beige or white	$d>170°$	(45)
cis-$[Pt(\eta^2-B_5H_8)Me(PMe_2Ph)_2]$	Beige or white	$d>170°$	(45)
cis-$[Pt(\eta^2-B_5H_8)Cl(AsPh_3)_2]$	Pale yellow	$d=125°$	(45)
trans-$[Pt(\eta^2-B_5H_8)I(PMe_2Ph)_2]$	White	$d=125°$	(45)
trans-$[Pt(\eta^2-B_5H_8)Me(PMe_2Ph)_2]$	White	$d=125°$	(45)
cis-$[\{Pt(\eta^2-B_5H_8)(PMe_2Ph)(\mu-SMe)\}_2]$	Very pale yellow	$d=125°$	(45)
$[Cu(\eta^2-B_5H_8)(PPh_3)_2]$	Cream white		(39–42)
$[Cu(\eta^2-B_5H_7-1-Me)(PPh_3)_2]$	Cream white		(39)
$[Cu(\eta^2-B_5H_7-4-Me)(PPh_3)_2]$	Cream white		(39)
$[Cu(\eta^2-B_5H_7-1-Br)(PPh_3)_2]$	White	$d=110°$	(40)
$[Cu(\eta^2-B_5H_8)(dppe)]$[a]	White		(40)
$[Ag(\eta^2-B_5H_8)(PPh_3)_2]$	Colorless (photosensitive)		(40)
$[Au(\eta^2-B_5H_8)(PPh_3)_n]$ (?)	Colorless	d below RT	(40)
$[Cd(\eta^2-B_5H_8)Cl(PPh_3)]$	White	83°	(46)
$[Cd(\eta^2-B_5H_7-1-Br)Cl(PPh_3)]$	White	98°	(46)
$[Hg(\eta^2-B_5H_8)_2]$	White	78°	(47)
$[Be(\eta^2-B_5H_8)(\eta^5-C_5H_5)]$	Colorless	ca. 38°	(48)
$[B(\eta^5-B_5H_8)Me_2]$	Colorless liquid		(49)
$[Si(\eta^2-B_5H_8)R_3]$ (R = H, Me, Et)	Colorless liquids or low mp solids		(50)
$[Ge(\eta^2-B_5H_8)R_3]$ (R = H, Me, Et)	Colorless liquids or low mp solids		(50)
$[Sn(\eta^2-B_5H_8)Me_3]$	Colorless liquid		(50)
$[Pb(\eta^2-B_5H_8)Me_3]$	Colorless	ca. $-5°$ (d.)	(50)

(continued)

Table 2 *(cont.)*

Complex	Color	mp	Reference
$[Si(\eta^2\text{-}B_5H_8)(\eta^1\text{-}B_5H_8)Me_2]$	Colorless liquid		(51)
$[Ge(\eta^2\text{-}B_5H_8)(\eta^1\text{-}B_5H_8)Me_2]$	Colorless liquid		(51)
$[P(\eta^2\text{-}B_5H_8)(CF_3)Me]$	Colorless	14°	(52)
$[P(\eta^2\text{-}B_5H_8)Me_2]$	Colorless	17°	(52)

[a] dppe = 1,2–bis (diphenylphosphino) ethane.

[b] dppp = 1,3–bis (diphenylphosphino) propane.

A second, more interesting, divergence from the simple general metathesis between an $\{\eta^2\text{-}B_5H_8\}^-$ and a halide ion at a metal center is exemplified by the reaction of $B_5H_8^-$ with the Ir(I) species *trans*-$[Ir(CO)Cl(PPh_3)_2]$. Although this reactant is isoelectronic with $[PtCl_2PPh_3)_2]$ and might, therefore, be expected to react analogously to give $[Ir^{(I)}(\eta^2\text{-}B_5H_8)(CO)(PPh_3)_2]$, the product is, in face, the *nido*-metallahexaborane $[(Ir^{(III)}B_5H_8)(CO)(PPh_3)_2]$ in which further reaction of the bridge-bonded intermediate has presumably occurred by internal oxidative addition and cluster-expansion (Figure 28).[56] The reaction occurs smoothly in THF/CH_2Cl_2 at $-70°$ to $-20°$ to give an 18% yield of the yellow product, mp (decomposition) $\sim180°$. The results of a single-crystal X-ray structure determination confirm the insertion of the Ir atom into the borane cluster, as shown in Figure 29: the structural relation to B_6H_{10} is clear and other noteworthy features are the B—H—Ir bridging bond and the *cis*-configuration of the phosphine ligands (Ph groups omitted for clarity). The topological similarity of B_6H_{10} and $[Ir(B_5H_8)(CO)(PPh_3)_2]$ is emphasized in Figure 30(a) and (b) which indicates replacement of $\{BH_tH_\mu\}$ by the neutral fragment $\{Ir(CO)$-$(PPh_3)_2\}$. An alternative topology in Figure 30(c) emphasizes the close relation with the $\{\eta^3\text{-}B_3H_7\}$ complexes of platinum and iridium on p. 48 (Figure 8). The bonding about the Ir atom is essentially octahedral. One of the P atoms is *trans* to an Ir—B two-center two-electron bond (Ir—P 240 pm) whereas the other is *trans* to an Ir—B—B three-center two-electron bond (Ir—P 235 pm).

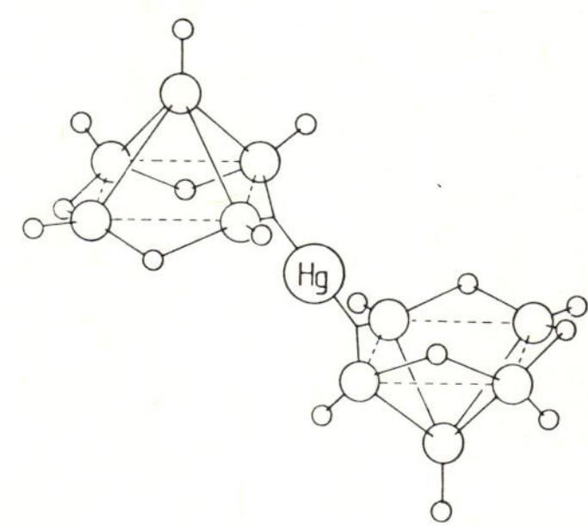

Figure 26. Proposed structure of $[Hg(\eta^2\text{-}B_5H_8)_2]$.

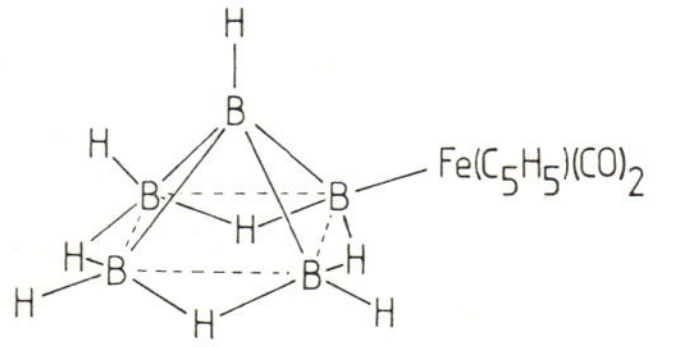

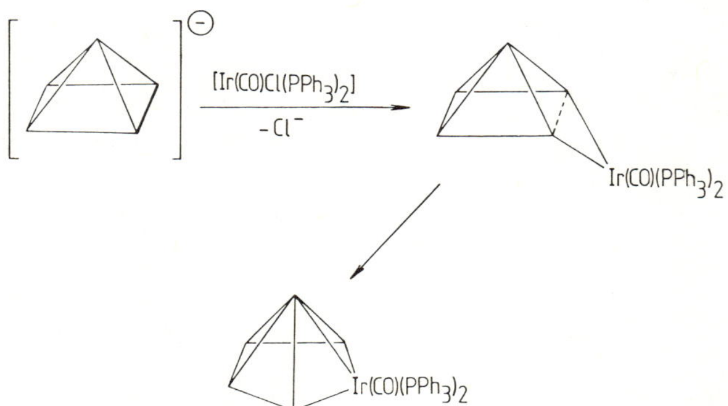

Figure 27. Proposed structure of the σ-bonded complex $[Fe(2-B_5H_8)(\eta^5-C_5H_5)(CO)_2]$.

Thus, the *trans* lengthening effect is greater for the σ-Ir—B bond and the sequence of σ-electron demand on p. 62 can be extended to give σ-B$\geqslant\sigma$-C$>$ μ-BB$\gg$I$>$Br$>$Cl. Other interatomic distances to the Ir atom are: B(1) 228, B(3) 225, B(6) 218, CO 187, and H 173 pm. The shortest B—B distance, as expected, is B(4)—B(5) (165 pm) and the other two basal B—B distances are 171 pm; B(apex)—B(base) are in the normal range 175–179 pm.

The geometrical disposition of the five boron atoms in the 2-IrB$_5$ cluster (Figures 28–30) is similar to that in *arachno*-pentaborane(11), B$_5$H$_{11}$. This has synthetic implications which have recently been realized in the reaction of the *arachno*-B$_5$H$_{10}^-$ anion with FeCl$_2$ in ether at $-45°$:[35]

$$B_5H_{11} + B_5H_8^- \longrightarrow B_5H_{10}^- + B_5H_9$$

$$2B_5H_{10}^- + FeCl_2 \longrightarrow [Fe(B_5H_{10})_2] + 2Cl^-$$

The product has been tentatively formulated as a *spiro*-metallaundecaborane which is believed to have the structure shown in Figure 31, similar to that established by X-ray crystallography for $[Be(B_5H_{10})_2]$.[57] It is interesting to note in this context that the 2-ferrahexaborane $[Fe(B_5H_{10})(\eta^5-C_5H_5)]$ (see the following) rearranges on heating to the semianalog of ferrocene $[1-\{Fe(\eta^5-C_5H_5)\}$

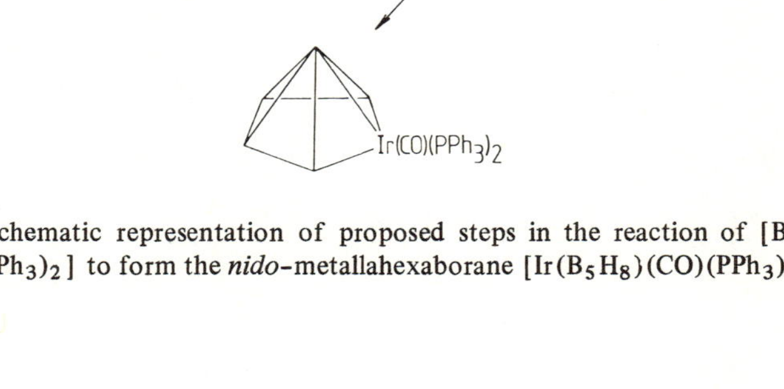

Figure 28. Schematic representation of proposed steps in the reaction of $[B_5H_8]^-$ with $[Ir(CO)Cl(PPh_3)_2]$ to form the *nido*-metallahexaborane $[Ir(B_5H_8)(CO)(PPh_3)_2]$.

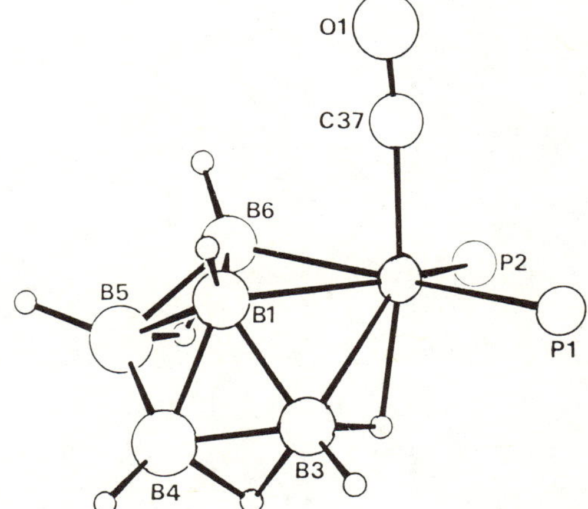

Figure 29. Structure of the metalla-hexaborane cluster in $[Ir(B_5H_8)(CO)(PPh_3)_2]$. The Ir atom is almost coplanar with the basal $B(3)$—$B(6)$ atoms (deviation 12 pm), in contrast to the position of the Cu atom in $[Cu(B_5H_8)(PPh_3)_2]$ (Figure 25).

$(B_5H_{10})]$; a similar process on $[Fe(B_5H_{10})_2]$ would, in principle, yield the complete borane analog of ferrocene (Figure 32).

Finally, in this section, it is convenient to note the thermolytic reaction between $B_5H_8^-$ and manganese pentacarbonyl hydride:[2]

$$LiB_5H_8 + [Mn(CO)_5H] \xrightarrow{\Delta} Li^+[Mn(B_5H_5)(CO)_3]^- + 2H_2 + 2CO$$

Few details of this reaction have been reported except that the yields are low. The product anion was tentatively formulated (without supporting data) as having *closo* six-vertex octahedral geometry which would correspond to a *tetrahapto* disposition of the $\{\eta^4\text{-}B_5H_5\}^-$ ligand. However, the subsequently developed electron counting rules[25] point to an η^3-*capped-closo* structure based on a trigonal bipyramidal $B_5H_5^{2-}$ capped by $\{Mn(CO)_3\}^+$, and a continued investigation of this system would be interesting.

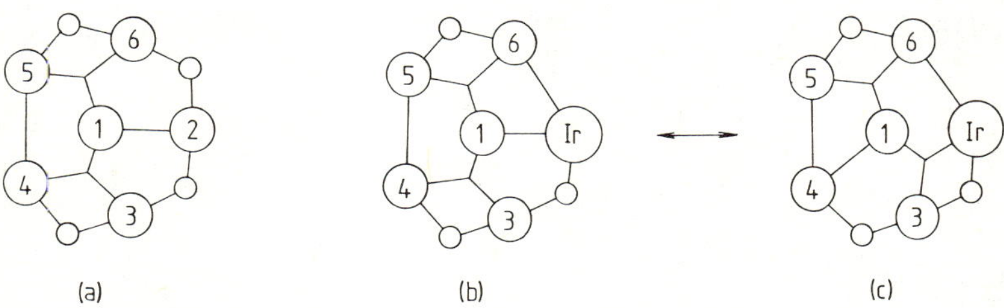

Figure 30. Valence-bond topological relationship between (a) B_6H_{10}, and (b), (c) $[(IrB_5H_8)(CO)(PPh_3)_2]$.

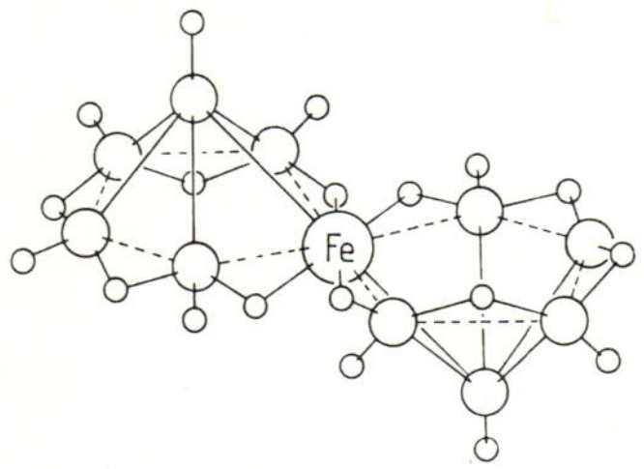

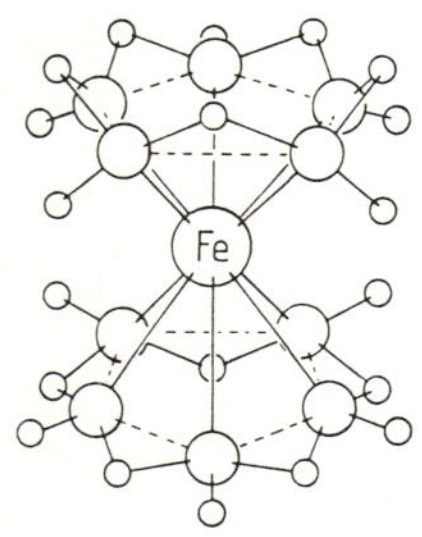

Figure 31. Proposed structure of $[Fe(B_5H_{10})_2]$.

5.4. Reactions of Halopentaboranes with Anionic Metal Complexes

The converse of the reactions on pp. 60-68, *viz.* the displacement of a halide ion from a haloborane by a metal anion, should also have great synthetic potential:

$$B_5H_8X + [ML_n]^- \longrightarrow [M(B_5H_8)L_n] + X^-$$

However, perhaps surprisingly, it has so far been limited in transition-metal pentaborane chemistry to the reactions of $2\text{-}ClB_5H_8$ and $2\text{-}BrB_5H_8$ with $Na[Mn(CO)_5]$ and $Na[Re(CO)_5]$.[58] The reactions are carried out in ether solution at or below room temperature and the products are brown liquids $[2\text{-}\{Mn(CO)_5\}B_5H_8]$ (mp $-11°$) and $[2\text{-}\{Re(CO)_5\}B_5H_8]$ (mp $10°$). Interestingly, no apex substituted derivatives $[1\text{-}\{M(CO)_5\}B_5H_8]$ are obtained, even when $1\text{-}ClB_5H_8$ or $1\text{-}BrB_5H_8$ is used in the reaction. Spectroscopic properties of the compounds indicate a σ-bonded basal-boron substituted structure analogous to that formed for the isoelectronic complex $[2\text{-}\{Fe(\eta^5\text{-}C_5H_5)\text{-}(CO)_2\}B_5H_8]$ (Figure 27).

5.5 Insertion by Oxidative Addition

Oxidative addition to a low-valent late transition metal is a well established inorganic reaction type, but it has been little exploited in the synthesis of metallaboranes. We have already noted examples of internal oxidative additions to Ir(I) resulting in cluster expansion by metal insertion (pp. 45, 67). In penta-

Figure 32. The (as yet?) hypothetical isomer of $[Fe(B_5H_{10})_2]$, "decaboraferrocene."

borane chemistry, *trans*-$[Ir^{(I)}Cl(CO)(PMe_3)_2]$ has been found to react rapidly with an excess of $2-BrB_5H_8$ to give the colorless, air-stable, crystalline complex $[2-\{Ir(Br)_2(CO)(PMe_3)_2\}B_5H_8]$, mp 148°(decomposition).[59] The molecular-structure as determined by X-ray diffraction, is shown in Figure 33.[59,60] Notable features are the almost regular octahedral coordination about the Ir atom with two identical Ir—P distances (236 pm), but significantly different Ir—Br distances (264 pm *trans* to σ-B_5H_8 and 252 pm *trans* to CO). It is clear that the $\{\eta^1$-$B_5H_8\}$ ligand exerts a straonger *trans* lengthening effect than does the CO group. Distances within the square-pyramidal B_5H_8 moiety are B(apex)—B(base) 164–169 pm (cf. 166 pm for B_5H_9), and B—B(base) 181–191 pm (cf. 177 pm for B_5H_9). The Ir—B(2) distance is 207 pm, somewhat shorter than the supposed two-electron two-center Ir—B(6) bond in $[(IrB_5H_8)(CO)(PPh_3)_2]$ discussed previously (Figure 29).

Interestingly, reaction of *trans*-$[IrCl(CO)(PMe_3)_2]$ with an excess of the apex-substituted borane $1-BrB_5H_8$ gives a 62% yield of the same, base-substituted dibromoiridium product whereas with an excess of B_5H_9 itself at $-45°$ there is an 80% yield of the less-stable, colorless simple oxidative addition product $[2-\{Ir-(CO)ClH(PMe_3)_2\}B_5H_8]$ [mp 64° (decomposition)].[59] The analogous PPh_3 complex *trans*-$[Ir(CO)Cl(PPh_3)_2]$ undergoes none of these reactions. It has also been shown that *trans*-$[Ir(CO)Cl(PMe_3)_2]$ undergoes oxidative B—H addition with $1-ClB_5H_8$ at $-22°$ to yield the thermally unstable $[1-Cl-2-\{Ir(CO)ClH-(PMe_3)_2\}B_5H_7]$ and that this isomerizes at 25° by interchange of Cl(—B) with H(—Ir) to give the more stable $[2-\{Ir(CO)Cl_2(PMe_3)_2\}B_5H_8]$ (see Figure 34).[61] It seems that the initial reaction is always the insertion of Ir into a basal B—H bond, the site expected to be most susceptible to nucleophilic attack. Qualitatively the relative order of reactivity is $2-BrB_5H_8 \approx 2-ClB_5H_8 > 1-BrB_5H_8 > 1-ClB_5H_8 \approx B_5H_9 \gg 1-MeB_5H_8$. As with the manganese and rhenium carbonyl anions mentioned in Section 5.4., no apically metallated intermediates have been detected.

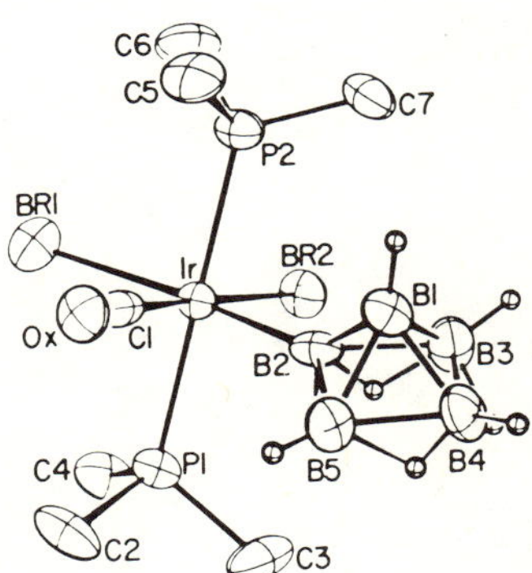

Figure 33. Molecular structure of $[Ir(B_5H_8)Br_2(CO)-(PMe_3)_2]$.

5.6. *Reaction of Pentaborane with Transition-Metal Complexes in the Presence of Reductive Bases*

The term reductive base refers to reagents such as $C_5H_5^-$, AlH_4^-, $C_{10}H_8^{\cdot-}$ (naphthalene radical anion), Na/K, and Na/Hg amalgam. When allowed to react with B_5H_9 (or $B_5H_8^-$) in the presence of transition-metal compounds they have yielded a fascinating variety of compounds which often have fewer than five boron atoms in the resulting cluster, although sometimes they contain five or more borons. This category of reaction includes the reactions of $CoCl_2$ with B_5H_9 in the presence of an excess of $C_5H_5^-$ (i.e., $CoCl_2/B_5H_8^-/C_5H_5^-$) which are extensively discussed in Section 4 (pp. 50–54) and elsewhere (p. 86), although this particular system does not appear to generate clusters having five boron atoms. Reactions of $B_5H_8^-$ with $[Ni(\eta^5\text{-}C_5H_5)_2]$ in the presence of Na/Hg have also been discussed (p. 54). Among the products is the green, air-stable *nido*-complex $[\{Ni(\eta^5\text{-}C_5H_5)\}_4(B_5H_5)]$. The structure is shown in Figure 35.[27] An unexpected feature is the occupation of vertices of high connectivity by boron and of low connectivity by nickel (see also p. 55). There has also been a rather severe disruption of the original B_5 cluster.

Attempts to extend the reaction to the preparation of ferraborane clusters lead to rather different products. Thus, the reaction of $FeCl_2$ with NaB_5H_8 and NaC_5H_5 in THF at 25° gives violet crystals of $[2\text{-}\{Fe(\eta^5\text{-}C_5H_5)\}B_5H_{10}]$ in low yield (2.5%).[62,63] The compound is moderately air-sensitive and the proposed structure is in Figure 36—it is a *nido*-2-ferrahexaborane and represents the insertion of an $\{Fe(\eta^5\text{-}C_5H_5)\}$ unit and two bridging H atoms into the base of $B_5H_8^-$ (cf. the *nido*-2-iridahexaborane cluster in Figure 29 and the proposed *nido*-2-manganahexaborane cluster in Figure 24). Thermal isomerization to the violet 1-ferra–isomer (isostructural with ferrocene) can be effected in a sealed tube at 175–180°, about 75 hr being required for complete conversion.

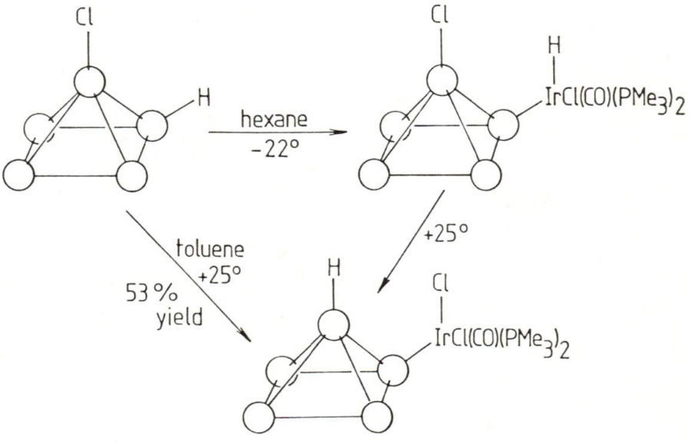

Figure 34. Reaction between $1\text{-}ClB_5H_8$ and *trans*-$[Ir(CO)Cl(PMe_3)_2]$.

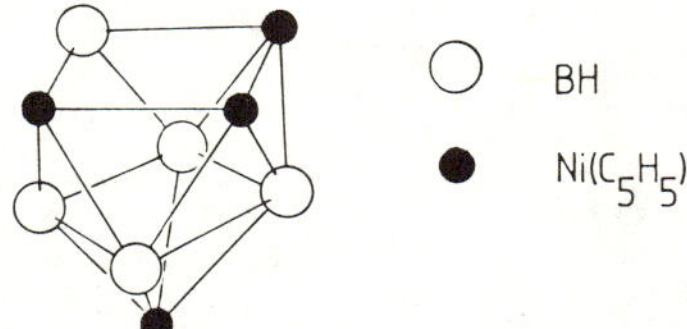

Figure 35. Schematic structure of the central cluster in [{Ni(η^5-C$_5$H$_5$)}$_4$(B$_5$H$_5$)].

The reaction of [Fe(CO)$_5$] and B$_5$H$_9$ in the presence of LiAlH$_4$ leads to degradation of the borane cluster and the production of [1-{Fe(CO)$_3$}B$_4$H$_8$] (p. 50), [1,2-{Fe(CO)$_3$}$_2$B$_3$H$_7$] (p. 49), and [{Fe(CO)$_3$}$_2$B$_2$H$_6$] (p. 45), depending on the conditions used.[11,15]

5.7. Metal Insertion into a Preformed Nido-Metallaborane

It is convenient at this point to mention the deprotonation of the *nido*-2-ferrahexaborane [2-{Fe(CO)$_3$}B$_5$H$_9$] (pp. 58–59) and the reaction of the resulting anion with a copper complex to form a mixed dimetallaborane:[64]

$$[2\text{-}\{Fe(CO)_3\}B_5H_9] + KH \xrightarrow{Me_2O/-78^\circ} K^+[2\text{-}\{Fe(CO)_3\}B_5H_8]^- + H_2$$

$$K^+[2\text{-}\{Fe(CO)_3\}B_5H_8]^- + [CuCl(PPh_3)_3] \xrightarrow{Me_2O/-45^\circ}$$
$$[2\text{-}\{Fe(CO)_3\}B_5H_8\text{-}\mu\text{-}4,5\text{-}\{Cu(PPh_3)_2\}] + KCl + PPh_3$$

The compound can be isolated as a yellow solid, stable at room temperature in the absence of air, and noticeably less photosensitive than its precursors [2-{Fe(CO)$_3$}B$_5$H$_9$] and K$^+$[2-{Fe(CO)$_3$}B$_5$H$_8$]$^-$. The molecular structure, obtained by X-ray analysis at -90° is shown in Figure 37. The remarkable feature of this structure is the very open bonding at the copper site: in contrast to [Cu(B$_5$H$_8$)(PPh$_3$)$_2$] (Figure 25) the Cu atom is only 43 pm below the basal plane and is, thus, within bonding distance of the (originally terminal) H atom on

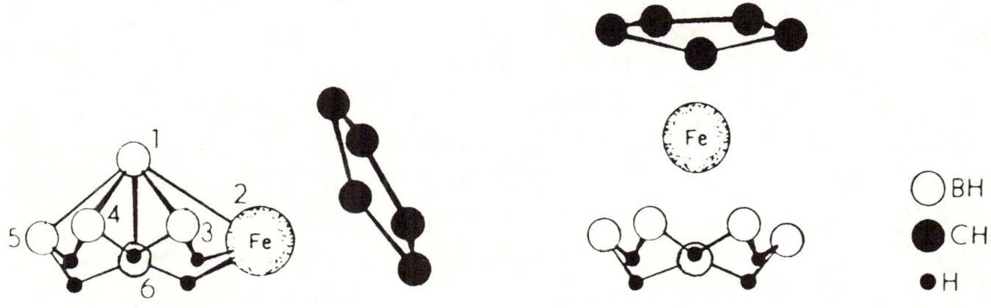

Figure 36. Proposed structures of [2-{Fe(η^5-C$_5$H$_5$)}(B$_5$H$_{10}$)] (left diagram) and [1-{Fe(η^5-C$_5$H$_5$)}(B$_5$H$_{10}$)] (right diagram). The 2-compound isomerizes to the 1-compound on being heated.

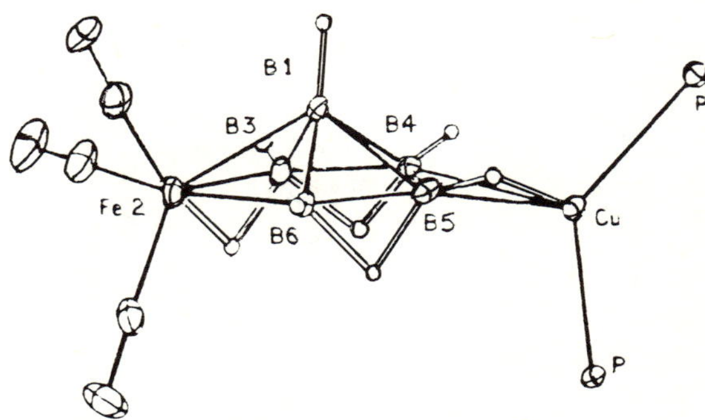

Figure 37. Molecular structure of $[2\text{-}\{Fe(CO)_3\}(B_5H_8)\text{-}\mu\text{-}4,5\text{-}\{Cu(PPh_3)_2\}]$. The Cu atom occupies a site relative to the pyramidal cage which is intermediate between that of the Cu atom in $[Cu(B_5H_8)(PPh_3)_2]$ (Figure 25) and that of the Ir atom in $[Ir(B_5H_8)\text{-}(CO)(PPh_3)_2]$ (Figure 29).

B(5), *viz.* 196 pm. This results in unsymmetrical bonding of the Cu atom to B(4)—B(5) with interatomic distances Cu—B(4) 227 pm and Cu—B(5) 216 pm. Other distances in the $\{Fe(B_5H_8)(CO)_3\}$ part of the molecule are normal, including B(4)—B(5), 166 pm, which establishes that the Cu atom has not been inserted into this bond to give a 2-ferra-5-cupra-heptaborane; it is also of interest that this latter site is not the site of the unbridged B(3)—B(4) bond on the parent anion $[Fe(B_5H_8)(CO)_3]^-$ (Figure 22).

6. HEXABORON COMPOUNDS

The topological similarity of $B_5H_8^-$ and *nido*-B_6H_{10}, and, in particular, the presence of a basal B—B bond in both species, suggests that there may be interesting parallels in their chemistry. This has to some extent been borne out, although the relative inaccessibility of B_6H_{10} has so far militated against any extensive exploration of its metal derivatives.

Reaction of B_6H_{10} with $[Fe_2(CO)_9]$ at room temperature results in the smooth elimination of $[Fe(CO)_5]$ and the formation of $[Fe(\eta^2\text{-}B_6H_{10})(CO)_4]$ as a volatile, yellow, crystalline solid, mp 37° (decomposition), yield 77%.[65,66] The compound can be handled for brief periods in air, but is thermally and photolytically unstable. Spectroscopic data and Mössbauer parameters are all consistent with the structure in Figure 38, which indicates that the basal B—B bond is acting as a two-electron *dihapto*-donor to form a three-center B—Fe—B bond; the local symmetry about the Fe atom is C_{2v}, thus, establishing an equatorial site for the borane ligand. In a similar reaction, displacement of ethene and Cl^- from Zeise's salt occurs smoothly at room temperature to give a 54% yield of *trans*-$[Pt(\eta^2\text{-}B_6H_{10})_2Cl_2]$ as yellow needles, unstable in air at room temperature, which decompose without melting at ca. 120°:[66]

$$K[Pt(\eta^2\text{-}C_2H_4)Cl_3] + 2B_6H_{10} \longrightarrow trans\text{-}[Pt(\eta^2\text{-}B_6H_{10})_2Cl_2] + C_2H_4 + KCl$$

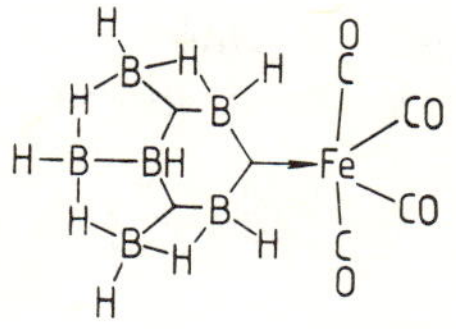

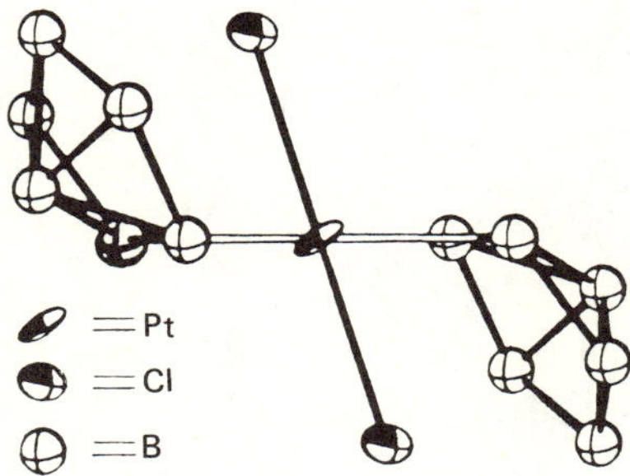

Figure 38. Proposed structure and bonding in $[Fe(\eta^2\text{-}B_6H_{10})\text{-}(CO)_4]$. The Fe atom is presumably in a bridging position below the basal plane of the hexaboron pyramidal cluster, and the actual electronic structure is presumably a hybrid of that depicted together with a number of other canonical forms.

The molecular structure (Figure 39), as determined by X-ray analysis at $-170°$, clearly shows the *dihapto*-bonding of each neutral B_6H_{10} ligand.[67] The most noteworthy feature of the B_6 fragment is that within the B—Pt—B three-center bond: the interatomic distance B(2)—B(3) of 182 pm is much longer than the unbridged B—B bond in the uncomplexed B_6H_{10} ligand (160 pm) and now has a value typical of B—B distances in triangulated polyhedral boranes. Other distances are Pt—B 227 pm and Pt—Cl 231 pm.

Rhodium and iridium complexes can be prepared similarly:[66]

$$[Rh(\eta^2\text{-}C_2H_4)_2(acac)] + 2B_6H_{10} \xrightarrow[15\%\ \text{yield}]{<0°} 2C_2H_4 + [Rh(\eta^2\text{-}B_6H_{10})_2(acac)]$$
yellow crystals

$$[Rh(\eta^2\text{-}C_2H_4)_2Cl]_2 + 4B_6H_{10} \xrightarrow[64\%\ \text{yield}]{<0°} 4C_2H_4 + [Rh(\eta^2\text{-}B_6H_{10})_2Cl]_2$$
orange-red solid

$$[Ir(C_8H_{14})_2Cl]_2 + 4B_6H_{10} \xrightarrow[38\%\ \text{yield}]{<0°} 4C_2H_4 + [Ir(\eta^2\text{-}B_6H_{10})_2Cl]_2$$
orange powder

In these reactions (acac) is the acetylacetonate anion $(MeCOCHCOMe)^-$, and C_8H_{14} is cyclo-octene. B_6H_{10} can also displace one CO ligand from $[Ni(CO)_4]$ to give a yellow liquid phase containing $[Ni(B_6H_{10})(CO)_3]$, but the complex is insufficiently robust to have so far been isolated.[66]

The numerous metathetical reactions in which halide ions are displaced from metal complexes by the $B_5H_8^-$ anion are mirrored in a few reactions involving $B_6H_9^-$ although the full potential of this reaction has not yet been ex-

Figure 39. Molecular structure of *trans*-$[Pt(\eta^2\text{-}B_6H_{10})_2Cl_2]$. All B—B distances are 182 ± 5 pm, angles about B atoms are $60\pm2°$, and internal angles of basal B atoms $108\pm2°$. The two B_6 units are, therefore, icosahedral fragments, with the Pt atom bridging the unique (4,5) position in both borane ligands. The structure may be viewed as involving *trans*-planar coordination about the Pt atom with the centers of the B(4)—B(5) vectors occupying two coordination sites.

ploited. One of the few examples is the reaction with $[CuCl(PPh_3)_3]$ to form a metallaheptaborane:[38,39]

$$[CuCl(PPh_3)_3] + K[B_6H_9] \xrightarrow[<0°]{THF/CH_2Cl_2} [Cu(B_6H_9)(PPh_3)_2] + PPh_3 + KCl$$

The product is a white crystalline solid, stable in air for about 4 days; it, therefore, closely resembles the $B_5H_8^-$ analog (p. 63) although it is somewhat less stable. Treatment with HCl in solution regenerates B_6H_{10} in 75% yield. The nickel complex $[NiCl_2(Ph_2PCH_2CH_2PPh_2)]$ has been reported to react similarly to yield $[Ni(\eta^2\text{-}B_6H_9)Cl(Ph_2PCH_2CH_2PPh_2)]$.[30] In both of these compounds it would be of interest to know whether the position of the metal with respect to the borane cluster resembles that in $[Cu(B_5H_8)(PPh_3)_2]$ (Figure 25) or that of the Cu atom in $[Cu\{B_5H_8Fe(CO)_3\}(PPh_3)_2]$ (Figure 37).

In principle, the $B_6H_9^-$ anion has two potential donor sites involving basal B—B bonds and these may both be used in the dimeric Ti(III) compound $[Ti(B_6H_9)(\eta^5\text{-}C_5H_5)_2]_2$ prepared by the following reaction:[30]

$$2[TiCl(\eta^5\text{-}C_5H_5)_2] + 2B_6H_9^- \longrightarrow [Ti(B_6H_9)(\eta^2\text{-}C_5H_5)_2]_2 + 2Cl^-$$

A plausible formulation for the structure of the dimer is in Figure 40 with a bidentate *tetrahapto* [i.e., bis(*dihapto*)] $B_6H_9^-$ ligand although the precise details of the stereochemistry and the orientation of the hexaborane clusters and the cyclopentadienyl rings must await an X-ray structure determination.

A perhaps unexpected route to a *trihapto*-hexaborane complex is the reaction of bis(*nido*-decaboranyl) oxide, $6,6'\text{-}(B_{10}H_{13})_2O$, with *cis*-$[PtCl_2(PMe_2Ph)_2]$ at room temperature.[68] Several metallaboranes are formed, one of the main ones being the yellow crystalline dimer $[Pt(\eta^3\text{-}B_6H_9)(PMe_2Ph)]_2$ (see Sections 8 and 10 for other products of this reaction). Single-crystal X-ray diffraction analysis revealed the structure shown in Figure 41 in which an essentially linear P—Pt—Pt—P group is coordinated by two bidentate bridging $2,3,4\text{-}\eta^3\text{-}nido\text{-}$ hexaboranate ligands. The Pt—Pt distance is 264.4 pm and the Pt—Pt—P angle 175.5°. Within each $B_6H_9^-$ unit the two adjacent basal B—B bonds coordinate to different Pt atoms, thus forming approximately square–planar bonding geometry

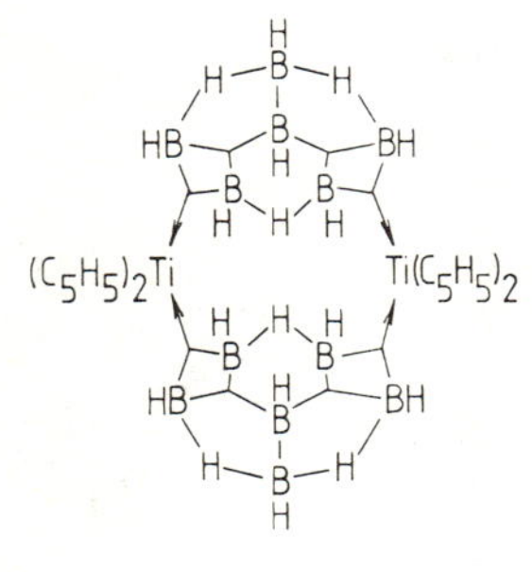

Figure 40. Proposed electronic structure for the metallaborane cluster in the dimer $[Ti\text{-}\mu\text{-}(\eta^2,\eta^2\text{-}B_6H_9)(\eta^5\text{-}C_5H_5)_2]_2$; a number of additional valence-bond structures may also be written down.

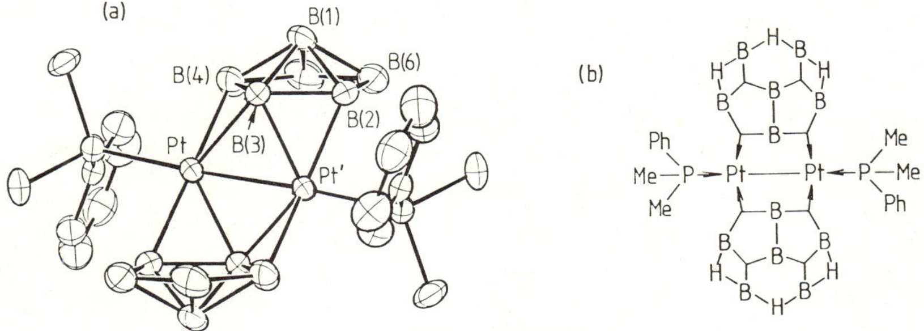

Figure 41. (a) Molecular structure of $[Pt(\eta^3\text{-}B_6H_9)(PMe_2Ph)]_2$ with H atoms omitted. Selected interatomic distances are Pt'—B(2) 222 pm, Pt'—B(3) 218 pm, Pt—B(3) 222 pm, and Pt—B(4) 223 pm. (b) Schematic representation of the proposed bonding, with bonds involving terminal H atoms omitted; a number of other canonical forms may be written down for the cluster valence-bond structure.

about each metal center [Figure 41(b)]. The metal atoms lie well below the basal plane of the borane cluster, as in $[Cu(\eta^2\text{-}B_5H_8)(PPh_3)_2]$ (Figure 25), but in contrast to $[Cu\{\eta^2\text{-}B_5H_8Fe(CO)_3\}(PPh_3)_2]$ (Figure 37), and thus take up positions analogous to bridging H atoms in $B_6H_{11}^+$. The mechanism by which the B_{10} cluster smoothly degrades even in modest yield to a B_6 cluster under mild conditions is not yet understood. It can also be noted that the formal replacement of the central "hinge" of $\{PhMe_2P\text{—}Pt\text{—}Pt\text{—}PMe_2Ph\}$ by the "isoelectronic" group $\{H\text{—}B\text{—}B\text{—}H\}$ would generate an as yet unknown isomer of $B_{14}H_{20}$ [Figure 42(a)] related to the known isomer[69] [Figure 42(b)] by an inversion of the orientation of one of the B_6 clusters. This encourages the belief that *iso*-$B_{14}H_{20}$ could also be prepared by a suitable route.

It has also been found that the *nido*-hexaborane cluster can act as a terdentate ligand as in the yellow trimetallaborane complex $[Pt_2(B_6H_9)\{B_6H_8\text{-}$

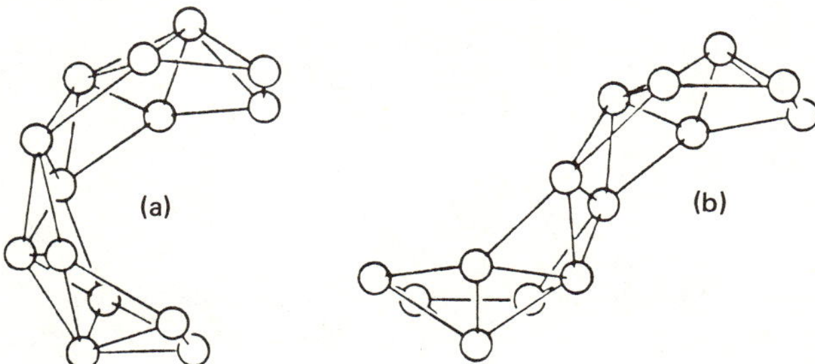

Figure 42. B atom skeletal structures of (left) the known tetradecaborane isomer $n\text{-}B_{14}H_{20}$ (Ref. 69) and (right) the as yet unsynthesized *iso*-$B_{14}H_{20}$ which is structurally analogous to $[Pt(B_6H_9)(PMe_2Ph)]_2$ (see Figure 41).

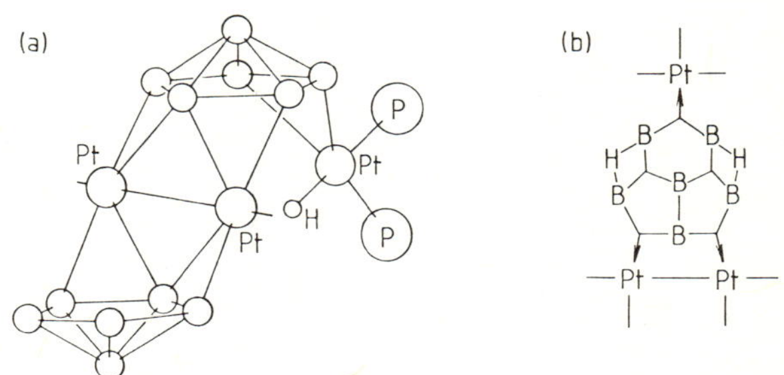

Figure 43. (a) Proposed metallaborane skeletal structure of $[Pt_2(B_6H_9)\{B_6H_8\text{-}cis\text{-}[PtH\text{-}(PMe_2Ph)_2]\}(PMe_2Ph)_2]$, formed by the replacement of a bridging H atom in one of the hexaboron clusters of $[Pt(B_6H_9)(PMe_2Ph)]_2$ by a three-center bond to the $cis\text{-}[PtH\text{-}(PMe_2Ph)_2]$ moiety. (b) A schematic electronic structure fo the terdentate $[B_6H_8]^{2-}$ ligand donation, with bonds involving terminal H atoms omitted.

$[PtH(PMe_2Ph)_2]\}(PMe_2Ph)_2]$ which is stable in air in the solid state and which is formed by deprotonation of $[Pt(B_6H_9)(PMe_2Ph)]_2$ with KH followed by the addition of $cis\text{-}[PtCl_2(PMe_2Ph)_2]$:[70]

$$[Pt_2(B_6H_9)_2(PMe_2Ph)_2] + 2KH + [PtCl_2(PMe_2Ph)_2] \longrightarrow$$
$$[Pt_2(B_6H_9)\{B_6H_8[PtH(PMe_2Ph)_2]\}(PMe_2Ph)_2] + 2KCl + H_2$$

^{1}H, ^{11}B, and ^{31}P nmr spectroscopy indicate that the gross structure of the product approximates to the structure of C_s symmetry shown in Figure 43(a), in which one of the *nido*-hexaborane clusters acts as a terdentate *pentahapto* $\{\eta^3,\eta^2\text{-}B_6H_8\}^{2-}$ ligand, i.e., simultaneously 2,3,4-*trihapto* to the Pt—Pt unit as in $[Pt(B_6H_9)(PMe_2Ph)]_2$ and 5,6-*dihapto* to the $\{PtH(PMe_2Ph)_2\}$ moiety.

So far, no σ-bonded metal hexaboranes have been reported that are analogous to those mentioned in Section 5 for $[L_nM\text{—}B_5H_8]$ and $[L_nM\text{—}B_5H_7\text{—}ML_n]$ although such compounds are known for main-group elements, e.g., the apex substituted derivatives $[1\text{-}(Me_3Si)B_6H_9]$ and $[1\text{-}(Me_3Ge)B_6H_9]$.[71] For completeness in this section it should also be noted that a metal hexaborane $[Fe\{B_6H_4(CO)_2\}(CO)_3]$ of unknown structure is believed to be among the products of the reaction between $[Fe(CO)_5]$ and B_5H_9 discussed in Section 5.2.[32]

7. HEPTABORON COMPOUNDS

No stable heptaboranes are known and the only examples of metallaheptaboron compounds are made by skeletal build-up from *nido*-hexaborane. Thus, the hexaborane–iron carbonyl adduct $[Fe(\eta^2\text{-}B_6H_{10})(CO)_4]$ mentioned in the preceding section can be rapidly and smoothly deprotonated at a bridging site by KH/ether and the B—B bond in the resulting anion can donate to a BH_3 group from diborane to form the moderately stable anion $[Fe(\eta^2\text{-}B_7H_{12})(CO)_4]^-$:[72]

$$[Fe(\eta^2\text{-}B_6H_{10})(CO)_4] + KH \longrightarrow K^+[Fe(\eta^2\text{-}B_6H_9)(CO)_4]^- + H_2$$

$$K^+[Fe(\eta^2\text{-}B_6H_9)(CO)_4]^- + [NBu_4]I \longrightarrow [NBu_4]^+[Fe(\eta^2\text{-}B_6H_9)(CO)_4]^- + KI$$

$$[NBu_4]^+[Fe(\eta^2\text{-}B_6H_9)(CO)_4]^- + \tfrac{1}{2}B_2H_6 \longrightarrow [NBu_4]^+[Fe(\eta^2\text{-}B_7H_{12})(CO)_4]^-$$

The anion can be isolated as its tetrabutylammonium salt and the results of an X-ray crystal structure determination at $-50°$ confirm the structure shown in Figure 44. The anion can be thought of as a bis-*dihapto*-$B_6H_9^-$ ligand donating from two nonadjacent basal B—B bonds into $\{BH_3\}$ and $\{Fe(CO)_4\}$, respectively [Figure 44(b)]. All the B—B and B—H distances fall within ranges previously observed for borohydride structures and the mean B—Fe distance is 220 pm. Treatment of this species with liquid anhydrous HCl at $-110°$ yielded 1 mole of H_2 to form the neutral complex $[Fe(B_7H_{11})(CO)_4]$, mp $20°$, which is stable at room temperature for short periods ($\sim$1 hr):

$$K^+[Fe(\eta^2\text{-}B_7H_{12})(CO)_4]^- + HCl \longrightarrow [Fe(B_7H_{11})(CO)_4] + H_2 + KCl$$

Although the structure of this interesting compound has not been definitively established, it seems highly probable that it is closely related to that of the gen-

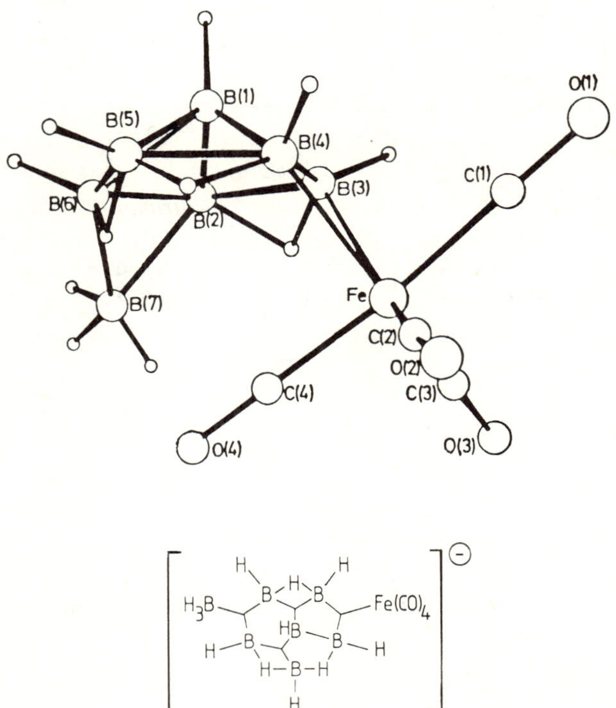

Figure 44. Top: the molecular structure of the $[Fe(\eta^2\text{-}B_7H_{12})(CO)_4]^-$ anion. Bottom: a schematic valence-bond structure; a number of canonical forms can be written for this.

erating anion; the compound would then be the first example of the stabilization of an otherwise nonexistent borane (*nido*-heptaborane, B_7H_{11}) by electron donation to an acceptor group $\{Fe(CO)_4\}$.

8. OCTABORON COMPOUNDS

Neutral and anionic *nido*-octaborane species are either unstable or difficult to prepare and the few non-*closo* metal octaboron compounds that are known are made by degradation of larger borane clusters.

Reaction of 6,6'-bis (*nido*-decaboranyl) oxide, $(B_{10}H_{13})_2O$, with *cis*-$[Pt(Cl)_2(PMe_2Ph)_2]$ yields a number of metallaboranes (p. 74) of which one is the very pale yellow *arachno*-platinanonaborane $[Pt(\eta^3\text{-}B_8H_{12})(PMe_2Ph)_2]$ having the structure shown in Figure 45.[70] The *trihapto* nature of the $\{B_8H_{12}\}^{2-}$ ligand is apparent, as is the analogy to the structures of the *arachno*-boranes *iso*-B_9H_{15} and $B_9H_{13}L$. Each B atom has an *exo*-terminal H atom bonded to it and, in addition, there are bridging H atoms in the B(2)—H—B(6) and B(8)—H—B(9) positions, together with *endo*-terminal "pseudobridging" H atoms on B(6) and B(9). The B(5)—B(6) and B(5)—B(9) distances average at 186 pm, somewhat longer than the B(2)—B(3) and B(3)—B(8) distances between the B atoms bonded to the Pt atom which average to 181 pm. The detailed geometry at the Pt atom is shown in Figure 46. A noteworthy feature is that the B(2) and B(3) atoms are contained in the Pt(II) bonding square plane which may imply two Pt—B two-center two-electron σ-bonds in contrast to the σ,μ structure found for the $\{\eta^3\text{-}B_3H_7\}^{2-}$ ligand in $[Pt(B_3H_7)(PMe_2Ph)_2]$ discussed previously (Figure 7).

This structure-type had previously been postulated on the basis of some preliminary work on the related compound $[Pt(B_8H_{12})(PEt_3)_2]$ which can be obtained by the ethanolic degradation of $[Pt(B_9H_{11}L)(PEt_3)_2]$ where L was an unspecified organic base:[73]

$$[Pt(B_9H_{11}L)(PEt_3)_2] + 3EtOH \longrightarrow [Pt(B_8H_{12})(PEt_3)_2] + B(OEt)_3 + H_2 + L$$

Other routes which also yield this type of compound involve the reaction of *cis*-$[PtCl_2(PMe_2Ph)_2]$ with either an excess of $[B_9H_{14}]^-$ or with base-deprotonated $[B_9H_{13}SMe_2]$ in neutral dipolar solvents at room temperature:[70]

$$[PtCl_2(PMe_2Ph)_2] + B_9H_{14}^- \xrightarrow[\text{yield}]{85\%} [Pt(\eta^3\text{-}B_8H_{12})(PMe_2Ph)_2] + \cdots$$

$$[PtCl_2(PMe_2Ph)_2] + B_9H_{13}SMe_2 \xrightarrow{\text{base}} [Pt(\eta^3\text{-}B_8H_{12})(PMe_2Ph)_2] + \cdots$$

The high yield (based on Pt) of the former reaction is particularly useful in securing access to this type of compound.

$[Pt(\eta^3\text{-}B_8H_{12})(PMe_2Ph)_2]$ can be readily deprotonated with KH, and the

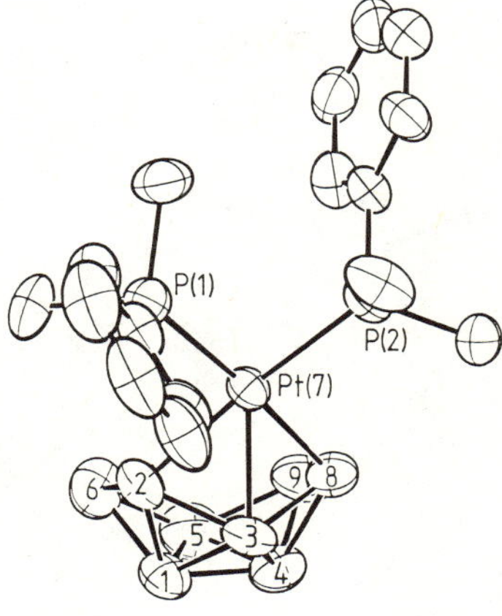

Figure 45. Molecular structure of $[Pt(\eta^3\text{-}B_8H_{12})\text{-}(PMe_2Ph)_2]$ (H atoms not shown). There are *exo*-terminal H atoms on each of the B atoms and bridging H atoms in the (2,6) and (8,9) positions; additionally there are "*endo*-terminal" (*pseudo*bridging) H atoms on B(6) and B(9). Details of the geometry are in Figure 46.

subsequent addition of $[PtCl_2(PMe_2Ph)_2]$ gives good yields of the colorless *arachno*-diplatinadecaborane $[\{Pt(PMe_2Ph)_2\}_2(B_8H_{10})]$:[70]

$$[Pt(B_8H_{12})(PMe_2Ph)_2] + [PtCl_2(PMe_2Ph)_2] + 2KH \longrightarrow$$
$$[\{Pt(PMe_2Ph)_2\}_2(B_8H_{10})] + 2KCl + H_2$$

The X-ray crystal structure of this compound (Figure 47) shows that the two *endo*-terminal/pseudobridging H atoms in the (5,6) and (5,9) positions of $[Pt(B_8H_{12})(PMe_2Ph)_2]$ have been replaced by the second $Pt^{(II)}$ moiety, but that the two (2,6) and (8,9) bridging hydrogens have been retained.[70] The compound can be regarded as a member of the series of *arachno*-platinahetero-boranes: $[Pt(B_{10}H_8S)(PR_3)_2]$, $[Pt(B_8H_{10}NH)(PR_3)_2]$, $[Pt(B_8H_{10}CH_2)\text{-}(PR_3)_2]$, $[Pt(B_8H_{10}BHL)(PR_3)_2]$, etc (see other chapters) in which the second $\{Pt(PR_3)_2\}$ group has now replaced the main-group heteroatom. Noteworthy interatomic distances in $[\{Pt(PMe_2Ph)_2\}_2B_8H_{10}]$ are $B(2)-B(3) = 179$ pm which is very similar to the corresponding distance (181 pm) in $[Pt(B_8H_{12})\text{-}$

Figure 46. Detailed coordination geometry about the Pt(7) atom in *arachno*-$[Pt(\eta^3\text{-}B_8H_{12})(PMe_2Ph)_2]$. That in $[\{Pt(PMe_2Ph)_2\}_2(B_8H_{10})]$ (Figure 47) is very similar. The η^3-bonding is essentially symmetrical. The B(2) and B(8) atoms are co-planar with Pt(7)−P(1)−P(2) and the Pt(7)−B(3) distance is 224 pm [compare with Figure 7].

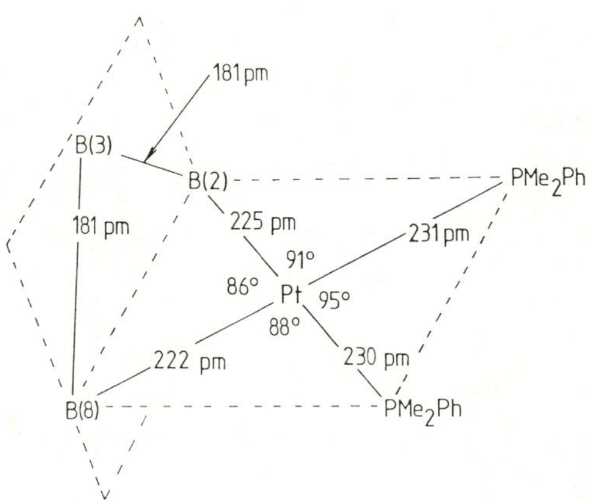

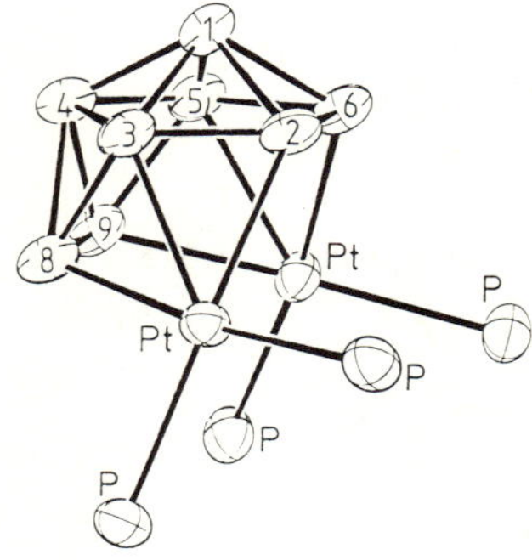

Figure 47. The structure of the $\{B_8Pt_2P_4\}$ cluster of *arachno*-[$\{Pt(PMe_2Ph)_2\}_2(B_8H_{10})$]. Selected distances are Pt–B(2) 225, Pt–B(3) 222, Pt–B(8) 225, and Pt–P (mean) 230.6 pm.

$(PMe_2Ph)_2$], and B(2)–B(6) = 189 pm which is somewhat shorter than the corresponding distance in *nido*-decaborane (197 pm) and compares to that in the *arachno*-decaborane anion $B_{10}H_{14}{}^{2-}$ (188 pm). The geometry about the Pt atom is almost identical to that in $[Pt(B_8H_{12})(PMe_2Ph)_2]$ (Figure 46).

The only other well-characterized metal-octaboron compound, other than the previously mentioned species $[1,6\text{-}\{Ni(\eta^5\text{-}C_5H_5)\}_2(B_8H_8)]$ [27] and those discussed in Section 14.4, is $[Mn(B_8H_{13})(CO)_3]$ in which the *arachno*-$B_8H_{13}{}^{-}$ ligand bonds to the $Mn(CO)_3$ group via three B—H—Mn bridges;[74] it is a minor product from the reaction between KB_9H_{14} and $[MnBr(CO)_5]$ in Et_2O and is discussed in more detail in Chapter 7. There is also a violet paramagnetic compound formulated as $[Fe(B_8H_8)(C_5H_5)_2]$; it can be isolated in trace amounts from the reaction of $FeCl_2$ with $B_5H_8{}^{-}$ and $C_5H_5{}^{-}$ (p. 70), but has not been structurally characterized. It may, in fact, be an 11-vertex *closo*-ferradicarbaborane with an exopolyhedral carbocycle, e.g., $[(\eta^5\text{-}C_5H_5)\{Fe^{(III)}H(C_3H_4)\text{-}C_2B_8H_8\}]$ rather than a true metallaborane.[63]

9. NONABORON COMPOUNDS

Nido-nonaborane,[13] B_9H_{13}, is unknown and the *arachno*-nonaboranes *n*- and *iso*-B_9H_{15}, although known, have so far been too inconvenient to use for synthetic purposes. However, the stable *nido*- and *arachno*-anions $B_9H_{12}{}^{-}$ and $B_9H_{14}{}^{-}$ are readily made and have both been used in the synthesis of metal-boron clusters. Thus, the reaction of *nido*-$B_9H_{12}{}^{-}$ with *trans*-$[PtClH(PEt_3)_2]$ in the presence of a number of ligands L (where L = amine, nitrile, phosphine, or sulfide) yields the *arachno*-species $[Pt(\eta^3\text{-}B_8H_{10}BHL)(PEt_3)_2]$ already mentioned briefly in the preceding section:[73]

$$\textit{trans-}[PtClH(PEt_3)_2] + CsB_9H_{12} \xrightarrow{\ L\ } [Pt(\eta^3\text{-}B_9H_{11}L)(PEt_3)_2] + CsCl + H_2$$

The proposed structure is in Figure 48. The compound can be degraded with ethanol to produce the *arachno* species $[Pt(B_8H_{12})(PEt_3)_2]$ as mentioned previously.

The only other example of the synthetic use of *nido*-B_9H_{12} in metal-

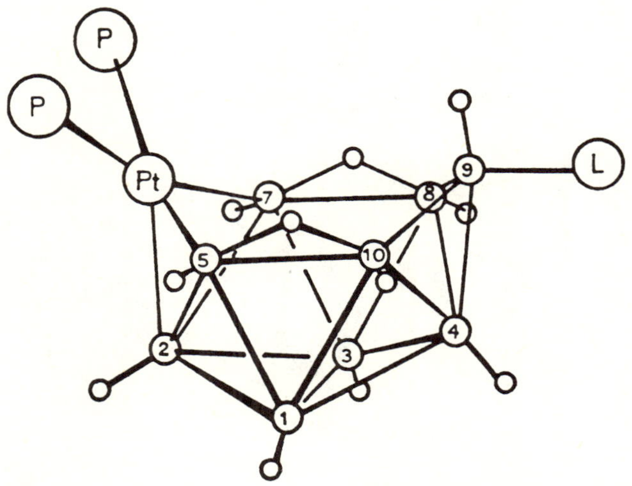

Figure 48. Proposed structure of $[Pt(\eta^3\text{-}B_9H_{11}L)(PEt_3)_2]$.

boron chemistry is its room-temperature reaction with nickelocene $[Ni(\eta^5\text{-}C_5H_5)_2]$ in the presence of sodium amalgam to give the *closo*-metalladecaborane anions 1- and 2-$[Ni(B_9H_9)(\eta^5\text{-}C_5H_5)]^-$; the same products are obtained in a slightly higher yield by nickel-insertion into *closo*-$B_9H_9^{2-}$ using $[Ni(\eta^5\text{-}C_5H_5)(CO)]_2$ as shown by the following scheme:[75,76]

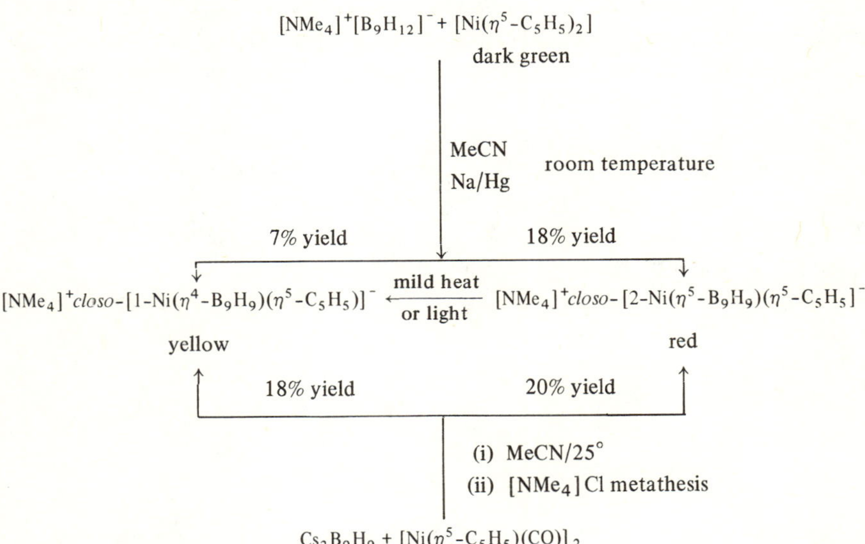

The proposed structures of the two product anions are shown in Figure 49 and the facile thermal isomerization (at $35°$) of the 2- to the 1-isomer is unusual in

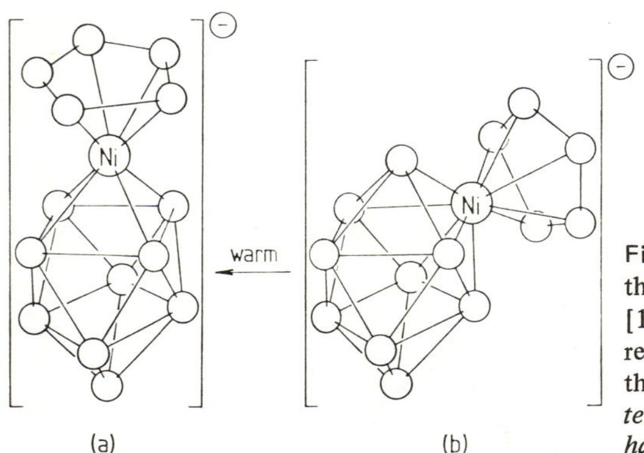

Figure 49. Proposed structures of the isomeric *closo*-anions (a) yellow [1-{Ni(η^5-C$_5$H$_5$)}(B$_9$H$_9$)] and (b) red [2-{Ni(η^5-C$_5$H$_5$)}(B$_9$H$_9$)]. In the 1–isomer the {B$_9$H$_9$} ligand is *tetrahapto*; in the 2–isomer *pentahapto*. (See also Figure 79).

that the metal atom moves from a site of higher to a site of lower connectivity within the borane cluster. The equatorial isomer [2-{Ni(η^5-C$_5$H$_5$)}B$_9$H$_9$]$^-$ is easily perchlorinated using Cl$_2$ at 5°C in acetonitrile solution to give the bright green [2-{Ni(η^5-C$_5$H$_5$)}B$_9$Cl$_9$]$^-$ in a yield of 80%. Chlorination of the apical 1-isomer does not proceed so successfully, but the deep yellow [1-{Ni(η^5-C$_5$H$_5$)}B$_9$Cl$_9$]$^-$ is obtained quantitatively by heating the perchlorinated 2-isomer at 130°C; interestingly, photolysis reverses this isomerization.[76]

A related neutral dichloro-*closo*-nickeladecaborane [2,4-Cl$_2$-1-{Ni-(PMe$_2$Ph)$_2$}B$_9$H$_7$] is obtained in a very small yield as air-stable red crystals from the reaction of 6,6′-bis(*nido*-decaboranyl) oxide with [NiCl$_2$(PMe$_2$Ph)$_2$] at room temperature.[70] The structure of the metallaborane cluster as determined by single-crystal X-ray diffraction analysis is that of a bicapped Archimedean square antiprism with the Ni atom in an apical position. There is essentially square–planar geometry about the Ni atom involving the two *cis*-P atoms and the B(3) and B(5) atoms.

An interesting series of *nido*-metallaboranes is produced by the reaction of *arachno*-B$_9$H$_{14}^-$ with manganese and rhenium carbonyl halides.[77,78] Thus, the reaction of KB$_9$H$_{14}$ with [MnBr(CO)$_5$] in THF under reflux gives [Mn(B$_9$H$_{13}$)-(CO)$_3$]$^-$ which can be isolated in a 30–50% yield as its air-stable reddish–orange crystalline tetramethylammonium salt, mp 200° (decomposition). The neutral complex [Mn(B$_9$H$_{12}$·OC$_4$H$_8$)(CO)$_3$], in which a molecule of solvent THF has replaced H$^-$, can also be isolated (5% yield) as air-stable, reddish–orange crystals which sublime in a high vacuum at 50–70° and melt with decomposition at 189°:

$$\text{B}_9\text{H}_{14}^- + [\text{BrMn(CO)}_5] \xrightarrow{\text{THF}} \begin{cases} [6\text{-}\{\text{Mn(CO)}_3\}(\text{B}_9\text{H}_{13})]^- + 2\text{CO} + \text{HBr} \\ \\ [6\text{-}\{\text{Mn(CO)}_3\}\{\text{B}_9\text{H}_{12}\text{-}2\text{-OC}_4\text{H}_8\}] + \cdots \end{cases}$$

The anion can also be transformed to the solvate in a separate reaction using I$_2$ or HgCl$_2$: the reaction can be viewed as a replacement of the ligand H$^-$ by THF followed by oxidation of H^- to H^+ as represented by the idealized equation:

$$[6\text{-}\{Mn(CO)_3\}(B_9H_{13})]^- + HgCl_2 + THF \longrightarrow$$
$$[6\text{-}\{Mn(CO)_3\}\{B_9H_{12}\text{-}2\text{-}OC_4H_8\}] + Hg + HCl + Cl$$

Under other conditions (see footnote 27b in Ref. 37) the anion $[6\text{-}\{Mn(CO)_3\}\text{-}(B_9H_{13})]^-$ reacts with I_2 in THF to give the monoiodo-derivative $[6\text{-}\{Mn(CO)_3\}\text{-}\{B_9H_{12}\text{-}6\text{-}I\}]^-$. The structure of the unusual solvate $[Mn(B_9H_{12}\cdot OC_4H_8)\text{-}(CO)_3]$ was determined by X-ray analysis and is shown in Figure 50. The B_9 cluster can be viewed as a *trihapto* ligand bonded to a six-coordinate Mn atom via two B—H—Mn bonds and a direct B(2)—Mn bond. The Mn—B(2) distance (219.6 pm) is substantially longer than the Mn—C(3) and Mn—H distances which are 179 and 175 pm, respectively. The only apparent substantial deviation from regular octahedral geometry at the Mn atom involves atom B(2) since the angle B(2)—Mn—C(3) is only 161.4°; this presumably derives from the tangential nature of the Mn and B(2) bonding orbitals. A third product isolated from the reaction mixture (7.2% yield) as red crystals has the THF coordinated at the B(5) position, i.e. $[6\text{-}\{Mn(CO)_3\}\{\eta^3\text{-}B_9H_{12}\text{-}5\text{-}OC_4H_8\}]$, as shown in Figure 51.[78] The dimensions of the metallaborane cage structure are very similar to those for the 2-substituted isomer (Figure 50) with some distortion because the site of substitution does not coincide with the cage mirror plane. Thus, the distance B(5)—B(10) is ca. 5 pm longer than B(7)—B(8), and the two Mn—H distances differ by ca. 16 pm. Use of Et_2O rather than THF as the reaction medium produces a small amount (0.6%) of the orange-red B(2)-coordinated etherate $[6\text{-}\{Mn(CO)_3\}\{B_9H_{12}\text{-}2\text{-}OEt_2\}]$ in addition to the anion and the bright

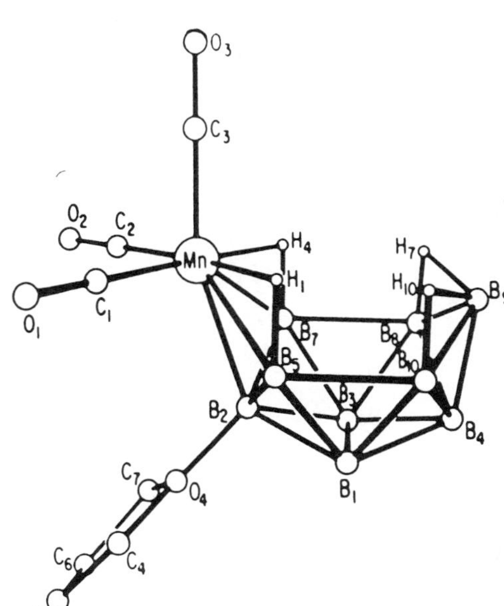

Figure 50. Molecular structure of $[6\text{-}\{Mn(CO)_3\}\{\eta^3\text{-}B_9H_{12}\text{-}2\text{-}OC_4H_8\}]$; for clarity the terminal H atoms have been omitted from the coordinated THF moiety and from B atoms B(1), B(3)–B(5), and B(7)–B(10).

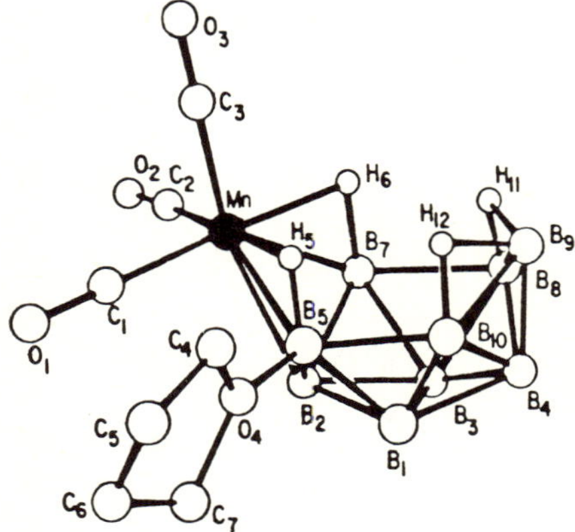

Figure 51.   Molecular structure of [6–{Mn(CO)$_3$} {η^3-B$_9$H$_{12}$-5-OC$_4$H$_8$}] . with B- and C-terminal H atoms omitted for clarity. Selected distances are B(5)–B(10) 199.3, B(7)–B(8) 203.7, Mn–H(5)*br* 167, Mn–H(6)*br* 183, Mn–B(2) 222.6, Mn–B(5) 223.2, Mn–B(7) 220.6, and B(5)–O(4) 152.0 pm.

yellow octaboron compound [Mn(B$_8$H$_{12}$)(CO)$_3$] mentioned in the preceding section.

The anion [Mn(B$_9$H$_{13}$)(CO)$_3$]$^-$ and the 2-tetrahydrofuranate adduct can both be used as the starting material for several other mangananonaboron complexes. Thus, photolysis in the presence of PPh$_3$ results in the replacement of one or more carbonyl ligands, and even the boron-complexed THF can be replaced. With NEt$_3$, however, the reaction takes a quite different and perhaps unexpected course involving isomerization and THF-ring opening to give the zwitterion [{Mn(CO)$_3$} {B$_9$H$_{12}$-O(CH$_2$)$_4$NEt$_3$}] as shown in Figure 52.[79,80] The compound is obtained in a 62% yield as air-stable orange-red crystals (mp 156-169°) (decomposition) and the *nido*-metalladecaborane cage dimensions are again very similar to those for the THF adducts (Figures 50 and 51). There

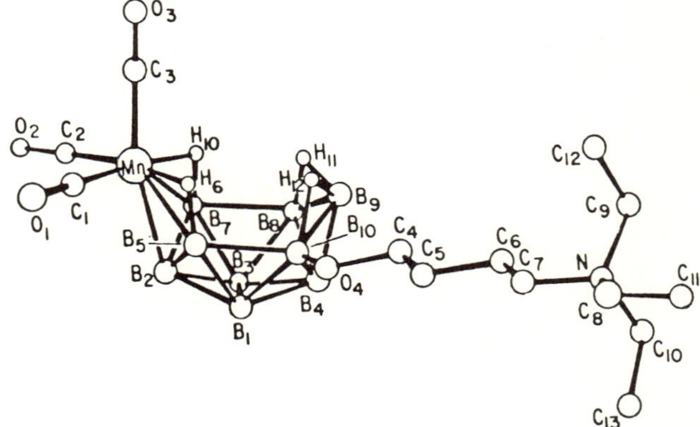

Figure 52.   Molecular structure of the *nido*-metallaborane zwitterion [6-{Mn(CO)$_3$} {B$_9$H$_{12}$-10-O(CH$_2$)$_4$NEt$_3$}] , with B- and C-terminal H atoms omitted for clarity. Selected distances are Mn–H(10)*br* 170, Mn–H(6)*br* 164, Mn–B(2) 221.6, Mn–B(5) 220.9, Mn–B(7) 226.2, and B(10)–O(4) 139.1 pm.

is, at present, no evidence as to how this novel reaction occurs although it has been suggested that it involves initial nucleophilic attack of NEt_3 at the α–C atom of THF followed by C—O bond scission and internal displacement of the Mn atom and bridging H-atoms as indicated in Figure 53.

When $[ReBr(CO)_5]$ rather than $[MnBr(CO)_5]$ reacts with KB_9H_{14} in boiling THF the main product is the anion $[6\text{-}\{Re(CO)_3\}B_9H_{13}]^-$ which can be isolated as its yellow tetramethylammonium salt in a 47% yield.[78] The unstable covalent yellow adduct $[6\text{-}\{Re(CO)_3\}\{B_9H_{12}\text{-}6\text{-}OC_4H_8\}]$ is also thought to be formed, but is too unstable to air and moisture to have so far been isolated or characterized.

The formal subrogation of a $\{BH\}$ group in the *nido-* $B_{10}H_{14}$ structure by a metal center to give an $\{MB_9\}$ cluster can, in principle, occur at any one of the four differing boron sites in decaborane, *viz.* the 1-, 2-, 5-, or 6-positions. Three of these four structural possibilities are known for the cyclopentadienylcobalta-decaboranes $[Co(B_9H_{13})(\eta^5\text{-}C_5H_5)]$. The 5-isomer is obtained in a small yield as red crystals from the reaction of $CoCl_2$ with NaB_5H_8 and NaC_5H_5 (see Table 1 in Section 4); its structure, established by X-ray diffraction,[81] is in Figure 54. The compound can be viewed as either a *nido-*cobaltadecaborane or as a complex in which a Co(III) atom is sandwiched between an $\{\eta^5\text{-}C_5H_5\}^-$ and an

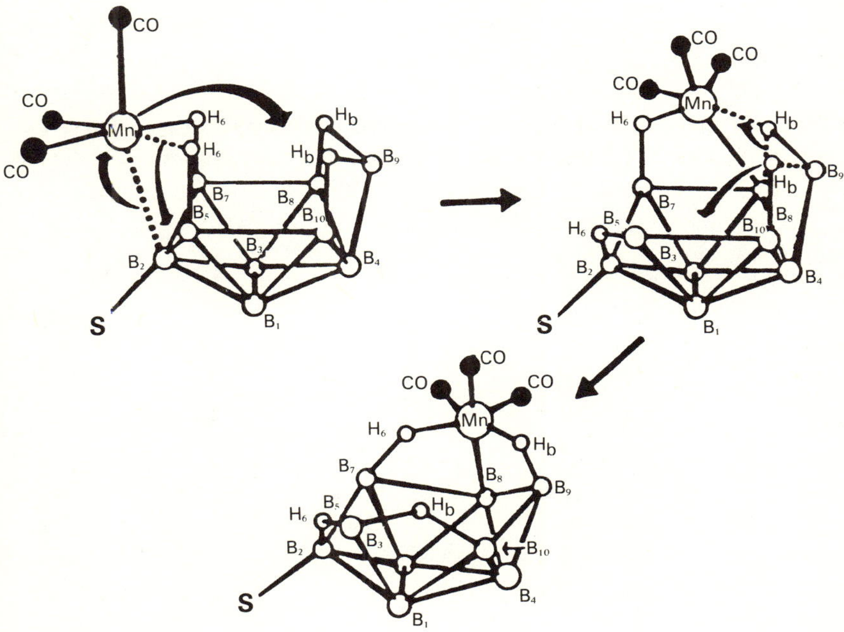

Figure 53. A possible rearrangement mechanism for the synthesis of $[6\text{-}\{Mn(CO)_3\}\{B_9H_{12}\text{-}10\text{-}O(CH_2)_4NEt_3\}]$. The S designates the substitutent, and to illustrate the rearrangement, the B atom numbers are not changed.

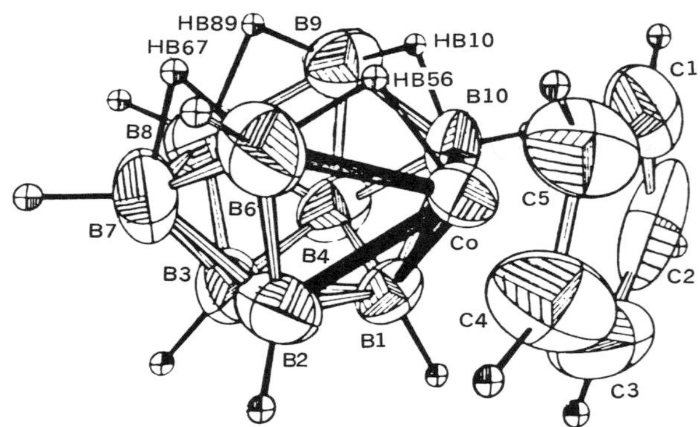

Figure 54. Molecular structure of $[5\text{-}\{Co(\eta^5\text{-}C_5H_5)\}(B_9H_{13})]$.

$\{\eta^5\text{-}B_9H_{13}\}^{2-}$ ligand. The distances between the Co atom and its five nearest neighbors in the B_9H_{13} moiety are Co—B(1) 208, Co—B(2) 209, Co—B(6) 207, Co—B(10) 218, and Co—H(5,6) 149 pm.

The 2-isomer (so assigned on the basis of nmr spectroscopy) has been prepared in a 0.9% yield as a coproduct of $[1\text{-}\{Co(\eta^5\text{-}C_5H_5)\}B_5H_9]$ (p. 60) in the direct reaction between B_5H_9 and $[Co(\eta^5\text{-}C_5H_5)(CO)_2]$ using hot/cold reactor conditions, e.g., 225°/75°C for 1 day.[36] The compound melts at 111-112°C and its yield can be increased to 2.5% by carrying out the reaction in the presence of an approximately equimolar amount of 2-butyne. The 6-isomer has been referred to briefly,[36] but no details of its preparation or characterization have yet appeared; however, the corresponding pentamethylcyclopentadienyl complex, $[6\text{-}\{Co(\eta^5\text{-}C_5Me_5)\}(B_9H_{13})]$, has been prepared and characterized by X-ray diffraction, as have the biscobalt species 5,7- and 6,9-$\{Co(\eta^5\text{-}C_5Me_5)\}_2$-$(B_8H_{12})$.[81a] Possible alternative syntheses include reactions with the $B_9H_{12}^-$ or $B_9H_{14}^-$ ions.

10. DECABORON COMPOUNDS

Nido-decaborane, $B_{10}H_{14}$, is the most readily available and easily handleable of the neutral boranes and its metallaborane chemistry is correspondingly quite extensive. Surprisingly, however, the variety of reaction types that have been utilized with *nido*-pentaborane (see Section 5) has not so far been paralleled by a similar diversity in the metallaborane chemistry of $B_{10}H_{14}$ and most of the reported examples can be thought of as being derived either directly or indirectly from the anionic ligands $B_{10}H_{13}^-$ and $B_{10}H_{12}^{2-}$.

Deprotonation of $B_{10}H_{14}$ [Figure 55(a)] occurs readily with a wide variety of bases from amines to alkali metal hydrides, and the resulting anion $B_{10}H_{13}^-$ [Figure 55(b)] has an unbridged B—B bond in the open face with a substantial two-center two-electron character. This site may then, in principle, act as a

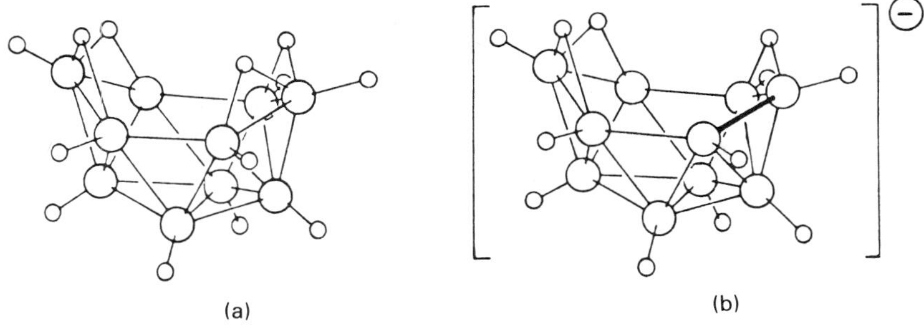

Figure 55. Schematic structures of (a) $B_{10}H_{14}$ and (b) the $B_{10}H_{13}^{-}$ anion.

dihapto-donor to a transition metal [or main-group metal (see other chapters)] as in the *nido*-pentaboranyl and *nido*-hexaboranyl compounds discussed in Sections 5 and 6. Examples of $B_{10}H_{13}^{-}$ acting in this *dihapto* mode have not, however, been unambiguously established, although possible candidates are the Cu(I) complexes $[Cu(B_{10}H_{13})(PPh_3)_2] \cdot CH_2Cl_2$, $[Cu(B_{10}H_{13})\{P(4\text{-}MeC_6H_4)_3\}_2]$, and $[Cu(B_{10}H_{13})Cl_2]^{2-}$. Thus, deprotonation of $B_{10}H_{14}$ with NaH in Et_2O followed by reaction with $[CuCl\{P(p\text{-}tol)_3\}_3]$ in CH_2Cl_2 at $0°$ affords a variable yield (up to 13%) of the air-sensitive yellow compound $[Cu(B_{10}H_{13})\text{-}\{P(p\text{-}tol)_3\}_2]$ (mp 272°): [82,83]

$$NaB_{10}H_{13} + [CuCl\{P(p\text{-}tol)_3\}_3] \longrightarrow [Cu(B_{10}H_{13})\{P(p\text{-}tol)_3\}_2] + NaCl + P(p\text{-}tol)_3$$

The corresponding solvated yellow complex $[Cu(B_{10}H_{13})(PPh_3)_2] \cdot CH_2Cl_2$ melts at 147–150°. Similarly, reaction of $NaB_{10}H_{13}$ with $[CuCl_2(OH_2)_2]$ in THF at $-10°$, followed by precipitation with $[NEt_4]Cl$ yields the bright yellow solid $[NEt_4]_2[Cu(B_{10}H_{13})Cl_2]$ (mp 145°).

The dianion $B_{10}H_{12}^{2-}$ can act either as an (η^2,η^2)-bis (*dihapto*) ligand or as an η^4-*tetrahapto* ligand and compounds containing both modes of bonding were, in fact, among the very first metal-polyhedral borane compounds to be synthesized, although their definitive structures were not established until some years later. Thus,[84] $B_{10}H_{14}$ reacts smoothly with $[ZnR_2]$ (R = Me, Et, Ph) and with $[CdEt_2]$ in ethereal solvents at room temperature to give air-sensitive crystalline products in yields of up to 83%. ($MgEt_2$ reacts similarly in Et_2O at $-78°$.) Typical products are yellow $[Zn(B_{10}H_{12})(OEt_2)_{1.5}]$ and $[Zn(B_{10}H_{12})\text{-}(THF)_{1.5}]$, white $[Cd(B_{10}H_{12})(OEt_2)_2]$, pale yellow $[Cd(B_{10}H_{12})(THF)_2]$, and colorless $[Zn(B_{10}H_{12})(OEt_2)_2]$.[84–86] Treatment of these complexes with ethereal HCl regenerates $B_{10}H_{14}$ essentially quantitatively. The physical and chemical properties of these compounds suggest that they are structurally more complex than the monomeric empirical formulas given above and an X-ray study of $[Cd(B_{10}H_{12})(OEt_2)_2]$, in fact, shows it to be dimeric with the unique structure shown in Figure 56.[87] Bridging H atoms at the 5,6 and 8,9 positions in $B_{10}H_{14}$ have been replaced by $\{Cd^{(II)}(OEt_2)_2\}$ groups which bridge two de-

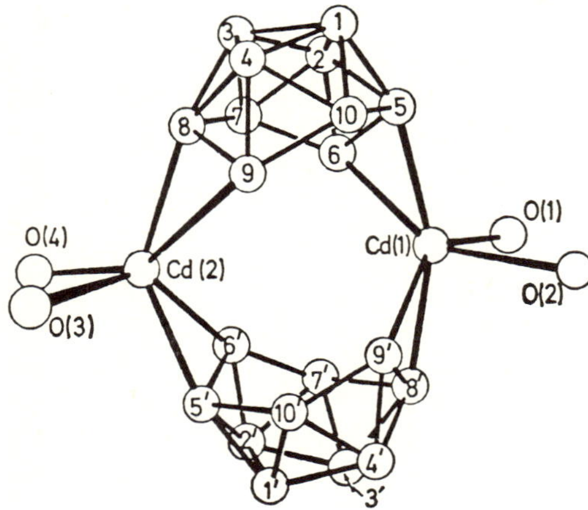

Figure 56. Structure of the dimeric complex $[Cd(\eta^2,\eta^2-B_{10}H_{12})(OEt_2)_2]_2$, showing the two bis (*dihapto*) $\{B_{10}H_{12}\}$ clusters bridging the two four-coordinate Cd atoms.

caboron clusters. Closest Cd—B approaches are 237-247 pm and the bonding at each Cd atom is essentially tetrahedral, with two three-center two-electron B—Cd—B bonds and two $Et_2O \longrightarrow Cd$ donor bonds with the angle $\angle O-Cd-O$ 88°.

The dimeric zinc and cadmium complexes react with water to yield the complex anions $[Zn(\eta^4-B_{10}H_{12})_2]^{2-}$ and $[Cd(\eta^4-B_{10}H_{12})_2]^{2-}$, and the mercury analog $[Hg(\eta^4-B_{10}H_{12})_2]^{2-}$ can be isolated by aqueous extraction of the reaction product of the decaboranyl "Grignard reagents" $B_{10}H_{13}MgX$ (X = Br, I) with alkylmercury(II) halides such as RHgCl (R = Me, Et, Pr^n) or MeHgX

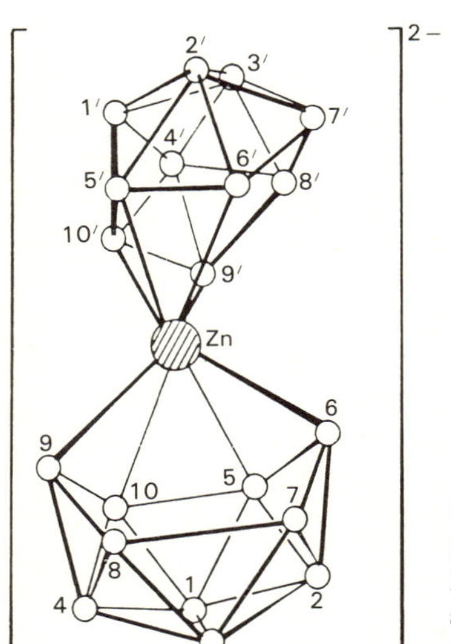

Figure 57. Structure of the anion in $[PMePH_3]_2^+$ $[Zn(\eta^4-B_{10}H_{12})_2]^{2-}$, showing the positions of the Zn and B atoms only. Each B atom has a terminal H atom attached, and there are bridging H atoms in the (6,7) and (8,9) positions in each cluster.

(X = Br, I).[88] The compounds $[NMe_4]_2[M(\eta^4\text{-}B_{10}H_{12})_2]$ and $[PMePh_3]_2$ $[M(\eta^4\text{-}B_{10}H_{12})_2]$ are all beautifully crystalline stable compounds with color deepening from pale yellow (Zn) to yellow-orange (Cd) and bright orange (Hg). The crystal structure of $[PMePh_3]_2^+[Zn(\eta^4\text{-}B_{10}H_{12})_2]^{2-}$ shows it to contain two bidentate *tetrahapto* $B_{10}H_{12}{}^{2-}$ ligands coordinated tetrahedrally about Zn(II) as in Figure 57.[89] The Zn—B contacts to B(5), B(6), B(9), and B(10) are 220, 245, 244, and 221 pm, respectively, indicating that the bridging H atoms at the 5,6 and 9,10 positions in $B_{10}H_{14}$ have been replaced by B—Zn—B three center two-electron bonds, and the disposition of the two B_{10} clusters ensures essentially tetrahedral bonding geometry about the Zn atom. Further spectroscopic and X-ray work established similar structures for the cadmium and mercury analogs.[89] The structures can also be described in terms of two metallaundecaborate units fused at a common metal ion which subrogates one {BH} unit in each of two *nido*-$B_{11}H_{13}{}^{2-}$ clusters.

An analogous structure has been found for the orange-red nickel(II) complex $[NMe_4]_2^+[Ni(\eta^4\text{-}B_{10}H_{12})_2]^{2-}$, the electronic requirements of the metal now dictating an essentially square-planar disposition of the bonding orbitals (Figure 58).[90] The interatomic distances of Ni to B(5), B(6), B(9), and B(10) are 211, 224, 222, and 218 pm, respectively, and the angle subtended at Ni by the midpoints of B(5)—B(6), and B(9)—B(10) is 80.8°. B—B distances within

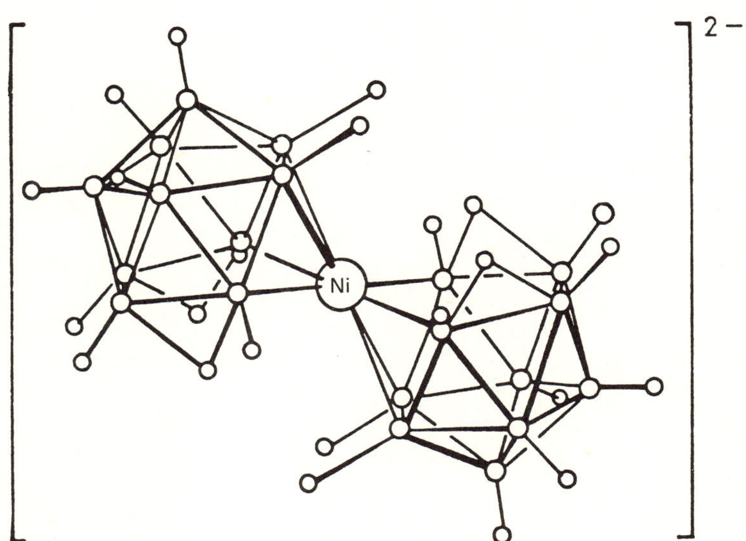

Figure 58. Structure of the anion in $[NMe_4]_2^+[Ni(\eta^4\text{-}B_{10}H_{12})_2]^{2-}$

Table 3. Properties of Some $[M(B_{10}H_{12})_2]^{-2}$ Salts

Compound	Color	mp	Yield (%)
$[NMe_4]_2[Ni(B_{10}H_{12})_2]$	Orange-red	180° (decomposition)	76
$[NMe_4]_2[Pd(B_{10}H_{12})_2]$	Yellow	240° (decomposition)	95
$[PHBu_3]_2[Pd(B_{10}H_{12})_2]$	Yellow	158° (decomposition)	20
$[NMe_4]_2[Pt(B_{10}H_{12})_2]$	Yellow	310° (decomposition)	96
$[NMe_4]_2[Co(B_{10}H_{12})_2]$	Red	Decomposition 140°	36
$[NMe_4]_2[Zn(B_{10}H_{12})_2]$	Yellow	Decomposition 300°	14

the η^4-B_4 unit are B(5)–B(6) 175, B(9)–B(10) 177, and B(5)–B(10) 187 pm; the latter is much shorter than that in uncomplexed $B_{10}H_{14}$ (201 pm).

The most general synthesis of $B_{10}H_{12}{}^{2-}$ metallaboranes is the reaction of excess $B_{10}H_{13}{}^-$ with transition-metal halide complexes. Three types of product can be distinguished.[91] With halo-complexes of Co(II), Ni(II), Pd(II), and Pt(II), salts of $[M(B_{10}H_{12})_2]^{2-}$ can be obtained in high yield, e.g,

$$4B_{10}H_{13}{}^- + [NiCl_2(PPh_3)_2] \xrightarrow{\text{THF}} 2B_{10}H_{14} + [Ni(B_{10}H_{12})_2]^{2-} + 2Cl^- + 2PPh_3$$

Typical products are seen in Table 3. In addition, neutral complexes of palladium and platinum can be isolated with composition $[M(B_{10}H_{12})(PR_3)_2]$, as can anionic complexes of composition $[M(B_{10}H_{12})L_3]^-$ where M = Co, Rh, Ir, and L = CO or tertiary phosphine:

$$2B_{10}H_{13}{}^- + [PtCl_2(PEt_3)_2] \xrightarrow{C_6H_6} B_{10}H_{14} + [Pt(B_{10}H_{12})(PEt_3)_2] + 2Cl^-$$

$$2B_{10}H_{13}{}^- + [Ir(CO)Cl(PPh_3)_2] \xrightarrow[\text{MeCN/THF}]{Et_2O} B_{10}H_{14} + [Ir(B_{10}H_{12})(CO)(PPh_3)_2]^- + Cl^-$$

The products (30–55% yield) are all yellow crystalline solids which tend to decompose without melting in the range 140-220°, e.g. $[Pd(B_{10}H_{12})(PPh_3)_2]$, $[Pd(B_{10}H_{12})(Ph_2PCH_2CH_2PPh_2)]$, $[Pt(B_{10}H_{12})L_2]$ (L = PEt_3, PBu_3, PPh_3), $[NMe_4][Co(B_{10}H_{12})(CO)_3]$, $[NMe_4][Rh(B_{10}H_{12})(CO)(PPh_3)_2]$, and $[NMe_4]$ $[Ir(B_{10}H_{12})(CO)(PPh_3)_2]$.[91] The spectroscopic properties of these various complexes all point towards the presence of one (or two) $\{\eta^4$-$B_{10}H_{12}\}^{2-}$ ligands of the type present in the *commo*-metallaborane anions illustrated in Figures 57 and 58. Similar structures have been proposed for the anion in $[PMePh_3][Hg(B_{10}H_{12})Me]$[92] and for the main-group metal *nido*-borane complexes $[In(B_{10}H_{12})Me]$,[92] $[In(B_{10}H_{12})Me_2]^-$,[92] $[Tl(B_{10}H_{12})Me_2]^{2-}$,[92-94] $[Sn^{(II)}(B_{10}H_{12})Cl_2]^{2-}$,[95] and $[Sn^{(IV)}(B_{10}H_{12})Cl_2Me_2]^{2-}$[95] which are discussed in more detail in Chapter 4.

The platinum species $[Pt(\eta^4$-$B_{10}H_{12})(PR_3)_2]$ can also be prepared by direct reaction of *nido*-decaborane with the platinum chlorohydrides $[PtClH(PR_3)]$,[91] e.g,

$$2[PtClH(PEt_3)_2] + B_{10}H_{14} \xrightarrow[9\%]{C_6H_6,\ 70°} [Pt(B_{10}H_{12})(PEt_3)_2] + [PtCl_2(PEt_3)_2] + H_2$$

A convenient route for their preparation involves the use of the nonnucleophilic base N,N,N',N'-tetramethylnaphthalene-1,8-diamine ("proton sponge") as a deprotonating agent:[70]

$$\textit{cis-}[PtCl_2(PMe_2Ph)_2] + B_{10}H_{14} + 2C_{10}H_6(NMe_2)_2 \xrightarrow[55\%]{CH_2Cl_2,\ 20°}$$

$$[Pt(B_{10}H_{12})(PMe_2Ph)_2] + 2[C_{10}H_6(NMe_2)(NHMe_2)]^+[Cl]^-$$

The yellow compound $[Pt(B_{10}H_{12})(PMe_2Ph)_2]$ undergoes a novel type of non-dissociative rotation about an axis in the Pt(II) coordination plane, with $\Delta G^{\ddagger}$ 79±5 kJ-mol^{-1} at 71°.[96] This mode of fluxionality is shown schematically in Figure 59(a) and is presumably general for platinaboranes of this type. The X-ray crystal structure of $[Pt(B_{10}H_{12})(PMe_2Ph)_2]$ is of particular interest in this connection as it reveals a rotation of the PtP$_2$ plane with respect to $\{\eta^4$-

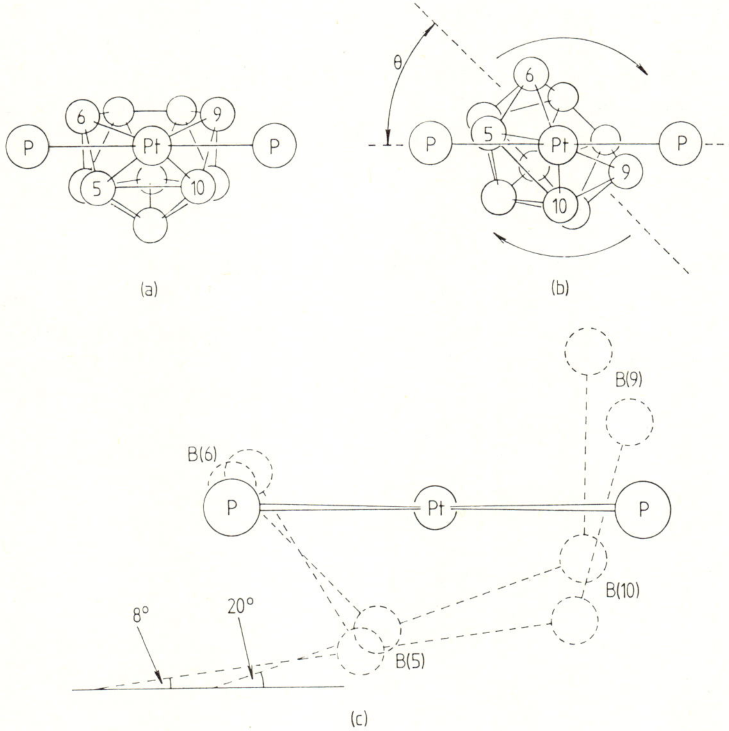

Figure 59. Diagrams (a) and (b): schematic diagrams of the $B_{10}PtP_2$ cluster of $[Pt(B_{10}H_{12})$-$(PMe_2Ph)_2]$ showing (a) the supposed minimum energy rotamer and (b) the nature of the rotational fluxionality. Diagram (c): superposition of the PtB(5)B(6)B(9)B(10)P(2) atoms of $[Pt(B_{10}H_{12})(PMe_2Ph)_2]$ (Figure 60) and $[Pt\{B_{10}H_{11}\text{-}2\text{-}(2'\text{-}B_{10}H_{13})\}(PMe_2Ph)_2]$ (Figure 68), viewed normal to the plane which contains the two P atoms and which is perpendicular to the PtP$_2$ plane.

$B_{10}H_{12}$} of about 21° as shown in Figure 59(b), which also includes data obtained for the analogous monoplatinum derivative of $[2,2'-(B_{10}H_{13})(B_{10}H_{11})]^{2-}$ (see Figure 68).[70] In the icosaborane compound the twist is approximately 8° and the superposition of these two effective rotations nicely defines the initial nature of the rotation and shows that it may also be associated with a movement of the Pt atom towards one corner of the B(5)B(6)B(9)B(10) grouping. It may be that both a *tetrahapto*-ligand and the Pt(II) electronic structure are required for the rotation to occur; certainly such fluxionality has not been observed for the *trihapto*-ligands in $[Pt(B_3H_7)(PMe_2Ph)_2]$ (Figure 6), $[Pt(B_8H_{12})(PMe_2Ph)_2]$ (Figure 45), or $[\{Pt(PMe_2Ph)_2\}_2(B_8H_{10})]$ (Figure 47).[70]

The structure of the complete $[Pt(B_{10}H_{12})(PMe_2Ph)_2]$ molecule is shown in Figure 60. Interatomic distances within the {Pt-η^4-B_4} unit are: PtB(5) 222, PtB(6) 231, PtB(9) 229, and PtB(10) 224 pm, and selected interboron distances are: B(5)—B(10) 182, B(5)—B(6) 177, B(9)—B(10) 178, B(1)—B(2) 179, and B(1)—B(4) 173 pm which illustrate the distortion associated with the rotation, as do the two PtP interatomic distances of 234.0 and 230.8 pm. The interatomic B—B distances of the atoms bonded to the metal atoms are somewhat shorter than those in $B_{10}H_{14}$, as found also for the $[Ni(B_{10}H_{12})_2]^{2-}$ anion (Figure 58).

An alternative route to $[M(B_{10}H_{12})_2]^{2-}$ complexes is by direct reaction of the *arachno*-decaborane salt $Cs_2B_{10}H_{14}$ with MCl_2, e.g.,[97]

$$2Cs_2B_{10}H_{14} + NiCl_2 \xrightarrow[\text{(ii) [NMe}_4]\text{ Cl}]{\text{(i) Me}_2CO(aq)} [NMe_4]_2[NiB_{10}H_{12})_2] + 4CsCl + 2H_2$$

The yield is 56% with $NiCl_2$ and 54% with $CoCl_2$. $PdCl_2$ is completely reduced under similar conditions and $ZnCl_2$ is unreactive. $PtCl_2$ affords a 7% yield of $[NMe_4]_2[Pt(B_{10}H_{12})_2]$, but the yield can be increased to 30% by starting with $[Pt(1,5\text{-cyclo-octadiene})I_2]$.

The {η^4-$B_{10}H_{12}$} mode of complexing results in particularly stable complexes and several other routes lead to such compounds. Thus, $B_{10}H_{14}$ reacts

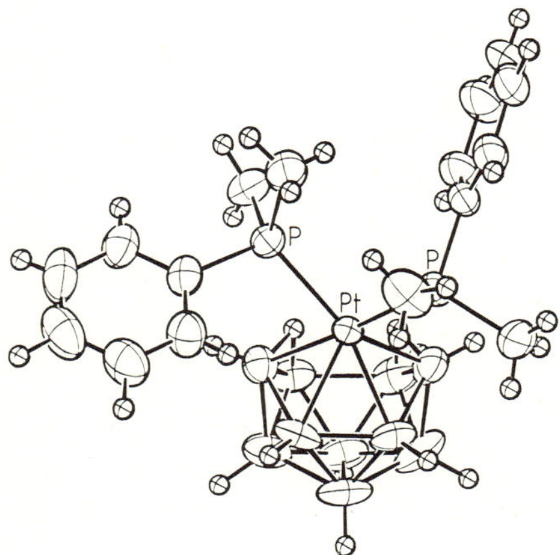

Figure 60. All-atom molecular structure of the yellow compound *nido*-$[Pt(\eta^4$-$B_{10}H_{12})$-$(PMe_2Ph)_2]$.

directly with nickelocene in either monoglyme ($MeOCH_2CH_2OMe$) or benzene to yield $[Ni(B_{10}H_{12})_2]^{2-}$ as the main product of prolonged reaction, but if the reaction is stopped after 1 hr a 60% yield of the red–brown intermediate product *nido*-$[Ni(B_{10}H_{13})(\eta^5-C_5H_5)]$ is obtained.[76] This can be reversibly deprotonated (see scheme) to give the red *nido*-anion $[Ni(B_{10}H_{12})(\eta^5-C_5H_5)]^-$ which can itself be prepared from reaction of $B_{10}H_{13}^-$ and nickelocene in monoglyme in the presence of a catalytic amount of sodium amalgam. This anion can react further with $[Co(\eta^5-C_5H_5)_2]$ to give the green *closo*-anion $[\{Co-(\eta^5-C_5H_5)\}\{Ni(\eta^5-C_5H_5)\}(B_{10}H_{10})]^-$ isolated as its $[AsPh_4]^+$ salt in a 2% yield. The golden-brown neutral bis(cyclopentadienylnickel) compound *closo*-$[\{Ni(\eta^5-C_5H_5)\}_2(B_{10}H_{10})]$, mentioned earlier,[27] is obtained in a 45% yield by direct insertion of $[Ni(\eta^5-C_5H_5)(CO)_2]$ or $[Ni(\eta^5-C_5H_5)_2]$ into *closo*-$B_{10}H_{10}^{2-}$ dissolved in monoglyme.[76,98] Proposed structures for these various complexes are in Figure 61; they are all stable to air, water, and heat. For example, $Cs[Ni(B_{10}H_{12})(\eta^5-C_5H_5)]$ is unchanged when heated to 180°C under a vacuum and even at 230° there is little pyrolytic damage after 1 hr; likewise $[\{Ni(\eta^5-C_5H_5)\}_2(B_{10}H_{10})]$ can be heated to 350° and subsequently sublimed at 265° over a period of hours with only partial decomposition.[98]

Derivatives of the *nido*-metallaborane complexes $[Pt(B_{10}H_{12})(PR_3)_2]$ are also produced when the *closo*-anion of anhydrous $K_2B_{10}H_{10}$ reacts with *cis*-$[PtCl_2(PR_3)_2]$ in ethanolic chloroform solution under reflux:[99] e.g.,

$$K_2B_{10}H_{10} + [PtCl_2(PPh_3)_2] \xrightarrow{\text{EtOH/CHCl}_3} [Pt(B_{10}H_{11}OEt)(PPh_3)_2] + 2KCl$$

$[PtCl_2(Ph_2PCH_2CH_2PPh_2)]$ reacted similarly. The yellow products can be separated chromatographically to give various isomers with ethoxy substituents in the 5-, 6-, or 7- positions in yields which vary from 3.5–74.1%; decomposition temperatures are in the range of 225–275°. A number of mechanisms may be written for this interesting cage-opening reaction[99] although the actual course of the reaction has yet to be established (see also Section 14.6).

Another derivative $[Pt(B_{10}H_{11}\text{-}6\text{-}Cl)(PMe_2Ph)_2]$ is formed by reaction of $[PtCl_2(PMe_2Ph)_2]$ with $[B_{10}H_{12}(SMe_2)_2]$.[70] The parent compound $[Pt(B_{10}H_{12})(PMe_2Ph)_2]$ is also formed in small amounts. The structure of yellow crystalline $[Pt(B_{10}H_{11}Cl)(PMe_2Ph)_2]$ has been determined by X-ray diffractometry and found to contain molecular units as shown in Figure 62. The twist

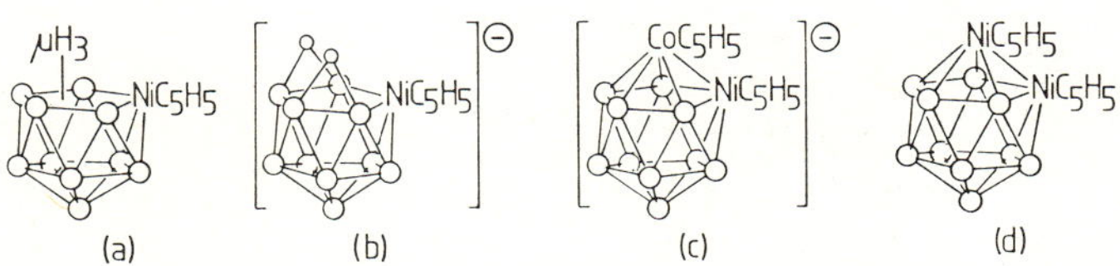

Figure 61. Proposed structure for (a) *nido*-$[Ni(B_{10}H_{13})(\eta^5-C_5H_5)]$, (b) *nido*-$[Ni(B_{10}H_{12})-(\eta^5-C_5H_5)]^-$, (c) *closo*-$[1\text{-}\{Co(\eta^5-C_5H_5)\}\text{-}2\text{-}\{Ni(\eta^5-C_5H_5)\}(B_{10}H_{10})]^-$, and (d) *closo*-$[1,2\text{-}\{Ni(\eta^5-C_5H_5)\}_2(B_{10}H_{10})]$.

$$[Ni(\eta^5\text{-}C_5H_5)_2] + B_{10}H_{14} \xrightarrow[\text{C}_6\text{H}_6/1\ \text{Hr}]{\text{glyme or}} nido\text{-}[Ni(B_{10}H_{13})(\eta^5\text{-}C_5H_5)]$$

bright green   red-brown

$$H^+/(aq) \Big\updownarrow OH^-(aq)$$

$$[Ni(\eta^5\text{-}C_5H_5)_2] + B_{10}H_{13}{}^- \xrightarrow[\text{(Na/Hg catalyst)}]{\text{glyme}} [NMe_4][nido\text{-}Ni(B_{10}H_{12})(\eta^5\text{-}C_5H_5)]$$

red

$$\Big\downarrow [Co(\eta^5\text{-}C_5H_5)_2]/(Na/Hg)$$

$$[AsPh_4][closo\text{-}\{Co(\eta^5\text{-}C_5H_5)\}\{Ni(\eta^5\text{-}C_5H_5)\}(B_{10}H_{10})]$$

green

$$\left.\begin{array}{c}[Ni(\eta^5\text{-}C_5H_5)_2]\\ \text{or}\\ [Ni(\eta^5\text{-}C_5H_5)(CO)_2]\end{array}\right\} + closo\text{-}B_{10}H_{10}{}^{2-} \xrightarrow[\text{or } C_6H_6]{\text{glyme}} closo\text{-}[\{Ni(\eta^5\text{-}C_5H_5)\}_2(B_{10}H_{10})]$$

golden brown

angle as defined in Figure 59 is ca. 24°, and like the parent compound, the 6-chloro derivative is fluxional, but with a much lower activation energy (56 kJ mol^{-1} at −6°C).

These *nido*-platinaundecaborane species, in general, seem to be particularly stable species and are formed as major or minor products in a number of reactions. Apart from the reaction discussed in the previous paragraph, the compound $[Pt(B_{10}H_{12})(PMe_2Ph)_2]$ is formed in small amounts in the reaction of $6,6'\text{-}(B_{10}H_{13})_2O$ to produce $[Pt(B_6H_9)(PMe_2Ph)]_2$ discussed above (Section 6, p. 75), and also when $[PtCl_2(PMe_2Ph)_2]$ and $B_{10}H_{14}$ are allowed to stand in contact in neutral solvents.[70] The reaction of the Pt^0 compound $[Pt(PPh_3)_3]$ with *nido*-$B_{10}H_{14}$ in refluxing benzene gives a high yield (85%) of $[Pt(B_{10}H_{12})(PPh_3)_2]$, but with *arachno*-$B_{10}H_{12}(SEt_2)_2$ the yield is <10%.[99]

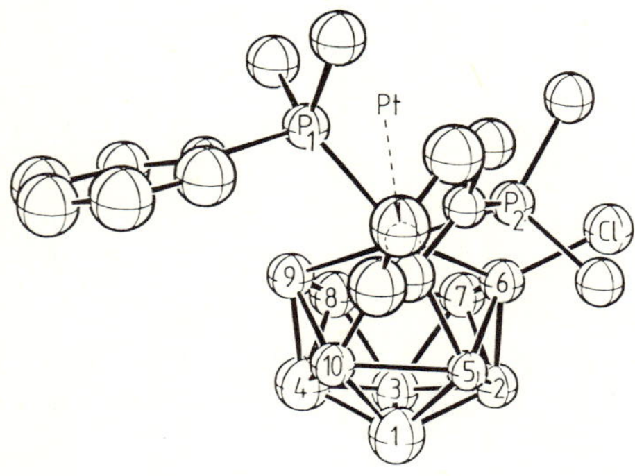

Figure 62. Molecular structure of $[Pt(5,6,9,10\text{-}\eta^4\text{-}B_{10}H_{11}\text{-}6\text{-}Cl)(PMe_2Ph)_2]$, with H atoms omitted. In this projection the Pt atom is obscured by the *para-C* atom of one of the *P*-phenyl groups. Distances from the Pt atom are B(5) 228, B(6) 222, B(9) 233, B(10) 225, P(1) 235, and P(2) 232 pm and the angle P(1)—Pt—P(2) is 96°. Other selected distances are B(5)—B(6) 175, B(5)—B(10) 206, B(9)—B(10) 176, and B(6)—Cl 184 pm.

A further route to metal–decaboron derivatives is by ligand exchange on a preformed complex as exemplified by the following reactions:[100]

$$[Pd(B_{10}H_{12})(PPh_3)_2] + 2NaCN \xrightarrow{\text{MeCN}} \xrightarrow{[NMe_4]Cl} [NMe_4]_2^+[Pd(B_{10}H_{12})(CN)_2]^{2-}$$

pale yellow-orange, 45% yield

$$[Pd(B_{10}H_{12})(PPh_3)_2] + Na_2[S_2C_2(CN)_2] \xrightarrow[\text{RT}]{\text{MeCN/Me}_2\text{CO}} \xrightarrow{[AsPh_4]Cl}$$

$$[AsPh_4]_2^+[Pd(B_{10}H_{12})\{S_2C_2(CN)_2\}]^{2-} \text{ yellow, 55\% yield, mp 205--207°}$$

$$[Pd(B_{10}H_{12})(PPh_3)_2] + 1,10\text{--phenanthroline} \xrightarrow[\text{RT}]{(ClCH_2)_2} [Pd(B_{10}H_{11})(1,10\text{--}C_{10}H_8N_2)]$$

yellow, 75% yield

$$\xrightarrow[\text{(ii) }[NMe_4]Cl]{\text{(i) LiBu}} [NMe_4]_2^+[\{Pd(B_{10}H_{11})\}_2(1,10\text{--}C_{10}H_8N_2)]^{2-} \text{ red needles, 18\% yield}$$

Metathesis has also yielded Cs^+ and $[AsPh_4]^+$ salts of this last binuclear complex. Again, reaction of $[NEt_4][Ir(B_{10}H_{12})(CO)(PPh_3)_2]$ with $Na_2[S_2C_2(CN)_2]$ in MeCN under reflux followed by metathesis with $[PMePh_3]Br$ affords low yields of the yellow 7-coordinate complex $[PMePh_3][Ir(B_{10}H_{12})(CO)(PPh_3)_2-\{S_2C_2(CN)_2\}]$ (mp 149-151°), and of the yellow-orange 6-coordinate complex $[PMePh_3][Ir(B_{10}H_{12})(PPh_3)_2\{S_2C_2(CN)_2\}]$ (mp 131-133°) (decomposition).[100]

This section concludes with a brief summary of various other miscellaneous reactions which have led to metal-*nido*-borane complexes. Thus, one of the products of the previously mentioned reaction between Fe^{2+}, $B_5H_8^-$, and $C_5H_5^-$ in THF at −78° followed by work-up in air[63] is the yellow compound $[2-\{Fe-(\eta^5-C_5H_5)\}(B_{10}H_{15})]$. This has a *nido*-electron count and has been assigned the structure in Figure 63; the $\{Fe(\eta^5-C_5H_5)\}$ group occupies the 2-position next to the apical boron atom. The open face has five B—H—B bridging bonds, and the compound is isoelectronic with the hypothetical *nido*-borane $B_{11}H_{15}$. Complex reactions have also been reported between $B_{10}H_{14}$ and $[FeBr_2(py)_4]$ in boiling benzene:[101] the first product, obtained in 95% yield, is a light-brown iron(III) complex of empirical formula $[Fe_2(B_{20}H_{24})Br_4(py)_5]$ of which postulated features include $\{\eta^4-B_{10}H_{11}py_2\}$ and $\{\eta^1-B_{10}H_{13}py\}$ groups coordinated to different Fe atoms which are themselves further coordinated by Br

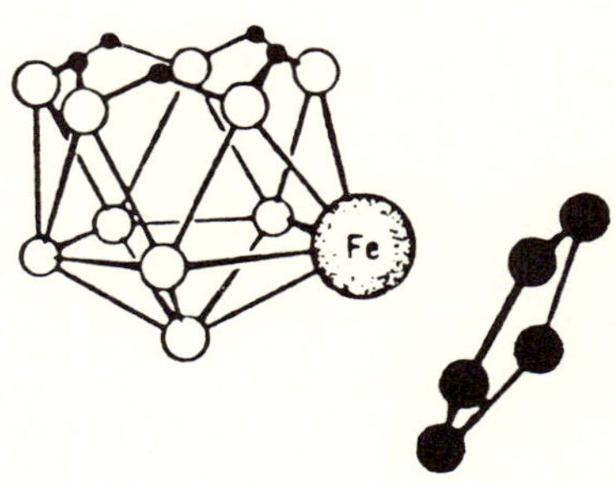

Figure 63. Proposed structure of $[2-\{Fe(C_5H_5)\}(B_{10}H_{15})]$.

and py groups and joined into a dinuclear complex by bridging Br atoms as in Figure 64. These bridges are thought to be cleaved by THF to give a quantitative yield of two new mononuclear complexes $[Fe\{B_{10}H_{11}(py)_2\}Br_2py]$ and the brown tetrahedral high-spin σ-bonded Fe(III) complex $[Fe(\sigma-B_{10}H_{13}py)Br_2py]$. With water the initial product yields a compound which has been tentatively formulated as $[Fe^{(III)}(\eta^4-B_{10}H_{11}py_2)_2][B_{10}H_{13}py]$. Further structural characterization of these curious species is clearly desirable.

Reaction of $[NMe_4][B_{10}H_{13}]$ with $HgCl_2$ in THF gives moderate yields (31%) of the bright yellow complex $[NMe_4]_2[Hg(B_{10}H_{12})_2]$ already described on p. 89. However, with $[CoCl_2(py)_2]$ and $[CoBr_2(py)_2]$ the reaction takes a different course to yield brown paramagnetic complexes believed to be $[NMe_4]$ $[Co(B_{10}H_{11}py)X_2(py)]$ and $[NEt_3H][Co(B_{10}H_{11}py)X_2(py)]$ (X = Cl, Br). The structure proposed for these novel *nido*-anions is depicted in Figure 65.[102] Yet another reaction course occurs when $NaB_{10}H_{13}$ reacts with the cation $[Fe(\eta^5-C_5H_5)(cyclohexene)(CO)_2]^+$ in Et_2O at room temperature:[103] a 40% yield of the yellow σ-bonded ferraundecaborane $[6-\{Fe(\eta^5-C_5H_5)(CO)_2\}-B_{10}H_{13}]$ results. The reaction, therefore, parallels that of $B_5H_8^-$ with $[Fe-(\eta^5-C_5H_5)(CO)_2)i]$ to give the σ-bonded species $[2-\{Fe(\eta^5-C_5H_5)(CO)_2\}-B_5H_8]$ and $[2,4-\{Fe(\eta^5-C_5H_5)(CO)_2\}_2B_5H_7]$ mentioned in Section 5.3.[54] Photolysis of $[6-\{Fe(\eta^5-C_5H_5)(CO)_2\}B_{10}H_{13}]$ in ethereal solvent leads to small yields of icosahedral metallacarboranes of the type $[Fe(\eta^5-B_{10}H_{10}CL)-(\eta^5-C_5H_5)]$ (L = Et_2O or THF) which are discussed further in another chapter.[104] Treatment of $[6-\{Fe(\eta^5-C_5H_5)(CO)_2\}B_{10}H_{13}]$ with Br_2 in CH_2Cl_2 below room temperature gives a 96% yield of the red–brown complex $[Fe(\eta^5-C_5H_4-6-B_{10}H_{13})(CO)_2Br]$ (mp 144–145°) (decomposition) which is a derivative of the well-known cyclopentadienyl compound $[Fe(\eta^5-C_5H_5)(CO)_2Br]$ with one H atom substituted by a $\{6-B_{10}H_{13}\}$ group.[103] A similar compound, but with a $\{2-B_5H_8\}$ substituent instead of $\{B_{10}H_{13}\}$, is produced when the pentaboron analog (Figure 27, Section 5) is treated similarly[70]; the reaction is also known in carborane chemistry. However, these products cannot be considered as metallaboranes.

Finally $B_{10}H_{13}^-$ reacts with the hexacarbonyls of chromium, molybdenum, and tungsten in THF under photolytic conditions to give curious icosahedral

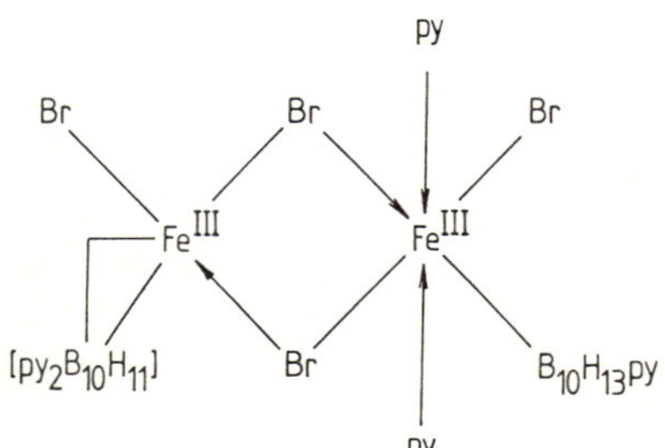

Figure 64. Possible structure for the supposed binuclear complex of empirical formula $[Fe_2(B_{20}H_{24})Br_4py_5]$.

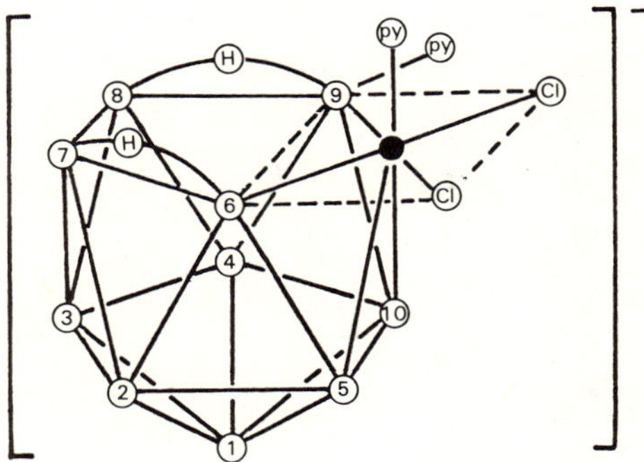

Figure 65. Proposed structure of the *nido*-anion $[Co(B_{10}H_{11}py)Cl_2py]^-$; the B(9) atom is believed to carry a pendant py ligand and each of the nine other B atoms is covalently bonded to a terminal H atom (not shown).

closo-metallacarborane anions $[M(\eta^5\text{-}B_{10}H_{10}COH)(CO)_4]^-$ (M = Cr, Mo, or W) which can be deprotonated with KH to give the exopolyhedral heterocyclic icosahedral *closo*-metallacarborane dianions $[B_{10}H_{10}COMCO(CO)_3]^{2-}$.[105]

11. UNDECABORON COMPOUNDS

Since the *nido*-undecaborate anions $B_{11}H_{14}^-$ and $B_{11}H_{13}^{2-}$ can be readily made in good yield from *nido*-decaborane one might have expected the development of a rich field of metal-undecaboron compounds. Surprisingly, there is only one report of such a compound, prepared by *nido*-cluster closure using nickelocene:[98]

$$[Ni(\eta^5\text{-}C_5H_5)_2] + Na_2B_{11}H_{13} \xrightarrow[\text{(ii) } H_2O/[NMe_4]Cl]{\text{(i) MeCN(Na/Hg)}}$$

$$[NMe_4][Ni(\eta^5\text{-}B_{11}H_{11})(\eta^5\text{-}C_5H_5)] + H_2 + \cdots$$

The bright yellow tetramethylammonium salt of the *closo*-anion $[Ni(\eta^5\text{-}B_{11}H_{11})(\eta^5\text{-}C_5H_5)]^-$ can be obtained in 64% yield and has been assigned the structure shown in Figure 66. (The same anion can be made in a 23% yield by the reaction of *closo*-$B_{11}H_{11}^{2-}$ with $[\{Ni(\eta^5\text{-}C_5H_5)(CO)\}_2]$ in refluxing THF.) An interesting extension of this reaction is the preparation of the mixed *closo*-dimetalla-

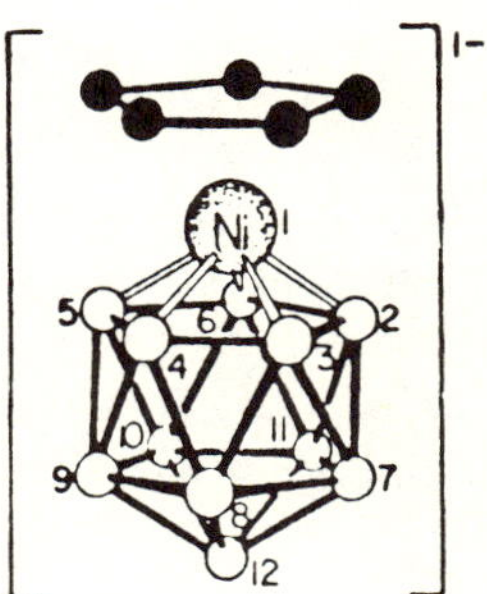

Figure 66. Proposed structure for the *closo*-anion $[Ni(\eta^5\text{-}B_{11}H_{11})(\eta^5\text{-}C_5H_5)]^-$.

dodecaborane anion $[\{Co(\eta^5\text{-}C_5H_5)\}\{Ni(\eta^5\text{-}C_5H_5)\}(B_{10}H_{10})]^-$ from cobaltocene and $[Ni(B_{10}H_{12})(\eta^5\text{-}C_5H_5)]^-$ as mentioned in the preceding section.

12. OCTADECABORON AND ICOSABORON COMPOUNDS

Of the higher boranes, only n- and iso-$B_{18}H_{22}$, $B_{20}H_{16}$, and the various isomers of bi($nido$-decaboranyl), $(B_{10}H_{13})_2$, are of sufficient stability and/or accessibility to permit their derivative chemistry to have been explored, and of these the isomers of $B_{18}H_{22}$ and $(B_{10}H_{13})_2$ fall within the scope of this chapter.

The isomers of $B_{18}H_{22}$ comprise edge-fused pairs of $nido$-decaborane clusters and, as such, can be readily deprotonated by NaH in THF to give n- and iso-$B_{18}H_{20}{}^{2-}$; these anions react with a variety of phosphine- and carbonyl-substituted derivatives of cobalt, rhodium, nickel, palladium, and platinum to give complexes of the general type $[M(B_{18}H_{22})L_n]$ [106] analogous to the decaboranyl complexes described previously. Typical reactions are:

$$\text{Na}_2[n\text{-}B_{18}H_{20}] + [Co_2(CO)_8] \xrightarrow[\text{(ii) } H_2O/Me_2CO/ \atop [NMe_4]Cl]{\text{(i) THF at RT}} [NMe_4]^+[Co(n\text{-}B_{18}H_{20})(CO)_3]^-$$

$$\text{Na}_2[n\text{-}B_{18}H_{20}] + [NiCl_2(PPh_3)_2] \xrightarrow[\text{reflux}]{\text{THF}} [Ni(n\text{-}B_{18}H_{20})(PPh_3)_2]$$

Table 4 summarizes the products and yields obtained and Figure 67 shows the proposed structure of these complexes in which the metal atom is bonded to one of the open faces of the $\{B_{18}H_{20}\}^{2-}$ ligand which acts as a bidentate *tetrahapto* donor as does $\{B_{10}H_{12}\}^{2-}$ to similar metal centers. [106] (See also Section 14.7).

The $(B_{10}H_{13})_2$ isomers consist of two $nido$-decaboranyl clusters joined by a stable two-electron two-center B—B σ-bond and so their metallaborane chemistry so far established is again similar to that of $nido$-$B_{10}H_{14}$. Thus, the reaction of $2,2'$-$(B_{10}H_{13})_2$ with cis-$[PtCl_2(PMe_2Ph)_2]$ and N,N,N',N'-tetramethyl-naphthalene-1,8-diamine in CH_2Cl_2 at ambient temperatures gives a 40% yield of the yellow crystalline $nido$-decaboranyl-$nido$-platinaundecaborane [Pt-$\{B_{10}H_{11}(B_{10}H_{13})\}(PMe_2Ph)_2]$: [70]

$$[PtCl_2(PMe_2Ph)_2] + (B_{10}H_{13})_2 + 2C_{10}H_6(NMe_2)_2 \longrightarrow$$
$$[Pt\{B_{10}H_{11}(B_{10}H_{13})\}(PMe_2Ph)_2] + 2[C_{10}H_6(NMe_2)(NHMe_2)]^+[Cl]^-$$

The X-ray crystal structure (Figure 68) shows that the molecule is similar to $[Pt(B_{10}H_{12})(PMe_2Ph)_2]$ (Figure 60), but with an additional $2'$-$\{nido$-$B_{10}H_{13}\}$ substituent on the 2-position. In the crystal the ligand rotational "twist" (Figure 59) is ca. $8°$, and the compound exhibits a similar ligand fluxionality to that of $[Pt(B_{10}H_{12})(PMe_2Ph)_2]$ and $[Pt(B_{10}H_{11}\text{-}6\text{-}Cl)(PMe_2Ph)_2]$ discussed previously. The interatomic distances from the Pt atom are B(5) 223, B(10 223.5, B(9) 232.5, B(6) 232, P(1) 233.5, and P(2) 234.1 pm, and the *conjuncto* B(2)—

Table 4. Complexes Derived from the $B_{18}H_{20}{}^{2-}$ Anions

Compound	Color	Yield (%)
$[NMe_4]^+[Co(n\text{-}B_{18}H_{20})(CO)_3]^-$	Red–orange	20
$[NMe_4]^+[Rh(n\text{-}B_{18}H_{20})(CO)(PPh_3)]^-$	Red	12
$[Ni(n\text{-}B_{18}H_{20})(PPh_3)_2]$	Red	27
$[Ni(n\text{-}B_{18}H_{20})(diphos)]^a$	Dark purple	32
$[Pd(n\text{-}B_{18}H_{20})(diphos)]$	Bright yellow	11
$[Ni(i\text{-}B_{18}H_{20})(diphos)]$	Red	26
$[Pd(i\text{-}B_{18}H_{20})(diphos)]$	Yellow	12.5
$[Pt(i\text{-}B_{18}H_{20})(diphos)]$	Bright yellow	18.6
$[Ni(i\text{-}B_{18}H_{18}CNH_2C_6H_{11})(diphos)]$	Red	16.6

^{a}Diphos = $Ph_2PCH_2CH_2PPh_2$

B(2′) intercluster link is 171 pm which is similar to the distance of 169.2 pm in unsubstituted $2,2'\text{-}(B_{10}H_{13})_2$. The icosaborane $2,6'\text{-}(B_{10}H_{13})_2$ reacts similarly to give a similar yellow compound, but now with a $6'\text{-}\{nido\text{-}B_{10}H_{13}\}$ substituent on the 2-position (so assigned on the basis of nmr spectroscopy); the isomer with a $2'\text{-}\{nido\text{-}B_{10}H_{13}\}$ substituent on the 6-position does not appear to be formed under these conditions. A second metal center can be added by further deprotonation followed by addition of more $[PtCl_2(PMe_2Ph)_2]$:

$$[Pt\{B_{10}H_{11}(B_{10}H_{13})\}(PMe_2Ph)_2] + 2C_{10}H_6(NMe_2)_2 + [PtCl_2(PMe_2Ph)_2] \longrightarrow$$
$$[\{Pt(PMe_2Ph)_2\}_2(B_{10}H_{11})_2] + 2[C_{10}H_6(NMe_2)(NHMe_2)]^+Cl^-$$

From the 2,2′-compound two isomeric yellow air-stable products are formed in a total yield of ca. 30%. These can be separated chromatographically to give what are thought to be the *cis* and *trans* isomers of $2,2'\text{-}[\{Pt(PMe_2Ph)_2\}(5,6,\text{-}9,10\text{-}\eta^4\text{-}B_{10}H_{11})]_2$ (Figure 69):[70] an X-ray diffraction analysis to show which is which would be of interest.

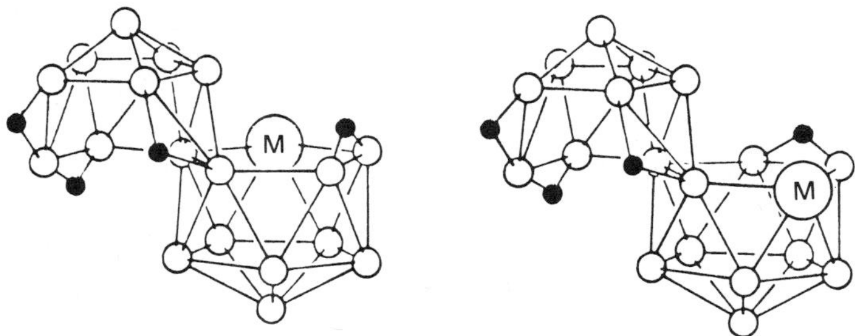

Figure 67. Presumed metallaborane cage structures of (left) $[Ni(n\text{-}B_{18}H_{20})(PPh_3)_2]$ and (right) $[Ni(iso\text{-}B_{18}H_{20})(diphos)]$.

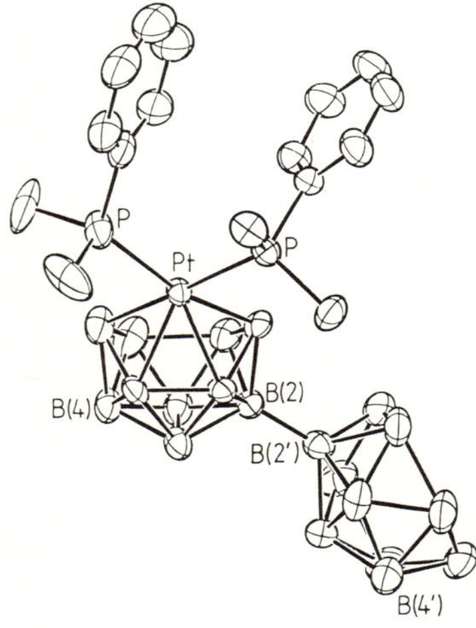

Figure 68. Molecular structure of $[Pt\{5,6,9,10-\eta^4-B_{10}H_{11}-2-(2'-B_{10}H_{13})\}(PMe_2Ph)_2]$ with H atoms omitted. The crystal unit cell also contains the $4-(4'-B_{10}H_{13})$ enantiomer.

13. SUMMARY AND CONCLUSIONS

The great versatility of metallaborane chemistry is apparent in the preceding sections. Some compounds can be prepared in high yield by rational syntheses whereas others are formed in minute yields by obscure and little understood pathways. Often, unstable and/or otherwise unknown cluster types are stabilized by the presence of metal centers. The field, while a comparatively new branch of borane chemistry, is already sufficiently extensive to reveal structural and bonding principles which relate it in a satisfying manner to the older and more thoroughly studied areas of organometallic chemistry, metallacarborane chemistry, and the chemistry of metal-metal cluster compounds. Future work will undoubtedly generate still further unsuspected structure types by unpredicted reaction paths and, thus, prove a rich hunting ground for the adventurous chemist for many years to come, but it is also likely that the growing corpus of knowledge in this area will permit increasing control of reactions to produce desired compounds in good yield. The flexibility afforded by varying the ligands coordinated to the transition-metal substrate, by modifying the solvent and other reaction conditions, and by using the chemical characteristics of the indi-

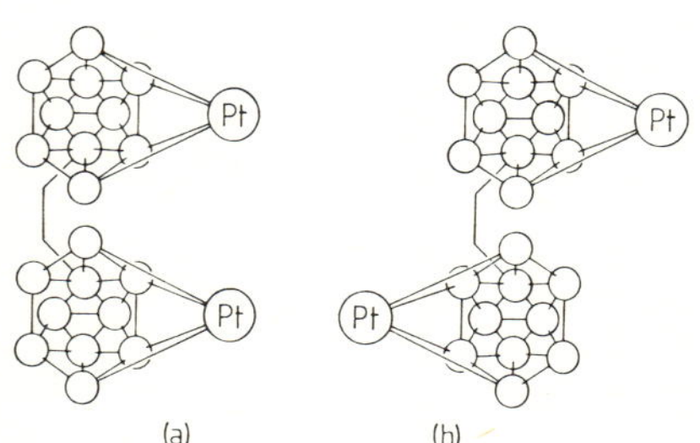

Figure 69. Schematic representation of the *conjuncto*-bis-(*nido*-metallaundecaboranyl) cluster structure of (a) the "*cis*" and (b) the "*trans*" isomers of $2,2'-[\{Pt(PMe_2Ph)_2\}-(B_{10}H_{11})]_2$. The *cis* compound (a) is a unique diastereoisomer whereas the structure depicted for the *trans* compound (b) is one of an enantiomeric pair.

vidual metals and boranes themselves, augurs well for the preparation of metalla-borane complexes in which there is an almost continuous variation in the extent to which the metals and the borane clusters interact.

14. APPENDIX (March 1982)

Since the submission of the main body of the text, a substantial number of results from further work have become available; these are now summarized.

14.1 Triboron Compounds

The compound $[Mn_2(B_3H_8)Br(CO)_6]$ has been reported; [107] this is a structural analogue of $[Mn_3(B_2H_6)(CO)_{10}H]$ discussed above in Section 2 (Figure 1). A more complete account of the work[15] on the *nido* five-vertex species $[Fe_2(B_3H_7)(CO)_6]$ (Section 3, Figure 9, above) has also appeared,[108] and X-ray crystallographic details of $[\{Co(C_5H_5)\}_3(CO)(B_3H_3)]$ have been reported.[109]

The results of single-crystal X-ray diffraction analysis on the iridatetraborane species $[Ir(B_3H_7)(CO)H(PPh_3)_2]$ (Section 3, Figure 8) are now available.[110] The structure (Figure 70) suggests that the compound is perhaps now best regarded as an iridium(V) analogue of *arachno*-B_4H_{10}, with the metal atom subrogating the B(1) vertex. The valence bonding geometry about the iridium atom is essentially seven-coordinate capped octahedral.

Another *arachno*-iridatetraborane has also been isolated;[110] this is the colorless iridium(III) species $[Ir(B_3H_8)H_2(PPh_3)_2]$, in which the metal atom subrogates now the B(2) vertex of B_4H_{10}, i.e., the $B_3H_8^-$ anion is acting as a bidentate η^2 ligand as in the numerous examples in Chapter 3. The geometry about iridium is presumably octahedral, with two mutually *cis* hydride ligands *trans* to the two Ir—H—B bridge bonds; these are all *cis* to the mutually *trans* phosphine ligands. The compound is the principal product in the reaction of $[IrCl(PPh_3)_3]$ with *arachno*-$B_3H_8^-$ or with *nido*-$B_5H_8^-$.[110]

There is a marked difference in the formal role of the metal center in these two *arachno*-B_4H_{10} analogues: in $[Ir(B_3H_8)H_2(PPh_3)_2]$ an iridium(III) center replaces a B_4H_{10} boron(III) center whereas in $[Ir(B_3H_7)(CO)H(PPh_3)_2]$ an iridium(V) center replaces a boron(III) center together with two bridging H

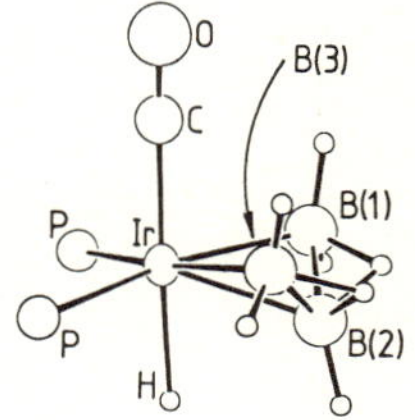

Figure 70. Cluster structure of $[Ir(B_3H_7)(CO)H(PPh_3)_2]$. Interatomic distances from the Ir atom to B(1), B(2) and B(3) are ca. 230, 220, and 230 pm respectively. The stereochemistry suggests a capped octahedral disposition of bonding orbitals about the metal atom with three 2-center 2-electron Ir—B bonds.

atoms. A further example of this $Ir^{(V)} \equiv \{Ir^{(III)} + 2H\}$ equivalence is discussed below in Section 14.5.

14.2 Tetraboron Compounds

Some theoretical work has been reported on the *closo* eight-vertex species $[3,4,5,6-\{Co(\eta^5-C_5H_5)\}_4B_4H_4]^{(24)}$ (Figure 14 above) and $[1,2,7,8-\{Ni-(C_5H_5)\}_4B_4H_4]^{(27)}$ (Section 4, Figure 16 above) which have $2n$ and $2n + 4$ framework electron counts respectively.$^{(111,112)}$

A *closo* di-irida(III)hexaborane species which has an idealized octahedral cluster structure has been structurally characterized: the yellow compound $[1,1,2-(CO)_3-1-(PPh_3)-2,2-(PPh_2\overline{C_6H_4})_2(Ir_2B_4H_2)]$ (Figure 71) has been isolated as an advanced degradation product of the reaction of $[Ir(CO)Cl(PPh_3)_2]$ with $B_{10}H_{10}^{2-}$ in the presence of MeOH.$^{(113)}$ It also exhibits a double incidence of an exocyclic *ortho*-boronation of the P-phenyl groups of the metal ligands.

The compound has no facial H atoms, in contrast to the structurally similar dicobaltahexaborane $[1,2-\{(C_5H_5)Co\}_2B_4H_6]^{(23)}$ (Section 4 above Figure 13), which has two extra H atoms to compensate for the two electrons involved in binding the $\{C_5H_5\}$ groups to the cobalt(III) atoms. In the iridium compound the *exo*-polyhedral ligands on the metal atoms are all neutral two-electron donors and so each iridium(III) center can contribute all three of its valence electrons to the cluster bonding.

Another iridium(III) species containing four B atoms is the pale yellow *arachno* compound $[Ir(B_4H_9)(CO)(PMe_3)_2]$ which may be regarded as an *arachno*-B_5H_{11} analogue.$^{(114)}$ In this compound the $\{Ir(CO)(PMe_3)_2\}$ center formally subrogates the unique (apical) $BH_2(1)$ group of B_5H_{11}. The borane-to-metal bonding has similarities to butadiene-to-metal bonding, and in accord with this the molecule is fluxional, as indicated in Figure 72. The compound was obtained as a minor by-product from the reaction of *nido*-$B_9H_{12}^-$ with $[Ir(CO)Cl-(PMe_3)_2]$, and it is also formed in low yield in the reaction of *nido*-$B_6H_9^-$ with the same metal complex.$^{(110)}$ The PMe_2Ph analogue $[Ir(B_4H_9)(CO)PMe_2Ph]$ has been more logically synthesized in good yield by the reaction of $[Ir(CO)Cl-$

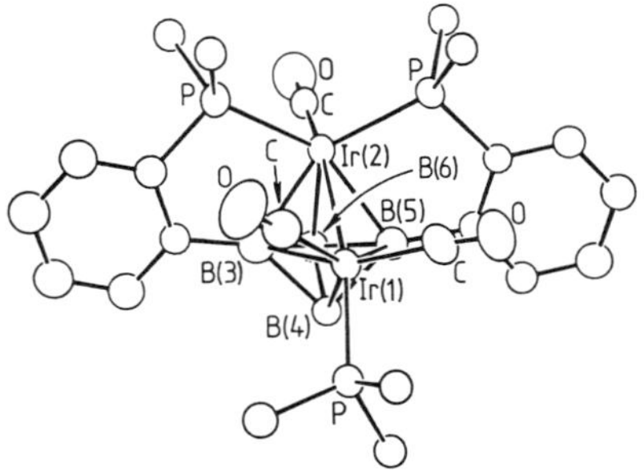

Figure 71. Molecular structure of $[(CO)_3-(Ph_3P)(Ph_2PC_6H_4)_2(Ir_2B_4H_2)]$ with H atoms and phenyl C atoms omitted. Selected interatomic distances are: from Ir(1) to Ir(2) 279.9(1), to B(3) 230.4(11), to B(4) 219.7(10), to B(5) 223.7(11), and from Ir(2) to B(3) 218.3(11), B(5) 218.7(10) and B(6) 219.0(10) pm.

Figure 72. Proposed structure of *arachno*-$[Ir(B_4H_9)(CO)(PMe_3)_2]$; the alternative view of the molecule (right) indicates the nature of the mutual pseudorotational fluxionality of the ligands ($\Delta G^{\ddagger} = 63$ kJ mol^{-1} at +17.5 and at +46°C) and shows the asymmetric distribution of the $\{(CO)$-$(PMe_3)_2\}$ ligands with respect to the formally C_s $\{IrB_4\}$ cluster; this proposed structure is analogous to that found by X-ray diffraction work on $[Ir(B_4H_9)(CO)(PMe_2Ph)_2]$ for which details are not yet available.

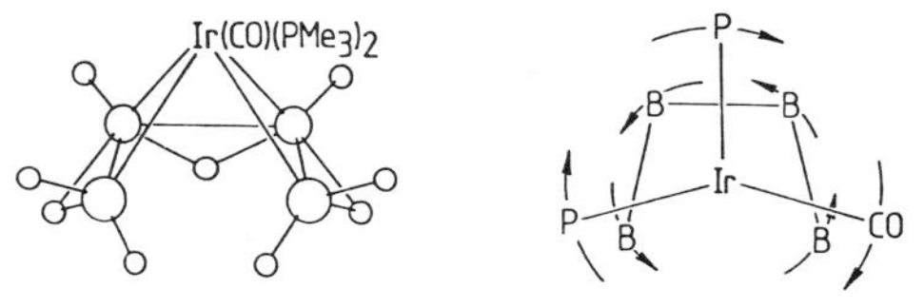

$(PMe_2Ph)_2]$ with *arachno*-$B_4H_9{}^-$; preliminary structural data have been reported for this species,[115] but full details are not yet available.

14.3 Pentaboron, Hexaboron, and Heptaboron Compounds

A more definitive and complete account of the reactions of metal atoms with *nido*-B_5H_9 (Section 5.1 above) has now appeared;[116] this also includes further details of the compound $[6-\{(C_5H_5)Co\}(B_9H_{13})]$ (Section 9 above). There has been some theoretical work on the relative stabilities of simple *nido*-metallahexaboranes, such as $[2-\{Fe(\eta^5-C_5H_5)\} B_5H_{10}]$ (e.g., Section 5.6 and Figure 36), which are structural analogues of B_6H_{10}.

There is still a dearth of examples of non-*closo* metallaboranes which have contiguous seven- and eight-vertex structures, although the macropolyhedral species $[Pt\{\eta^6-B_{16}H_{18}(PMe_2Ph)\}(PMe_2Ph)]$ (see Figures 90 and 91 in Section 14.7 below) does contain a $\{PtB_7\}$ sub-cluster which is topographically similar to *nido*-B_8H_{12}.[118]

14.4 Octaboron Compounds

A full account of the *arachno* nine- and ten-vertex species $[Pt(\eta^3-B_8H_{12})-(PMe_2Ph)_2]$ and $[\{Pt(PMe_2Ph)_2\}_2(\eta^3,\eta^3-B_8H_{10})]$ (Section 8 above, Figures 45–47) has now appeared.[119] Mixed metal species such as $[\{Pt(PMe_2Ph)_2\}-\{Pd(PMe_2Ph)_2\}(\eta^3,\eta^3-B_8H_{10})]$ may be similarly prepared,[120] and $[\{Pt-(PMe_3)_2\}\{Ir(CO)H(PMe_3)_2\}(\eta^3,\eta^3-B_8H_{10})]$ has been synthesized in a ca. 10% yield in an analogous manner from $[PtCl_2(PMe_3)_2]$ and *arachno*-$[Ir(\eta^3-B_8H_{12})-(CO)H(PMe_3)_2]$.[110]

This last iridanonaborane may be isolated from the reaction of *nido*-$B_9H_{12}{}^-$ or of *arachno*-$B_9H_{14}{}^-$ with *trans*-$[Ir(CO)Cl(PMe_3)_2]$. It is formed along with a second product, the yellow 2-chlorinated derivative $[Ir(\eta^3-B_8H_{11}Cl)(CO)H(PMe_3)_2]$, for which the molecular structure has been reported (Figure 73).[121] The structure is similar to that of $[Pt(B_8H_{12})(PMe_2Ph)_2]$ discussed above (Sec-

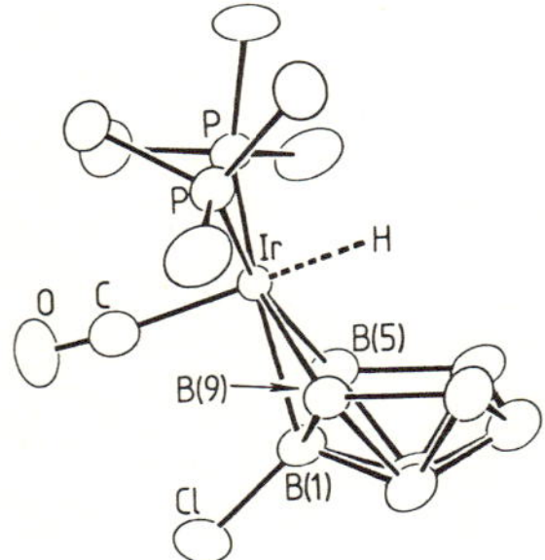

Figure 73. Molecular structure of *arachno*-[Ir(B$_8$H$_{11}$Cl)-(CO)H(PMe$_3$)$_2$]; H atoms were not located but NMR spectroscopy shows seven expected *exo*-terminal H atoms, *endo*-terminal H atoms on B(6) and B(8), bridging H atoms at B(5)B(6) and B(8)B(9) and a terminal H atom on the metal as indicated. Distances from the Ir(4) atom to B(1), B(5), and B(9) are 226.1(10), 228.4(11), and 229.7(8) pm respectively.

tion 8, Figure 45), although it is claimed that the compounds exhibit important differences in the detail of the metal-to-borane bonding.

On gentle heating these *arachno*-iridanonaborane species decompose smoothly with first order kinetics to give quantitatively the corresponding yellow *nido* compounds, as exemplified in the following equation

$$[\text{Ir}(\text{B}_8\text{H}_{12})(\text{CO})\text{H}(\text{PMe}_3)_2] \longrightarrow [\text{Ir}(\text{B}_8\text{H}_{11})(\text{CO})(\text{PMe}_3)_2] + \text{H}_2$$

Of these products, *nido*-[Ir(B$_8$H$_{11}$)(CO)(PMe$_3$)$_2$] has been structurally characterized by X-ray diffraction work (Figure 74). The geometry of the metallaborane cluster is that of a bicapped square antiprism with one five-connected vertex missing; the iridium(III) atom occupies one of the vertices on the open face.[121]

On stronger heating, these *nido* species in turn decompose to give moderate yields (1–45%) of poppy-red *closo* species such as [Ir(B$_8$H$_7$Cl)H(PMe$_3$)$_2$]:

$$[\text{Ir}(\text{B}_8\text{H}_{10}\text{Cl})(\text{CO})(\text{PMe}_3)_2] \longrightarrow [\text{Ir}(\text{B}_8\text{H}_7\text{Cl})\text{H}(\text{PMe}_3)_2] + \text{CO} + \text{H}_2$$

This reaction, however, is by no means quantitative, and the disposition of the chloro-substituent in the product has been taken to indicate that the mechanism is not a straightforward cluster closure accompanied by elimination.[121] The molecular structure of *closo*-[Ir(B$_8$H$_7$Cl)H(PMe$_3$)$_2$] is given in Figure 75. The nine-vertex metallaborane cluster has an *iso-closo* structure of idealized C$_{2v}$ symmetry, as distinct from the otherwise expected tricapped trigonal prismatic structure typified by *closo*-B$_9$H$_9$$^{2-}$. The application of electron-counting rules

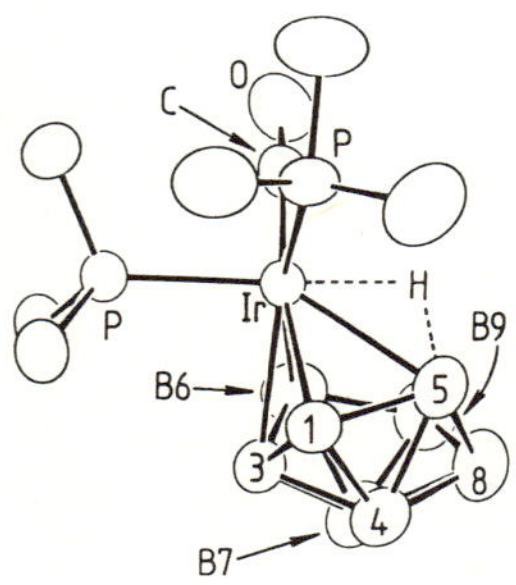

Figure 74. Molecular structure of *nido*-[Ir(B$_8$H$_{11}$)(CO)-(PMe$_3$)$_2$]. There is an *exo*-terminal H atom on each B atom, and there are bridging H atoms at B(6)B(9) and B(8)B(9) and also at Ir(2)B(5) *trans* to P(1) as indicated. Distances from Ir(2) to B(1), B(3), B(5) and B(6) are 218.5(8), 230.4(7), 250.0(8), and 233.0(10) pm respectively.

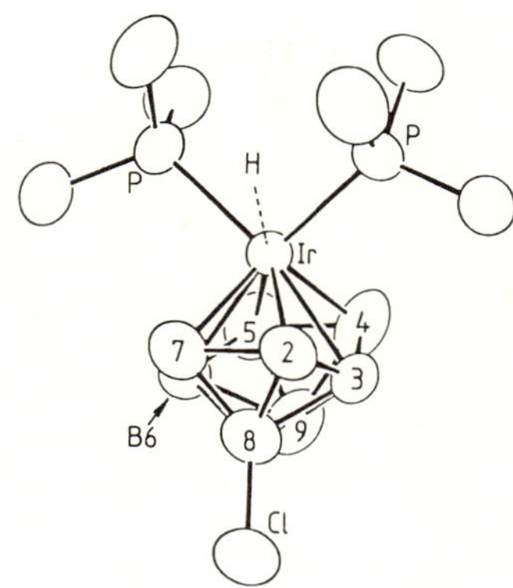

Figure 75. Molecular structure of *iso-closo-* $[Ir(B_8H_7Cl)-H(PMe_3)_2]$. There is an *exo*-terminal H atom on each B atom except B(8) and there is a terminal H atom on the metal as indicated. Two exchanging tautomers exist in solution with the H atom *syn* or *anti* to the Cl substituent. The borane-to-metal coordination mode is inverted boat η^6; distances from Ir(1) to B(2,4,5,7) are in the range 217.7(10) to 219.4(9) pm and from Ir(1) to B(3) and B(6) are 231.5(8) and 230.7(9) pm. Ir(1)–B(8) and Ir(1)–B(9) are 282.5(10) and 285.2(10) pm respectively.

suggests that the metal may be best regarded as iridium(V) (see also the *iso-closo* ten-vertex iridadecaboranes [e.g., Figure 81] in section 14.5 below). The compound is pseudorotationally fluxional about the metal atom with $\Delta G^{\ddagger}$ ca. 64 kJ mol^{-1} at –10 and at +45°C.[110]

These iridanonaborane cluster processes constitute a key series of reactions. Although the *formal* conversion sequence *arachno* to *nido* to *closo*, by the successive loss of electron pairs as 2H atoms, has been recognized for some years, these species are the first examples reported that actually show this in practice (Figure 76).

Some insight has been gained into the mechanism of the production of the eight-boron *arachno*-metallanonaborane species, such as those in Figures 45 and 73, from nine-boron *arachno* species related to $B_9H_{14}^-$. Substituent position analysis on the products of the reactions of $[PtCl_2(PMe_2Ph)_2]$ with various substituted *arachno* species $B_9H_{13}L$ indicateds that the preferred site of attack is as in Figure 77. However, other species such as the previously mentioned metallatetraborane $[Pt(B_3H_7)(PMe_2Ph)_2]$ (Section 3 above, Figures 6 and 7) and dimetalla-tetradecaborane $[Pt_2(B_6H_9)_2(PMe_2Ph)_2]$ (Section 6 above, Figure 41) are also produced in significant quantities in these reactions.[110]

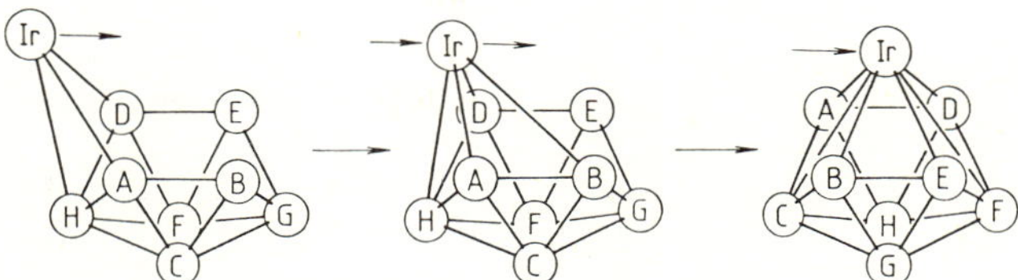

Figure 76. Schematic representation of the *arachno* to *nido* to *closo* cluster sequence exhibited by the iridanonaboranes. The position of the Cl substituent in $[Ir(B_8H_7Cl)H(PMe_3)_2]$ (Figure 75) indicates that the *nido* to *closo* process may involve a diamond–square–diamond transition state as implied by the lettering of the B atom positions in the structure.

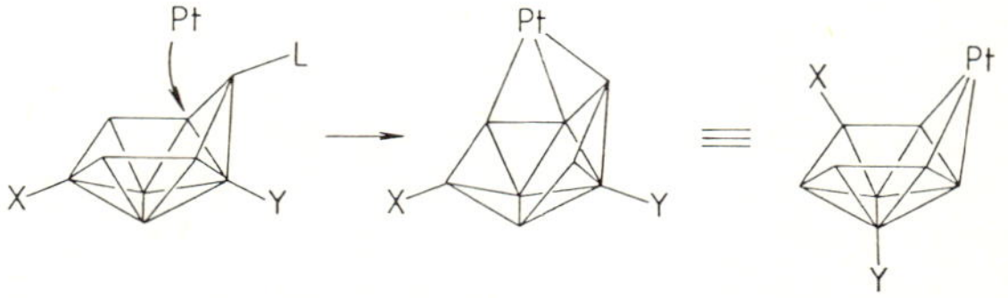

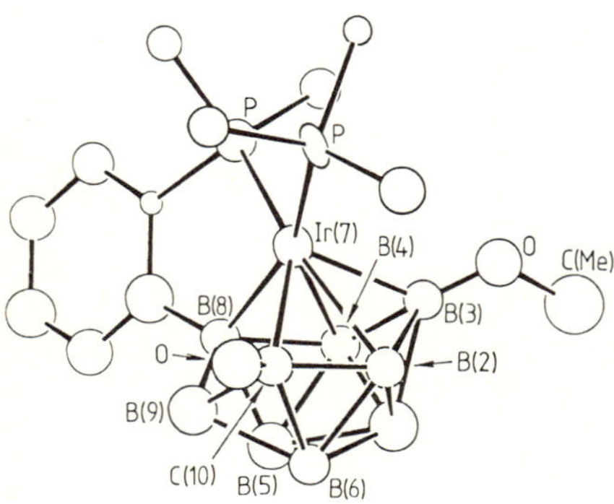

Figure 77. Presumed course of the reaction of $[PtCl_2(PMe_2Ph)_2]$ with substituted $B_9H_{13} \cdot L$ species in the presence of base. Substituents XY at the positions indicated in the starting *arachno*-nonaborane-ligand species are found at the positions indicated in the *arachno*-platinanonaborane product species. These products are substituted derivatives of the compound $[Pt(B_8H_{12})(PMe_2Ph)_2]$ depicted in Figures 45 and 46.

The intensely deep-purple ten-vertex *iso-nido* iridium(V) compound $[Ir\{\eta^5\text{-}B_8H_6(OMe)C(OH)\}(PPh_2C_6H_4)(PPh_3)]$ should also be mentioned (Figure 78). This compound is formed in ca. 3% yield from the reaction of *trans*-$[Ir(CO)Cl(PPh_3)_2]$ with *closo*-$B_{10}H_{10}{}^{2-}$ in the presence of MeOH. This reaction also produces a number of other unusual species (e.g., as in Section 14.2 above, Figure 71). The ten-vertex iridacarbadecaborane structure is based on that of *closo*-$B_{11}H_{11}{}^{2-}$ from which a 4-connected vertex has been formally removed. The compound exhibits an *ortho*-cycloboronation of a P-phenyl ligand (see also Section 14.2 above, Figure 71) and Section 14.5 below, Figure 81), together with a *clusto*-boronation of an Ir-carbonyl group to form a hydroxy-substituted carbon cluster vertex.[122]

14.5 Nonaboron Compounds

Full structural data for the neutral dichloro-*closo*-nickeladecaborane [2,4-Cl_2-1$\{Ni(PMe_2Ph)_2\}B_9H_7$] (Section 9 above) are now available.[120] The structure (Figure 79) is that of a bicapped square antiprism with the Ni atom in a four-connected axial position. The formal *closo* electron-count requires the metal center to contribute four electrons to the cluster bonding, so the compound can be regarded formally as a nickel(IV) species. The eclipsed position of the ligand

Figure 78. Molecular structure of *iso-nido*-$[Ir\{B_8H_6(OMe)C(OH)\}(PPh_2C_6H_4)(PPh_3)]$ with H and selected phenyl C atoms omitted. There are *exo*-terminal H atoms on B(1), B(2), B(4), B(5), B(6), and B(9), and no bridging or Ir-terminal H atoms. Distances from Ir(7) are as follows: to B(2) 241(3), to B(3) 213(4), to B(4) 231(3), to B(8) 215(3), to B(9) 282(4), and to C(10) 215(3) pm. The distance from B(8) to C(10) is ca. 266 pm.

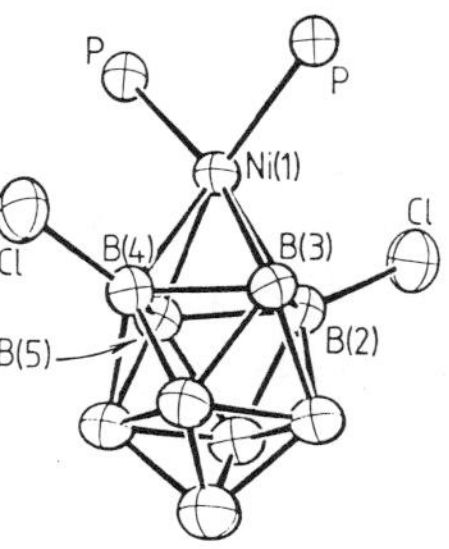

Figure 79. Molecular structure of *closo*-$[Ni(B_9H_7Cl_2)(PMe_2\text{-}Ph)_2]$. The borane-to-metal coordination mode is square η^4 with distances from Ni(1) to B(2), B(3), B(4), and B(5) at 198.7(6), 208.4(7), 198.7(6) and 208.4(7) pm respectively (compare Figures 49a and 81).

P atoms with respect to the B(3) and B(5) atoms contrasts to that in otherwise analogous metalladicarbaboranes in which the metal center is formally a two-electron contributor to the cluster bonding.

It has been reported that the yellow ten-vertex *nido*-6-iridadecaboranes $[Ir(B_9H_{13})H(PPh_3)_2]$ and $[Ir(B_9H_{13})H(PMe_3)_2]$ are produced in low yield ($\leqslant$ ca. 2%) in the reaction of the *arachno*-$B_9H_{14}^-$ anion with *trans*-$[Ir(CO)Cl(PR_3)_2]$:[123]

$$[Ir(CO)Cl(PR_3)_2] + B_9H_{14}^- \longrightarrow [Ir(B_9H_{13})H(PR_3)_2] + Cl^- + CO$$

Use of $[IrCl(PPh_3)_3]$ gives a much improved yield ($> 85\%$),[124] whereas use of $[Rh(cod)Cl]_2$ or $[Ir(cod)Cl]_2$ gives the analogous (cod) (cod = cyclooctadiene) complexes $[Rh(B_9H_{13})(cod)Cl]$ and $[Ir(B_9H_{13})(cod)Cl]$ respectively.[123] The molecular structure of $[Ir(B_9H_{13})H(PPh_3)_2]$ (Figure 80) shows that it is a direct analogue of *nido*-$B_{10}H_{14}$, with the $\{IrH(PPh_3)_2\}$ center subrogating the BH(6) group of the borane. As such the compound is very similar to the *nido*-6-mangana- and *nido*-6-cobalta-decaboranes[77-81] discussed above (Section 9, structures 50–54), a more complete description of $[6\text{-}\{Co(C_5H_5)\}B_9H_{13}]$ has recently appeared,[116] and X-ray crystallographic details of its $2\text{-}\{Co(C_5H_5)\}$ isomer have also been reported.[125]

Use of the *nido*-anion $B_9H_{12}^-$ instead of *arachno*-$B_9H_{14}^-$ in the reaction with $[IrCl(PPh_3)_3]$ yields (85%) derivatives of the *nido*-6-iridadecaborane depicted in Figure 80 which exhibit *ortho*-cycloboronation of ligand P-phenyl groups to the 5-position of the *nido* ten-vertex cluster:[124]

$$[IrCl(PPh_3)_3] + B_9H_{12}^- \longrightarrow [Ir(B_9H_{12})H(PPh_2C_6H_4)(PPh_3)] + Cl^- + PPh_3$$

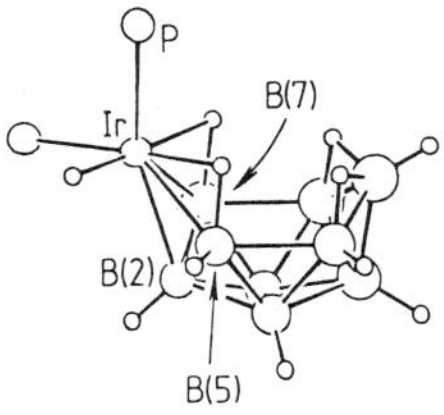

Figure 80. Molecular structure of $[Ir(B_9H_{13})H(PPh_3)_2]$. The structure is directly analogous to that of $B_{10}H_{14}$ (see also Figures 50–52 above). Distances from Ir(6) are as follows: to B(2) 226.9(8), to B(5) 228.9(9), to B(7) 228.1(11), to H(6) 155(11), to H(5,6) 181(8), to H(6,7) 184(10), to P(1) 230.3(2), and to P(2) 230.5(2) pm.

Two isomers are formed which differ in the disposition of the various ligands about the metal atom. The more stable has the disposition exhibited by the acyclic compound in Figure 80, and the less stable readily converts to this at +65°C. Stronger heating of either isomer results in quantitative decomposition to give the orange-yellow *iso-closo* ten-vertex irida(V) decaborane $[Ir(\overline{B_9H_8})H(PPh_2C_6H_4)$-$(PPh_3)]$ (Figure 81):[124]

$$[Ir(\overline{B_9H_{12}})H(PPh_2C_6H_4)(PPh_3)] \longrightarrow [Ir(\overline{B_9H_8})H(PPh_2C_6H_4)(PPh_3)] + 2H_2$$

This exocyclic *iso-closo* species is also produced quantitatively when the acyclic analogue $[Ir(B_9H_{13})H(PPh_3)_2]$ (Figure 80) is heated under similar conditions:[124]

$$[Ir(B_9H_{13})H(PPh_3)_2] \longrightarrow [Ir(\overline{B_9H_8})H(PPh_2C_6H_4)(PPh_3)] + 3H_2$$

The descriptor *iso-closo* has been used since the structure differs from that of the otherwise expected ten-vertex bicapped square antiprismatic *closo* geometry (e.g., Figure 77), in that the metallaborane cluster has an idealized C_{3v} geometry with a chair-η^6 borane-to-metal coordination mode.[124]

The acyclic *iso-closo* species $[Ir(B_9H_9)(cod)Cl]$ and $[Ir(B_9H_9)H(PMe_3)_2]$ are also formed similarly by mild thermolysis of the corresponding *nido* compounds mentioned above. These acyclic species appear to exhibit pseudo-rotation in the borane-to-metal bonding with a very low activation energy;[110,124] pseudo-rotation is precluded for the cyclic species of Figure 81 because of the *ortho*-phenylene link.

The formal valence state of iridium(V) in systems such as this and the eight-boron species described above (Section 14.4, Figures 75 and 78) implies that the species may be regarded as "non-Wade's rules" compounds in the sense that the metals contribute more than three orbitals to the cluster bonding. The *iso-closo* and *iso-nido* cluster geometries may perhaps reflect this to some extent.

Of interest in this regard is the isolation of the pale-violet *iso-nido* iridade-caborane $[Ir\{B_9H_{10}(PPh_3)\}(PPh_3)_2]$ as a product in trace quantities from the

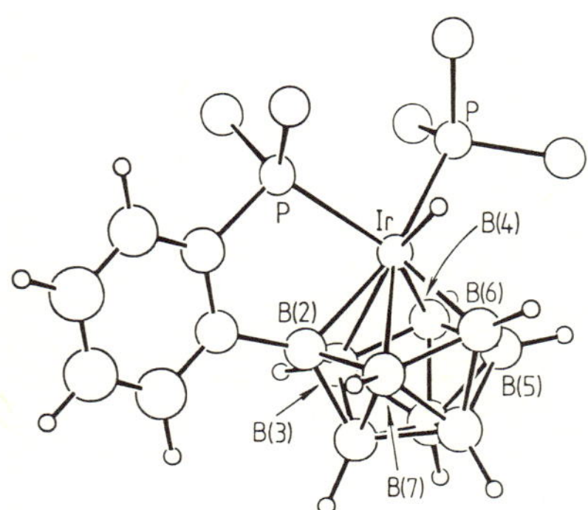

Figure 81. Molecular structure of the *exo-cyclic iso-closo* compound $[Ir(B_9H_8)H(PPh_2$-$C_6H_4)(PPh_3)]$. The borane-to-metal coordination mode is chair η^6 (compare to Figure 75 and contrast to Figure 79), and distances from Ir(1) are as follows: to B(2) 216.3(9), B(3) 246.0(9), to B(4) 218.8(9), to B(5) 238.8(10), to B(6) 214.8(9), to B(7) 237.8(9) and to H(1) 157(8) pm. Distances B(2)B(4), B(4)B(6), and B(6)B(2) are in the range 295–301 pm.

reaction of *arachno*-$B_9H_{14}^-$ with [IrCl(PPh$_3$)$_3$] discussed above.[110] The cluster geometry of this compound (Figure 82) can in the first instance be regarded as essentially isostructural with the deep-purple *iso-nido* iridium(V) species [Ir$\{B_8H_6(OMe)C(OH)\}(PPh_2C_6H_4)(PPh_3)$] mentioned above (Section 14.4, Figure 78), except that this species (of Figure 82) now has two additional H atoms associated with the metal atom and the open face; the formal *nido* cluster electron count then indicates that this is best regarded as an iridium(III) species. This contrasts to the {IrB$_8$C} compound and indicates that the energetic differences between {Ir$^{(III)}$ + 2H} and Ir$^{(V)}$ + H$_2$ in these systems may be very small, which may have catalytic significance.[110] An additional example of this type of equivalence is mentioned above in Section 14.1.

14.6 Decaboron Compounds

Complete descriptions of the *nido*-platinaundecaborane [Pt(B$_{10}$H$_{12}$)-(PMe$_2$Ph)$_2$] and its 4-(2′-B$_{10}$H$_{13}$) substituted derivative [Pt(B$_{20}$H$_{24}$)(PMe$_2$Ph)$_2$] (Section 10, Figures 59 and 60, and Section 12, Figure 68) have now appeared,[126] and the NMR properties of the {PtB$_{10}$H$_{12}$} cluster of [Pt(B$_{10}$H$_{12}$)(PMe$_2$Ph)$_2$] have also been discussed in some detail.[127] The reaction between *cis*-[PtCl$_2$-(PMe$_2$Ph)$_2$] and B$_{10}$H$_{10}{}^{2-}$ in the presence of alcohols to give alkoxy-substituted *nido*-platinaundecaboranes [Pt(B$_{10}$H$_{11}$OR)(PMe$_2$Ph)$_2$], as discussed above (Section 10),[99] has been extended in two ways. Firstly, the *ortho*-hydroxy ligand [PPh$_2$C$_6$H$_4$OH] has been used as the alcohol to give quantitatively a cyclized compound:[110]

$$[PtCl_2(PMe_2Ph)(PPh_2C_6H_4OH)] + B_{10}H_{10}{}^{2-} \longrightarrow$$

$$[Pt(B_{10}H_{11}O)(PPh_2C_6H_4)(PMe_2Ph)] + 2Cl$$

The molecular structure of this product (Figure 83) shows that the *nido*-platinaundecaborane cage is again produced and that an *ortho*-cycloboronation of the hydroxy group has occurred at B(8) to give a six-membered Pt-P-C-C-O-B ring which has an approximate twist-boat configuration.[110] The second extension has been to the reaction of the iridium complex *trans*-[Ir(CO)Cl(PPh$_3$)$_2$]

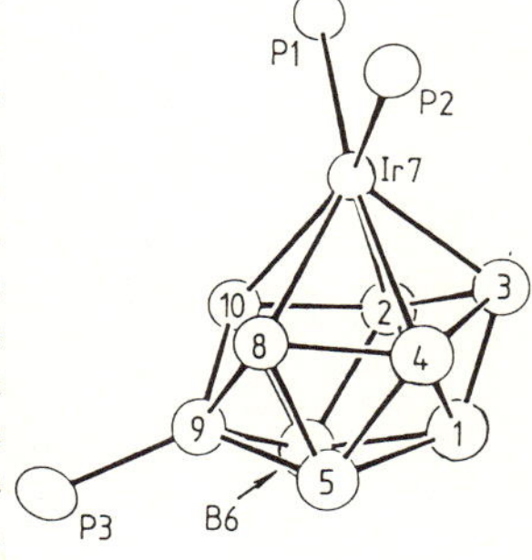

Figure 82. Preliminary molecular structure of the iridium-(III) species [Ir(B$_9$H$_{10}$·PPh$_3$)(PPh$_3$)$_2$]; each B atom except B(9) has a terminal H atom associated with it, and distances from Ir(7) are as follows: to B(2) 237.1(15) to B(3) 211.3(16), to B(4) 232.2(15), to B(5) 233.4(14), and to B(10) 233.4(14) pm. The distance B(8)B(10) is ca. 211 pm. There are two inequivalent H atoms associated with the metal atom and the Ir(1)B(8)B(9)B(10) face which undergo mutual exchange with ΔG‡ ca. 45 kJ mol^{-1} at -10°C. The cluster configuration is perhaps at present best regarded as intermediate between a *nido* (which would require an Ir-H-B bridging H atom) and a *closo* one (which would require an Ir-H terminal H atom).

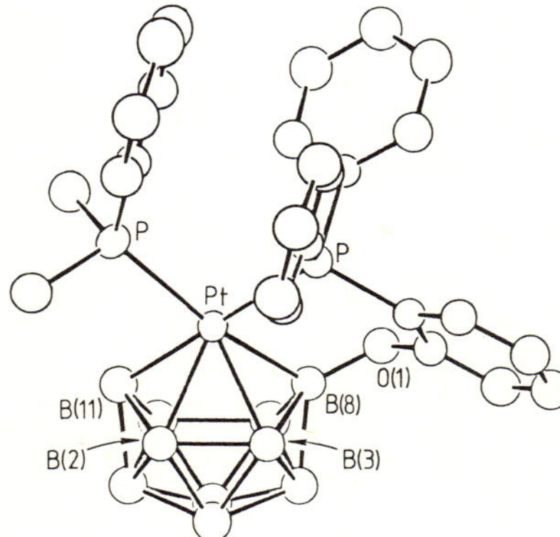

Figure 83. Molecular structure of the *exo*-cyclic platinaundecaborane [Pt(B$_{10}$H$_{11}$O)(PPh$_2$C$_6$H$_4$)(PMe$_2$Ph)]. Selected distances from Pt(7) are as follows: to B(2) 221.9(8), to B(3) 222.9(8), to B(8) 232.4(8), and to B(11) 231.3(9) pm. B(8)–0(1) is 140.6(9) pm.

with B$_{10}$H$_{10}$$^{2-}$ in the presence of MeOH which has been reported to yield a variety of novel metallaborane species including the *closo*-diiridahexaborane and *iso-nido*-iridacarbadecaborane species discussed above (Section 14.2, Figure 71, and Section 14.4, Figure 78).[113,121] This reaction is believed to proceed principally *via* the initial formation of methoxy-substituted *nido*-iridaundecaborane species followed by their subsequent solvolytic and/or thermolytic degradation.[113]

When the reaction between *cis*-[PtCl$_2$(PMe$_2$Ph)$_2$] and [B$_{10}$H$_{12}$(SMe$_2$)$_2$], which yields [Pt(B$_{10}$H$_{11}$Cl)(PMe$_2$Ph)$_2$] (Section 10 above, Figure 62), is modified by using [PtCl$_2$(PMe$_2$Ph)]$_2$ as the metal species, then a product formulated as [Pt{η^4-B$_{10}$H$_{11}$(SMe$_2$)}Cl(PMe$_2$Ph)] is formed instead of the {η^4-B$_{10}$H$_{11}$Cl} derivative; the SMe$_2$ group is coordinated to the 8-position on the *nido*–7–platinaundecaborane cluster.[110] In another variant of this reaction, use of [2-ClB$_{10}$H$_{11}$-(SMe$_2$)$_2$] instead of [B$_{10}$H$_{12}$(SMe$_2$)$_2$] with [PtCl$_2$(PMe$_2$Ph)$_2$] results in the formation of the 3,8-dichloro substituted species [Pt(η^4-B$_{10}$H$_{10}$-3,8-Cl$_2$)(PMe$_2$Ph)$_2$] rather than the expected 4,8- or 6,8-substituted species (undecaboranyl numbering system) indicating that these reactions may be far from straightforward.[110]

The influence of cluster halogen substituents on the pseudorotation fluxionality exhibited by *nido*–platinaundecarboranes (see Section 10 above, Figure 59)

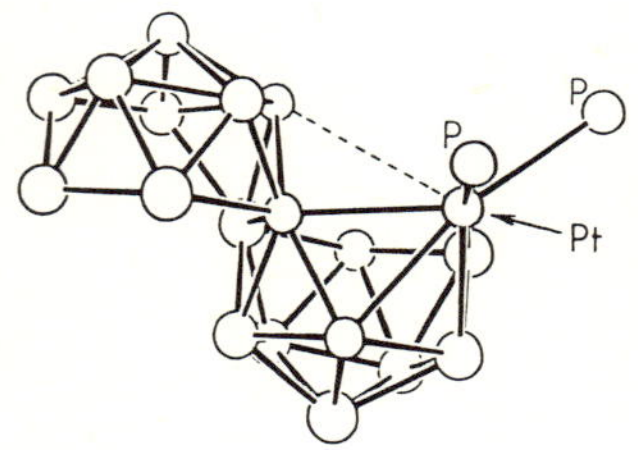

Figure 84. Molecular structure of [Pt(η^4-*n*-B$_{18}$H$_{20}$)(PMe$_2$Ph)$_2$] with H and C atoms omitted. Platinum-boron distances within the η^4 coordination mode are 236, 227, 224, and 234 pm, and the distance from the Pt atom to the next nearest B atom (broken line on diagram is 304 pm.

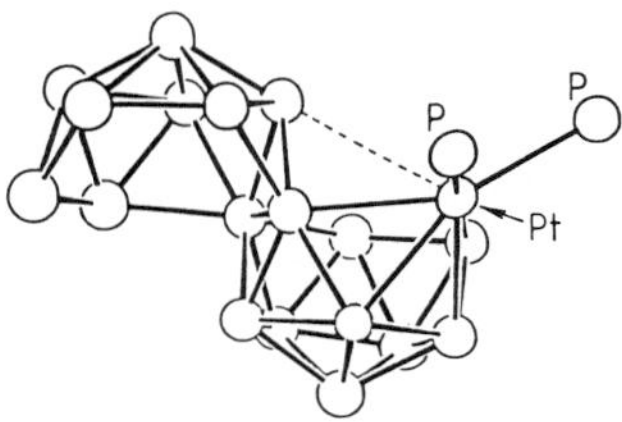

Figure 85. Molecular structure of $[Pt(\eta^4\text{-}iso\text{-}B_{18}H_{20})$-$(PMe_2Ph)_2]$ with H and C atoms omitted. Platinum-boron distances within the η^4 coordination mode are 231, 225, 224, and 235 pm, and the distance from the Pt atom to the next nearest B atom (broken line on diagram) is 296 pm.

has been investigated.[110] Single B(4) or double B(4)B(6) halogenation (*nido*-undecaboranyl numbering system) does not significantly affect $\Delta G^{\ddagger}$ at ca. 75 kJ mol^{-1} for this process, whereas an 8–chloro substituent decreases it to ca. 55 kJ mol^{-1} and a 3–chloro substituent increases it to ca. 80 kJ mol^{-1}. An 8–phenoxy substituent results in a value of $\Delta G^{\ddagger}$ of ca. 67 kJ mol^{-1}. It has also been noted in this work that the halogenated *nido*-platinaundecaboranes are readily isomerized under basic conditions to give ultimately the 3-substituted derivatives (again the *nido*-undecaboranyl numbering system[126] is used here).[110]

14.7 Derivatives of Higher Boranes: Hexadecaboron, Octadecaboron, and Icosaboron Compounds

More complete details on the reactions of *cis*–$[PtCl_2(PMe_2Ph)_2]$ with the icosaboranes $(B_{10}H_{13})_2$ in the presence of base have now appeared.[126] These include structural and NMR information on the platinahenicosaboranes and di-platinadocosaboranes mentioned above in Section 12 (Figures 68 and 69).

Advances have been made in the chemistry of metal derivatives of *n*- and *iso*-$B_{18}H_{22}$, and structural data are now available for five different platino derivatives.[110,128] The molecular structures of the straightforward bright yellow bis(dimethylphenylphosphine) platinum derivatives of *n*- and *iso*-$B_{18}H_{22}$ (Figures 84 and 85, respectively) are as previously predicted,[106] although there are in addition reasonably close interactions between the metal atom and the adjacent apical atoms of the unmetallated borane sub-cluster as indicated by the broken lines in the figures. These two compounds do not exhibit fluxionality of

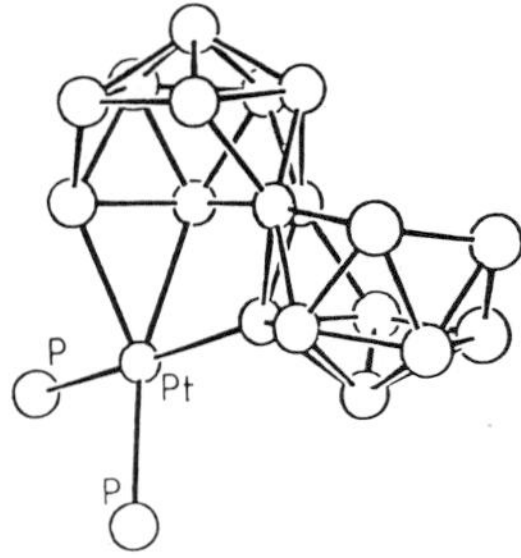

Figure 86. Molecular structure (H and C atoms omitted) of the $\{Pt(PMe_2Ph)_2\}$ derivative of *n*–$B_{18}H_{22}$ showing an η^2,η^1 borane-to-metal coordination mode, *viz.* $[Pt(\eta^2,\eta^1\text{-}B_{18}H_{20})(PMe_2Ph)_2]$. Distances from the Pt atom to the η^2 B atoms are 229 and 232 pm, and to the η^1 B atom 212 pm.

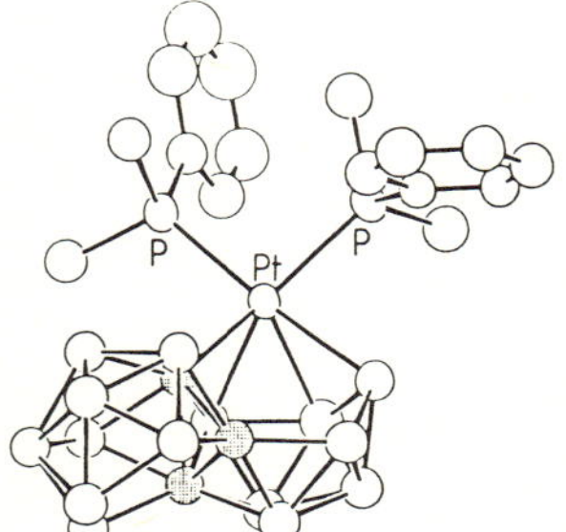

Figure 87. Molecular structure of $[Pt(\eta^4\text{-}n\text{-}B_{18}H_{18})$-$(PMe_2Ph)_2]$ with H and C atoms omitted. Platinum-boron distances within the η^4 coordination mode are 233.5(12), 226.7(12), 219.6(11), and 223.6(10) pm. Atoms in the B–B–B *triangulo-conjuncto* link are shaded.

the type summarized above (Section 10, Figure 59), even at reasonably high temperatures. The unmetallated borane sub-cluster presumably provides inhibitory steric presence. The compounds are readily formed by the treatment of the appropriate $B_{18}H_{22}$ isomer with $[PtCl_2(PMe_2Ph)_2]$ in the presence of base:

$$B_{18}H_{22} + [PtCl_2(PMe_2Ph)_2] + 2\text{ base} \longrightarrow [Pt(\eta^4\text{-}B_{18}H_{20})(PMe_2Ph)_2] + 2\text{ baseH}^+Cl^-$$

Other products are also formed in these reactions. With $n\text{-}B_{18}H_{22}$ a major product is a yellow $[Pt(B_{18}H_{20})(PMe_2Ph)_2]$ isomer in which the borane ligand can be regarded as coordinating η^1,η^2 to the platinum center rather than η^4 (Figure 86). The geometry about the metal atom has been taken to suggest a formally square planar platinum(II) environment rather than the platinum(IV) valency state which is now believed to best describe that more generally found in the metallaboranes.[128]

    This species is thermolytically unstable, giving readily upon mild thermolysis the η^4-bonded platinum(IV) isomer (Figure 84), together with, via loss of dihydrogen, a second orange-yellow product $[Pt(\eta^4\text{-}B_{18}H_{18})(PMe_2Ph)_2]$ (Figure 87) which can be regarded as a bis(dimethylphenylphosphine)platinum derivative of

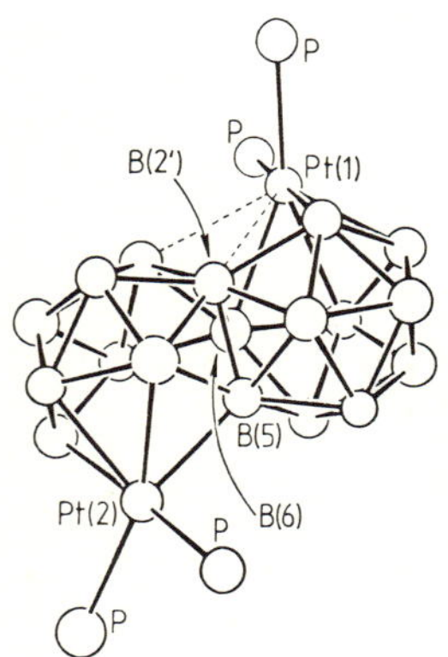

Figure 88. Molecular structure of the dimetallaicosaborane $[Pt_2(\eta^4,(\eta^4+\eta^2)\text{-}n\text{-}B_{18}H_{16})(PMe_2Ph)_2]$ with H and C atoms omitted. Distances from the Pt(2) atom to the straightforwardly bonded sub-cluster are 227, 223, 223, and 230 pm, and the three B atoms in the B–B–B *triangulo-conjuncto* link are those designated B(5), B(6), and B(2′). The bonding detail at Pt(1) is given in Figure 89.

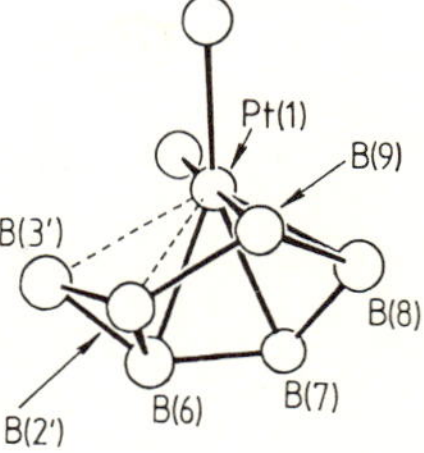

Figure 89. Bonding detail at the $(\eta^4 + \eta^2)$ coordinated Pt(1) atom in $[Pt_2(\eta^4,(\eta^4+\eta^2)-n-B_{18}H_{16})(PMe_2Ph)_4]$. Distances from Pt(1) are as follows: to B(6) 233 pm, to B(7) 226 pm, to B(8) 224 pm, to B(9) 227 pm, to B(2') 269 pm, and to B(3') 254 pm.

the as yet unknown higher borane $B_{18}H_{20}$;[110] this last species would have the *triangulo-conjuncto* structure derived from the fusion of a *nido*-decaborane cluster and a *nido*-undecaborane cluster with a common triangular face. The borane-to-metal bonding is again η^4 as typified by $[Pt(B_{10}H_{12})(PMe_2Ph)_2]$ (Section 10 above, Figures 59 and 60). In accord with this the compound is pseudorotationally fluxional albeit with a higher activation energy than the unsubstituted platinaundecaboranyl species. Presumably the formation of the *triangulo-conjuncto* link relieves somewhat the steric inhibitions to rotation that prevail in the straightforward $[Pt(\eta^4-B_{18}H_{20})(PMe_2Ph)_2]$ isomers (Figures 84 and 85).

Additional higher platinaborane products are present in these reaction mixtures; one that has been reported[128] is the bottle-green dimetallaicosaborane $[Pt_2(\eta^4,\eta^4-B_{18}H_{16})(PMe_2Ph)_4]$, which is more readily obtained in quantitative yield from the treatment of $[Pt(\eta^4-B_{16}H_{18})(PMe_2Ph)_2]$ (Figure 87) with an equivalent of $[PtCl_2(PMe_2Ph)_2]$ in the presence of base:[110]

$$[Pt(\eta^4-B_{18}H_{18}(PMe_2Ph)_2] + [PtCl_2(PMe_2Ph)_2] + 2\ \text{base} \longrightarrow$$
$$[Pt_2(\eta^4,\eta^6-B_{18}H_{16})(PMe_2Ph)_4] + 2\ \text{baseH}^+Cl^-$$

The molecular structure (Figure 88) shows that the second platinum atom is bonded η^5 to the undecaboranyl sub-cluster in a similar manner to that proposed[98] for the borane-to-metal bonding in the *closo*-nickeladodecaborane species $[Ni(\eta^5-B_{11}H_{11})(\eta^5-C_5H_5)]^-$ (Section 11 above, Figure 66), but that there is also interaction, perhaps sterically forced, between this metal atom and the adjacent apical B and H atoms of the other sub-cluster as indicated in Figure 89.

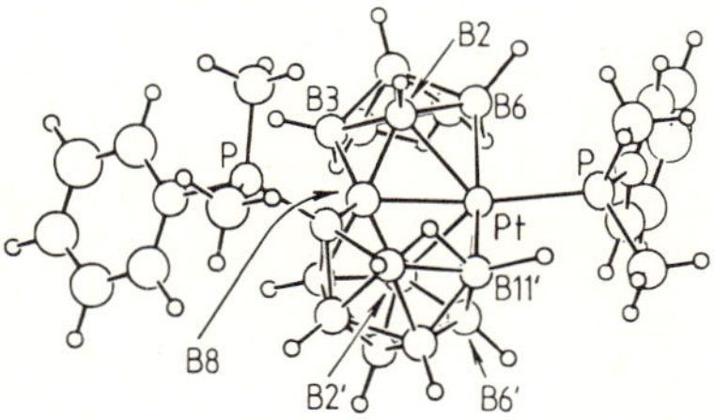

Figure 90. Molecular structure of the 17-vertex species $[Pt(\eta^6-B_{16}H_{18} \cdot PMe_2Ph)(PMe_2Ph)]$. Distances from Pt(7) are as follows: to B(2) 223.9(8), to B(6) 227.4(11), to B(8) 223.0(12), to B(2') 223.7(11), to B(6') 224.7(20), to B(11') 229.4(11), and to P(1) 232.2(4) pm. The distance B(9')–P(2) is 191.8(11) pm.

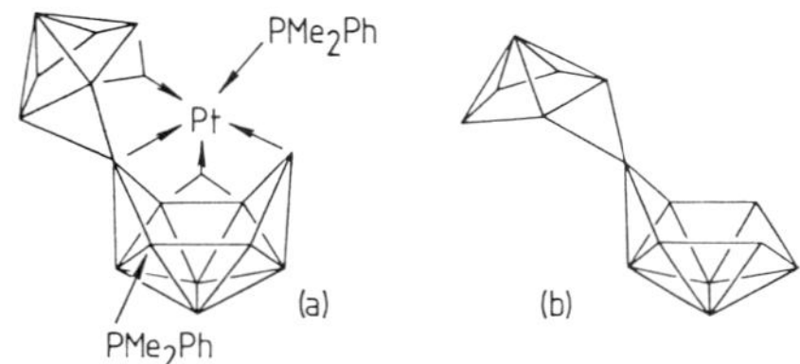

Figure 91. (a) Representation of the structure of $[Pt(B_{16}H_{18} \cdot PMe_2 Ph)(PMe_2Ph)]$ as a derivative of *arachno*-$B_{10}H_{12}LL'$, where $L = PMe_2Ph$ and $L' = B_6H_{10}$. (b) Representation of the supposed structure of $B_{15}H_{23}$ which may be regarded as a derivative of *arachno*-$B_9H_{13}L$, where $L = B_6H_{10}$.

Both borane-to-platinum coordination configurations in this compound exhibit pseudorotational fluxionality, and there appears to be additionally a low energy configurational flip in the η^6 borane-to-metal bonding.[110] It has also been claimed that other products from these reactions probably contain further variations of borane-to-metal coordination modes,[128] and further information will be of interest.

Finally, the 17-vertex flame-red platinaheptadecaborane $[Pt(\eta^6-B_{16}H_{18}-\{PMe_2Ph\}(PMe_2Ph)]$ merits mention.[118] This compound was obtained in small yield (<2%) from the controlled thermolysis of *arachno*-$[Pt(B_8H_{12})-(PMe_2Ph)_2]$ (Section 8 above, Figure 45). The structure (Figure 90) is that of a *nido* eight-vertex platinaoctaborane and a *nido* eleven-vertex platinaundecaborane conjoined with a Pt–B vector as a common edge. The borane ligand may be regarded as being formally based on the as yet hypothetical species $B_{16}H_{22}L$ (Figure 91a), formally derived from an *arachno*-$B_{10}H_{12}L_2$ structure by the replacement of one of the ligands L by the Lewis base B_6H_{10}. This is analogous to the interpretation[129] of the structure of the known higher borane $B_{15}H_{23}$ as an *iso-arachno*-nonaborane species $B_9H_{13}L$, where L is the borane ligand B_6H_{10} (Figure 91b). It is likely that this platinaheptadecaborane species arises from the conjoining of two octaboron units about the metal center, and more details on the other products of this reaction, tentatively claimed to be further higher metallaborane species,[118] will be of obvious interest.

REFERENCES

1. N. N. Greenwood, in: *Comprehensive Inorganic Chemistry*, J. C. Bailar, H. J. Eméleus, R. Nyholm, and A. F. Trotman–Dickenson, Eds. Pergamon: Oxford, Vol. 1, Chapter 11, 665–991 (1973).
2. E. L. Muetterties, *Pure Appl. Chem.* **29**, 585–595 (1972).
3. N. N. Greenwood and I. M. Ward, *Chem. Soc. Rev.,* **3**, 231–271 (1974).
4. *Gmelin Handbuch der Anorganische Chemie, Supplementary Works, Vol. 19. Boron Compounds Part 3*, Springer–Verlag Berlin 182–291 (1975).
5. P. A. Wegner, in: *Boron Hydride Chemistry*, E. L. Muetterties, Ed., Academic Press, New York, Chapter 12, 431–480 (1975).

6. N. N. Greenwood, *Pure Appl. Chem.*, **49**, 791–801 (1977).

7. R. N. Grimes, *Acc. Chem. Res.*, **11**, 420–427 (1978).

8. H. D. Kaesz, W. Fellmann, G. R. Wilkes, and L. F. Dahl, *J. Am. Chem. Soc.*, **87**, 2753–2755 (1965).

9. G. Medford and S. G. Shore, *J. Am. Chem. Soc.*, **100**, 3953–3954 (1978).

10. J. S. Plotkin and S. G. Shore, *J. Oranometallic Chem.*, **182**, C15–C19 (1979).

11. E. L. Andersen and T. P. Fehlner, *J. Am. Chem. Soc.*, **100**, 4606–4607 (1978).

12. A. R. Kane and E. L. Muetterties, *J. Am. Chem. Soc.*, **93**, 1041–1042 (1971).

13. L. J. Guggenberger, A. R. Kane, and E. L. Muetterties, *J. Am. Chem. Soc.*, **94**, 5665–5673 (1972).

14. N. N. Greenwood, J. D. Kennedy, and D. Reed, *J. C. S. Dalton.* 196–200 (1980).

15. E. L. Andersen, K. J. Haller, and T. P. Fehlner, *J. Am. Chem. Soc.*, **101**, 4390–4391 (1979).

16. J. R. Pipal and R. N. Grimes, *Inorg. Chem.*, **16**, 3255–3261 (1977).

17. N. N. Greenwood, C. G. Savory, R. N. Grimes, L. G. Sneddon, A. Davison, and S. S. Wreford, *J. C. S. Chem. Comm.*, 718 (1974).

18. D. R. Salahub, *J. C. S. Chem. Comm.*, 385–386 (1978).

19. P. Brint and T. R. Spalding, *Inorg. Nucl. Chem. Lett.*, **15**, 355–359 (1979).

20. V. R. Miller and R. N. Grimes, *J. Am. Chem. Soc.*, **95**, 5078–5080 (1973).

21. V. R. Miller, R. Weiss, and R. N. Grimes, *J. Am. Chem. Soc.*, **99**, 5646–5651 (1977).

22. L. G. Sneddon and D. Voet, *J. C. S. Chem. Comm.*, 118–119 (1976).

23. J. R. Pipal and R. N. Grimes, *Inorg. Chem.*, **18**, 252–257 (1979).

24. J. R. Pipal and R. N. Grimes, *Inorg. Chem.*, **18**, 257–263 (1979).

25. K. Wade, *Adv. Inorg. Chem. Radiochem.*, **18**, 1–66 (1976).

26. R. Weiss, J R. Bowser, and R. N. Grimes, *Inorg. Chem.*, **17**, 1522–1527 (1978).

27. J. R. Bowser, A. Bonny, J. R. Pipal, and R. N. Grimes, *J. Am. Chem. Soc.*, **101**, 6229–6236 (1979).

28. J. R. Bowser and R. N. Grimes, *J. Am. Chem. Soc.*, **100**, 4623–4624 (1978).

29. K. E. Inkrott and S. G. Shore, *J. C. S. Chem. Comm.*, 866–867 (1978).

30. S. G. Shore, *Abstracts of Papers XIX Int. Conf. on Coord. Chem. Prague*, September (1978).

31. L. W. Hall, G. J. Zimmerman, and L. G. Sneddon, *J. C. S. Chem. Comm.*, 45–46 (1977).

32. J. A. Ulmann and T. P. Fehlner, *J. C. S. Chem. Comm.*, 632–633 (1976).

33. T. P. Fehlner, J. Ragaini, M. Mangion, and S. G. Shore, *J. Am. Chem. Soc.*, **98**, 7085–7086 (1976).

34. S. G. Shore, J. D. Ragaini, R. L. Smith, C. E. Cottrell, and T. P. Fehlner, *Inorg. Chem.*, **18**, 670–673 (1979).

35. S. G. Shore, *Fourth Internat. Meeting Boron Chem. (IMEBORON IV), Salt Lake City/Snowbird, Utah*, July (1979).

36. R. Wilcyznski and L. G. Sneddon, *Inorg. Chem.*, **18**, 864–865 (1979).

37. M. B. Fischer and D. F. Gaines, *Inorg. Chem.*, **18**, 3200–3205 (1979).

38. V. T. Brice and S. G. Shore, *J. C. S. Chem. Comm.*, 1312–1313 (1970).

39. V. T. Brice and S. G. Shore, *J. C. S. Dalton*, 334–336 (1975).

40. N. N. Greenwood, and J. Staves, *J. C. S. Dalton*, 1144–1145 (1977).

41. G. G. Outterson, V. T. Brice, and S. G. Shore, *Inorg. Chem.*, **15**, 1456–1457 (1976).

42. N. N. Greenwood, J. A. Howard, and W. S. McDonald, *J. C. S. Dalton*, 37–39 (1976).

43. N. N. Greenwood and J. Staves, *J. C. S. Dalton*, 1788–1791 (1977).

44. J. D. Kennedy and J. Staves, *Z. Naturforsch.*, **34B**, 808–813 (1979).

45. N. N. Greenwood, J. D. Kennedy, and J. Staves, *J. C. S. Dalton*, 1146–1152 (1978).

46. N. N. Greenwood and J. Staves, *J. C. S. Dalton*, 1786–1788 (1977).

47. N. S. Hosmane and R. N. Grimes, *Inorg. Chem.*, **18**, 2886–2891 (1979).

48. D. F. Gaines, K. M. Coleson, and J. C. Calabrese, *J. Am. Chem. Soc.*, **101**, 3979–3980 (1979).

49. D. F. Gaines and T. V. Iorns, *J. Am. Chem. Soc.*, **92**, 4571–4574 (1979).

50. D. F. Gaines and T. V. Iorns, *J. Am. Chem. Soc.*, **90**, 6617–6621 (1968).

51. D. F. Gaines and J. Ulman, *Inorg. Chem.*, **13**, 2792–2796 (1974).

52. A. B. Burg and N. Heinen, *Inorg. Chem.*, **7**, 1021–1025 (1968).

51. D. F. Gaines, and J. Ulman, *Inorg. Chem.*, **13**, 1792–2796 (1974).

52. A. B. Burg, and N. Heinen, *Inorg. Chem.*, **7**, 1021–1025 (1968).

53. D. F. Gaines, M. B. Fischer, S. J. Hildebrandt, J. A. Ulman, and J. W. Lott, Paper 26 in *Inorganic Compounds with Unusual Properties*, R. B. King, Ed., Advances in Chem. Series, **150**, 311–317 (1979).

54. N. N. Greenwood, J. D. Kennedy, C. G. Savory, J. Staves, and K. R. Trigwell, *J. C. S. Dalton*, 237–244 (1978).

55. L. G. Sneddon and R. N. Grimes, *J. Am. Chem. Soc.*, **94**, 7161–7171 (1972); L. G. Sneddon, D. C. Beer, and R. N. Grimes, *J. Am. Chem. Soc.*, **95**, 6623–6629 (1973).

56. N. N. Greenwood, J. D. Kennedy, W. S. McDonald, D. Reed, and J. Staves, *J. C. S. Dalton*, 117–123 (1979).

57. D. F. Gaines, J. L. Walsh, and J. C. Calabrese, *Inorg. Chem.*, **17**, 1242–1248 (1978).

58. D. F. Gaines and T. V. Iorns, *Inorg. Chem.*, **7**, 1041–1043 (1968).

59. M. R. Churchill, J. J. Hackbarth, A. Davison, D. D. Traficante, and S. S. Wreford, *J. Am. Chem. Soc.*, **96**, 4041–4042 (1974).

60. M. R. Churchill, and J. J. Hackbarth, *Inorg. Chem.*, **14**, 2047–2051 (1975).

61. R. W. Marks, S. S. Wreford, and D. D. Traficante, *Inorg. Chem.*, **17**, 756–759 (1978).

62. R. Weiss and R. N. Grimes, *J. Am. Chem. Soc.*, **99**, 8087–8088 (1977).

63. R. Weiss and R. N. Grimes, *Inorg. Chem.*, **18**, 3291–3295 (1979).

64. M. Mangion, J. D. Ragaini, T. A. Schmitkons, and S. G. Shore, *J. Am. Chem. Soc.*, **101**, 754–755 (1979).

65. A. Davison, D. D. Traficante, and S. S. Wreford, *J. C. S. Chem. Comm.*, 1155–1156 (1972).

66. A. Davison, D. D. Traficante, and S. S. Wreford, *J. Am. Chem. Soc.*, **96**, 2802–2805 (1974).

67. J. P. Brennan, R. Schaeffer, A. Davison, and S. S. Wreford, *J. C. S. Chem. Comm.*, 354 (1973).

68. N. N. Greenwood, M. J. Hails, J. D. Kennedy, and W. S. McDonald, *J. C. S. Chem. Comm.*, 37–38 (1980).

69. J. C. Huffman, D. C. Moody, and R. Schaeffer, *J. Am. Chem. Soc.*, **97**, 1621–1622 (1975).

70. S. K. Boocock, Y. M. Cheek, J. E. Crook, N. N. Greenwood, M. J. Hails, J. D. Kennedy, W. S. McDonald, and J. Staves, unpublished observations made in the Department of Inorganic and Structural Chemistry, University of Leeds 1979–80.

71. D. F. Gaines, S. Hildebrandt, and J. Ulman, *Inorg. Chem.*, **13**, 1217–1219 (1974).

72. O. Hollander, W. R. Clayton, and S. G. Shore, *J. C. S. Chem. Comm.*, 604–605 (1974).

73. A. R. Kane, L. J. Guggenberger, and E. L. Muetterties, *J. Am. Chem. Soc.*, **92**, 2571–2572 (1970).

74. J. C. Calabrese, M. B. Fischer, D. F. Gaines, and J. W. Lott, *J. Am. Chem. Soc.*, **96**, 6318–6323 (1974).

75. R. N. Leyden and M. F. Hawthorne, *J. C. S. Chem. Comm.*, 310–311 (1975).

76. R. N. Leyden, B. P. Sullivan, R. T. Baker, and M. F. Hawthorne, *J. Am. Chem. Soc.*, **100**, 3758–3765 (1978).

77. J. W. Lott, D. F. Gaines, H. Shenhav, and R. Schaeffer, *J. Am. Chem. Soc.*, **95**, 3042–3043 (1973).

78. J. W. Lott, and D. F. Gaines, *Inorg. Chem.*, **13**, 2261–2267 (1974).

79. D. F. Gaines, J. W. Lott, and J. C. Calabrese, *J. C. S. Chem. Comm.*, 295–296 (1973).

80. D. F. Gaines, J. W. Lott, and J. C. Calabrese, *Inorg. Chem.*, **13**, 2419–2423 (1974).

81. J. R. Pipal and R. N. Grimes, *Inorg. Chem.*, **16**, 3251–3255 (1977).

81a. T. L. Venable and R. N. Grimes, *Inorg. Chem.*, **21**, 887 (1982).

82. E. L. Muetterties, W. G. Peet, P. A. Wegner, and C. W. Alegranti, *Inorg. Chem.*, **9**, 2447–2451 (1970).

83. F. Klanberg, E. L. Muetterties, and L. J. Guggenberger, *Inorg. Chem.*, **7**, 2272–2278 (1968).

84. N. N. Greenwood and N. F. Travers, *Inorg. Nucl. Chem. Lett.*, **2**, 169–171 (1966).

85. N. N. Greenwood and N. F. Travers, *J. Chem. Soc. (A)* 880–884 (1967).

86. N. N. Greenwood and N. F. Travers, *J. Chem. Soc. (A)* 15–20 (1968).

87. N. N. Greenwood, J. A. McGinnety, and J. D. Owen, *J. C. S. Dalton*, 989–992 (1972).

88. N. N. Greenwood and N. F. Travers, *J. Chem. Soc. (A)*, 3257–3264 (1971).

89. N. N. Greenwood, J. A. McGinnety, and J. D. Owen, *J. Chem. Soc. (A)*, 809–813 (1971).

90. L. J. Guggenberger, *J. Am. Chem. Soc.*, **94**, 114–119 (1972).

91. F. Klanberg, P. A. Wegner, G. W. Parshall, and E. L. Muetterties, *Inorg. Chem.*, **7**, 2072–2077 (1968).

92. N. N. Greenwood, B. S. Thomas, and D. W. Waite, *J. C. S. Dalton*, 299–304 (1975).

93. N. N. Greenwood, N. F. Travers, and D. W. Waite, *Chem. Comm.*, 1027 (1971).

94. N. N. Greenwood and J. A. Howard, *J. C. S. Dalton*, 177–180 (1976).

95. N. N. Greenwood and B. Youll, *J. C. S. Dalton*, 158–162 (1975).

96. S. K. Boocock, N. N. Greenwood, and J. D. Kennedy, *J. C. S. Chem. Comm.*, 305–306 (1980).

97. A. R. Siedle and T. A. Hill, *J. Inorg. Nucl. Chem.*, **31**, 3874–3875 (1969); A. R. Siedle, G. M. Bodner, A. R. Garber, R. F. Wright, and L. J. Todd, *J. Mag. Res.*, **31**, 203–206 (1978).

98. B. P. Sullivan, R. N. Leyden, and M. F. Hawthorne, *J. Am. Chem. Soc.*, **97**, 455–456 (1975).

99. T. E. Paxson and M. F. Hawthorne, *Inorg. Chem.*, **14**, 1604–1607 (1975).

100. A. R. Siedle and L. J. Todd, *Inorg. Chem.*, **11**, 2838–2842 (1976).

101. N. N. Greenwood and H. Schick, *Chem. Comm.*, 935–936 (1969).

102. N. N. Greenwood and D. N. Sharrocks, *J. Chem. Soc. (A)*, 2334–2338 (1969).

103. F. Sato, T. Yamamoto, J. R. Wilkinson, and L. J. Todd, *J. Organomet. Chem.*, **86**, 243–251 (1975).

104. R. V. Schultz, F. Sato, and L. J. Todd, *J. Organomet. Chem.*, **125**, 115–118 (1977).

105. P. A. Wegner, L. J. Guggenberger, and E. L. Muetterties, *J. Am. Chem. Soc.*, **92**, 3473–3474 (1970).

106. R. L. Sneath and L. J. Todd, *Inorg. Chem.*, **12**, 44–48 (1973).

107. M. W. Chen, D. F. Gaines, and L. G. Hoard, *Inorg. Chem.*, **19**, 2989–2993 (1980).

108. K. J. Haller, E. L. Andersen, and T. P. Fehlner, *Inorg. Chem.*, **20**, 309–313 (1981).

109. J. M. Gromek and J. Donohue, *Crystal Struct. Commun.*, **10**, 849–854 (1981).

110. R. Ahmad, M. A. Beckett, J. Bould, Y. M. Cheek, J. E. Crook, N. N. Greenwood, J. D. Kennedy, and W. S. McDonald. Unpublished observations made in the Department of Inorganic and Structural Chemistry, University of Leeds, 1980–82.

111. D. N. Cox, D. M. P. Mingos, and R. Hoffman, *J.C.S., Dalton*, 1788–1797 (1981).

112. M. E. O'Neill and K. Wade, *Inorg. Chem.*, **21**, 461–464 (1982).

113. J. E. Crook, N. N. Greenwood, J. D. Kennedy, and W. S. McDonald, *J.C.S. Chem. Comm.* (in press).

114. J. Bould, N. N. Greenwood, and J. D. Kennedy, *J.C.S. Dalton*, 481–483 (1982).

115. S. K. Boocock, M. J. Toft, and S. G. Shore, *182nd Amer. Chem. Soc. National Meeting, New York, 23–28th August 1981*, Abstract INOR 149 (1981).

116. G. J. Zimmerman, L. W. Hall, and L. G. Sneddon, *Inorg. Chem.*, **19**, 3642–3650 (1980).

117. P. Brint, W. K. Pelin, and T. R. Spalding, *J.C.S., Dalton*, 546–551 (1981).

118. M. A. Beckett, J. E. Crook, N. N. Greenwood, J. D. Kennedy, and W. S. McDonald, *J.C.S. Chem. Comm.*, submitted, (1982).

119. S. K. Boocock, N. N. Greenwood, M. J. Hails, J. D. Kennedy, and W. S. McDonald, *J.C.S. Dalton*, 1415–1429 (1981).
120. M. J. Hails, Thesis, University of Leeds (1981).
121. J. Bould, J. E. Crook, N. N. Greenwood, J. D. Kennedy, and W. S. McDonald, *J.C.S. Chem. Comm.* (in press).
122. J. E. Crook, N. N. Greenwood, J. D. Kennedy, and W. S. McDonald, *J.C.S. Chem. Comm.*, 933–934 (1981).
123. S. K. Boocock, J. Bould, N. N. Greenwood, J. D. Kennedy, and W. S. McDonald, *J.C.S. Dalton* (in press).
124. J. Bould, N. N. Greenwood, J. D. Kennedy, and W. S. McDonald, *J.C.S. Chem. Comm.* (in press).
125. J. M. Gromek and J. Donohue, *Crystal Struct. Commun.*, **10**, 871–877 (1981).
126. S. K. Boocock, N. N. Greenwood, J. D. Kennedy, W. S. McDonald, and J. Staves, *J.C.S. Dalton*, 2573–2584 (1981).
127. J. D. Kennedy and B. Wrackmeyer, *J. Magnet. Reson.*, **38**, 529–535 (1980).
128. Y. M. Cheek, N. N. Greenwood, J. D. Kennedy, and W. S. McDonald, *J.C.S. Chem. Comm.*, 80–81 (1982).
129. J. Rathke and R. Schaeffer, *J. Am. Chem. Soc.*, **95**, 3402 (1973); and *Inorg. Chem.*, **13**, 3008–3011 (1974).

3

Interactions of Metal Groups with the Octahydrotriborate(1 –) Anion, $B_3H_8{}^-$

Donald F. Gaines and Steven J. Hildebrandt

1. INTRODUCTION

During the last decade interest in metal-containing cluster molecules has increased dramatically. This interest is in large part due to a growing mass of evidence supporting the importance of metal-containing clusters as catalysts and reaction centers in biochemical and industrial systems. The use of preformed borane and carborane clusters in the synthesis of metal–borane and metal–carborane clusters provides unique access to the study of metals in clusters. The formation of clusters containing the $B_3H_8{}^-$ anion as a bi-, tri-, or tetradentate ligand constitutes an increasingly revealing part of the overall area of metal–borane clusters.

1.1. Ionic $B_3H_8{}^-$ Salts

Alfred Stock was probably the first to synthesize $B_3H_8{}^-$, which he characterized as $Na_2B_2H_6$, by direct reaction of sodium metal and diborane.[1] In a typical preparation, ether is included to enhance the rate of reaction between sodium and diborane,[2] as shown in reaction (1). This reaction is

$$2Na/Hg + 2B_2H_6 \xrightarrow{\text{ether}} NaB_3H_8 + NaBH_4 + 2Hg \qquad (1)$$

Donald F. Gaines • Department of Chemistry, University of Wisconsin, Madison, Wisconsin. ***Steven J. Hildebrandt*** • Minerals and Chemicals Division, Engelhard Minerals and Chemicals Corporation, Edison, New Jersey.

complete in 1 or 2 days at room temperature, producing up to 80% yields of NaB_3H_8. Addition of a carrier (ϕ_3B or naphthalene) enhances the rate of reaction such that it is completed in a matter of seconds at $0°$ in tetrahydrofuran,[3] [reaction (2)].

$$2\ Na/Hg - carrier + 2B_2H_6 \xrightarrow{\text{ether}} NaB_3H_8 + NaBH_4 + carrier + 2Hg \qquad (2)$$

The reaction is easily followed by the gradual disappearance of color from the Na-carrier species as the sodium is consumed. Sodium amalgam has also been found to react with B_4H_{10}[4,5] and B_5H_{11}[4,6] in ether producing NaB_3H_8.

A synthesis of triborohydride salts adaptable to large-scale preparations involves the oxidation of sodium borohydride with iodine as shown in reaction (3).[7,8]

$$3NaBH_4 + I_2 \xrightarrow[100°]{\text{diglyme}} NaB_3H_8 + 2H_2 + 2NaI \qquad (3)$$

The mechanism of this reaction most likely involves the oxidation of BH_4^- to B_2H_6 and the subsequent reaction between B_2H_6 and BH_4^- to give $B_3H_8^-$. Direct reaction between diborane and borohydride ion in ethereal solutions, as shown in reaction (4), has also proven to be a useful large-scale synthesis.[3,9,10]

$$B_2H_6 + NaBH_4 \xrightarrow[\text{diglyme}]{100°} NaB_3H_8 + H_2 \quad \sim 65{-}80\% \qquad (4)$$

This reaction can be carried out using conventional equipment at atmospheric pressure in an inert atmosphere. The diborane can be generated simultaneously or *in situ* by the reaction shown in reaction (5).

$$4BF_3 \cdot O(C_2H_5)_2 + 3NaBH_4 \xrightarrow{\text{diglyme}} 3NaBH_4 + 2B_2H_6 + 4(C_2H_5)_2O \qquad (5)$$

In these preparations the alkali metal $B_3H_8^-$ salts are solvated[5] and have limited air stability. A large number of other cations have been used to produce salts of enhanced stability. For example, the dioxane adduct of the sodium salt can be prepared in $\sim 65\%$ yield and can then be used to synthesize the unsolvated, air-stable cesium salt [reaction (6)].[9-12]

$$CsBr + NaB_3H_8 \cdot 3C_4H_8O_2 \xrightarrow{H_2O} CsB_3H_8 + NaBr + 3C_4H_8O_2 \qquad (6)$$

Tetraalkylammonium triborohydride salts have proven to be very useful because of their high thermal and air stability, low moisture sensitivity, increased solubility in organic solvents, and value as starting reagents for metathesis to other $B_3H_8^-$ salts. The general method of preparation involves the addition of a saturated aqueous solution of the tetraalkylammonium halide or hydroxide salt to a NaB_3H_8/ether solution [reaction (7)].[3,7-11,13]

$$NaB_3H_8 + R_4NX \xrightarrow{\hspace{2cm}} R_4NB_3H_8 + NaX \qquad (7)$$

Current nomenclature rules require that $B_3H_8^-$ be called the octahydrotriborate(1–) ion, although it is commonly referred to as the triborohydride ion.

The mechanistic details of the syntheses of $B_3H_8^-$ from the reactions outlined in the text have not been deciphered, although several possible schemes have been postulated.

Heřmánek and Plešek[14] have proposed a mechanism for the sodium-diborane reaction that involves a radical reaction resulting in the formation of the $B_2H_6^{2-}$ ion [reaction (8)].[1,15]

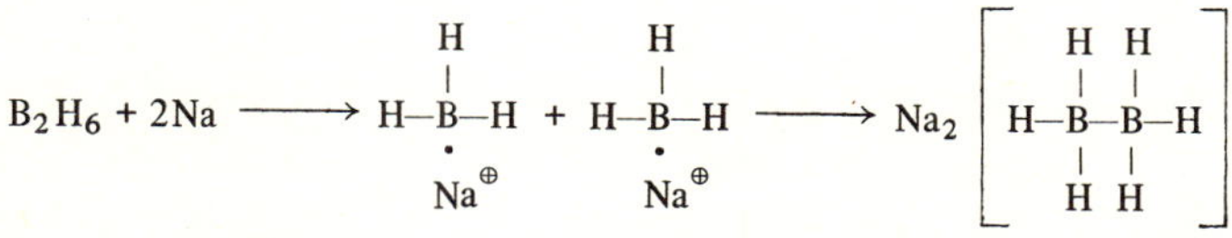

$$B_2H_6 + 2Na \longrightarrow \underset{\underset{Na^{\oplus}}{\bullet}}{H-\overset{\overset{\displaystyle H}{|}}{B}-H} + \underset{\underset{Na^{\oplus}}{\bullet}}{H-\overset{\overset{\displaystyle H}{|}}{B}-H} \longrightarrow Na_2 \left[H-\overset{\overset{\displaystyle H}{|}}{\underset{\underset{\displaystyle H}{|}}{B}}-\overset{\overset{\displaystyle H}{|}}{\underset{\underset{\displaystyle H}{|}}{B}}-H \right] \tag{8}$$

Subsequent reaction of $Na_2B_2H_6$ with diborane results in the formation of the complex ion $Na_2B_4H_{12}$, which decomposes to $B_3H_8^-$ [reaction (9)].

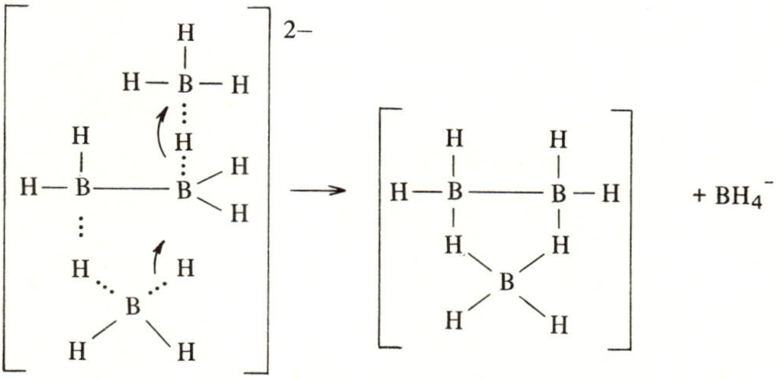

$$\tag{9}$$

Another approach is to consider the $B_3H_8^-$ ion as a salt of a relatively strong acid B_3H_9,[16] which is formed from B_2H_6 and deprotonated by BH_4^- [reaction (10)].

$$\tfrac{3}{2}B_2H_6 \longrightarrow [B_3H_9] \xrightarrow{\ BH_4^-\ } B_3H_8^- + H_2 + \tfrac{1}{2}B_2H_6 \tag{10}$$

Alternately the diborohydride(7) ion, $B_2H_7^-$,[11] may add BH_3 to form the intermediate (I) which could then lose molecular hydrogen and form $B_3H_8^-$.

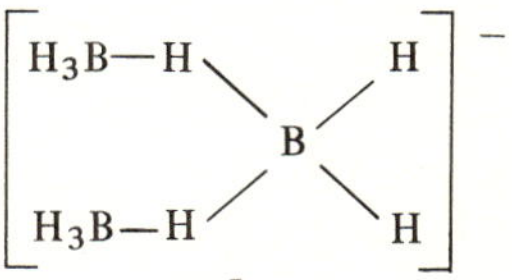

$$\left[\begin{array}{c} H_3B-H \\ H_3B-H \end{array} \diagdown\!\!\diagup \overset{\displaystyle H}{\underset{\displaystyle H}{B}} \diagup\!\!\diagdown \begin{array}{c} H \\ H \end{array} \right]^-$$

I

The $B_2H_7^-$ ion may lose H_2 directly to form $B_2H_5^-$ which could add BH_3 to form $B_3H_8^-$,[3] or two $B_2H_7^-$ ions may react to form $B_3H_8^-$ and BH_4^- ions [reaction (10)].[10]

$$2NaB_2H_7 \xrightarrow{\ \Delta\ } NaB_3H_8 + NaBH_4 + H_2 \tag{11}$$

1.2. Structural and Bonding Characteristics of the Free $B_3H_8^-$ Ion

The various topological representations for the localized bonding in $B_3H_8^-$ are shown in Figure 1.[17]

The solid-state structure of $B_3H_8^-$ (II) was determined by an X-ray structural study on $[H_2B(NH_3)_2][B_3H_8]$ [18] to be the predicted 2013 structure, where the digits 2013 in Lipscomb's styx system represent, respectively, the numbers of B—H—B bridges (s), B—B—B three-center bonds (t), B—B two-center bonds (y), and BH_2 groups (x). The three boron atoms form an isosceles triangle and the hydrogens are oriented so as to give apparent C_{2v} symmetry to the ion. The bridge hydrogen bonds are not symmetrical. The reasons for this have been discussed in terms of molecular orbital theory using several models.[19,20] Indications are that atoms B(1) and B(3) are more negatively charged than the unique boron atom, B(2). On this basis is is not surprising that the relatively positively charged bridge hydrogen atoms are closer to B(1) and B(3) than to B(2). Further calculations, using orbital vacancy considerations, indicate that the 2013 valence structure [Figure 1(a)] describes the major bonding in $B_3H_8^-$, but that other resonance contributions are also important.[17] For example, the 2103 structure [Figure 1(d)] involves an unsymmetrical B—B—B three-center bond in which the two electrons originally localized in the B(1)—B(3) bond of the 2013 structure are delocalized over all three boron atoms. While the major electron density resides in the basal B—B region of the ion, there is also a bonding interaction with the apex boron. The effect of this interaction is that the bridging hydrogen atoms move away from the apex boron. The 0105 structure [Figure 1(c)] is considered to be an unoptimized form of the double-bridged 2013 structure.[17]

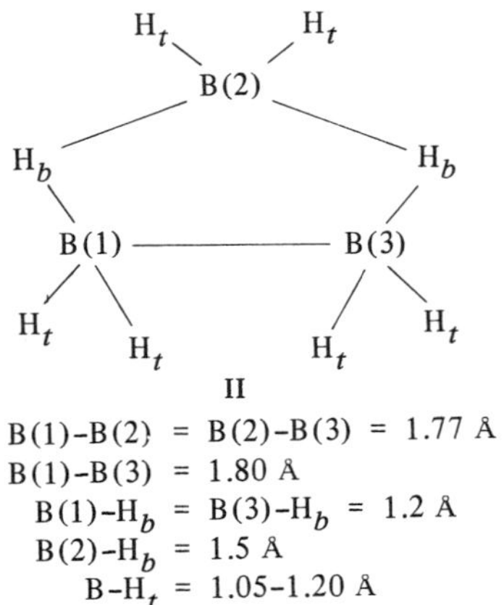

B(1)–B(2) = B(2)–B(3) = 1.77 Å
B(1)–B(3) = 1.80 Å
B(1)–H_b = B(3)–H_b = 1.2 Å
B(2)–H_b = 1.5 Å
B–H_t = 1.05–1.20 Å

1.3. Covalent Metal Derivatives of the $B_3H_8^-$ Ion

The covalent metal derivatives of $B_3H_8^-$ are electron deficient and the attachment between metal and the borane anion takes the form of metal–hydrogen–

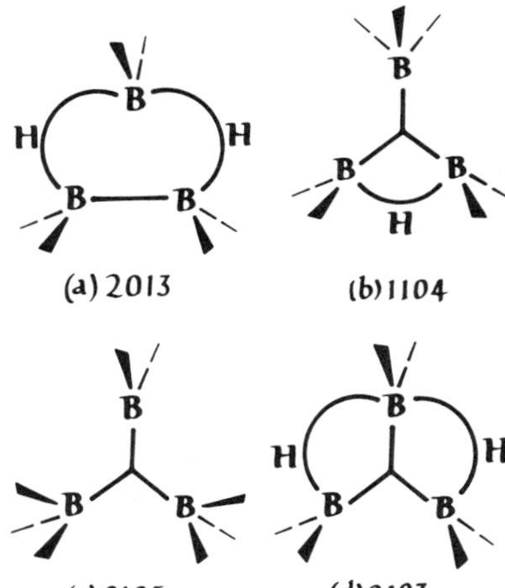

Figure 1. Possible topological structures for $B_3H_8^-$; (a) 2013 topologically allowed double bridge;(b) 1104 topologically allowed single bridge;(c) 0105 vacant-orbital zero bridge;(d) 2103 topologically forbidden double bridge.

boron three-center two-electron bridge bonds. There are examples of $B_3H_8^-$ forming two and three M—H—B bonds to a single metal. In one case a $B_3H_8^-$ moiety is bound to two metals via four M—H—B bonds, two to each metal. The following compilation of metallaboranes containing the $B_3H_8^-$ ligand is organized in accordance with the position of the metal in the Periodic Table.

1.3.1. Group IIA Complexes

1.3.1.1. Be(B$_3$H$_8$)$_2$. This colorless volatile compound is best prepared in a reliable high-vacuum system by the stirred solid–solid reaction shown in reaction (1). The $Be(B_3H_8)_2$ is removed from the reaction mixture as it forms.[21]

$$BeCl_2 + 2CsB_3H_8 \xrightarrow[\text{3 days}]{60^\circ} Be(B_3H_8)_2 + 2CsCl \qquad (12)$$
$$75\%$$

The X-ray determined low-temperature static molecular structure of $Be(B_3H_8)_2$ is shown in Figure 2.[22] The molecule exhibits C_2 point group symmetry and is the only known covalent metal–B_3H_8 compound in which a metal is bonded only to B_3H_8 units.

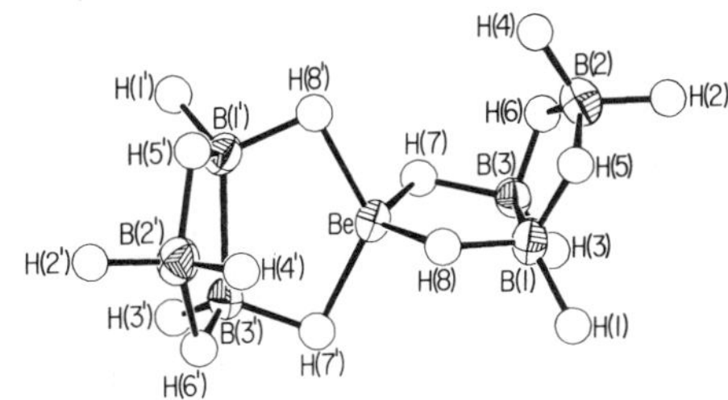

Figure 2. The low-temperature static molecular structure of $Be(B_3H_8)_2$.

1.3.1.2. $C_5H_5BeB_3H_8$. This compound is prepared by the reaction shown in reaction (13).[21]

$$Be(B_3H_8)_2 + NaC_5H_5 \longrightarrow C_5H_5BeB_3H_8 + NaB_3H_8 \qquad (13)$$

1.3.1.3. $(CH_3BeB_3H_8)_2$. Reaction of $Zn(CH_3)_2$ with $Be(B_3H_8)$ produces $(CH_3BeB_3H_8)_2$ along with a more volatile product thought to be $CH_3ZnB_3H_8$.[21] The ^{11}B nmr spectra indicate that the B_3H_8 units are equivalent and nearly static. The ^1H spectra, however, suggest the presence of three types of CH_3 groups. At $7°C$ the spectrum contains three relatively sharp CH_3 resonances in a ratio of $\sim 1:2:1$. As the solution is warmed the CH_3 resonances broaden, as do the B_3H_8 resonances. By $60°$ the CH_3 resonances have merged to a single resonance and have sharpened significantly, while the B_3H_8 resonances have broadened such that no clear maxima are present. These spectral results are consistent with the presence of two geometric isomers of $(CH_3BeB_3H_8)_2$ in a methyl bridged dimer in which there are two different relative orientations of the B_3H_8 groups, as shown in (III) and (IV).

III   IV

1.3.1.4. $Mg(B_3H_8)_2$. Two triborohydride derivatives of magnesium have been described.[14] The bis(octahydrotriborate) complex is prepared in an autoclave by the reaction shown in reaction (14), producing a 42% yield of the white

$$B_2H_6 + Mg(BH_4)_2 \cdot 3THF \xrightarrow[\text{THF}]{100°} Mg(B_3H_8)_2 \cdot 6THF \qquad (14)$$

crystalline product. A mixed borohydride–triborohydride magnesium metallaborane may be prepared by the reaction of diborane with magnesium hydride as shown in reaction (15).

$$B_2H_6 + MgH_2 \xrightarrow[\text{THF}]{85°} Mg(BH_4)(B_3H_8) \cdot 5THF \qquad (15)$$

Elevated pressure and temperature are necessary to promote this reaction. The product is a yellow crystalline solid which enflames in air and is decomposed explosively by alcohols after a variable induction period.

1.3.2. Group IIIA Complexes

1.3.2.1. $(CH_3)_2AlB_3H_8$ and $(CH_3)_2GaB_3H_8$. Dimethylaluminum and dimethylgallium triborane(8) have been prepared by the reaction of the appropriate

Group IIIA organometallic halide and salts of the octahydrotriborate ion,[23] [reaction (16)] .

$$(CH_3)_2MCl + B_3H_8^- \longrightarrow (CH_3)_2MB_3H_8 + Cl^- \quad (M = Al, Ga) \quad (16)$$

Direct reaction of the reagents at room temperature in the absence of solvent afforded approximately 60% yields of the product when a slight excess of Na^+, $(CH_3)_4N^+$, or $(n\text{--}C_4H_9)_4N^+$ salt of the $B_3H_8^-$ ion was used. The products are volatile, air-sensitive liquids that have limited thermal stability at room temperature.

1.3.2.2. $Al(BH_4)_2B_3H_8$. This compound has been prepared by the route shown in reaction (17).[24]

$$Al(BH_4)_3 + (CH_3)_2AlB_3H_8 \longrightarrow Al(BH_4)_2(B_3H_8) + (CH_3)_2AlBH_4 \quad (17)$$

$Al(BH_4)_2(B_3H_8)$ has a vapor pressure of 4 Torr at $0°C$. The molecule is fluxional on the nmr time scale.

1.3.3. Group VIB—Chromium, Molybdenum, and Tungsten

Reaction between the Group VIB metal hexacarbonyls, $M(CO)_6$ (M = Cr, Mo, W), and CsB_3H_8 in refluxing monoglyme/diglyme solutions proceeds with the formation of transition-metal octahydrotriborate complexes in approximately 70% yield [reaction (18)] .[25,26]

$$M(CO)_6 + CsB_3H_8 \xrightarrow[]{\Delta} \xrightarrow{(CH_3)_4NCl} [(CH_3)_4N^+][(CO)_4MB_3H_8^-] + CsCl + 2CO \quad (18)$$

The tetraalkylammonium salts are air-stable as solids, but are air-sensitive in solution.

An X-ray crystal structure of $[(CH_3)_4N^+] (CO)_4CrB_3H_8^-]$ [27] (Figure 3), shows the $B_3H_8^-$ ligand bonded to the metal in a bidentate fashion through two hydrogen bridges. The octahedral geometry about the chromium atom is preserved, and the anion possesses C_s point symmetry. Molecular orbital calculations suggest that the degree of σ-bonding is about the same for all Cr—C carbonyl bonds, but that there is a 28% greater back-donation for the equatorial carbonyl groups. The unique boron atom carries a higher positive charge than do the boron atoms adjacent to the metal.

Unlike the coinage metal complexes, discussed later, these group VIB complexes appear to be static on the nmr time scale, and the ^{11}B nmr spectra[26,27] consist of two resonances: a low-field resonance of area 1 [$\delta \approx -4$ ppm versus $BF_3 \cdot O(C_2H_5)_2$] and a high-field resonance of area 2 [$\delta \approx -43$ ppm]. The ^1H nmr spectra[27] contain an absorption at approximately $-7.3 \, \delta$ (versus TMS), appropriate for those hydrogens bridging between the metal and the boron atoms adjacent to the metal.

The reaction chemistry of these group VIB complexes has not been extensively studied, but one general reaction, reaction (19), describes the formation of several otherwise inaccessible organometallic complexes.[28]

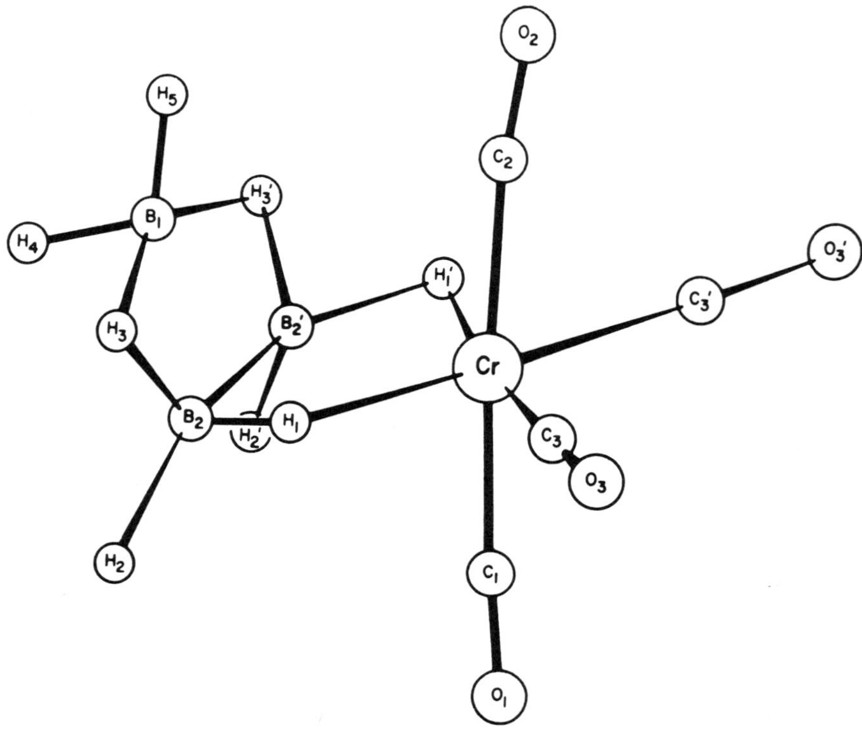

Figure 3. The molecular structure of the $(CO)_4CrB_3H_8^-$ anion.

$$(CO)_4MB_3H_8^- + 2L \xrightarrow{\text{THF}} cis\text{-}(CO)_4ML_2 \qquad M = Cr, Mo, W; L = PH_3, P\phi_3, RNC \qquad (19)$$

1.3.4. Group VIIB—Manganese and Rhenium

The most investigated $B_3H_8^-$ complexes are those of manganese. The parent complex, $(CO)_4MnB_3H_8$, is prepared in yields up to 80% from $(CO)_5MnBr$ and $(CH_3)_4NB_3H_8$ [reaction (20)].

$$(CH_3)_4NB_3H_8 + (CO)_5MnBr \longrightarrow (CO)_4MnB_3H_8 + CO + (CH_3)_4NBr \qquad (20)$$

Other ionic and covalent main-group B_3H_8 compounds can also be used.[29] The isostructural $(CO)_4ReB_3H_8$ is prepared using an analogous reaction. At room temperature the $B_3H_8^-$ group in these complexes is bonded to the metal in the same way as it is in $(CO)_4CrB_3H_8^{1-}$. Several representative reactions of $(CO)_4MnB_3H_8$ provide insights into the mutual effects of the metal group and the $B_3H_8^-$ ligand on one another.[30] The reaction of halogens with $(CO)_4MnB_3H_8$ [reaction (21)], results in stereospecific *exo* substitution at the unique boron atom, B(2), which is not adjacent to the manganese. The low-temperature X-ray determined structure of $(CO)_4MnB_3H_7Br$ is shown in Figure 4.[31]

$$(CO)_4MnB_3H_8 + X_2 \longrightarrow (CO)_4MnB_3H_7X + HX \qquad X = Cl, Br, I \qquad (21)$$

Photolytic or thermal decarbonylation of $(CO)_4MnB_3H_8$ produces $(CO)_3MnB_3H_8$, as shown in reaction (22).

$$(CO)_4MnB_3H_8 \underset{\sim 95\%}{\overset{\text{UV or } \Delta, \, >80\%}{\rightleftharpoons}} (CO)_3MnB_3H_8 + CO \qquad (22)$$

The low-temperature X-ray determined structure of $(CO)_3MnB_3H_8$ is shown in Figure 5.[32] This is the only known tridentate complex of $B_3H_8{}^-$. The tridentate ligation of the $B_3H_8{}^-$ ion reverts to the bidentate when $(CO)_3MnB_3H_8$ is treated with CO, PF_3, or NH_3. In the case of PF_3 and NH_3, however, a mixture of the $(CO)_3(L)MnB_3H_8$ isomers is produced.

The reactions of Br_2 or HCl with $(CO)_3MnB_3H_8$ in the presence of $AlBr_3$ or $AlCl_3$ produces μ-$Br(CO)_6(B_3H_8)Mn_2$ and μ-$Cl(CO)_6(B_3H_8)Mn_2$, respectively. The low-temperature X-ray determined structure of the former is shown in Figure 6.[33] These complexes are the first examples of a new class in which a $B_3H_8{}^-$ ligand is simultaneously bound to two metal centers through four M—H—B bridge bonds from two of the $B_3H_8{}^-$ boron atoms. The two metal bound boron atoms are bonded to both manganese atoms by bridging Mn—H—B bonds.

Small yields of μ-$Br(CO)_6(B_3H_8)Mn_2$ are also obtained in the reaction of $Mn_2(CO)_8Br_2$ with salts of the $B_3H_8{}^-$ anion. A second type of M2—B_3 cluster, $B_3H_7Fe_2(CO)_6$, is described in the group VIII section.

1.3.5. Group VIII—Iron and Ruthenium

Reactions of $(\eta^5$-$C_5H_5)(CO)_2FeI$[30] and $(\eta^5$-$C_5H_5)(CO)_2RuCl$[34] with $B_3H_8{}^-$ salts produce the sublimable complexes $(\eta^5$-$C_5H_5)(CO)_2FeB_3H_8$ and $(\eta^5$-$C_5H_5)(CO)_2RuB_3H_8$, whose structures are analogous to those of $(CO)_4MnB_3H_8$. In addition, reaction of $B_3H_8{}^-$ salts with $Fe(CO)_4Br_2$ produces

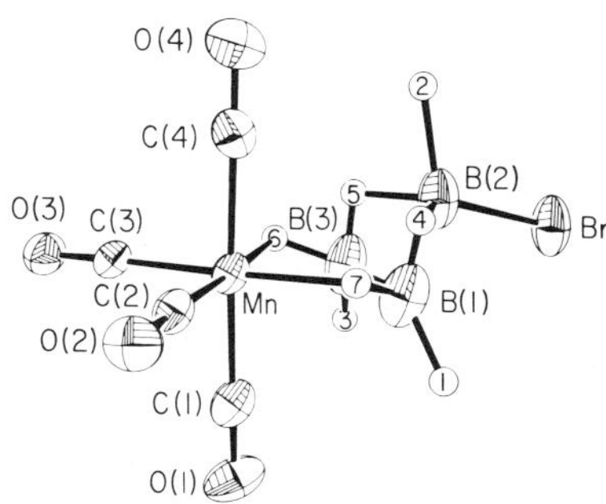

Figure 4. The low-temperature static structure of $(CO)_4MnB_3H_7Br$.

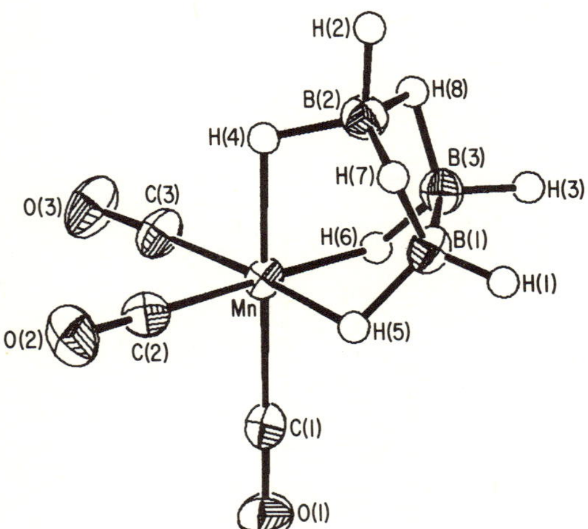

Figure 5.   The low-temperature static structure of $(CO)_3MnB_3H_8$.

$(H)(CO)_3FeB_3H_8$.[30] It is the only known metal-hydride $B_3H_8^-$ complex which exists as a single isomer in which the terminal hydrogen on iron is in the same plane as the two Fe—H—B bonds, thus eliminating all molecular symmetry. This complex is very unstable, decomposing primarily to $1\text{-}(CO)_3FeB_4H_8$, B_2H_6, and B_4H_{10}.

A second type of M_2B_3 cluster is produced by reaction of B_5H_9 with $Fe(CO)_5$ and $LiAlH_4$ at 60-70° in 1,2-dimethoxyethane.[35] Low yields of $B_3H_7Fe_2(CO)_6$, a very different class of cluster, are obtained from this reaction. Figure 7[36] shows the X-ray determined structure of this new metallaborane cluster in which the B_3 unit is open rather than triangular, and the cluster may be viewed as a pentaborane(9) molecule in which the apical BH unit and the

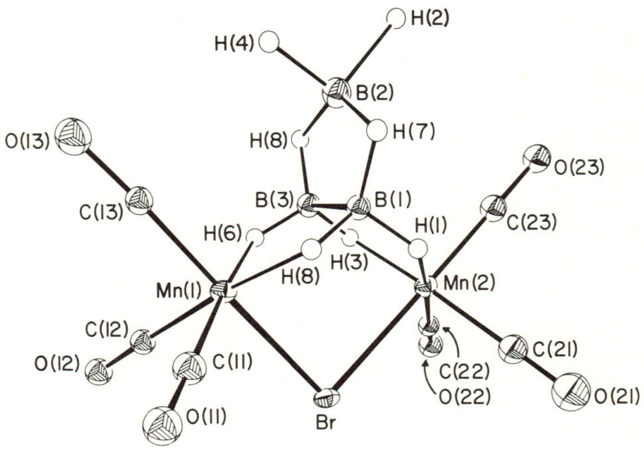

Figure 6. The structure of $\mu\text{-Br-}(CO)_6(B_3H_8)Mn_2$.

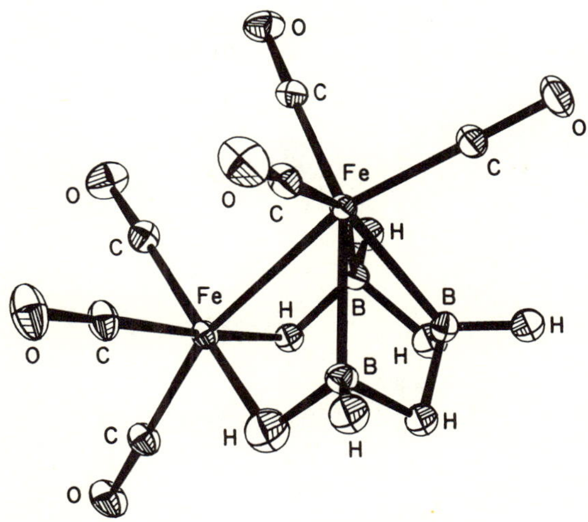

Figure 7. The structure of $(B_3H_7)Fe_2(CO)_6$.

basal BH unit have been replaced by two $Fe(CO)_3$ units. An alternate view is that a $B_3H_7^{2-}$ (borallyl) unit is added to $Fe_2(CO)_6^{2+}$.

Platinum complexes of the heptahydrotriborate(2−), $B_3H_7^{2-}$, ligand are formed according to reaction (23).[37,38]

$$\textit{cis-}L_2PtCl_2 + CsB_3H_8 + (C_2H_6)_3N \xrightarrow{CH_3CN} L_2Pt(\eta^3\text{-}B_3H_7) + CsCl + (C_2H_5)_3NHCl$$

$$L = (C_2H_5)_3P, (C_2H_5)P\phi_2, \phi_3P, (CH_3)_2P\phi, (p\text{-}CH_3C_6H_4)_3P \qquad (23)$$

The corresponding nickel and palladium complexes could only be prepared using the $\phi_2PCH_2P\phi_2$ ligand, and the nickel complex proved too unstable for isolation. The stability of these complexes with respect to heat, oxidation, and hydrolysis decreases in the order $Pt > Pd > Ni$.

The ^1H and ^{11}B nmr spectra of these complexes indicate that the molecules are rigid on the nmr time scale. The ^{11}B nmr spectrum is broad and structureless while the ^1H nmr spectrum consists of four resonances. The X-ray determined structure of the $[(CH_3)_2(C_6H_5)P]_2PtB_3H_7$ complex, shown in Figure 8, shows a markedly different mode of coordination for the $B_3H_7^{2-}$ ligand compared to that observed for the $B_3H_8^-$ ligand, in that there are apparently no hydrogen atoms bridging between the metal and the adjacent boron atoms.[37]

The π-borallyl, $B_3H_7^{2-}$, ligand can be considered a four-electron donor in this platinum complex, with the platinum atom in the +2 oxidation state. On the basis of ESCA studies, this formalism is favored over the alternative view that the complex is a Pt° adduct of the neutral B_3H_7 ligand.

Reaction of TlB_3H_8 with $\textit{trans-}Ir(I)(CO)Cl(PPh_3)_2$ produces Ir(III) $(\eta^3\text{-}B_3H_7)(CO)H(PPh_3)_2$ according to reaction (24).[39]

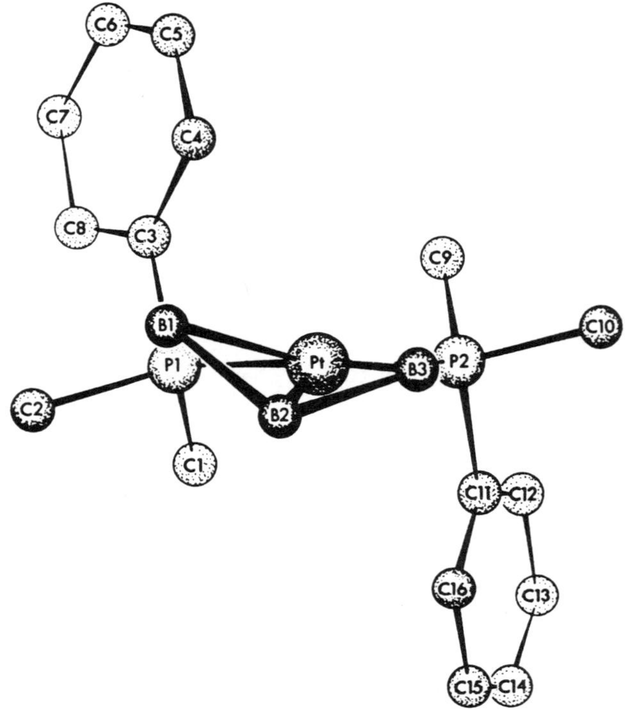

Figure 8. The structure of $[(CH_3)_2(C_6H_5)P]_2PtB_3H_7$.

$$B_3H_8^- + Ir(I)(CO)Cl(PPh_3)_2 \longrightarrow Cl^- + Ir(III)(\eta^3\text{-}B_3H_7)(CO)H(PPh_3)_2 \quad (24)$$

Spectroscopic evidence suggests that the $B_3H_7{}^{2-}$ ligand is bound to the iridium in the same η^3-allyl manner as in the platinum complex, and may be regarded as an *arachno*-$B_3H_7{}^{2-}$ ligand. Alternately, these "borallyl" complexes can be derived by replacement of H—B(1) (in the 2112 vacant orbital configuration) in B_4H_8 with an $Ir(CO)H(PPh_3)_2$ or PtL_2 moiety.

1.3.6. Group IB

The coinage metal $B_3H_8^-$ complexes were first prepared according to reaction (25):[40,41]

$$L_nCuCl + CsB_3H_8 \xrightarrow{\text{acetone}} L_mCuB_3H_8 + CsCl + (n-m)L \quad (25)$$

for $L = \phi_3P$, $n = 3$, $m = 2$; $L = \phi_3As$, $n = 1$ or 2, $m = 2$; $L = \phi_3Sb$, $n = 3$, $m = 3$. Because of dissociation, the exact stoichiometry of the L_nCuCl complexes in solution is not known. The $(\phi_3As)_2CuB_3H_8$ and $(\phi_3Sb)_3CuB_3H_8$ complexes decompose slowly in chloroform/acetone solutions, but the $(\phi_3P)_2CuB_3H_8$ complex is stable. Lippard and Ucko[41] postulated that the decreased rate of reactivity of $B_3H_8^-$ with $Cu(I)$, as compared to the analogous BH_4^- reaction, may indicate that the $B_3H_8^-$ ion prefers a specific stereochemical mode of attachment.

The crystallographically determined structure of the $(\phi_3P)_2CuB_3H_8$ complex shows the $B_3H_8^-$ ion bonded to the copper atom through two $Cu-H-B$ bonds as shown in Figure 9. The tetrahedral coordination of the copper atom is completed by the two phosphine ligands and the two $Cu-H-B$ bridging hydrogen atoms. The $H-Cu-H$ angle is $103°$ and the $P-Cu-P$ angle is $120°$.[42] $L_2CuB_3H_8$ complexes can also be prepared from $Cu(II)$ salts as shown in reaction (26).

$$2R_3P + CuSO_4 \cdot 5H_2O + CsB_3H_8 \xrightarrow{\quad (C_2H_5)_2O/H_2O \quad} (R_3P)_2CuB_3H_8 \qquad (26)$$
$$R = \phi,^{(25)} \; p\text{-}CH_3C_6H_4^{(43)}$$

The silver complexes $(\phi_3P)_2AgB_3H_8$[25] and $[(p\text{-}CH_3C_6H_4)_3P]_2AgB_3H_8$[43] can be prepared by substituting $AgNO_3$ for $CuSO_4$ in the reaction above. They are photosensitive in solution, producing silver metal. Gold derivatives of the triborohydride ion proved to be too unstable to isolate as crystalline materials due to an even more pronounced tendency to reduction to the metal.

The solution-state structures of the octahydrotriborate derivatives of these coinage metals appear to be complex and have been studied in at least nine investigations.[40,41,43-49] Molecular weight and conductivity measurements on the copper complexes indicate monomeric behavior in benzene[41] and chloroform,[40] but behavior as a 2:1 electrolyte in acetonitrile.[40,41] The explanation proposed for this behavior is the formation of the trinuclear $[(L_2Cu)_3B_3H_8]^{2+}$ ion which is postulated to have the structure shown in scheme (V).

$$\left[\begin{array}{c} \text{structure} \end{array} \right]^{2+} \quad [B_3H_8^-]_2$$

V

^{31}P nmr spectra and conductivity measurements on CH_2Cl_2 and $CH_2Cl_2/$ $CH_3C_6H_5$ solutions of copper, silver, and gold $B_3H_8^-$ complexes, with $(p\text{-}CH_3C_6H_4)_3P$ ligands on the metal, showed the existence of a variety of metal species in solution.[49] The solution structures are dependent on (1) temperature, (2) the metal atom, and (3) the concentration of free phosphine ligand when it is added to a solution containing the metal complex. All of the following species were observed in solution: $L_2CuB_3H_8$, $L_2AgB_3H_8$, $L_3AgB_3H_8$ (or $[L_3Ag^+]$

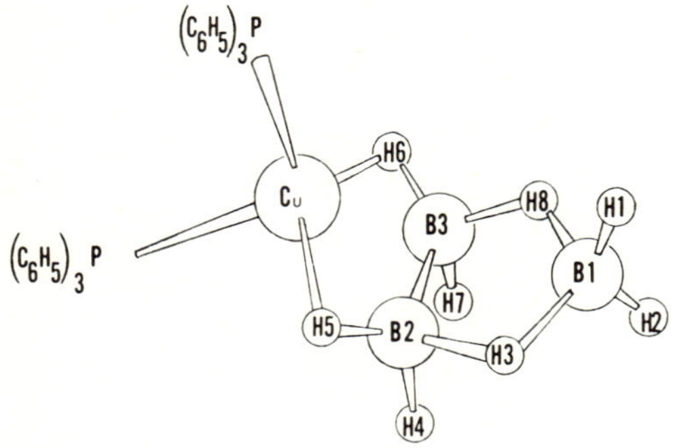

Figure 9. The structure of $[(C_6H_5)_3P]_2CuB_3H_8$.

$[B_3H_8^-])$, $[L_4Ag^+][B_3H_8^-]$, $[L_2Au^+][B_3H_8^-]$, $[L_3Au^+][B_3H_8^-]$ and perhaps $[L_4Au^+][B_3H_8^-]$.

1.4. Internal Exchange in Metal-B_3H_8 Complexes

Internal exchange of hydrogen atoms is often observed in metal complexes of the $B_3H_8^-$ anion. In many cases the barriers to such exchange are in the range accessible via nmr spectroscopy, and several of these have been studied in some detail. The following selected descriptions illustrate the types of internal hydrogen and boron exchange exhibited by the various metal-B_3H_8 complexes.

By way of introduction, the free $B_3H_8^-$ ion provides the least constrained, lowest energy barrier to internal exchange. The room temperature ^1H nmr spectrum of the $B_3H_8^-$ ion consists of a symmetrical ten-lined pattern at ~+0.3 ppm (versus TMS, $J_{^{11}B-^1H} = 33$ Hz.[9,50-52] The corresponding ^{11}B nmr spectrum is a symmetrical nonet at ~ −30.0 ppm (versus $BF_3 \cdot OEt_2$, $J_{^{11}B-^1H} = 33$ Hz.[3,9,50,51,53-57] These data indicate that all hydrogen and all boron environments are equivalent owing to rapid internal exchange in solution. Low-temperature nmr studies[45,46,52] suggest an upper limit of ~8 kcal/mole for the free energy of activation, $\Delta G^{\ddagger}$, for a hydrogen pseudorotation exchange process. A recent calculated barrier to hydrogen exchange in the gas phase ion is about 1 kcal/mole.[17] The localized molecular orbital structures and topological structures are consistent with a hydrogen tautomerism mechanism as shown in Figure 10.[17]

Metal-B_3H_8 complexes and their derivatives exhibit a broad range of internal exchange phenomena with activation energies covering the range accessible via nmr. The exchange mechanism is, in some cases, associated with other exchange processes occurring at the metal center, as demonstrated by external phosphine ligand exchange in $(\phi_3P)_2CuB_3H_8$ and internal CH_3 group exchange in $(CH_3)_2AlB_3H_8$. In other cases the whole ligand-metal-B_3H_8 system appears to be static except at such elevated temperatures that decomposition complicates detailed analyses. In between these extremes lie a group of very interesting complexes that exhibit internal hydrogen (and in some cases boron) exchange of

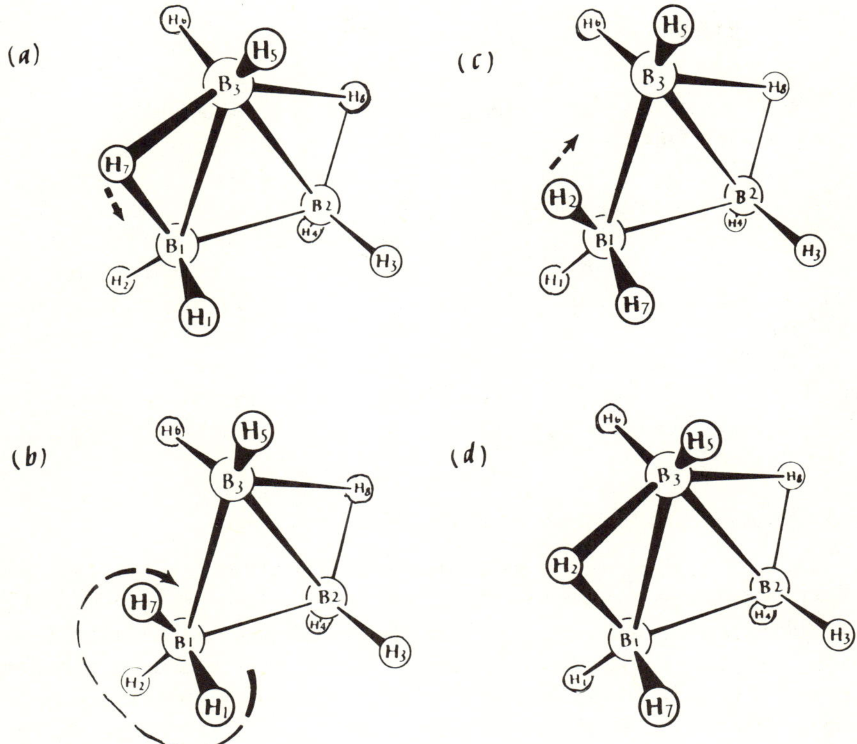

Figure 10. Proposed tautomerism mechanism for B_3H_8 :(a) interconversion of double bridge to single bridge; (b) single bridge with rotation; (c) single to double bridge conversion; (d) double bridge structure with interconverted hydrogens.

several different types, the mechanism and activation energies of which are dependent on the metal and its other ligands, substituents on the B_3 ligand, and the way in which the B_3 ligand is bonded to the metal. Specific examples to illustrate each of the various types of exchange follow.

1.4.1. $(CH_3)_2AlB_3H_8$ and $(CH_3)_2GaB_3H_8$

The ^1H and ^{11}B nmr spectra of $(CH_3)_2AlB_3H_8$ and $(CH_3)_2GaB_3H_8$ are temperature dependent and indicate that rapid internal exchanges occur at room temperature that equilibrate the methyl groups, the boron atoms, and all their attached hydrogen atoms. At lower temperatures the ^{11}B nmr spectra show that boron exchange slows to reveal two resonances characteristic of a B_4H_{10}-like structure in which the $B_3H_8^-$ is bonded to the Al or Ga through two Al—H—B bridge hydrogen bonds. By $-70°$ the ^1H nmr spectra of both complexes indicate that the methyl groups are nonequivalent and by $-90°$ four environments are visible for the $B_3H_8^-$ hydrogens. These low-temperature nmr spectra are consistent with static structures for $(CH_3)_2AlB_3H_8$ and $(CH_3)_2GaB_3H_8$ in which the $B_3H_8^-$ ligand is bidentate. The fluxional behavior of these molecules has been explained on the basis of the proposed mechanism shown in Figure 11.[23]

The lowest energy process is the opening of an Al—H—B bond in (a) to form (b), which allows rotation about the remaining Al—H—B bond to equilibrate the methyl groups. Further opening of a B—H—B bond allows equilibration of the three hydrogens in (c) associated with each boron atom adjacent to aluminum in the static structure. Finally, hydrogen migration and rotation of the B_3 unit shown in (d) allow exchange between all hydrogen and boron environments in the $B_3H_8^-$ group. The barriers to these exchanges increase in the order (a)⇌(b)< (b)⇌(c)<(c)⇌(d). In the presence of diethyl ether these barriers to internal exchange are substantially decreased. For example, the ^{11}B spectrum of $(CH_3)_2$-AlB_3H_8 in diethyl ether is a symmetrical nonet at room temperature, whereas in the absence of ether the spectrum is a slightly asymmetric singlet devoid of fine structure.

1.4.2. $Al(BH_4)_2(B_3H_8)$

At 30° the ^{11}B nmr spectrum consists of a broad featureless hump, suggestive of the equivalence of all of the boron atoms in the molecule. By –50°, the hydrogen atom exchange within the $B_3H_8^-$ fragment of the molecule is slowed and two boron resonances appear, suggesting bidentate fluxionality for the ligand. The borohydride groups, however, are still equivalent and fluxional at –50°, as indicated by the presence of a single quintet in the ^{11}B nmr spectrum.[24]

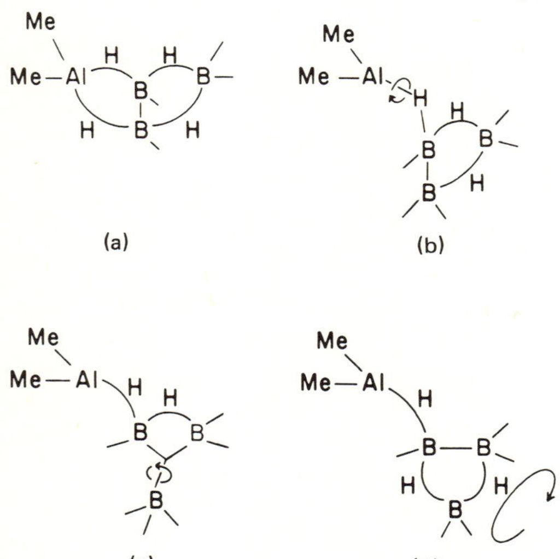

Figure 11. Proposed mechanism for internal hydrogen and methyl group exchange in $(CH_3)_2AlB_3H_8$.

1.4.3. $Be(B_3H_8)_2$

This molecule is unique in that the C_2 symmetry of the low-temperature static structure gives rise to nonequivalence of B(1) and B(3), as shown in Figure 2,[22] which can be distinguished in the low-temperature [11]B nmr spectra shown in Figure 12.[21,58] As the temperature is raised, however, the molecule becomes increasingly fluxional. At room temperature, for example, B(1) and B(2) become equivalent (the molecular symmetry may be thought of as changing from C_2 to C_{2v}) and the three hydrogen atoms bonded to each of these boron atoms become equivalent. At higher temperatures the [11]B nmr resonances broaden and near 80° they merge, indicating that all atoms in the B_3H_8 unit are involved in the internal exchange process similar to that proposed for $(CH_3)_2$ AlB_3H_8 (Figure 11).

1.4.4. $C_5H_5BeB_3H_8$

In this compound all atoms in the C_5H_5 and B_3H_8 groups appear to be rapidly but internally fluxional in solution on the nmr time scale even at –87°C.[21] The equivalence of the C—H hydrogens suggests that the C_5H_5 ring is η^5, as it is in $C_5H_5BeBH_4$. The rapid exchange within the B_3H_8 moiety precludes determination of its mode of bonding to Be on the basis of the nmr evidence.

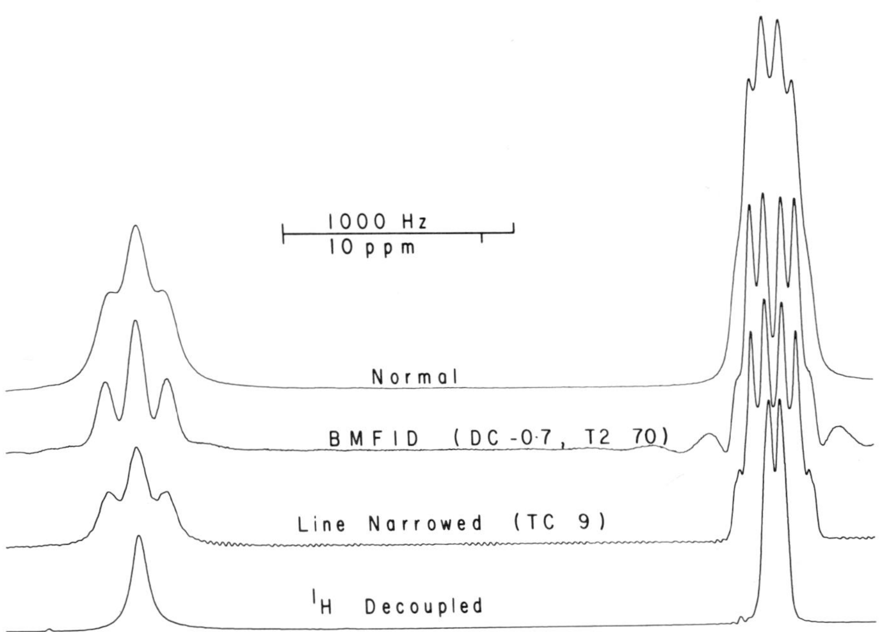

Figure 12. The [11]B nmr spectrum of $Be(B_3H_8)_2$ at 86.6 MHz in toluene-D_8 at –21°C.

1.4.5. $(CO)_4MnB_3H_7X$ and $(CO)_3MnB_3H_8$

Nearly all of the bidentate chromium, manganese, and iron group $B_3H_8^-$ complexes studied to date are static at room temperature on the nmr time scale.[30] The ^1H nmr spectrum shown in Figure 13 is representative of these static complexes. The halo derivatives of $(CO)_4MnB_3H_7X$ (X = Cl, Br, I), however, are partially fluxional at room temperature, as illustrated by their reversible temperature-dependent nmr spectra. Figure 14 shows the temperature dependance of the ^1H spectrum of $(CO)_4MnB_3H_7Cl$, from which several important conclusions can be extracted. First, it is clear that the highest field ^1H resonance, arising from the two Mn—H—B bridge hydrogens, is temperature independent

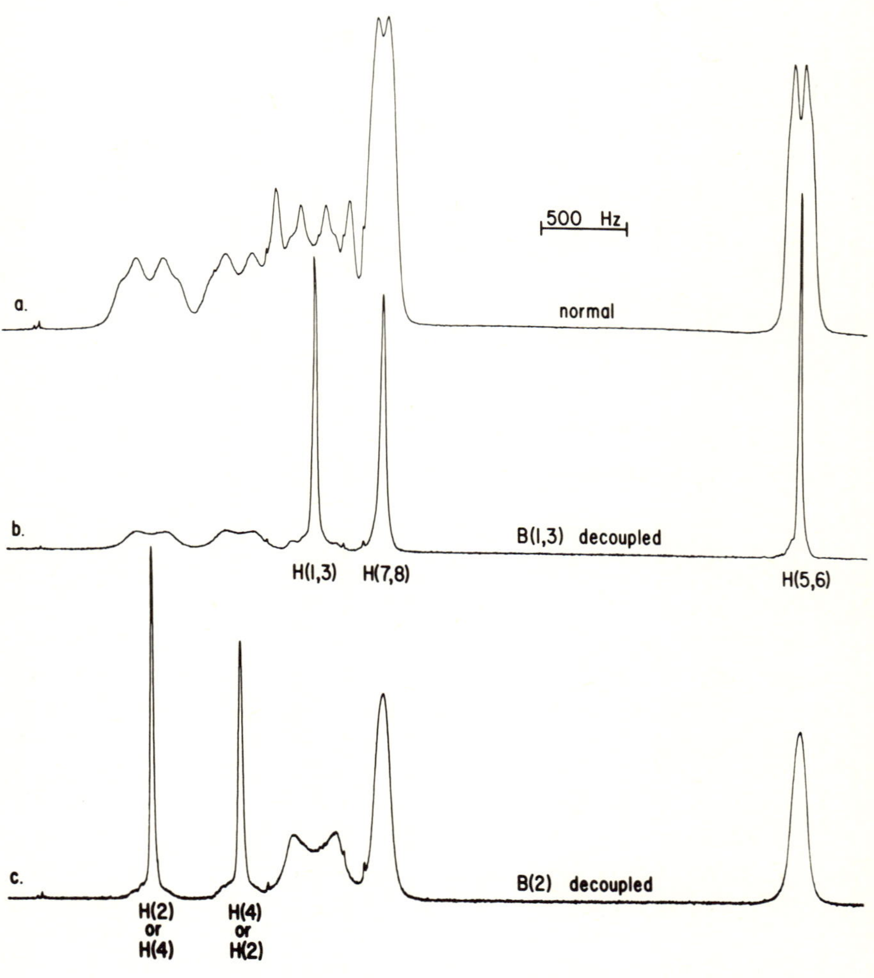

Figure 13. The ^1H nmr spectrum of $(CO)_4ReB_3H_8$ at 270 MHz.

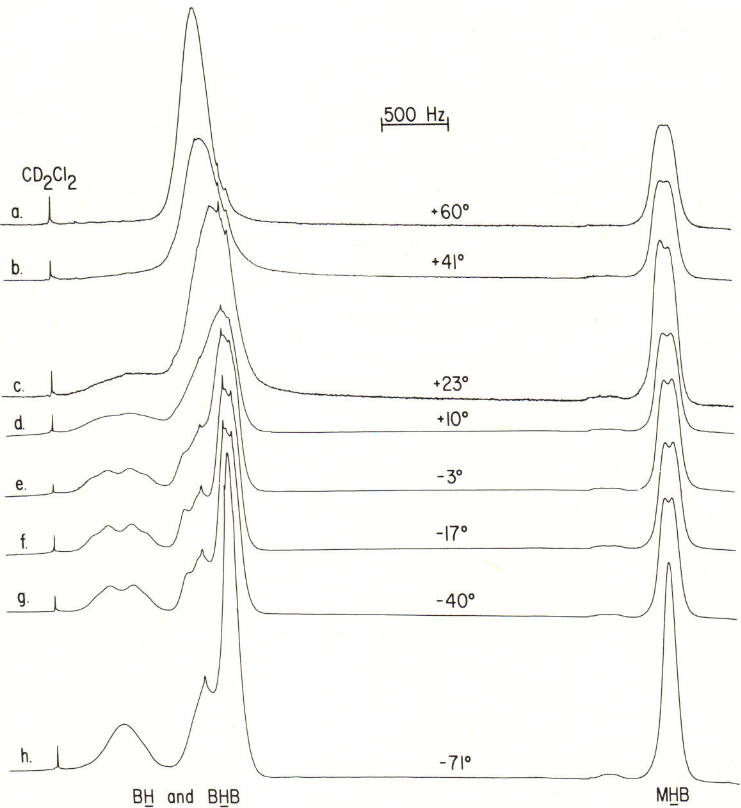

Figure 14. The variable temperature nmr spectra of $(CO)_4MnB_3H_7Cl$ at 270 MHz.

and that those hydrogens, and the boron atoms to which they are bound, are not subject to exchange in the temperature range investigated. Second, only one of the terminal positions on the unique boron B(2) (Figure 4) is involved in the exchange process, and that position is apparently fixed by the stereospecific location of the halogen atom. Thus, only the five boron bound hydrogen atoms are involved in the internal exchange around an essentially static boron framework. In order to study this exchange in more detail, the variable ^1H nmr spectra of $(CO)_4MnB_3H_7Br$ were measured under conditions of complete ^{11}B decoupling.[31] Line shape analysis of these spectra produced the results shown in Figure 15, and the activation parameters shown in Table 1. The exchange process was described by a "jump" model which is consistent with a process initiated by the rotation of hydrogens about the B—Br bond axis coupled with hydrogen rotations about the B—H—M bond axes. The overall atom movement proposed is shown in Figure 16.

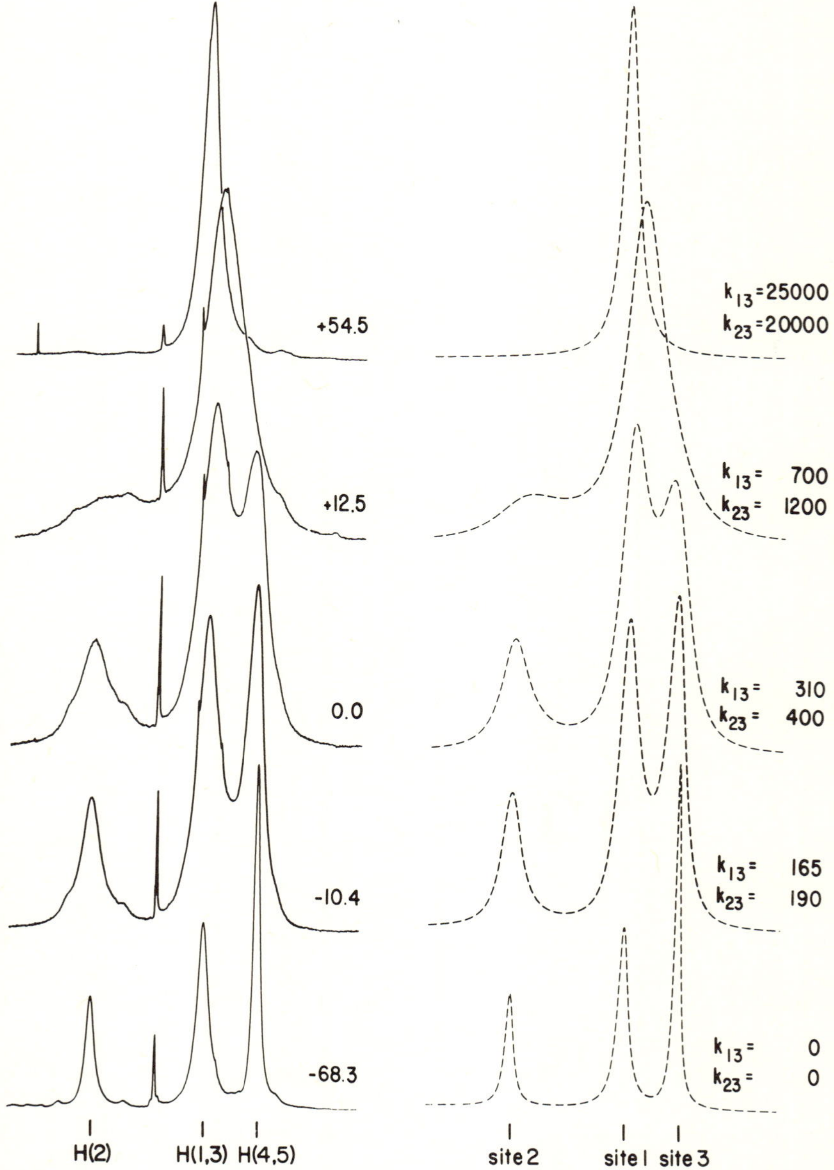

Figure 15. A comparison of selected experimental and theoretical spectra obtained in the total line shape analysis of $(CO)_4MnB_3H_7Br$.

Nmr studies of the tridentate $B_3H_8^-$ complex $(CO)_3MnB_3H_8$ show that it is also partially fluxional, and that a static configuration cannot be observed at temperatures as low as $-80°C$. The spectra indicate that three equivalent M—H—B hydrogens are not exchanging, but that the five remaining boron bound hydro-

Table 1. Rate Constants from the Total NMR Line Shape Analysis
of $(CO)_4MnB_3H_7Br$ at Various Temperatures

Temperature K	k_{13} sec^{-1}	k_{23} sec^{-1}
211	4.5	2
222	25	8
231	36	11
245	41	26
252	95	75
263	165	190
273	310	400
281	550	900
286	700	1 200
296	1 500	3 000
304	11 000	5 500
310	14 000	12 000
320	25 000	12 000
328	25 000	20 000

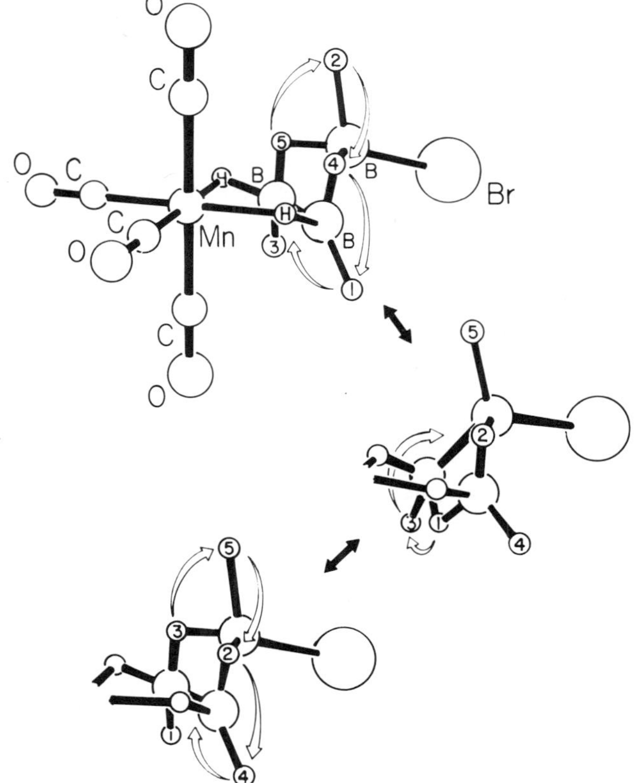

Figure 16. A proposed mechanism of intramolecular hydrogen exchange in $(CO)_4MnB_3H_7Br$.

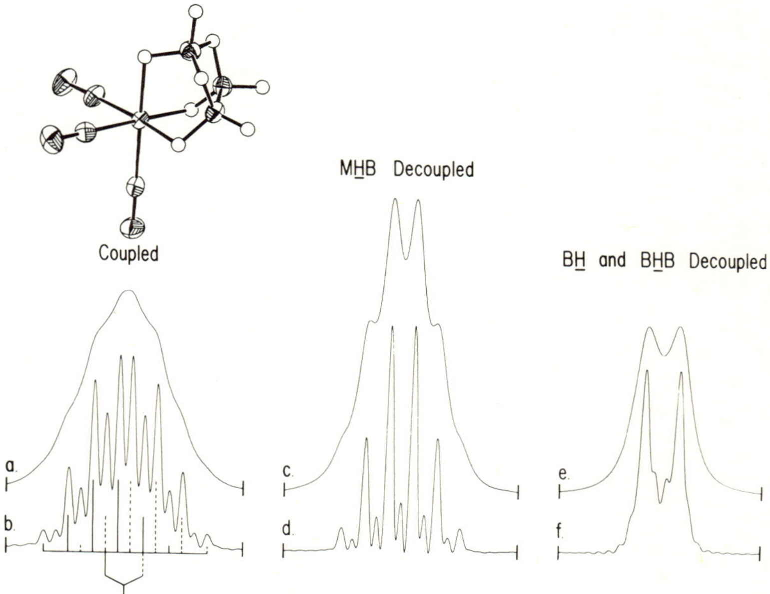

Figure 17. The ^{11}B nmr spectrum of $(CO)_3MnB_3H_8$ at 86.6 MHz and −10°C.

gens are equivalently coupled to the three equivalent boron atoms. The ^{11}B nmr spectra shown in Figure 17 show several of the diagnostic features. We have proposed that the most reasonable internal exchange mechanism is a simple lower energy varient of the $(CO)_4MnB_3H_7Br$ mechanism and this is shown in Figure 18.

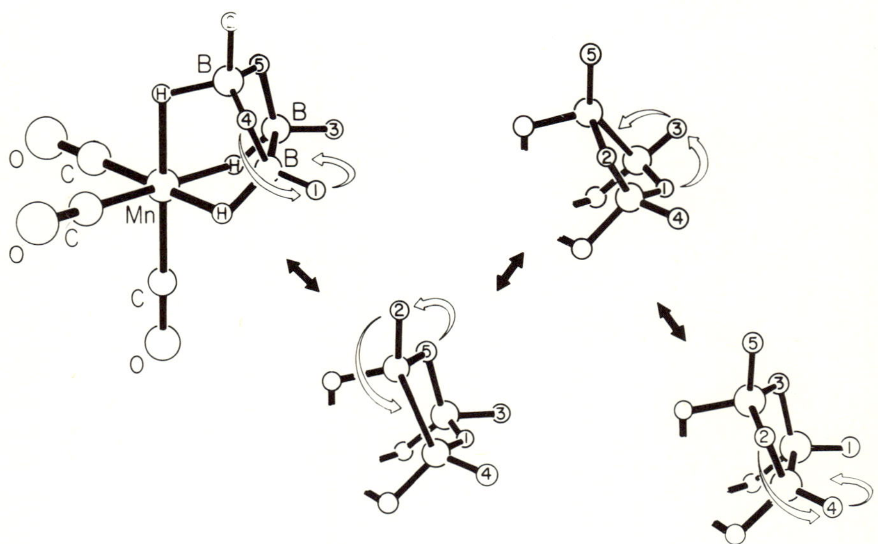

Figure 18. A proposed mechanism of intramolecular hydrogen exchange in $(CO)_3MnB_3H_8$.

The proposed exchange mechanisms for $(CO)_4MnB_3H_7Br$ and $(CO)_3MnB_3H_8$ are compared in Figure 19. As evident in the figure, the exchanging hydrogen atoms in $(CO)_3MnB_3H_8$ are all on one side of the B_3 triangle whereas in $(CO)_4MnB_3H_7Br$ the B(2) hydrogen position and the bridge hydrogens are on the opposite side of the B_3 triangle. The required movement of hydrogen from one side of the B_3 triangle to the other in the $(CO)_4MnB_3H_7X$ complexes is considered a primary factor in the observed higher activation barriers.

While both $(CO)_3MnB_3H_8$ and $(CO)_4MnB_3H_7Br$ have rapidly fluxional hydrogens at room temperature, the parent $(CO)_4MnB_3H_8$ is static. At temperatures above $80°C$, however, $(CO)_4MnB_3H_8$ undergoes hydrogen exchange that appears qualitatively similar to that in $(CO)_4MnB_3H_7Br$. In $(CO)_4MnB_3H_8$ the *exo* hydrogen of the BH_2 group is most likely excluded in this exchange just as the halogen is in $(CO)_4MnB_3H_7X$.

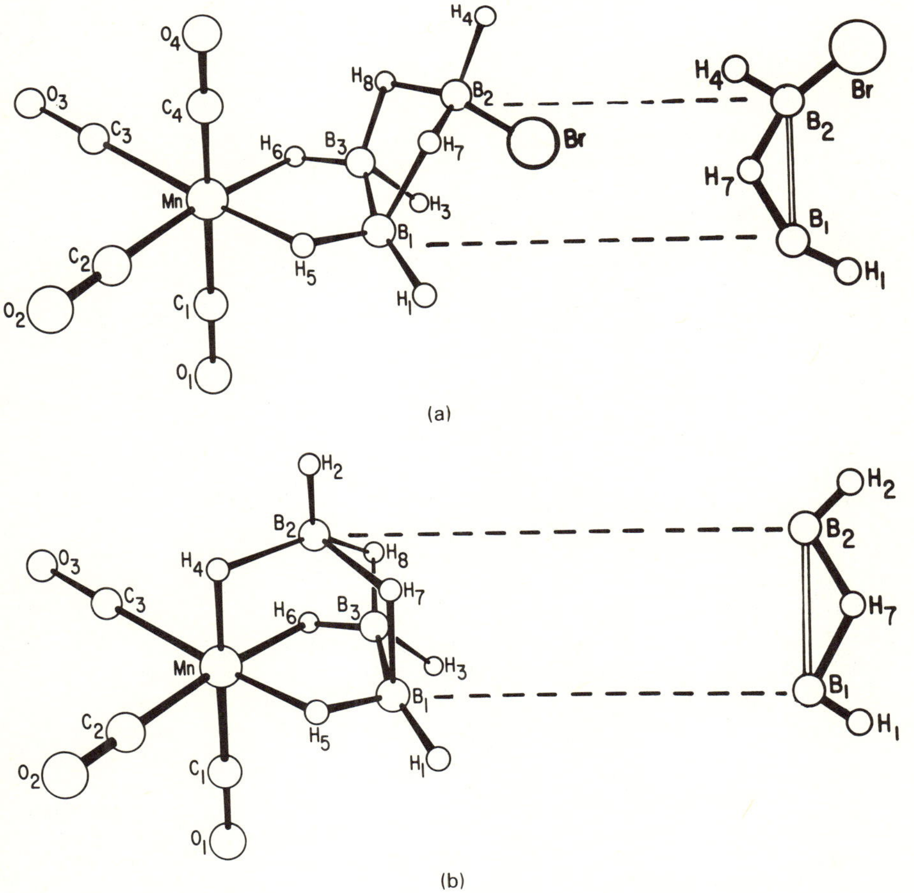

Figure 19. A comparison of the proposed mechanistic exchange processes in $(CO)_4MnB_3H_7Br$ and $(CO)_3MnB_3H_8$.

1.4.6. $L_2CuB_3H_8$

Several copper complexes of formula $L_2CuB_3H_8$ [L = $(C_6H_5)_3P$, $(CH_3C_6H_4)_3P$, $(C_6H_5O)_3P$] exhibit broad structureless ^1H nmr singlets for the B_3H_8 hydrogens at room temperature.[44] At lower temperatures the singlet broadens and two (or possibly three) broad resonances become somewhat resolved. On the basis of the X-ray determined static structure of $[(C_6H_5)_3P]_2$ CuB_3H_8, which is similar to that of other bidentate metal B_3H_8 complexes, these spectral changes have been interpreted as evidence for slowing B_3H_8 pseudorotation to a static structure at $-97°$. Facile intermolecular phosphine exchange has also been observed in the studies of these complexes.

REFERENCES

1. A. E. Stock, *Hydrides of Boron and Silicon*, Cornell University Press, Ithaca, New York (1933).
2. W. V. Hough, L. J. Edwards, and A. D. McElroy, *J. Am. Chem. Soc.*, 78, 689 (1956).
3. D. F. Gaines, R. Schaeffer, and F. Tebbe, *Inorg. Chem.*, 2, 526–528 (1963).
4. *Boron, Metallo-Boron Compounds and Boranes* (R. M. Adams, ed.), p. 470, Interscience, New York (1964).
5. W. V. Hough, M. D. Ford, and L. J. Edwards, Callery Chemical Company, 1024–TR–214, Dec. 18, 1956.
6. R. F. Gould, ed., *Adv. Chem. Ser.*, 32, 191 (1961).
7. G. E. Ryschkewitsch, and K. C. Nainan, *Inorg. Synth.*, 15, 113–114 (1974).
8. K. C. Nainan, G. E. Ryschkewitsch, *Inorg. Nucl. Chem. Lett.*, 6, 765–766 (1970).
9. W. J. Dewkett, M. Grace, and H. Beall, *Inorg. Synth.*, 15, 115–118 (1974).
10. W. J. Dewkett, M. Grace, and H. Beall, *J. Inorg. Nucl. Chem.*, 33, 1279–1280 (1971).
11. H. C. Miller, N. E. Miller, and E. L. Muetterties, *Inorg. Chem.*, 3, 1456–1463 (1964).
12. H. C. Miller, and E. L. Muetterties, *Inorg. Synth.*, 10, 81–91 (1967); E. Amberger, and E. Gut, *Chem. Ber.*, 101, 1200–1204, (1968).
13. *Production of the Boranes and Related Research* (R. T. Holzmann, ed.), p. 252, Academic Press, New York (1967).
14. S. Heřmánek, and J. Plešek, *Coll. Czech. Chem. Commun.*, 31, 177–189 (1966).
15. W. V. Hough, L. J. Edwards, and A. D. McElroy, *J. Am. Chem. Soc.*, 80, 1828–1829 (1958).
16. R. E. Enrione, and R. Schaeffer, *J. Inorg. Nucl. Chem.*, 18, 103–107 (1961).
17. I. M. Pepperberg, D. A. Dixon, W. N. Lipscomb, and T. A. Halgren, *Inorg. Chem.*, 17, 587–593 (1978).
18. C. R. Peters, and C. E. Nordman, *J. Am. Chem. Soc.*, 82, 5758 (1960).
19. W. N. Lipscomb, *Boron Hydrides*, p. 110, W. A. Benjamin, New York (1963).
20. R. Hoffman, and W. N. Lipscomb, *J. Chem. Phys.*, 37, 2872–2883 (1962).
21. D. F. Gaines, J. Morris, D. Hillenbrand, and J. L. Walsh, *Inorg. Chem.*, 17, 1516–1522 (1978).
22. J. C. Calabrese, D. F. Gaines, S. J. Hildebrandt, and J. H. Morris, *J. Am. Chem. Soc.*, 98, 5489–5492 (1976).
23. J. Borlin, and D. F. Gaines, *J. Am. Chem. Soc.*, 94, 1367–1369 (1972).
24. P. R. Oddy, D. L. Shaw, and M. G. H. Wallbridge, reported at the *Third International Meeting on Boron Chemistry*, Munich, July 1976.
25. F. Klanberg, E. L. Muetterties, and L. J. Guggenberger, *Inorg. Chem.*, 7, 2272–2278 (1970).

26. F. Klanberg, and L. J. Guggenberger, *J. C. S., Chem. Commun.*, 1293–1294 (1967).
27. L. J. Guggenberger, *Inorg. Chem.*, 9, 367–373 (1970).
28. F. Klanberg, and E. L. Muetterties, *J. Am. Chem. Soc.*, 90, 3296–3297 (1968).
29. D. F. Gaines, and S. J. Hildebrandt, *J. Am. Chem. Soc.*, 96, 5574–5576 (1974).
30. D. F. Gaines, and S. J. Hildebrandt, *Inorg. Chem.*, 17, 794–806 (1978).
31. M. W. Chen, J. C. Calabrese, D. F. Gaines, and D. F. Hillenbrand, *J. Am. Chem. Soc.*, 102, 4928–4933 (1980).
32. D. F. Gaines, S. J. Hildebrandt, and J. C. Calabrese, *Inorg. Chem.*, 17, 790–794 (1978).
33. M. W. Chen, D. F. Gaines, and L. G. Hoard, *Inorg. Chem.*, in press.
34. M. W. Chen, and D. F. Gaines, unpublished results.
35. E. L. Anderson, K. J. Haller, and T. P. Fehlner, *J. Am. Chem. Soc.*, 101, 4390–4391 (1979).
36. K. J. Haller, E. L. Andersen, and T. P. Fehlner, *Inorg. Chem.*, in press.
37. L. J. Guggenberger, A. R. Kane, and E. L. Muetterties, *J. Am. Chem. Soc.*, 94, 5665–5673 (1972).
38. A. R. Kane, and E. L. Muetterties, *J. Am. Chem. Soc.*, 93, 1041–1042 (1971).
39. N. N. Greenwood, J. D. Kennedy, and D. Reed, *J. Chem. Soc. Dalton*, 196–200 (1980).
40. S. J. Lippard, and D. Ucko, *J. C. S., Chem. Commun.*, 983 (1967).
41. S. J. Lippard, and D. Ucko, *Inorg. Chem.*, 7, 1051–1056 (1968).
42. S. J. Lippard, and K. M. Melmed, *Inorg. Chem.*, 8, 2755–2762 (1969).
43. E. L. Muetterties, W. G. Peet, P. A. Wegner, and C. W. Alegranti, *Inorg. Chem.*, 9, 2447–2451 (1970).
44. H. Beall, and C. H. Bushweller, *Chem. Rev.*, 73, 465–486 (1973).
45. H. Beall, C. H. Bushweller, W. Dewkett, and M. Grace, *J. Am. Chem. Soc.*, 92, 3484–3486 (1970).
46. C. H. Bushweller, H. Beall, M. Grace, W. Dewkett, and H. Bilofsky, *J. Am. Chem. Soc.*, 93, 2145–2149 (1971).
47. C. H. Bushweller, H. Beall, and W. J. Dewkett, *Inorg. Chem.*, 15, 1739–1740 (1976).
48. H. Beall, C. H. Bushweller, and M. Grace, *Inorg. Nucl. Chem. Lett.*, 7, 641–645 (1971).
49. E. L. Muetterties, and C. W. Alegranti, *J. Am. Chem. Soc.*, 92, 4114–4115 (1970).
50. E. L. Muetterties, and W. D. Phillips, *Adv. Chem. Radiochem.*, 4, 231–292 (1962).
51. G. R. Eaton, and W. N. Lipscomb, *NMR Studies of Boron Hydrides and Related Compounds*, p. 73 and references therein, W. A. Benjamin, New York (1969).
52. D. Marynick, and T. Onak, *J. Chem. Soc. (A)*, 1160–1161 (1970).
53. R. Schaeffer, F. Tebbe, and C. Phillips, *Inorg. Chem.*, 3, 1475–1749 (1964).
54. B. M. Graybull, J. K. Ruff, and M. F. Hawthorne, *J. Am. Chem. Soc.*, 83, 2669–2670 (1961).
55. D. F. Gaines, *Inorg. Chem.*, 2, 523–526 (1963).
56. W. D. Phillips, H. C. Miller, and E. L. Muetterties, *J. Am. Chem. Soc.*, 81, 4496–4500 (1959).
57. A. D. Norman, and R. Schaeffer, *J. Phys. Chem.*, 70, 1662–1664 (1966).
58. D. F. Gaines, and J. H. Morris, *J. C. S., Chem. Commun.*, 626–627 (1975).

4

Metallaboron Cage Compounds of the Main Group Metals

Lee J. Todd

1. INTRODUCTION

The first carborane compounds were reported in 1962. About 5 years later the first boron cage compounds containing main group metals appeared in the literature. During the past 15 years considerable progress has been made in the synthesis and structure elucidation of these heteroatom boranes, but the research activity has been far less intense than that experienced in the carborane area.

This review covers mainly boron hydride derivatives with five to eleven boron atoms in the framework along with one or more heteroatoms. Compounds continuing only one or two boron atoms in the structure are not covered in this review of metallaboron "cage" compounds. Metal–B_3H_8 derivatives are not covered here because they are treated in detail in a separate chapter of this book (see Chapter 3 by D. F. Gaines and S. J. Hildebrandt). Main group metallaboron compounds with 12 or more boron atoms have not been reported to date and represent an interesting area for future investigation.

In addition to main group metals, some metalloid derivatives are also included. In most cases, derivatives of the alkali metals have been used primarily as reactive intermediates and have not been fully characterized. Some of these materials most likely have interesting structural features (similar to organolithium reagents), but a discussion of this area must await future definitive structural investigations.

The review is divided into a discussion of the syntheses and properties of

Lee J. Todd • Department of Chemistry, Indiana University, Bloomington, Indiana.

appropriate group IIA, IIIA, IVA, VA, and VIA derivatives. Each section is further divided into carborane and borane subsections. Boron-11 nmr remains as a rapid method for compound identification and as a check on compound purity. For this reason detailed ^{11}B nmr data for most derivatives are included in the review.

2. COMPOUNDS CONTAINING BERYLLIUM AND MAGNESIUM

2.1. Carborane Derivatives

Diethyl- or dimethylberyllium diethyl etherates reacted with $7,8\text{-}B_9C_2H_{13}$ in benzene/ether at room temperature to produce 1 mole of alkane. The second mole of alkane was evolved during removal of the solvent at $50°$ to form crystalline $(C_2H_5)_2O\cdot BeC_2B_9H_{11}$.[1,2] This compound was very sensitive to moist air and hydrolyzed to form $7,8\text{-}B_9C_2H_{12}^-$. Treatment of the ether complex with trimethylamine formed $Me_3N\cdot BeC_2B_9H_{11}$ which was considerably more stable in moist air than the ether complex. Ethanolic KOH degraded the trimethylamine complex to form $7,8\text{-}B_9C_2H_{12}^-$. These beryllium derivatives are proposed to be *closo*-icosahedral analogs of $B_{10}C_2H_{12}$ or $B_{12}H_{12}^{2-}$.

2.2. Borane Derivatives

The reaction of $Be(BH_4)_2$ with $1\text{-}ClB_5H_8$ resulted in the opening of the pentaborane cage and insertion of the beryllium atom into a basal position to form $B_5H_{10}BeBH_4$.[3] The mechanism of this reaction is complex and not well understood. The X-ray structure of this unusual molecule has been determined[4,5] (see Figure 1) and the cage is a distorted pentagonal pyramid with the beryllium atom in a basal position. Treatment of $B_5H_{10}BeBH_4$ with HCl or HBr at $-80°$ formed $B_5H_{10}BeCl$ or $B_5H_{10}BeBr$, both of which are isoelectronic with B_6H_{10}. Reaction of $B_5H_{10}BeBr$ with $Al_2(CH_3)_6$ or NaC_5H_5 formed $B_5H_{10}BeCH_3$

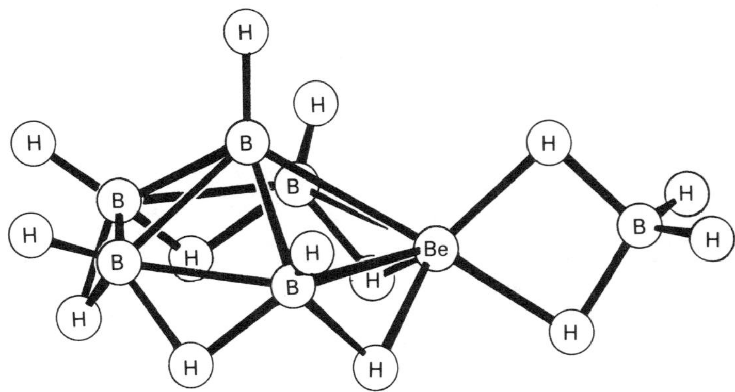

Figure 1. The structure of $B_5H_{10}HeBH_4$.

and $B_5H_{10}BeC_5H_5$, respectively. In an unexpected result, $B_5H_{10}BeBr$ reacted with CsB_3H_8 to form $B_5H_{10}BeB_5H_{10}$.[3] The structure of this molecule has been determined[5] and consists of two $B_5H_{10}Be$ fragments, similar to that of $B_5H_{10}BeBH_4$, linked at a common beryllium atom. The dihedral angle between the planes defined by the basal boron atoms of the two cages is 66°. The 86.6 MHz ^{11}B nmr spectrum (toluene-d_8 solvent) of $B_5H_{10}BeBH_4$ contained four resonances in a 2:2:1:1 area ratio (see Table 1). The signal at −9.8 ppm which was assigned to the two basal boron atoms adjacent to the beryllium, was resolved by line narrowing into a doublet of doublets due to terminal hydrogen and bridge hydrogen (BHBe) coupling. The quintet observed for the −33.4 ppm resonance indicates that the BH_4 unit is probably undergoing rapid internal exchange relative to the nmr time scale.

Reaction of KB_5H_8 with excess C_5H_5BeCl between −40° and ambient temperature formed $\mu[\eta^5\text{-}(C_5H_5)Be]B_5H_8$ in good yield.[6] The compound is thermally stable at room temperature, but is air and moisture sensitive. The ^{11}B nmr spectrum consisted of three doublets at −13.4(161)2B, −21.8(141)2B and −54.6 ppm (170 Hz)1B. A single-crystal X-ray study[6] confirmed that a C_5H_5Be unit occupies a bridge hydrogen position of B_5H_9.

In a 2:1 mole ratio, B_6H_{10} and $Mg(CH_3)_2$ in THF at −78° reacted to form $Mg(THF)_2(B_6H_9)_2$ and methane after 6–8 hr.[7] One mole of methane was evolved within minutes at −78° to possibly form $CH_3MgB_6H_9$. The X-ray structure of $Mg(THF)_2(B_6H_9)_2$ was determined.[7] The magnesium atom appears to be inserted into one of the basal boron-boron bonds of each $B_6H_9^-$ framework to form three-center B—Mg—B bonds. The Mg—B distances are 2.38 and 2.48 Å. The THF rings are distorted from planarity and are attached to the magnesium atom. The ^{11}B nmr spectrum of $Mg(THF)_2(B_6H_9)_2$ in CH_2Cl_2 at room temperature consisted of a low-field doublet of area 5 at 9.6 ppm and a high-field doublet of area 1 at −49.3 ppm. The magnesium atom is believed to be involved in an exchange process. The boron-11 spin-decoupled proton nmr spectrum of

Table 1. 86.6 MHz ^{11}B NMR Data for $B_5H_{10}BeX$
(X = BH_4, B_5H_{10}, Cl, Br, CH_3, C_5H_5)

Compound	Assignment[a] {chemical shift (ppm) [$J_{^{11}B-H}$ (Hz)]}			
	B(1)	B(3), B(6)	B(4), B(5)	B(7)
$B_5H_{10}BeBH_4$	−53.5(146)	−9.8(135,83)	8.2(164)	−33.4(86)
$B_5H_{10}BeB_5H_{10}$	−54.6(139)	−8.1(127,98)	5.3(159)	
$B_5H_{10}BeBr$	−54.0(146)	−1.6(138,84)	10.0(161)	
$B_5H_{10}BeCl$	−54.5(146)	−2.9(138,84)	9.2(164)	
$B_5H_{10}BeCH_3$	−54.6(146)	−2.0(138,83)	9.6(161)	
$B_5H_{10}BeC_5H_5$	−64.0(117)	−3.6(117)	−0.3(156)	

[a]Benzene or toluene used as solvent.

this compound in CD_2Cl_2 at 20° displayed three resonances at 3.86 ppm (basal terminal H), -1.65 ppm (apical H), and -3.22 ppm (bridge H). At $-100°$, the proton nmr spectrum contained two resonances in both the basal terminal and bridge hydrogen regions which indicated that the exchange process had slowed down considerably. This compound exhibited thermal and air stability. Evidence for decomposition of the solid was noted in the infrared spectrum after standing for several weeks, under vacuum, at room temperature.

It has been reported[8] that the Grignard reagent (RMgX) reacted with $B_{10}H_{14}$ in diethyl ether to form hydrocarbon and a material of the composition $B_{10}H_{13}MgX(Et_2O)_2$. Other work indicates that some nucleophilic attack of $B_{10}H_{14}$ by the RMgX to form $RB_{10}H_{13}$ also occurs.[9] Diethylmagnesium reacted with decaborane in ether at $-78°$ to form a white solid with the overall composition $MgB_{10}H_{12}(Et_2O)_2$.[10] This compound decomposed rapidly in air and was insoluble in most organic solvents. It did dissolve in ethanol and water, but decomposed with evolution of hydrogen. Treatment of the magnesium compound in HCl did not give decaborane (as was the case with related zinc and cadmium derivatives), but decomposed with evolution of hydrogen.

3. COMPOUNDS CONTAINING ALUMINUM, GALLIUM, INDIUM, OR THALLIUM

3.1. Carborane Derivatives

The reaction of $Ga(CH_3)_3$ with $2,3$-$C_2B_4H_8$ in the gas phase at 180–215°C formed $CH_3GaC_2B_4H_6$ in 20% yield.[11] A similar reaction employing $(CH_3)_3In$ and a temperature of 95–110°C formed $CH_3InC_2B_4H_6$ in 50% yield.[12] A single-crystal X-ray study of the gallacarborane established the molecular geometry indicated in Figure 2.[12] One distinctive feature of the structure is that the $Ga-CH_3$ axis is tilted 23° with respect to the perpendicular to the equatorial plane. The 32.1 MHz ^{11}B nmr spectrum of the gallium derivative (CDCl$_3$ solution) contains resonances at 8.9, 4,5, and -39.2 ppm (181 Hz) in a 2:1:1 area ratio. The ^{11}B nmr spectrum of $CH_3InC_2B_4H_6$ contains resonances at 4.5, 0.3, and -46.3 ppm (173 Hz) in a 2:1:1 area ratio. At greater than 100°C both the gallium and indium compounds decompose generating $2,3$-$C_2B_4H_8$. Both HCl and Br_2 readily attack the gallacarborane and destroy the cage structure.

The reaction of lithium or sodium salts of $2,3$-$C_2B_4H_7^-$ anion with $(CH_3)_2AlCl$ or $(CH_3)_2GaCl$ formed unstable $(CH_3)_2AlC_2B_4H_7$ and stable $(CH_3)_2GaC_2B_4H_7$, respectively.[60] In both compounds the metal is apparently linked to the open face of the cage by a B—M—B three-center, two-electron bond. The 32 MHz ^{11}B nmr spectrum of the gallium derivative contained doublets at 2.72(138) and -51.9 ppm ($J_{B-H} = 182$ Hz) with relative areas of 3 and 1.

When $7,8$-$B_9C_2H_{13}$ was mixed with triethylaluminum in benzene at room temperature, ethane was evolved and *nido*-$(C_2H_5)_2AlB_9C_2H_{12}$ was formed

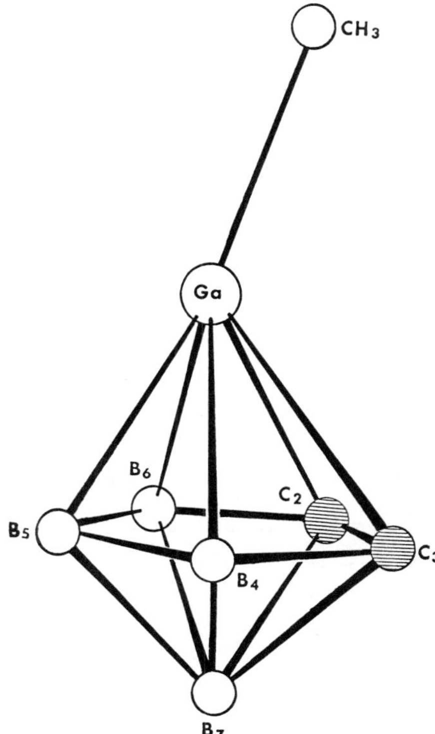

Figure 2. The structure of $CH_3GaC_2B_4H_6$.

in good yield.[13] Similar reactions were used to form $(CH_3)_2AlB_9C_2H_{12}$ and $(C_2H_5)_2GaB_9C_2H_{12}$. A single-crystal X-ray diffraction study of $(CH_3)_2AlB_9C_2H_{12}$ indicated that the $(CH_3)_2Al$ group is in a bridging position.[14] The variable temperature ^{11}B nmr spectrum of $(CH_3)_2AlB_9C_2H_{12}$ demonstrated that the moelcule is fluxional at room temperature. The fluxional process presumably involves interchange of the bridging hydrogen and $(CH_3)_2Al$ group. The room temperature ^{11}B nmr data of the aluminum and gallium derivatives are given in Table 2. It is quite probable that the structures and ^{11}B nmr spectra of such formal $7,8\text{-}B_9C_2H_{13}$ derivatives as $R_2MB_9C_2H_{12}$ (M = Al or Ga), $TlB_9C_2H_{12}$, and $(R_2As)_2B_9C_2H_{11}$ are closely related with a heteroatom unit occupying a three-center two-electron bridging position between two boron atoms on the open face of the B_9C_2 cage.

When $(C_2H_5)_2AlB_9C_2H_{12}$ in benzene was heated at reflux for 25 hr ethane was evolved and the *closo*-derivative $(C_2H_5)AlB_9C_2H_{11}$ was produced in good yield.[13] $CH_3AlB_9C_2H_{11}$ and $(C_2H_5)GaB_9C_2H_{11}$ were formed in a similar manner. The *closo*-icosahedral nature of this type of molecule was demonstrated by an X-ray structure.[15]

Sublimation of 3–ethyl–3–alumino–1,2–dicarbadodecaborane(12) through a glass-wool filled hot tube at 410° produced a new isomer believed to be 3–

Table 2. 80 MHz Boron NMR Spectra of $R_2MB_9C_2H_{12}$ and
$RMB_9C_2H_{11}$ Derivatives (M = Al or Ga)

Compound	δ_B (ppm) [relative intensity, J_{B-H} (Hz)]
$1,2-B_9C_2H_{12}Al(C_2H_5)_2$	$-7.7(4,138); -12.8(1,160); -18.0(2,107); -26.2(1,59);$ $-31.7(1,142)$
$1,2-B_9C_2H_{12}Al(CH_3)_2$	$-9.6(4,134); -15.5(1,170); -21.1(2,97); -28.6(1,58);$ $-33.8(1,143)$
$1,2-B_9C_2H_{12}Ga(C_2H_5)_2$	$-9.8(4,134); -15.4(1,145); -19.7(2,109); -28.7(1,74);$ $-33.2(1,141)$
$1,2-B_9C_2H_{11}Al(C_2H_5)$	$-7.3(2,146); -13.5(1,150); -15.7(3,150); -17.8(2,152);$ $-24.0(1,152)$
$1,2-B_9C_2H_{11}Al(CH_3)$	$-8.5(2,154); -14.7(1,150); -17.0(3,151); -19.2(2,147);$ $-25.5(1,151)$
$1,2-B_9C_2H_{11}Ga(C_2H_5)$	$-9.5(2,140); -17.1(3,150); -19.2(3,145); -23.0(1,143)$
$1,7-B_9C_2H_{11}Al(C_2H_5)$	$-8.6(2,154); -15.5(3,149); -18.5(2,151); -20.6(1,146);$ $-25.9(1,146)$

ethyl–3–alumino–1,7–dicarbadodecaborane (12).[13] A similar thermal treatment of $3-(C_2H_5)-3Ga-1,2-C_2B_9H_{11}$ at 400° formed mainly $B_9C_2H_{11}$, however, a small amount of a new isomer was detected. Controlled hydrolysis of the very reactive $3-C_2H_5Al-1,2-C_2B_9H_{11}$ and $3-C_2H_5Al-1,7-C_2B_9H_{11}$ formed 7,8- and $7,9-B_9C_2H_{12}^-$, respectively. When $3-C_2H_5Al-1,2-C_2B_9H_{11}$ was dissolved in tetrahydrofuran a crystalline adduct identified as $C_2H_5AlC_2B_9H_{11}(THF)_2$ was obtained. This adduct was also prepared from $7,8-B_9C_2H_{11}^{2-}$ and $C_2H_5AlCl_2$ in THF.[19] Trialkylaluminum compounds also readily expand their coordination sphere to form adducts with Lewis bases.[20]

When thallium(I) acetate reacted with $7,8-B_9C_2H_{12}^-$ in aqueous alkaline solution light yellow $(Tl)_2B_9C_2H_{11}$ precipitated from solution.[21] This compound appears to be very stable to air and moisture and is useful for the syntheses of metallacarborane compounds. One thallium ion is readily replaced by other large cations to form $Ph_3PMe[TlB_9C_2H_{11}]$. An X-ray structure of this material has been determined.[22] The thallium atom is fairly symmetrically located over the open face of the $7,8-B_9C_2H_{11}^{2-}$ unit. However, the thallium cage atom distances considerably exceed the sum of the covalent radii so that the $TlB_9C_2H_{11}$ unit appears to be an ion pair. The yellow color of the compound suggests some charge transfer in the ion pair, however. Treatment of $Tl_2[7-CH_3-7,8-B_9C_2H_{10}]$ with acetic acid or heating the thallium compound to 300° formed relatively volatile $Tl[7-CH_3-7,8-B_9C_2H_{11}]$ [23] and thallium metal. Further heating produced the *closo*-carborane, $CH_3CCHB_9H_9$. The 80 MHz [11]B nmr spectrum (THF solvent) of $Tl[7-CH_3-7,8-B_9C_2H_{11}]$ contained resonances at $-10.3, -11.1, -13.8, -17.8(2B), -18.7, -22.0, -33.4,$ and -37 ppm and the spectrum looked very much like that of a C-substituted $7,8-B_9C_2H_{12}^-$ ion.

3.2. Borane Derivatives

Decaborane-14 reacted with trimethylamine-alane in ether to form $NHMe_3$ $[AlB_{10}H_{14}] \cdot (OEt_2)_2$ [16] which may be similar to the known all-boron species $B_{11}H_{14}^-$. More recently decaborane was found to react with two equivalents of trimethylindium in benzene at room temperature to form the air-sensitive ionic compound $InMe_2[Me_2InB_{10}H_{12}]$.[17] The compound is a 1:1 electrolyte in acetone and shows the presence of two types of $InMe_2$ groups in the proton nmr spectrum. When the reaction is carried out using a 1:1 mole ratio of $B_{10}H_{14}$ and $InMe_3$, a nonelectrolyte product $MeInB_{10}H_{12}$ and two equivalents of methane are formed.[17] Two equivalents of trimethylthallium reacted similarly with decaborane to form $TlMe_2[Me_2TlB_{10}H_{12}]$.[17] With 1:1 stoichiometry $[TlMe_2]$ $[B_{10}H_{13}]$ was obtained. An X-ray structural analysis of $PMePh_3[Me_2TlB_{10}H_{12}]$ revealed a distorted eleven-atom *nido*-icosahedral structure.[18] Treatment of $TlMe_2[Me_2TlB_{10}H_{12}]$ with CH_3HgCl at room temperature in tetrahydrofuran lead to formation of $Me_2Tl[MeHgB_{10}H_{12}]$ and Me_2TlCl.[17] Metathesis of the borane product with $[MePPh_3]Br$ gave orange air-stable $MePPh_3[MeHgB_{10}H_{12}]$.

4. COMPOUNDS CONTAINING GERMANIUM, TIN, OR LEAD

4.1. Carborane Derivatives

The $7,8-B_9C_2H_{11}^{2-}$ ion reacts with germanium diiodide, stannous chloride, and lead acetate to form $1,2,3-C_2GeB_9H_{11}$, $1,2,3-C_2SnB_9H_{11}$, and $1,2,3-C_2PbB_9H_{11}$, respectively.[24,25] Similarly $7,9-B_9C_2H_{11}^{2-}$ can be reacted with germanium diiodide to produce $1,2,7-C_2GeB_9H_{11}$. The metal is easily removed from these derivatives with ethanolic potassium hydroxide. The product from reaction of $7,9-B_9C_2H_{11}^{2-}$ and stannous chloride is thermally unstable and decomposes to form tin metal and *closo*-$B_9C_2H_{11}$.[26] The ^{11}B nmr data available on these compounds are not very informative because of extreme overlap of resonances measured at 32 MHz. These compounds are assumed to be *closo*-icosahedral molecules containing an unsubstituted germanium, tin, or lead atom in the cage framework. The ^{119}Sn Mössbauer resonance of $1,2,3-C_2SnB_9H_{11}$ at 77 K was found well above that of β-tin.[29] The isomer shift of β-tin is considered as the dividing line between Sn(II) and Sn(IV) compounds with tin(II) derivatives being above this value.

Insertion of germanium into $7,8-$ or $7,9-B_9CH_{10}E^{2-}$ (E = P or As) by reactions similar to those described above afforded $1,2,3-$ and $1,2,7-GeB_9H_9CHE$ in which three different heteroatoms are contained in the icosahedral framework.[27] Pyrolysis of $1,2,7-GeB_9H_9CHP$ generated a new isomer of undetermined structure. The 70.6 MHz ^{11}B nmr spectrum in benzene contains resonances at 1.8, 1.2, -0.4, -3.1, and -5.9 ppm with 1:2:2:2:2 relative areas, respectively. Charge distribution in the $1,2,7-$isomers is probably more even than in the $1,2,3-$derivatives and this appears to be reflected in the smaller span of

Table 3. 70.6 MHz ^{11}B NMR Data of 1,2,3- and 1,2,7–GeB$_9$H$_9$CHP Derivatives

Compound	δ_B (ppm) [relative intensity, J_{B-H}, (Hz)][a]
1,2,3-GeB$_9$H$_9$CHP	14.2(1B,150); 7.9(1B, ~127); 5.8(1B, ~117); 2.5(1B, 155); −1.3(2B, 155); −4.3(1B, ~140); −6.4(1B, ~123); −8.4(1B, ~137)
1,2,7–GeB$_9$H$_9$CHP	7.5(1B, ~155); 6.0(1B, ~130); −0.0(2B, ~150); −1.9(1B, ~130); −3.0(2B, ~150); −6.3(2B, ~165)
1,2,3-GeB$_9$H$_9$CHAS	16.7(1B, 140); 7.3(1B, ~150); 5.2(1B, ~140); 2.2(1B, ~150); 1.1(1B, ~140); 0.1(1B, ~145); −6.7(2B, ~145); −8.6(1B, ~135)
1,2,7–GeB$_9$H$_9$CHAs	9.4(1B, 160); 4.8(1B, ~160); 2.5(2B, ~165); −0.7(1B, ~150); −2.9(2B, ~153); −4.4(1B, ~130); −8.4(1B, ~165)

[a]Benzene solvent.

^{11}B chemical shifts (see Table 3). The still smaller range of shifts in the product formed by high-temperature isomerization indicates that the heteroatoms have, as expected, moved further apart.

Smaller two-carbon carboranes have also been found to undergo reactions like those described above for icosahedral systems. Addition of SnCl$_2$ or PbBr$_2$ to a THF solution of NaC$_2$B$_4$H$_7$ formed off-white SnC$_2$B$_4$H$_6$ and light yellow PbC$_2$B$_4$H$_6$, respectively, in low yields.[28] Treatment of C$_2$B$_4$H$_7^-$ with GeI$_2$ gave a product whose mass spectrum contained peaks assigned to GeC$_2$B$_4$H$_6$, but thermal instability precluded isolation of the product. Reactions of SnCl$_2$ and PbBr$_2$ with Na[(CH$_3$C)$_2$B$_4$H$_5$] produced the corresponding sublimable M(CH$_3$C)$_2$B$_4$H$_4$ species (M = Sn or Pb). The C,C'-dimethyl derivatives appear to be a little more stable than the parent compounds. The lead derivatives have ^{11}B nmr spectra (see Table 4) each containing three doublets with 1:2:1 area ratios. These nmr results are consistent with the *closo*-pentagonal-bipyramidal structure (Figure 3) proposed for these compounds.

Reaction of Sn(CH$_3$C)$_2$B$_4$H$_4$ with (η^5-C$_5$H$_5$)Co(CO)$_2$ in a hot-cold reactor maintained at 70 and 150°C formed (C$_5$H$_5$)CoSn(CH$_3$C)$_2$B$_4$H$_4$ in low yield.[28] In solution this compound loses tin to form1-(C$_5$H$_5$)Co-2,3-

Table 4. 32.1 MHz ^{11}B NMR Data of C$_2$B$_4$H$_6$E (E = Sn or Pb)

Compound	Solvent	δ (ppm)[(J (Hz)]	Relative area
SnC$_2$B$_4$H$_6$	CDCl$_3$	0.1(165), −6.2(180)	3:1
PbC$_2$B$_4$H$_6$	d_6-acetone	35.5(151), 19.7(132), −6.7(166)	1:2:1
SnB$_4$H$_4$C$_2$(Me)$_2$	CDCl$_3$	17.6(144), −1.6(168)	3:1
PbB$_4$H$_4$C$_2$(Me)$_2$	d_6-acetone	33.3(130), 20.1(140), 1.0(180)	1:2:1

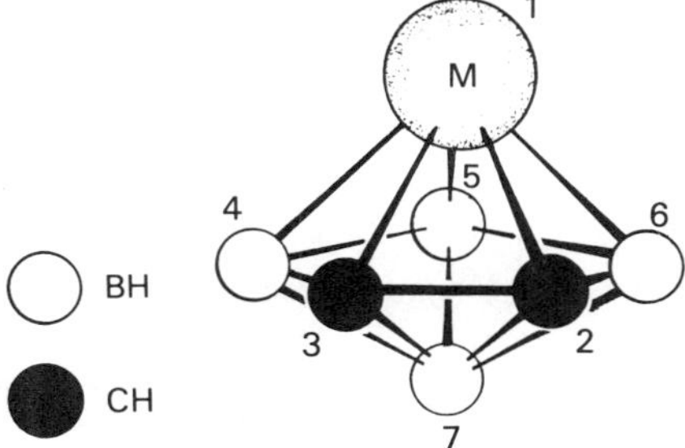

Figure 3. The proposed structure for 1,2,3–M(CH₃)₂C₂B₄H₄ and 1,2,3–MC₂B₄H₆ (M = Sn or Pb). Note: ●, CH; ○, BH.

$(CH_3C)_2B_4H_4$. The two-metal compound is most likely a *closo* molecule (Figure 4) which may be fluxional in solution.

Reduction of $2,4-C_2B_5H_7$ with sodium naphthalide in tetrahydrofuran formed $C_2B_5H_7{}^{2-}$; treatment of the dianion with $SnCl_2$ formed mainly $C_2B_5H_7$, although traces of $SnC_2B_5H_7$ were observed in the mass spectrum. Analogous experiments using $PbBr_2$ or GeI_2 did not give isolable metallacarboranes.[28]

Reaction of $Na_3B_{10}H_{10}CH(THF)_2$ with $GeCl_4$ at $0°C$ produced $1,2-B_{10}H_{10}CHGe^-$.[30] Treatment of the crude anionic product with methyl iodide formed sublimable $1,2-B_{10}H_{10}CHGeCH_3$ in 27% overall yield. This neutral compound can be stored in air for several months with negligible decomposition. Like $1,2-B_{10}H_{10}C_2H_2$, the neutral germanacarborane has a large melting point depression constant. The ^{11}B nmr spectrum of the methyl derivative (see Table 5) is very similar to that of $1,2-B_{10}H_{10}CHP$. There are two unit area doublets at low field, which must be due to the two unique boron atoms which are antipodal to the carbon and germanium atoms in the icosahedral structure. When $1,2-B_{10}H_{10}CHGeCH_3$ is reacted with piperidine at reflux, the methyl group is removed to form $1,2-B_{10}H_{10}CHGe^-$; in addition, some germanium atom abstraction is observed to form $7-B_{10}H_{12}CH^-$.[31] No isomerized compounds were obtained by heating $1,2-B_{10}H_{10}CHGeCH_3$ at $450°C$ or $1,2-B_{10}H_{10}CHGe^-$ at $475°C$, however, thermal decomposition was extensive. Photochemical irradi-

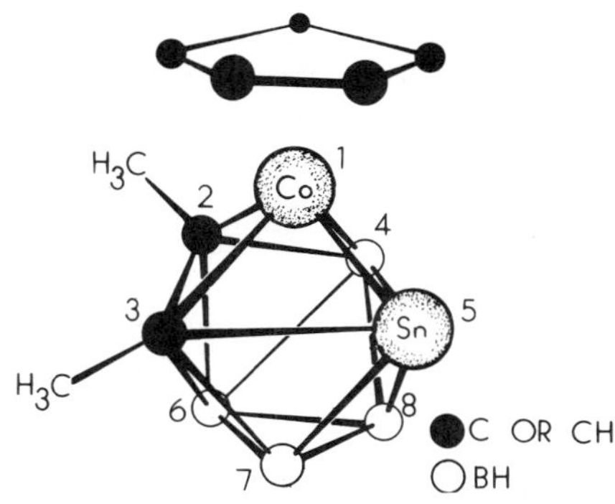

Figure 4. The proposed structure for $(\eta^5-C_5H_5)CoSn(CH_3)_2C_2B_4H_4$. Note: ●, C or CH; ○, BH.

Table 5.   70.6 MHz ^{11}B NMR Spectra of Germanacarborane Derivatives

Compound[a]	Chemical shift[b] (relative intensity) [J_{B-H} (Hz)][c]
$1,2-B_{10}H_{10}CHGeCH_3$	$-5.9(1)[148], -10.9(1), -14.0(6), -14.9(2)$
$1,2-B_{10}H_{10}CHGeC_2H_5$	$-5.6(1)[142], -8.6(1), -13.7(6), -15.3(2)$
$N(CH_3)_4[1,2-B_{10}H_{10}CHGe]$	$-3.7(2), -6.7(2), -8.2(2), -12.7(4)$
$N(CH_3)_4[1,2-B_{10}H_{10}CHGe \cdot Cr(CO)_5]$	$-4.2(1)[142], -8.0(1), -10.0(4), -13.1(2),$ $-13.7(2)$
$N(CH_3)_4[1,2-B_{10}H_{10}CHGe \cdot Mo(CO)_5]$	$-4.2(1), -7.3(1), -9.7(4), -13.2(4)$
$N(CH_3)_4[1,2-B_{10}H_{10}CHGe \cdot W(CO)_5]$	$-4.5(1), -7.3(1), -10.2(4), -13.3(4)$
$(C_5H_5)Fe(CO)_2GeCHB_{10}H_{10}$[d]	$-3.6(1), -9.6(1), -11.0(4), -12.8(2), -14.1(2)$
$(C_7H_7)Mo(CO)_2GeCHB_{10}H_{10}$	$-4.1(1), -7.3(1), -10.0(4), -13.0(2), -13.9(2)$

[a] Acetone solvent.

[b] ppm versus $BF_3 \cdot O(C_2H_5)_2$.

[c] All signals are doublets with J_{B-H} 150 ± 15 Hz. Due to the overlap of signals, measurement of accurate values of some coupling constants is not possible.

[d] CH_2Cl_2 solvent.

ation of a tetrahydrofuran solution of $(CH_3)_4N[1,2-B_{10}H_{10}CHGe]$ and $Cr(CO)_6$ in a 1:1 mole ratio formed the σ-bonded complex $(CH_3)_4N[1,2-B_{10}H_{10}CHGe \cdot Cr(CO)_5]$ in moderate yield.[30] In a different synthetic method, the $1,2-B_{10}H_{10}CHGe^-$ ion was treated with $[(C_5H_5)Fe(CO)_2(\eta^2\text{-cyclohexene})]PF_6$ to produce the netural σ-bonded complex $(C_5H_5)Fe(CO)_2GeCHB_{10}H_{10}$.[52] In a similar manner the compound $(\eta^7\text{-}C_7H_7)Mo(CO)_2GeCHB_{10}H_{10}$ was formed. The σ-bonded complexes and $1,2-B_{10}H_{10}CHGeR$ have very similar ^{11}B nmr spectra. In the case of $1,2-B_{10}H_{10}CHGe^-$ the available lone pair electrons on germanium are more accessible for interaction with cage orbital electrons and this is reflected in the different ^{11}B nmr spectrum of this anion.

The anion $[(CH_3)_2C_2B_4H_4]_2FeH^-$ reacted with germanium(II) iodide or tin(II) chloride to form $MFe(CH_3)_4C_4B_8H_8$ (M = Ge or Sn).[32] The reaction of the ferracarborane anion with lead(II) bromide gave $(CH_3)_4C_4B_8H_8$ as the only isolated product. The ^{11}B nmr spectrum $(CDCl_3)$ of $GeFe(CH_3)_4C_4B_8H_8$ contains resonances at 13.8(166, 2B), 3.7(156, 4B), and −9.9 ppm (126 Hz, 2B) and the tin derivative spectrum $(CDCl_3)$ has resonances at 11.4(175, 2B), 2.3 (170, 4B), and −10.9 ppm (160 Hz, 2B). It is possible that the simple nature of

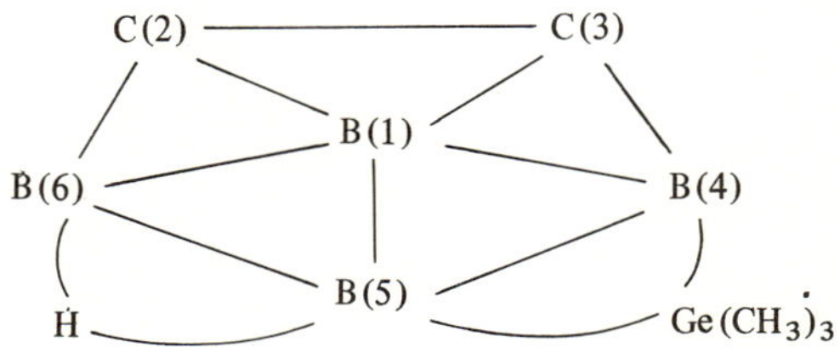

these spectra is due to their fluxional nature in solution. The tin species exhibits a single-methyl resonance in the proton nmr spectrum as low as $-66°C$. In many cluster systems having fewer than $2n + 2$ framework valence electrons in an n-vertex polyhedron, the geometry adopted is that of a capped polyhedron, as in the case of $(\eta^5\text{-}C_5H_5)CoFe(CH_3)_4C_4B_8H_8$, whose capped polyhedral structure was established by X-ray diffraction.[70] The proposed structure for the germanium derivative is shown in Figure 5.

When $Na[2,3\text{-}C_2B_4H_7]$ is reacted with organometallic reagents such as $(CH_3)_3GeCl$, the metal atom is inserted into a bridging position on the base of the pentgonal–pyramidal carborane cage as indicated below.[33] The ^{11}B nmr spectrum of the related derivative, $\mu,\mu'\text{-}SiH_2(C_2B_4H_7)_2$ consists of four doublets of equal area at 12.8(163), 1.6(167), -7.0(137), and -45.7 ppm (184 Hz). The -7.0 ppm doublet is unusually sharp and may be assigned to B(4) which would experience less $^{11}B-^{11}B$ and bridge hydrogen coupling than other boron atoms in this type of structure (see Table 6 for ^{11}B nmr data of other related derivatives.[33-37] The high-field signal is readily assigned to the apex boron atom. The trimethylgermyl derivative is stable at $25°C$, but at higher temperatures isomerizes irreversibly to a terminally bound species with substitution at B(4).

4.2. Borane Derivatives

When $NaB_{10}H_{13}$ is refluxed with $(CH_3)_3GeCl$ or $(CH_3)_3SnCl$ in diethyl ether, low yields of $B_{10}H_{12}Ge(CH_3)_2$ and $B_{10}H_{12}Sn(CH_3)_2$, respectively, are formed.[38] These metallaborane derivatives are thermally stable and exhibit reasonable stability in the air. The 32 MHz ^{11}B nmr spectrum (benzene solution) of $B_{10}H_{12}Ge(CH_3)_2$ contains resonances at 13.3(145, 2B), 5.5(160, 2B), -0.7(2B), -6.9(170, 2B), and -29.4 ppm (160 Hz, 2B) and the spectrum of $B_{10}H_{12}Sn(CH_3)_2$ contains resonances at 10.4, 2.4, -5.4(148), and -29.1 ppm

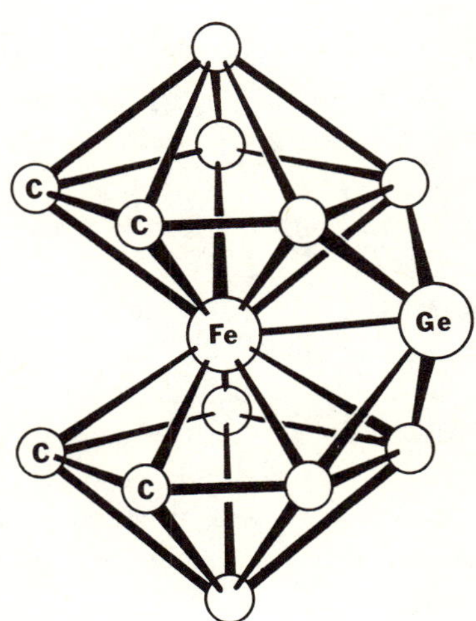

Figure 5. The proposed structure for $GeFe(CH_3)_4$-$C_4B_8H_8$.

Table 6. ^{11}B NMR Parameters of $(\mu\text{-MR}_n)C_2B_4H_7$ Derivatives

μ-MR$_n$ group	^{11}B NMR shifts [(J$_{B-H}$ (Hz))]			
	B(1)		B(4, 5, 6)	
SiH$_3$[a]	−49.5(179)	−8.2(138)	−0.5(170)	5.9(163)
Si(CH$_3$)$_3$[a]	−47.2(181)	−0.2	−0.2	8.7(148)
GeH$_3$[a]	−48.9(177)	−7.3(148)	−0.7(183)	6.2(177)
Ge(CH$_3$)$_3$[a]	−46.4(179)	−0.2	−0.3	8.4(155)
Sn(CH$_3$)$_3$[b]	−51.3(182)	−2.3(137)	2.1(158)	5.4(147)
Pb(CH$_3$)$_3$[b]	−51.0(183)	−0.9(138)	2.5(157)	5.3(148)

[a]Neat liquid.

[b]CS$_2$ solution.

(157 Hz). The structure proposed for these compounds is a *nido* eleven-atom icosahedral fragment with the $M(CH_3)_2$ unit in the open face of the molecule. In this structure the two methyl groups would lie on a mirror plane through the molecule and would be nonequivalent. The proton nmr spectrum of $B_{10}H_{12}Ge$ $(CH_3)_2$ contains resonances at 0.46 and 0.27 ppm assigned to the methyl groups and a high-field broad signal at −5.4 ppm assigned to the two bridge hydrogen atoms of the molecule. Treatment of $B_{10}H_{12}Sn(CH_3)_2$ with HCl generates $B_{10}H_{14}$ and $(CH_3)_2SnCl_2$ in greater than 95% yield. Base [(CH$_3$)$_4$NOH] removes the heteroatom from the tin derivative to give $B_{10}H_{13}{}^-$ and $B_{10}H_{10}{}^{2-}$

When $(AsPh_4)_2B_{10}H_{12}$ was reacted with tin(II) chloride both $B_{10}H_{12}{}^{2-}$ and pale yellow $[Cl_2SnB_{10}H_{12}]^{2-}$ were formed.[39] The tin-containing compound could be considered as a Sn(II) *arachno* structure or as a Sn(IV) *nido* compound. The ^{119}Sn Mössbauer spectrum at liquid nitrogen temperature showed that $[Cl_2SnB_{10}H_{12}]^{2-}$ had a chemical isomer shift δ + 3.17 mms^{-1} relative to BaSnO$_3$ which is typical of Sn(II). Reaction of $(AsPh_4)_2B_{10}H_{12}$ with $(CH_3)_2SnCl_2$ produced $B_{10}H_{10}{}^{2-}$ and $[(CH_3)_2SnCl_2(B_{10}H_{12})]^{2-}$,[39] whose ^{119}Sn Mössbauer spectrum suggested the presence of Sn(IV). It should be noted that the tetraphenylarsonium, but not the sodium salt of $B_{10}H_{12}{}^{2-}$, reacts with the Sn(II) and Sn(IV) reagents.

Low-temperature reaction (−30°C) of LiB$_5$H$_8$ and R$_3$MX (X = Cl, Br, or I; R = C$_2$H$_5$, CH$_3$, or H when M = Ge; R = CH$_3$ when M = Sn or Pb) formed μ-R$_3$MB$_5$H$_8$ in 60% or greater yield.[40,41] The 32 MHz ^{11}B nmr spectrum (see Table 7) of μ-(CH$_3$)$_3$GeB$_5$H$_8$ contains three doublets in a 2:2:1 area ratio. This spectrum is consistent with a B$_5$H$_9$ structure in which one bridge hydrogen is replaced by the (CH$_3$)$_3$Ge group. A single-crystal X-ray study of 1-Br-μ-(CH$_3$)$_3$SiB$_5$H$_7$ confirmed the proposed structure for these derivatives. When μ-(CH$_3$)$_3$GeB$_5$H$_8$ is allowed to stand in diethyl ether solution at room temperature for several hours, the molecule isomerizes to 2-(CH$_3$)$_3$GeB$_5$H$_8$ in which the R$_3$M

Table 7. ^{11}B NMR Data of μ-$R_3MB_5H_8$ Derivatives

μ-R_3M	Chemical shifts (ppm)			
	$B(2,3)$	$B(2,5)$	$B(4,5)$	$B(1)$
H_3Ge	–	–12.1	–	–51.2
$(CH_3)_3Ge$	–8.7	–	–12.9	–48.4
$(C_2H_5)_3Ge$	–	–11.6	–	–49.7
$(CH_3)_3Sn$	–12.0	–	–13.7	–50.6
$(CH_3)_3Pb$	–	–13.3	–	–52.4

group replaces a terminal hydrogen atom. Isomerization of μ-$(CH_3)_3SnB_5H_8$ did not occur under these conditions and μ-$(CH_3)_3PbB_5H_8$ was thermally unstable at room temperature.

5. COMPOUNDS CONTAINING ARSENIC OR ANTIMONY

5.1. Carborane Derivatives

The first cage molecules prepared in this group were $1,2$-$B_{10}H_{10}CHAs$[42] and $1,2$-$B_{10}H_{10}CHSb$.[42] These *closo*-icosahedral molecules were prepared by reaction of $Na_3B_{10}H_{10}CH\cdot(THF)_n$ with $AsCl_3$ (25% yield) and SbI_3 (41% yield), respectively. Thermal isomerization of the $1,2$–isomer at 495°C for 22 hr in a sealed tube formed $1,7$-$B_{10}H_{10}CHAs$[42] in 65% yield. The thermal stability of $1,2$-$B_{10}H_{10}CHSb$ is lower than the corresponding phosphorus and arsenic derivatives. Heating $1,2$-$B_{10}H_{10}CHSb$ at 450°C for 13 hr in a sealed tube gave mainly black decomposition products and a 20% yield of $1,7$-$B_{10}H_{10}CHSb$.[42] Heating the $1,2$–isomer to 575°C formed a 1:1 mixture of $1,7$-$B_{10}H_{10}CHAs$ and $1,12$-$B_{10}H_{10}CHAs$.[42] Similar thermal isomerization of $1,2$-$B_{10}H_{10}CHSb$ at 500°C probably formed a very small amount of $1,12$-$B_{10}H_{10}CHSb$ which was not isolated, but was detected by gas chromatography. Thermal isomerization of a mixture of $1,2$-$B_{10}H_{10}C_2H_2$ and $1,2$-$B_{10}H_{10}CHAs$ at 475°C indicated that $1,2$-$B_{10}H_{10}CHAs$ isomerizes to its $1,7$–isomer more easily than $1,2$-$B_{10}H_{10}C_2H_2$ does.

When $1,2$-$B_{10}H_{10}CHAs$ is reduced with sodium metal in liquid ammonia and the resulting anion treated with aqueous acid, the known 7-$B_{10}H_{12}CH^-$ ion was obtained as the major product.[47] Treatment of the reduced product of $1,7$-$B_{10}H_{10}CHAs$ with an aqueous solution led to boron abstraction and the formation of $7,9$-$B_9H_{10}CHAs^-$ as the major product. Reduction of $1,12$-$B_{10}H_{10}$-$CHAs$ and subsequent acid hydrolysis formed the new 1-$B_{10}H_{12}CH^-$ ion.[47] The ^{11}B nmr spectrum of the anion product contains two doublets of equal area. This suggests that the carbon atom is located on the opposite side of the *nido*-cage from the open face.

The boron-11 nmr data for the compounds described above are presented in Table 8. The boron spectra of the 1,2- and 1,7-isomers are consistent with a 1:1:2:2:2:2 pattern of doublets expected for these moelcules. In all the 1,2- and 1,7-derivatives, two doublets of area 1 are found at lowest field. These signals are due to the two unique boron atoms located on the mirror plane, on the opposite side of the molecule from the heteroatoms. It should be noted that the signals of the analogous boron atoms of 1,2- and $1,7-B_{10}H_{10}C_2H_2$ were also found at lowest field.[43] The proton nmr data of the CH groups in these derivatives are given in Table 9. The shielding values of the CH groups are strongly solvent dependent.

Reaction of 1,2- or $1,7-B_{10}H_{10}CHAs$ in neat piperidine at reflux formed 7,8- and $7,9-B_9H_{10}CHAs^-$, respectively, in good yields.[42] The 70.6 MHz ^{11}B nmr spectrum of the $7,8-B_9H_{10}CHAs^-$ ion is given in Figure 6. Seven of the nine expected doublets for this unsymmetrical ion are observed. Treatment of $1,2-B_{10}H_{10}CHSb$ with piperidine in dilute benzene solution at reflux formed $7,8-B_9H_{10}CHSb^-$ in good yield.[42] Use of neat piperidine in the case of the antimony derivative produced mainly the $B_{10}H_{12}CH^-$ ion. Treatment of $7,9-B_9H_{10}CHAs^-$ with methyl iodide produced $7,9-B_9H_{10}CHAsCH_3$ in 96% yield. Similar methylation of $7,8-B_9H_{10}CHAs^-$ gave a mixture of volatile products which were difficult to separate. Attempted methylation of $7,8-B_9H_{10}CHSb^-$ gave no observable reaction. The proton nmr data of these eleven-atom *nido* derivatives are given in Table 9.

Reaction of $7,9-B_9H_{10}CHAs^-$ with sodium hydride followed by reaction with anhydrous $FeCl_2$ gave red-violet $Fe(7,9-B_9H_9CHAs)_2^{2-}$.[44] Oxidation of this Fe(II) complex with $FeCl_3$ formed the green paramagnetic complex $Fe(7,9-B_9H_9CHAs)_2^-$, whose magnetic susceptibility as measured by the Faraday technique gave a value of 2.42 BM (Bohr magneton). Treatment of $Fe(7,9-B_9H_9-CHAs)_2^{2-}$ with excess methyl iodide formed red $[(7,9-B_9H_9CHAs)Fe(7,9-B_9H_9CHAsCH_3)]^-$ as the major product. No neutral dimethylated complex was obtained in this reaction. Reaction of $7,9-B_9H_{10}CHAsCH_3$, $FeCl_2$, C_5H_6, and triethylamine gave a low yield of orange-red sublimable $(C_5H_5)Fe(7,9-B_9H_9-CHAsCH_3)$. Compounds with other metals were also prepared using the $7,9-B_9H_9CHAs^{2-}$ ligand [i.e., yellow $(7,9-B_9H_9CHAs)Mn(CO)_3^-$ and yellow-orange

Table 8. ^{11}B NMR Data of $B_{10}H_{10}CHAs$, $B_{10}H_{10}CHSb$, and Related Derivatives

Compound (solvent)	Relative intensity	δ_B (ppm) [J_{B-H}(Hz)]
$1,2-B_{10}H_{10}CHAs(CH_3CN)$	1:1:2:2:2:2	10.3(150), −0.3, −2.7, −5.0, −8.9, −9.6
$1,12-B_{10}H_{10}CHAs$	1:1	−7.5(160), −8.2(175)
$1,2-B_{10}H_{10}CHSb$ (acetone)	1:1:4:2:2	6.2(141), −1.45(148), −5.8, −10.2(162), −11.1(155)

Table 9. Proton NMR Data of $B_{10}H_{10}CHAs$, $B_{10}H_{10}CHSb$, and Related Derivatives

Compound	Solvent	δ_H (ppm) (assignment)
$1,2-B_{10}H_{10}CHAs$	C_6H_6	2.08 (CH)
$1,7-B_{10}H_{10}CHAs$	C_6H_6	2.62 (CH)
$1,12-B_{10}H_{10}CHAs$	C_6H_6	3.24 (CH)
$1,2-B_{10}H_{10}CHSb$	C_6H_6	2.50 (CH)
$1,7-B_{10}H_{10}CHSb$	C_6H_6	3.12 (CH)
$(CH_3)_4N[7,8-B_9H_{10}CHAs]$	d_6-acetone	3.4 (Me_4N) 1.3 (CH)
$(CH_3)_4N[7,9-B_9H_{10}CHAs]$	d_6-acetone	3.4 (Me_4N) 1.8 (CH)
$7,9-B_9H_{10}CHAsCH_3$	d_6-acetone	2.5 (CH) 2.2 (CH_3)
$(7,9-B_9H_9CHAsCH_3)Fe(C_5H_5)$		4.66 (C_5H_5) 2.88 (CH) 2.55 (CH_3)
$(C_6H_5)_4As[Co(7,8-B_9H_9CHAs)_2]$		7.85 (tetraphenylarsonium ion) 2.2 (CH)

$Co(7,9-B_9H_9CHAs)_2^-]$. Using the synthetic procedures described above $Co(7,9-B_9H_9CHAs)_2^-$ and $(C_5H_5)Co(7,8-B_9H_9CHAs)$ were also prepared. A single-crystal X-ray study of the neutral cobalt complex has been completed[45] (see Figure 7) and showed a distorted icosahedral structure for this molecule.

Attempts to methylate or form metal complexes with $7,8-B_9H_{10}CHSb^-$ were not successful.

A tetrahydrofuran solution of $(CH_3)_4N[7,8-B_9H_{10}CHAs]$ and $Mo(CO)_6$ in a 1:1 mole ratio was irradiated with a high-pressure mercury-vapor lamp to form the sigma-bonded $[7,8-B_9H_{10}CHAs \rightarrow Mo(CO)_5]^-$ ion.[46] Similar complexes involving the $7,9-B_9H_{10}CHAs^-$ ion and chromium, molybdenum or tungsten

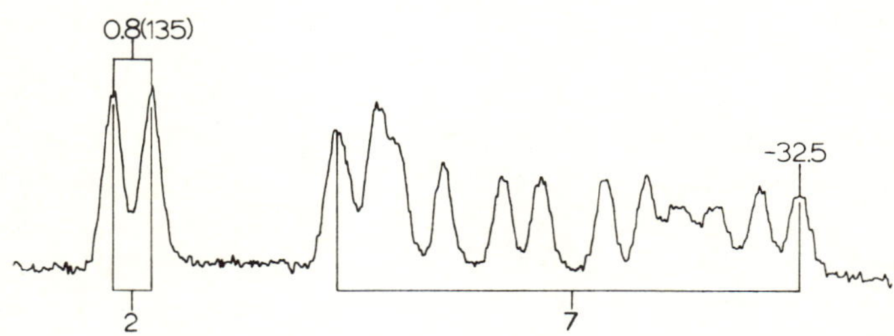

Figure 6. The 70.6 MHz ^{11}B nmr spectrum of $Cs[7,8-B_9H_{10}CHAs]$ in acetone solution referenced to $BF_3 \cdot (C_2H_5)_2O$.

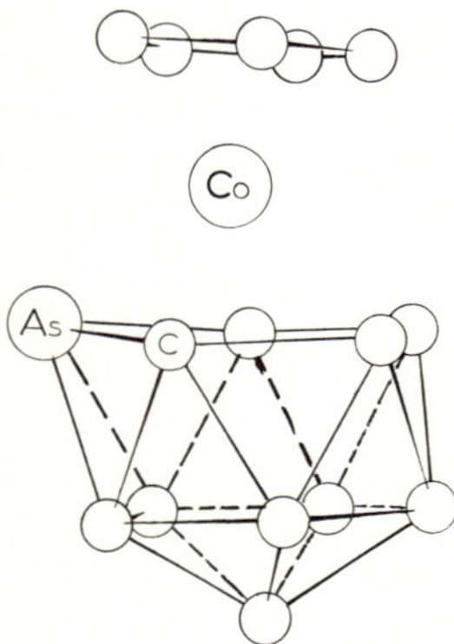

Figure 7. The structure of $(C_5H_5)Co(7,8-B_9H_9CHAs)$.

carbonyls were also prepared. The complexes are somewhat air sensitive in solution, but can be stored in the solid state in air at room temperature for 3 weeks without noticeable decomposition. Sigma coordination through the arsenic atom would not be expected to seriously change the electronic environment of the boron cage. Indeed the boron nmr spectra of the free ligand and the complexed ligand are very similar. An attempt to form a sigma complex with $7,8-B_9H_{10}CHSb^-$ was not successful.

The photochemical reaction of $(7,9-B_9H_{10}CHAs)_2Fe^{2-}$ with $Cr(CO)_6$ in tetrahydrofuran led to the formation of $[7,9-B_9H_{10}CHAs \cdot Cr(CO)_5]_2Fe^{2-}$. It is proposed that each heteroatom carborane ligand is π-bonded to the iron atom and σ-bonded through the arsenic atom to a $Cr(CO)_5$ unit.[48]

Other *nido* twelve-atom cage arsacarborane derivatives have been synthesized more recently. Reaction of $B_{10}H_{12}CNMe_3$ with sodium hydride in THF and then $PhAsCl_2$ formed *nido*-$Me_3NCB_{10}H_{10}AsPh$ in low yield.[55] The ^{11}B nmr spectrum (acetone) contained resonances at 2.8, 0.5, $-3.1(140)$, $-8.4(155)$, $-12.2(140)$, and -18.9 ppm ($J_{11_{B-H}} = 141$ Hz) in a 1:2:1:2:2:2 area ratio. A single-crystal X-ray structure determination of the corresponding phosphorus derivative, $Me_3NCB_{10}H_{10}PPh$, indicates that the "PPh" unit bridges two boron atoms in the open face of the $B_{10}H_{10}CNMe_3$ eleven-atom icosahedral fragment.[56] A compound of the formula $PhAsB_9C_2H_{11}$ has been prepared from $Na_2[7,8-B_9C_2H_{11}]$ and $PhAsCl_2$ in THF.[62] The ^{11}B nmr spectrum (see Table 10) consists of six doublets in a 1:2:1:2:2:1 area ratio. A study of the low-temperature ^1H nmr spectrum of this material indicates that the carborane CH protons become nonequivalent at $-40°C$ suggesting that the molecule has a fluxional, *nido* structure, similar to that of $(CH_3)_2AlB_9C_2H_{11}$. Another method for synthesis of $RAsB_9C_2H_{11}$ derivatives (R = Ph, CH_3, $n-C_4H_9$) involves reaction of $Tl_2[7,8-B_9C_2H_{11}]$ with $RAsCl_2$.[57] Treatment of $PhAsB_9C_2H_{11}$ with bromine caused cleavage of the phenyl-arsenic bond and formation of $BrAsB_9C_2H_{11}$. These arsacarboranes reacted rapidly with ethanolic potassium hydroxide producing the potassium salt of $7,8-B_9C_2H_{12}^-$ in good yield. When

Table 10. ^{11}B NMR Data of $RAsB_9C_2H_{11}$ and Related Derivatives

Compound	Solvent	δ_B (ppm) [(J_{B-H} (Hz))]
$CH_3AsC_2B_9H_{11}$	CCl_4	8.6(1), 3.2(2), −0.4(1), −4.8(2), −14.5(2), −17.0(1)
$CH_3AsC_2B_9H_9(CH_3)_2$	CCl_4	4.9(3), −3.7(1), −11.9(5)
$C_6H_5AsC_2B_9H_{11}$	CCl_4	5.7(1), 3.7(2), 0.0(1), −3.6(2), −14.1(2), −17.1(1)
n−$C_4H_9AsC_2B_9H_{11}$	CCl_4	6.6(1), 2.6(2), −1.1(1), −5.6(2), −16.2(2), −18.3(1)
$BrAsC_2B_9H_{11}$	CCl_4	8.8(1), 2.1(2), 1.9(1), 0.0(2), −12.4(2), −14.2(1)
$[(CH_3)_2As]_2C_2B_9H_{11}$	C_6H_6	−3.6(1), −10.0(1), −12.0s(1), −15.4(1), −16.8(1), −24.0(2), −27.9(1), −35.7(1)
$[(CH_3)_2As]_2C_2B_9H_9(CH_3)_2$	C_6H_6	−5.8(2), −9.0br, sh(1), −9.3s(1), −14.2(1), −18.1(1), −24.3(1), −28.5(1), −33.7(1)
$(CH_3)_2AsC_2B_9H_{11}(OC_2H_5)$	$(CH_3)_2CO/C_2H_5OH$	−9.0s(1), −16.0(1), −20.6(1), −23.9(2), −30.0br(2), −39.0(1), −44.0(1)

bromodimethylarsine was added to $Li[7,8-B_9C_2H_{12}]$ in hot benzene solution both $7,8-B_9C_2H_{13}$ and $(Me_2As)_2B_9C_2H_{11}$ were formed.[57] The ^{11}B nmr spectrum of this compound contains a singlet at −12 ppm (see Table 10) and a doublet at −28 ppm broadened by coupling with a bridge proton. This compound probably has one bridging Me_2As group and one Me_2As group single-bonded to a boron atom of the cage. The molecule is considered to be a disubstituted $B_9C_2H_{13}$ derivative. An acetone solution of $(Me_2As)_2B_9C_2H_{11}$ was found to react with ethanolic potassium hydroxide to form $(Me_2As)B_9C_2H_{11}$ (OC_2H_5). Available data are consistent with substitution of the terminally bound Me_2As group by ethoxide.

Treatment of $B_7C_2H_{13}$ with arsenic triiodide and triethylamine formed white $B_7C_2H_9As_2$ in moderate yield.[58] The ^{11}B nmr spectrum (hexane) contained resonances at −0.8(144), −1.0, −4.1, −14.3(164), and −32.2 ppm $(J_{B-H} = 146$ Hz) with 1:2:2:1:1 relative areas, respectively. The proton nmr spectrum contained one broadened singlet at 2.4 ppm which is assigned to the apparently equivalent CH groups. The molecule has $n + 2$ pairs of valence electrons suggesting that it has a *nido* eleven-atom structure (see Figure 8). This new type of molecule is isoelectronic with $B_7C_4H_{11}$, a member of the rapidly expanding tetracarbon carborane series. The $B_7C_4H_{13}^-$ system is known as a tetramethyl derivative, $B_7C_4(CH_3)_4H_9^{71}$, and several iron and cobalt complexes containing a C_4B_7 ligand system have been prepared.[32,59,61] Photolysis of $(\eta^5-C_5H_5)Co(CO)_2$ and $B_7C_2H_9As_2$ in THF solution formed $(\eta^5-C_5H_5)$

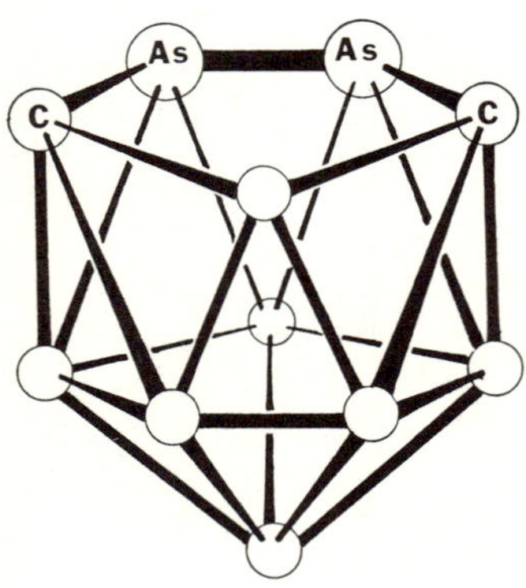

Figure 8. Proposed structure of $B_7C_2H_9As_2$.

$CoB_7C_2H_9As_2$ in low yield. The appearance of two distinct CH signals in the proton nmr spectrum and six resonances in the ^{11}B nmr spectrum of this metal derivative suggests that this twelve-atom *nido* structure is very unsymmetrical.

5.2. Arsenaboranes

Reaction of decaborane-14 with arsenic trichloride and sodium hydride or sodium borohydride formed $7\text{-}B_{10}H_{12}As^-$ in moderate yield.[49] The ^{11}B nmr spectrum of this ion (see Table 11) has a 1:2:2:2:1:2 pattern of doublets which is consistent with a *nido* eleven-atom icosahedral structure. In fact most compounds of the type $B_{10}H_{12}E$ (E = BH^{2-}, CH^-, $CNMe_3$, PR, AsR, As^-, S, Se, and Te) have very similar spectra. An assignment of the resonances in these molecules has been obtained by the use of labeled derivatives, proton decoupling, and spin-lattice relaxation studies.[50]

Treatment of $7\text{-}B_{10}H_{12}As^-$ with triethylamine borane in diglyme at $160°C$ formed the *closo*-icosahedron $B_{11}H_{11}As^-$ in 52% yield.[49] Heating of solid $CsB_{10}H_{12}As$ at $375°C$ also formed $B_{11}H_{11}As^-$. No $B_9H_9As^-$ ion was detected in the pyrolysis products. Methyl iodide in THF reacted with $(CH_3)_4N(7\text{-}B_{10}H_{12}As)$ to form sublimable $7\text{-}B_{10}H_{12}AsCH_3$ in 76% yield. The proton nmr spectrum (see Table 12) contains a sharp singlet at 2.1 ppm assigned to the methyl group bonded to the arsenic atom. Formation of $7\text{-}B_{10}H_{12}AsC_6H_5$ could not be obtained by the procedure used for the methyl derivative. Treatment of decaborane with triethylamine (excess) and $C_6H_5AsCl_2$ formed $7\text{-}B_{10}H_{12}C_6H_5$ in low yield. Both of these neutral arsaboranes are easily deprotonated in aqueous ammonia to form the $B_{10}H_{11}AsR^-$ (R = CH_3 or C_6H_5) ions. More vigorous treatment of $7\text{-}B_{10}H_{12}AsCH_3$ with sodium in liquid ammonia led to demethylation and formation of $7\text{-}B_{10}H_{12}As^-$ in good yield.

Reaction of $7\text{-}B_{10}H_{12}As^-$ with cobalt or nickel chloride in a 33% by weight aqueous sodium hydroxide solution formed $Co(B_{10}H_{10}As)_2^{3-}$ and

Table 11. 70.6 MHz ^{11}B NMR Data of Arsenaborane Derivatives

Compound (solvent)	Relative intensity	δ_B (ppm) [J_{B-H} (Hz)]
$(CH_3)_4N[7-B_{10}H_{12}As]$ (acetone)	1:2:2:2:1:2	7.9[140], −6.8[145], −8.5, −14.3[140], −21.4, −23.5[145]
$7-B_{10}H_{12}AsCH_3(CHCl_3)$	1:2:2:2:1:2	4.6[145], −2.3[170], −4.7[165], −14.3[155], −17.4[160], −23.4[150]
$7-B_{10}H_{12}AsPh(CHCl_3)$	1:2:2:2:1:2	4.4[140], −1.7[170], −5.6[165], −14.2[150], −17.7[155], −23.3[150]
$(CH_3)_4N[7-B_{10}H_{11}AsCH_3]$ (acetone)	2:2:2:2:1:1	−10.6, −11.7, −17.4[145], −22.8[130], −24.8[140], −37.1[140]
$(CH_3)_4N[7-B_{10}H_{11}AsPh]$ (acetone)	2:2:2:2:1:1	−10.6[145], −13.0, −17.2[150], −22.1, −24.5, −36.8[140]
$(C_5H_5)Fe(CO)_2AsB_{10}H_{12}$ (acetone)	1:4:2:3	4.8, −6.5, −14.6, −22.6
$1,2-B_{10}H_{10}As_2$ (DMSO)	2:2:4:2	15.5[139], 3.8, 0.9, −0.8
$(CH_3)_4N[7,8-B_9H_{10}As_2]$ (acetone)	2:2:1:2:1:1	1.4[139], −0.5[140], −6.9, −9.8[161], −16.2[140], −26.4[154]
$7,8-B_9H_{10}As_2CH_3$ (THF)	1:1:2:2:2:1	8.1[156], 5.0, −7.3, −16.8, −19.3, −30.2[159]
$(C_5H_5)Fe(CO)_2As_2B_9H_{10}(CH_2Cl_2)$	1:1:1:1:1:3:1	10.0, 8.1, −4.4, −6.1, −8.0, −16.4, −28.7
$(CH_3)_4N[B_{11}H_{11}As]$ (acetone)	1:5:5	8.0[135], −7.6[125], −8.6[140]
$Na[Co(B_{10}H_{10}As)_2]$ (water)	1:2:2:1:2:2	6.2[130], 4.3[130], 1.7, 0.0, −11.9[125], −13.4
$(CH_3)_4N[Ni(B_{10}H_{10}As)_2]$ (DMF)	1:4:1:2:2	20.9[130], 12.5, 8.3[145], −4.3[130], −6.1
$(CH_3)_4N[(C_5H_5)Co(B_{10}H_{10}As)]$ (acetone)	2:1:1:2:2:2	8.0[125], 6.7, 5.0, 3.2[140], −7.3[135], −12.7[145]
$(CH_3)_4N[Co(7,8-B_9H_9As_2)_2$ (acetone)	1:1:2:2:1:2	20.6[145], 16.1, 13,3, 11.5, −1.7, −2.6[145]
$(C_5H_5)Co(7,8-B_9H_9As_2)$ (acetone)	1:5:2:1	17.1[145], 12.3, −4.7[155], −8.8[165]

Table 12. Proton NMR Data of Arsenaboranes

Compound	Solvent	δ_H (ppm) (assignment)
$7\text{-}B_{10}H_{12}AsCH_3$	$CDCl_3$	2.1 (CH_3)
$7\text{-}B_{10}H_{12}AsC_6H_5$	$CDCl_3$	7.6, multiplet (C_6H_5)
$(CH_3)_4N[7\text{-}B_{10}H_{11}AsCH_3]$	d_6-acetone	3.38 (Me_4N) 2.01 (CH_3)
$(CH_3)_4N[7\text{-}B_{10}H_{11}AsC_6H_5]$	d_6-acetone	3.34 (Me_4N) 7.3; multiplet (C_6H_5)
$(CH_3)_4N[(C_5H_5)Co(B_{10}H_{10}As)]$	d_6-acetone	3.4 (Me_4N) 4.98 (C_5H_5)
$(C_5H_5)Co(7,8\text{-}B_9H_9As_2)$	d_6-acetone	5.83 (C_5H_5)
$(C_5H_5)Fe(CO)_2AsB_{10}H_{12}$	d_6-acetone	5.54 (C_5H_5) −4.2 (broad, bridge hydrogens)
$(C_5H_5)Fe(CO)_2As_2B_9H_{10}$	d_6-acetone	5.56 (C_5H_5) −4.2 (broad, bridge hydrogen)

$Ni(B_{10}H_{10}As)_2{}^{2-}$, respectively, in good yield.[51] The complex $(CH_3)_4N$ $[(C_5H_5)Co(B_{10}H_{10}As)]$ was prepared in moderate yield by reaction of $B_{10}H_{12}AsCH_3$, C_5H_6, and $CoCl_2$ in ethanolic KOH solution. Attempted methylation of the arsenic atom to form a neutral complex was not successful.

Reaction of $7\text{-}B_{10}H_{10}As^-$ with triethylamine and $AsCl_3$ in THF formed $1,2\text{-}B_{10}H_{10}As_2$ in 30% yield.[49] Reaction of $B_{10}H_{14}$ with As_2O_3 and aqueous sodium hydroxide produced a mixture of $7\text{-}B_{10}H_{12}As^-$ and $1,2\text{-}B_{10}H_{10}As_2$.[53] This molecule is isoelectronic and closely related in structure to $1,2\text{-}B_{10}H_{10}C_2H_2$ (*ortho*-carborane). The compound is surprisingly polar and is sparingly soluble in boiling benzene. The ^{11}B nmr spectrum was obtained in DMF solution (see Table 11). Thermal isomerization probably occurs at 575°C although the products were not completely identified.[49] Piperidine removes a boron atom from $1,2\text{-}B_{10}H_{10}As_2$ to form $7,8\text{-}B_9H_{10}As_2{}^-$ in good yield. The ^{11}B nmr spectrum of this ion is very similar to that of $7,8\text{-}B_9C_2H_{12}{}^-$. The doublet of area one at −16.2 ppm shows secondary splitting attributed to the bridging hydrogen. Reaction of $7,8\text{-}B_9H_{10}As_2{}^-$ with methyl iodide in THF at room temperature forms $7,8\text{-}B_9H_{10}As_2CH_3$.[54]

Formation of π-bonded metal complexes with this ligand are not as facile as with the carborane analog. To date only cobalt complexes have been fully characterized. Reaction of the piperidinium salt of $7,8\text{-}B_9H_{10}As_2{}^-$ with $CoCl_2$ in strong aqueous sodium hydroxide solution formed red $Co(7,8\text{-}B_9H_9As_2)_2{}^-$ in 46% yield.[51] In a similar reaction, in the presence of cyclopentadiene monomer the neutral complex $(C_5H_5)Co(7,8\text{-}B_9H_9As_2)$ was formed.

These arsaborane anions do function as monodentate ligands in transition metal complexes. The anions $7\text{-}B_{10}H_{12}As^-$ and $7,8\text{-}B_9H_{10}As_2{}^-$ were found to react readily with $[(\eta^5\text{-}C_5H_5)Fe(CO)_2(\text{cyclohexene})]PF_6$ to afford the yellow derivatives $(\eta^5\text{-}C_5H_5)Fe(CO)_2AsB_{10}H_{12}$ and $(\eta^5\text{-}C_5H_5)Fe(CO)_2As_2B_9H_{10}$.[52] It is proposed that iron–arsenic single bonds are present in these two complexes.

5.3. Stibaboranes

Two antimony atoms were inserted into a ten-boron cage by reaction of $B_{10}H_{14}$ with $SbCl_3$, triethylamine, and zinc dust.[54] The ^{11}B nmr spectrum (see Table 13) is similar to other icosahedral molecules such as $1,2\text{-}B_{10}H_{11}C_2H_2$. When decaborane was added to a slurry of excess NaH in ether followed by addition of $SbCl_3$, the unstable $B_{10}H_{12}Sb^-$ ion appeared to be formed (characterized by IR). This molecule has not been fully characterized.

Starting with $7\text{-}B_{10}H_{12}As^-$ it was possible to insert one antimony atom (using $SbCl_3$ and triethylamine) to form $1,2\text{-}B_{10}H_{10}AsSb$. Reaction of $B_{11}H_{14}^-$ with triethylamine and $SbCl_3$ produced the icosahedral ion, $B_{11}H_{11}Sb^-$, in low yield.

Reaction of $1,2\text{-}B_{10}H_{10}Sb_2$ or $1,2\text{-}B_{10}H_{10}AsSb$ with neat piperidine at $70°C$ formed the $7,8\text{-}B_9H_{10}Sb_2^-$ and $7,8\text{-}B_9H_{10}AsSb^-$ ions, respectively. On attempted purification the salts of these ions slowly decompose; however, the eleven-atom *nido* ions can be stabilized by the formation of cobalt complexes. Treatment of $7,8\text{-}B_9H_{10}Sb_2^-$ or $7,8\text{-}B_9H_{10}AsSb^-$ with cyclopentadiene monomer and $CoCl_2$ in piperidine formed $(C_5H_5)Co(7,8\text{-}B_9H_9Sb_2)$, mp $296\text{-}298°C$, and $(C_5H_5)Co(7,8\text{-}B_9H_9AsSb)$, mp $270\text{-}272°C$ respectively, in low yield.[54]

5.4 Arsenathia- and Arsenaselenaboranes

The first compound of this type was formed by the reaction of $K(B_9H_{12}S)$ with excess arsenic trichloride and triethylamine which formed *nido-* $B_8H_8As_2S$.[62] More recently it was discovered that this diarsenathiaborane could be formed in slightly better yield by the reaction of $K(B_9H_{12}S)$ with As_2O_3 in basic aqueous solution.[58] One doublet of area 2 at 0.3 ppm in the ^{11}B nmr

Table 13. 70.6 MHz ^{11}B NMR Data of Stibaborane Derivatives

Compound (solvent)	Relative intensity	δ_B (ppm) [J_{B-H} (Hz)]
$1,2\text{-}B_{10}H_{10}Sb_2$ (DMF)	2:2:4:2	13.9[140], 3.9[155], −1.7[155], −4.6
$1,2\text{-}B_{10}H_{10}AsSb$ (DMF)	1:1:2:2:2:2	14.9(140), 12.3, 3.0[160], −0.1[140], −2.0, −3.5
$(CH_3)_4N[B_{11}H_{11}Sb]$ (DMF)	1:5:5	9.4[135], −6.9, −9.3
$(C_5H_5)Co(B_9H_9AsSb)$ (acetone)	1:3:2:2:1	15.8, 13.7[150], 11.2, −5.7, −7.3
$(C_5H_5)Co(B_9H_9Sb_2)$ (acetone)	2:3:1:1:2	15.7[140], 13.8, 9.5[145], −4.0[160], −7.9[145]

Table 14. ^{11}B NMR Data for the Heteroatom Boranes
Containing Two Arsenic Atoms

Compound (solvent)	Chemical shift (ppm) $[(J_{B-H}$ (Hz)] (relative area)
$B_8H_8As_2S$ (CH_3CN)	10.3(161)[2], 8.0(156)[1], 1.8(140)[2], 0.3(156)[2], −30.0(171)[1]
$B_8H_8As_2Se$ (heptane)	10.6[2], 10.0[1], 1.8(156)[2], 0.7(158)[2], −30.2(161)[1]
$B_7H_7As_2SCo(C_5H_5)$ (CH_2Cl_2)	14.7(146)[1], 7.0(156)[2], 1.5[1], 0.16[1], −9.9(156)[1], −23.3(146)[1]
$B_7C_2H_9As_2$ (hexane)	−0.8(144)[1], −1.0[2], −4.1[2], −14.3(164)[1], −32.2(146)[1]
$B_7C_2H_9As_2Co(C_5H_5)$ (CH_2Cl_2)	14.6(142)[1], 5.6(156)[1], ~2[1], 1.5[2], 0.0[1], −18.9(132)[1]

spectrum (see Table 14) is unusually sharp. This signal is probably due to the two equivalent boron atoms on the open face of the molecule (see Figure 9 for the proposed structure). These two boron atoms would experience less $^{11}B-^{11}B$ coupling than other boron atoms in the molecule. When $K(B_{10}H_{11}Se)$ was reacted with As_2O_3 in basic solution a low yield of $B_8H_8As_2Se$ was formed.[58] Reaction of $B_8H_8As_2S$ with KOH in methanol and then complexation of the base degradation product by treatment of the mixture with triethylamine/cyclopentadiene monomer/$CoCl_2$ formed red $B_7H_7As_2SCo(C_5H_5)$.[58] This complex is another of a rapidly expanding class of neutral *nido* eleven-atom heteroatom boranes, and suggests that $B_7H_9As_2S$ might be isolated with carefully chosen experimental conditions.

6. COMPOUNDS CONTAINING TELLURIUM

The reaction of decaborane with Na_2Te_x in aqueous ammonia gave $NaB_{10}H_{11}Te$. The aqueous solution was acidified with HCl and the mixture extracted with diethyl ether to obtain $B_{10}H_{12}Te$ in good yield.[63] A pure sample of $(CH_3)_4N[B_{10}H_{11}Te]$ was obtained by treating $B_{10}H_{12}Te$ with aqueous potassium hydroxide and then saturated tetramethylammonium iodide solution.

The *closo*-icosahedral molecule $B_{11}H_{11}Te$ has been prepared by two different reactions. Treatment of $B_{10}H_{12}Te$ with trimethylamine borane in refluxing xylenes formed $B_{11}H_{11}Te$ in 46% yield.[63] In this reaction the salt $[(Me_3N)_2BH_2]B_{10}H_{11}Te$ was also formed in substantial yield. An attempt to thermally force this salt to convert to $B_{11}H_{11}Te$ was not successful. The second method for preparing $B_{11}H_{11}Te$ involved the reaction of $B_{11}H_{14}^-$ with TeO_2 in aqueous solution with the product being extracted into a heptane layer (25% yield).[64]

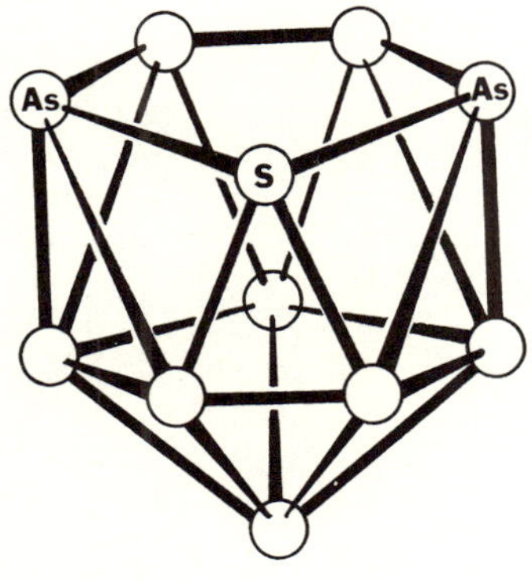

Figure 9. Proposed structure of $B_8 H_8 As_2 S$.

The ^{11}B nmr spectrum of $B_{11}H_{11}Te$ (see Table 15) has the 1:5:5 pattern of resonances expected for this type of *closo* structure.

Treatment of $B_{10}H_{12}Te$ with aqueous potassium hydroxide and $FeCl_2 \cdot 6H_2O$ formed green $[N(CH_3)_4][Fe(B_{10}H_{10}Te)_2]$ in low yield.[63] Prolonged air oxidation of a solution of this complex converted it to the paramagnetic yellow Fe(III) derivative, $N(CH_3)_4[Fe(B_{10}H_{10}Te)_2]$. With aqueous base, cobalt(II) chloride and $B_{10}H_{12}Te$ form orange $N(CH_3)_4[Co(B_{10}H_{10}Te)_2]$. Under anhydrous conditions, cyclopentadiene monomer, $B_{10}H_{12}Te$, cobalt(II) chloride, and triethylamine reacted to form yellow $(C_5H_5)Co(B_{10}H_{10}Te)$ in moderate yield. The 32 MHz ^{11}B nmr spectrum of $Co(B_{10}H_{10}S)_2^-$ had been reported previously.[66] Although not as well resolved, this spectrum is consistent with that obtained at 70.6 MHz for the tellurium analog.

When $B_9H_{13}(SMe_2)$ reacted with an aqueous solution of $K_2(Te)_n$ the ion $B_9H_{12}Te^-$ was formed in 70% yield.[65] This compound was isolated and characterized as the tetramethylammonium salt, but had limited stability even in the solid state. The ^{11}B nmr spectrum of this anion (see Table 15) is very similar to that of the well-known $B_9H_{12}S^-$ ion.[67] Treatment of a slurry of $N(CH_3)_4$ $B_9H_{12}Te$ in benzene with I_2 formed $B_9H_{11}Te$ as indicated by its ^{11}B nmr spectrum which was very similar to that of $B_9H_{11}S$. The neutral telluraborane was very unstable and the white solid turned black within hours. A much more stable derivative was obtained by reaction of freshly formed $B_9H_{11}Te$ with acetonitrile which produced the adduct $B_9H_{11}Te \cdot NCCH_3$. The proton nmr spectrum contained two solvent dependent singlet resonances in the spectral region where one methyl group signal was expected. A similar finding was observed previously for $B_9H_{11}S \cdot NCCH_3$[66] and this observation is still not understood.

Reaction of $K[B_9H_{12}S]$ with TeO_2 in water solution formed a low yield of *nido*-B_9H_9STe which was extracted into a hexane layer placed over the aqueous medium.[68] The ^{11}B nmr spectrum of the thiatelluraborane was very

Table 15. 70.6 MHz ^{11}B NMR Data of Telluraborane Derivatives

Compound (solvent)	Relative intensity	δ_B (ppm) [J_{B-H} (Hz)]
$B_{11}H_{11}Te(CHCl_3)$	1:5:5	25.9, −4.4, ~ −5
$B_{10}H_{12}Te(CHCl_3)$	1:4:2:1:2	21.4(149), −3.2(176), ~ −5.6, −14.0(151), −24.8(156)
$[N(CH_3)_4]_2[Fe(B_{10}H_{10}Te)_2]$ (acetonitrile)	1:2:1:2:2:2	12.4(134), 5.7(120), ~3.4, ~ −1.5, −12.9(137), −19.5(142)
$N(CH_3)_4[Co(B_{10}H_{10}Te)_2]$ (acetone)	1:1:2:2:2:2	18.9(144), 15.2(143), 10.4(135), 5.9(157), −6.1(144), −10.9(152)
$(C_5H_5)Co(B_{10}H_{10}Te)$ (acetone)	1:2:1:2:2:2	19.9(140), 13.7(144), 9.1(145), 4.2(160), −6.2(145), −13.6(162)
$N(CH_3)_4[B_{10}H_{11}Te]$ (acetonitrile)	2:3:2:2:1	−5.4(139), −12.4(139), ~ −16.9, ~ −18.3, −33.7(147)
$N(CH_3)_4B_9H_{12}Te$ (acetonitrile)	1:4:2:2	7.3(137), ~ −10.8, −25.3(137), −35.5(139)
$B_9H_{11}Te(C_6H_6)$	2:1:2:2:1:1	31.3(170), 15.7(158), 7.5(146), −7.4(144), −21.4(149), −31.0(177)
$B_9H_{11}Te \cdot CH_3CN(CH_2Cl_2)$	1:1:3:2:2	8.7(137), −6.9(159), ~ −14.9, −23.3(129), −36.2(149)

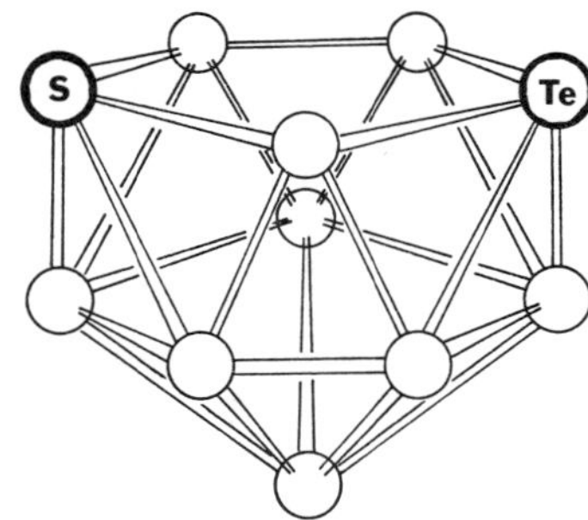

Figure 10. Proposed structure for B_9H_9STe.

similar to the spectra of B_9H_9SSe, $B_9H_9Se_2$, and $B_9H_9S_2$.[69] The *nido* structure given in Figure 10 has been proposed for this molecule.

REFERENCES

1. G. Popp and M. F. Hawthorne, *Inorg. Chem.*, **10**, 391 (1971).
2. G. Popp and M. F. Hawthorne, *J. Am. Chem. Soc.*, **90**, 6553 (1968).
3. D. F. Gaines and J. L. Walsh, *Inorg. Chem.*, **17**, 1238 (1978).
4. D. F. Gaines and J. L. Walsh, *Chem. Commun.*, 482 (1976).
5. D. F. Gaines, J. L. Walsh, and J. C. Calabrese, *Inorg. Chem.*, **17**, 1242 (1978).
6. D. F. Gaines, K. M. Coleson, and J. C. Calabrese, *J. Am. Chem. Soc.*, **101**, 3979 (1979).
7. D. L. Denton, W. R. Clayton, M. Mangion, S. G. Shore, and E. A. Meyers, *Inorg. Chem.*, **15**, 541 (1976).
8. B. Siegel, J. L. Mack, J. U. Lowe, and J. Gallaghan, *J. Am. Chem. Soc.*, **80**, 4523 (1958).
9. I. Dunstan, N. J. Blay, and R. L. Williams, *J. Chem. Soc.*, 5016 (1960).
10. N. N. Greenwood, and N. F. Travers, *J. Chem. Soc. A*, 15 (1968).
11. R. N. Grimes, and W. J. Rademaker, *J. Am. Chem. Soc.*, **91**, 6498 (1969).
12. R. N. Grimes, W. J. Rademaker, M. L. Denniston, R. F. Bryan, and P. T. Greene, *J. Am. Chem. Soc.*, **94**, 1865 (1972).
13. D. A. T. Young, R. J. Wiersema, and M. F. Hawthorne, *J. Am. Chem. Soc.*, **93**, 5687 (1971).
14. M. R. Churchill, A. H. Reis, D. A. T. Young, G. R. Willey, and M. F. Hawthorne, *Chem. Commun.*, 298 (1971).
15. D. A. T. Young, G. R. Willey, M. F. Hawthorne, A. H. Reis, and M. R. Churchill, *J. Am. Chem. Soc.*, **92**, 6663 (1970).
16. N. N. Greenwood, and J. A. McGinnety, *J. Chem. Soc., A*, 1090 (1966).
17. N. N. Greenwood, B. S. Thomas, and D. W. Waite, *J. Chem. Soc. Dalton*, 299 (1975).
18. N. N. Greenwood, and J. A. Howard, *J. Chem. Soc. Dalton*, **177** (1976).
19. B. M. Mikhailov, and T. V. Potapova, *Izvest. Akad. Nauk SSSR Ser. Khim.*, **5**, 1153 (1968).
20. G. E. Coates, M. L. H. Green, and K. Wade, *Organometallic Compounds*, Vol. 1, 3rd edn., p. 305, Methuen, London (1967).
21. J. L. Spencer, M. Green, and F. G. A. Stone, *Chem. Commun.*, 1178 (1972).
22. H. M. Colquhoun, T. J. Greenhough, and M. G. H. Wallbridge, *Chem. Commun.*, 737 (1977).
23. J. Smith, G. Allender, and H. D. Smith, *Inorg. Chem.*, **16**, 1814 (1977).

24. R. L. Voorhees, and R. W. Rudolph, *J. Am. Chem. Soc.*, **91**, 2173 (1969).
25. R. W. Rudolph, R. L. Voorhees, and R. E. Cochoy, *J. Am. Chem. Soc.*, **92**, 3351 (1970).
26. V. Chowdhry, W. R. Pretzer, D. N. Rai, and R. W. Rudolph, *J. Am. Chem. Soc.*, **95**, 4560 (1973).
27. D. C. Beer, and L. J. Todd, *J. Organometal. Chem.*, **50**, 93 (1973).
28. K-S. Wong, and R. N. Grimes, *Inorg. Chem.*, **16**, 2053 (1977).
29. R. W. Rudolph, and V. Chowdhry, *Inorg. Chem.*, **13**, 248 (1974).
30. G. S. Wikholm, and L. J. Todd, *J. Organometal. Chem.*, **71**, 219 (1974).
31. D. E. Hyatt, F. R. Scholer, L. J. Todd, and J. L. Warner, *Inorg. Chem.*, **6**, 2229 (1967).
32. W. M. Maxwell, K-S. Wong, and R. N. Grimes, *Inorg. Chem.*, **16**, 3094 (1977).
33. M. L. Thompson and R. N. Grimes, *Inorg. Chem.*, **11**, 1925 (1972).
34. A. Tabereaux and R. N. Grimes, *J. Am. Chem. Soc.*, **94**, 4768 (1972).
35. A. Tabereaux and R. N. Grimes, *Inorg. Chem.*, **12**, 792 (1973).
36. C. G. Savory and M. G. H. Wallbridge, *J. Chem. Soc. Dalton*, 918 (1972).
37. C. G. Savory and M. G. H. Wallbridge, *Chem. Commun.*, 622 (1971).
38. R. E. Loffredo and A. D. Norman, *J. Am. Chem. Soc.*, **93**, 5587 (1971).
39. N. N. Greenwood and B. Youll, *J. Chem. Soc. Dalton*, 158 (1975).
40. D. F. Gaines and T. V. Iorns, *J. Am. Chem. Soc.*, **89**, 4249 (1967).
41. D. F. Gaines and T. V. Iorns, *J. Am. Chem. Soc.*, **90**, 6617 (1968).
42. L. J. Todd, A. R. Burke, H. T. Silverstein, J. L. Little, and G. S. Wikholm, *J. Am. Chem. Soc.*, **91**, 3376 (1969).
43. J. A. Potenza, W. N. Lipscomb, G. D. Vickers, and H. Schroeder, *J. Am. Chem. Soc.*, **88**, 5340 (1966).
44. A. R. Burke, Ph.D. thesis, University of Illinois (1969).
45. W. E. Streib and C. Boss, Indiana University, unpublished results.
46. H. T. Silverstein, D. C. Beer, and L. J. Todd, *J. Organometal. Chem.*, **21**, 139 (1970).
47. D. C. Beer, A. R. Burke, T. R. Engelmann, B. N. Storhoff, and L. J. Todd, *J. Chem. Soc. Chem. Commun.*, 1611 (1971).
48. D. C. Beer and L. J. Todd, *J. Organometal. Chem.*, **36**, 77 (1972).
49. J. L. Little, S. S. Pao, and K. K. Sugathan, *Inorg. Chem.*, **13**, 1752 (1974).
50. W. F. Wright, A. R. Garber, and L. J. Todd, *J. Magn. Reson.*, **30**, 595 (1978).
51. J. L. Little and S. S. Pao, *Inorg. Chem.*, **17**, 584 (1978).
52. T. Yamamoto and L. J. Todd, *J. Organometal. Chem.*, **67**, 75 (1974).
53. G. D. Friesen and L. J. Todd, unpublished results.
54. J. L. Little, *Inorg. Chem.*, **18**, 1598 (1979).
55. W. F. Wright and L. J. Todd, unpublished results.
56. W. F. Wright, J. C. Huffman, and L. J. Todd, *J. Organometal. Chem.*, **148**, 7 (1978).
57. H. D. Smith, Jr., and M. F. Hawthorne, *Inorg. Chem.*, **13**, 2312 (1974).
58. A. M. Barriola, T. P. Hanusa, and L. J. Todd, *Inorg. Chem.*, **19**, 2801 (1980).
59. W. M. Maxwell, R. F. Bryan, and R. N. Grimes, *J. Am. Chem. Soc.*, **99**, 4008 (1977).
60. C. P. Magee, L. G. Sneddon, D. C. Beer, and R. N. Grimes, *J. Organometal. Chem.*, **86**, 1975 (1975).
61. K. S. Wong, J. R. Bowser, J. R. Pipal, and R. N. Grimes, *J. Am. Chem. Soc.*, **100**, 5045 (1978).
62. A. R. Siedle, and L. J. Todd, *J. Chem. Soc. Chem. Commun.*, 914 (1973).
63. J. L. Little, G. D. Friesen, and L. J. Todd, *Inorg. Chem.*, **16**, 869 (1977).
64. G. D. Friesen, and L. J. Todd, *Chem. Commun.*, 349 (1978).
65. G. D. Friesen, R. L. Kump, and L. J. Todd, *Inorg. Chem.*, **19**, 1485 (1980).
66. W. R. Hertler, F. Klanberg, and E. L. Muetterties, *Inorg. Chem.*, **6**, 1696 (1967).
67. A. R. Siedle, G. M. Bodner, A. R. Garber, and L. J. Todd, *Inorg. Chem.*, **13**, 1756 (1974).

68. T. Barriola, R. L. Kump, and L. J. Todd, unpublished results.
69. G. D. Friesen, A. Barriola, P. Daluga, P. Ragatz, J. C. Huffman, and L. J. Todd, *Inorg. Chem.*, **19**, 458 (1980).
70. W. M. Maxwell, E. Sinn, and R. N. Grimes, *J. Am. Chem. Soc.*, **98**, 3490 (1976).
71. D. C. Finster and R. N. Grimes, *J. Am. Chem. Soc.*, **103**, 1399 (1981).

5

closo-Carborane–Metal Complexes Containing Metal–Carbon and Metal–Boron σ-Bonds

Silvano Bresadola

1. INTRODUCTION

The carboranes are a class of compounds containing boron and carbon atoms situated at the vertices of polyhedral fragment frameworks.

The compounds of general formula $C_2B_nH_{n+2}$ constitute the class of the *closo*-carboranes in which carbon and boron atoms are arranged in a framework closed on itself in a polyhedral configuration. The *nido*-carboranes, $C_mB_nH_{n+4}$ (m = 1–4), and *arachno*-carboranes, $C_mB_nH_{n+6}$, exhibit "open" structures that are polyhedral fragments.

The scope of this chapter is restricted to the metal derivatives of the *closo*-carboranes formed through carborane carbon-metal and carborane boron-metal bonds. For a more comprehensive treatment of the chemistry and structural aspects of all the carborane compounds, the reader is invited to consult the excellent reviews regarding these topics.[1-7] In order to provide a better perspective of the *closo*-carborane-metal compounds, this discussion will also include derivatives containing carboranyl groups bonded to boron and silicon, which are normally classified as nonmetals.

Silvano Bresadola • Istituto di Chimica, Università di Trieste, Trieste (Italy), and CNR–Centro di Studio sulla Stabilità e Reattività dei Composti di Coordinazione, Padova, Italy.

2. STRUCTURES, NOMENCLATURE SYSTEM, AND GENERAL PREPARATIVE METHODS OF THE CLOSO-CARBORANES

2.1. Structures

The *closo*-carboranes exhibit closed polyhedral structures with triangulated faces (Figure 1). The polyhedral carboranes can be formally viewed as derived from $B_nH_n^{2-}$ ions by replacement of BH^- groups with isoelectronic and isostruc-

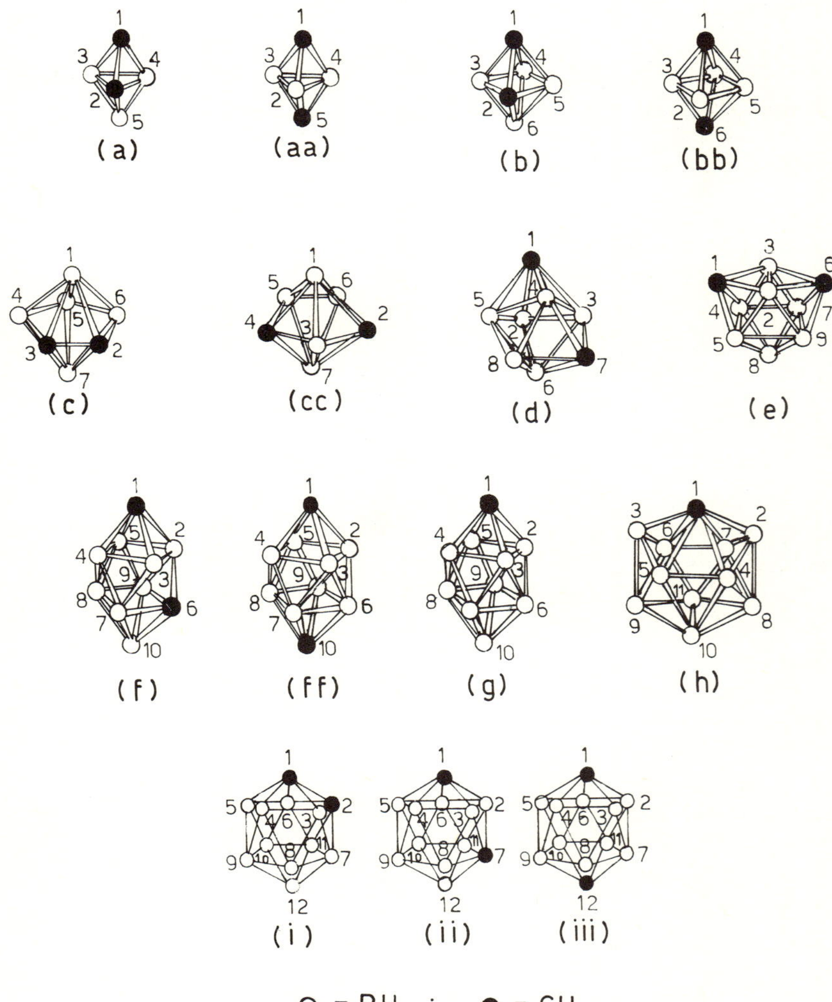

$$O = BH \quad ; \quad \bullet = CH$$

Figure 1. The polyhedral structures of the *closo*-carboranes: (a) $1,2\text{-}C_2B_3H_5$; (aa) $1,5\text{-}C_2B_3H_5$; (b) $1,2\text{-}C_2B_4H_6$; (bb) $1,6\text{-}C_2B_4H_6$; (c) $2,3\text{-}C_2B_5H_7$; (cc) $2,4\text{-}C_2B_5H_7$; (d) $1,7\text{-}C_2B_6H_8$; (e) $1,6\text{-}C_2B_7H_9$; (f) $1,6\text{-}C_2B_8H_{10}$; (ff) $1,10\text{-}C_2B_8H_{10}$; (g) $1\text{-}CB_9H_{10}^-$; (h) $1\text{-}CB_{10}H_{11}^-$; (i) $1,2\text{-}C_2B_{10}H_{12}$; (ii) $1,7\text{-}C_2B_{10}H_{12}$; (iii) $1,12\text{-}C_2B_{10}H_{12}$.

tural CH groups. Stereoisomerism based on the relative position of the carbon atoms in the polyhedral cage is possible for all the *closo*-carboranes and it is observed for several members of the series $C_2B_nH_{n+2}$.

The carboranes are stable, "electron deficient" molecules which have been the subject of detailed theoretical investigations reported in previous reviews.[1-7] In these systems the carbon atoms have unusual coordination numbers and exhibit electropositive behavior, in accordance with the general observation that the higher electropositive character is associated with the carbon atom with the higher coordination number when two carbon atoms differently situated in the same framework are compared.[8] Furthermore, the *closo*-carboranes in general can be easily deprotonated at the C—H groups, and this behavior is assumed to be a manifestation of the electropositive nature of the skeletal carbon atoms.[7]

2.2. Nomenclature System

The polyhedral (fully triangulated) compounds containing boron and carbon atoms in the skeletal framework are called *closo*-carboranes. The prefix *closo* (from the greek word for cage, κλωβός) indicates a closed network.

As proposed by the Nomenclature Committee of the Division of Inorganic Chemistry of the American Chemical Society,[9] the carboranes are named as polyboranes in which boron atoms have been replaced by carbon atoms by an adaptation of organic replacement nomenclature (oxa–aza convention). The cage atoms of the *closo*-carboranes are numbered, assigning the number 1 to an apical atom. When no apparent apex exists, the number 1 is attributed to a carbon atom, and the numbering procedure goes on with successive rings in a clockwise way. The number of hydrogen atoms in the carborane molecule is given in parentheses.

Thus, $1,6\text{-}C_2B_4H_6$ is named 1,6-dicarba-*closo*-hexaborane(6), indicating a closed cage molecule (*closo*) with two skeletal carbon atoms (dicarba) situated in positions 1 and 6.

For the icosahedral carboranes, $C_2B_{10}H_{12}$, trivial names and useful symbols are commonly used (Table 1).

Table 1. Trivial Names and Symbols of the Icosahedral Carboranes, $C_2B_{10}H_{12}$

Isomer	Trivial names	Symbol
$1,2\text{-}C_2B_{10}H_{12}$ 1,2-dicarba-*closo*-dodecarborane(12)	*o*-Carborane (barene in the Russian literature)	$H-C$⎯$C-H$ over $B_{10}H_{10}$
$1,7\text{-}C_2B_{10}H_{12}$ 1,7-dicarba-*closo*-dodecaborane(12)	*m*-Carborane Neocarborane (neobarene)	$H-CB_{10}H_{10}C-H$
$1,12\text{-}C_2B_{10}H_{12}$ 1,12-dicarba-*closo*-dodecaborane(12)	*p*-Carborane (*p*-Barene)	$H-CB_{10}H_{10}C-H$

2.3. Synthetic Methods

The smaller *closo*-carboranes, such as $C_2B_3H_5$ and $C_2B_4H_6$, and their alkyl derivatives can be obtained in low yields from circulation of mixtures of diborane and acetylene through an electric discharge.[10,11]

Small quantities of the medium-size *closo*-carborane $C_2B_6H_8$ and its C,C'-dimethyl derivative are formed in the reaction of hexaborane, B_6H_{10}, with acetylene and dimethylacetylene, respectively, in the presence of ultraviolet irradiation.[12]

The $C_2B_9H_{11}$ carborane has been obtained from pyrolysis of the $C_2B_9H_{13}$ intermediates formed in the protonation of $C_2B_9H_{12}^-$ salts.[13] By oxidation of the *closo*-carborane $C_2B_9H_{11}$, a $C_2B_7H_{13}$ series has also been obtained, which in turn gives $C_2B_6H_8$, $C_2B_7H_9$, and $1,6$-$C_2B_8H_{10}$ derivatives by pyrolysis at $215°C$. On heating to $350°C$, the $1,6$-$C_2B_8H_{10}$ species isomerizes to $1,10$-$C_2B_8H_{10}$.[13] Among the *closo*-carboranes, the icosahedral carboranes, $C_2B_{10}H_{12}$, exhibit the greater chemical stability and ease of preparation in large quantities. The *o*-carborane $1,2$-$C_2B_{10}H_{12}$, and its *C*-alkyl and *C*-aryl derivatives are easily prepared[14] from the reaction of $B_{10}H_{12}L_2$ with acetylene or substituted acetylenes.

$$B_{10}H_{14} + 2L \longrightarrow B_{10}H_{12}L_2 + H_2 \qquad\qquad (1)$$

$$B_{10}H_{12}L_2 + HC{\equiv}CR \longrightarrow HC{-}C{-}R + 2L + H_2 \qquad\qquad (2)$$

$$B_{10}H_{10}$$

$$L = \text{Lewis base } [(C_2H_5)_2S,\ CH_3CN,\ \text{etc.}]$$
$$R = H,\ \text{alkyl, aryl, alkenyl, alkynyl}$$

The *m*-carborane[15] and the *p*-carborane[16] ($1,7$- and $1,12$-$C_2B_{10}H_{12}$, respectively) are the products of thermal rearrangement of the parent *o*-carborane isomer on heating to $465–500°C$ and $500–620°C$, respectively, for several hours in an inert atmosphere.

3. NONTRANSITION-METAL CLOSO-CARBORANE DERIVATIVES

3.1. General Preparative Methods

The mildly acidic C–H bonds in the carboranes react easily with organolithium reagents such as alkyllithium or phenyllithium, to yield *C*-lithium- and C,C'-dilithium-substituted carboranes.[6,7]

Metallation of the carboranes can also be accomplished by reaction of the carboranes with alkali metal amides in liquid ammonia. Grignard carborane derivatives are formed by treating the carboranes with alkylmagnesium halides.

The alkali metal carboranes and the Grignard carborane derivatives are useful intermediates, undergoing many of the reactions common to the organometallic compounds.[6,7] On the other hand, *C*-metal carborane compounds formed through metal–carbon σ-bonds can generally be synthesized by reacting an alkali metalcarborane or a Grignard carborane derivative with the appropriate metal compound containing a metal–halogen bond.

3.2. Alkali Metal Derivatives

3.2.1. Small Carboranes: $1,6$-$C_2B_4H_6$; $2,4$-$C_2B_5H_7$; $1,6$-$C_2B_8H_{10}$; $1,10$-$C_2B_8H_{10}$; $1,8$-$C_2B_9H_{11}$; 1-$CB_9H_{10}^-$; 1-$CB_{11}H_{12}^-$

The 1,6-dicarba-*closo*-hexaborane(6), $1,6$-$C_2B_4H_6$, reacts with excess *n*-butyllithium in ether–hexane solvent mixtures at room temperature to give[17] the corresponding C,C'-dilithium derivative.

$$1,6\text{--}HCB_4H_4CH + 2LiR \longrightarrow 1,6\text{--}LiCB_4H_4CLi + 2RH \tag{3}$$

The reaction is extremely slow in comparison with the corresponding metallation of higher carboranes, such as $2,4$-$C_2B_5H_7$ or $C_2B_{10}H_{12}$ isomers. Conversion to the dilithium derivative is incomplete even after 24 hr, but glpc analysis of the solution after 48 hr indicates no remaining unreacted carborane. The dilithium derivative readily reacts with methyl iodide in diethyl ether, giving[17,18] the C,C'-dimethyl derivative, $1,6$-$(CH_3)_2$-$1,6$-$C_2B_4H_4$. On the other hand, monomethylation of $1,6$-Li_2-$1,6$-$C_2B_4H_4$ is complicated by side reactions.[17]

The stable and relatively accessible 2,4-dicarba-*closo*-heptaborane(7), $2,4$-$C_2B_5H_7$, undergoes metallation at the carbon atoms.[19,20] Thus $2,4$-dilithio-2,4-dicarba-*closo*-heptaborane(7) was obtained[21,22] by reacting $2,4$-$C_2B_5H_7$ with excess *n*-butyllithium (ca. 1:3) in solution. While in a 2:1 diethyl ether–hexane mixture the reaction proceeds to completion in 20 hr at room temperature, attempts to carry out this lithiation in pure ether produce an intractable gum.[21] Moreover, by working in nonpolar solvents, such as pentane or hexane, no reaction is observed when both reactant concentrations are below $1M$. However, at higher concentrations the white dilithiocarborane slowly precipitates at room temperature.

The relatively low reactivity toward proton abstraction shown by $C_2B_5H_7$ in comparison with the $C_2B_{10}H_{12}$ isomers may be due to a lower positive charge on the carbon atoms.[23] On the other hand, unexpected difficulty was encountered in preparing the *C*-monolithium derivative by reacting $2,4$-$C_2B_5H_7$ with an equimolecular amount of *n*-C_4H_9Li in hexane-ether. In this case the formation of a large quantity of solid carborane polymer occurs. To explain[21] this polymer formation, the following reactions have been suggested:

$$LiCB_5H_5C\text{--}H + LiCB_5H_5C\text{--}H \xrightarrow{-LiH}$$

$$Li\text{--}CB_5H_5C\text{--}CB_5H_5C\text{--}H \xrightarrow[-LiH]{n\text{-butyllithium}} Li\text{--}[\text{--}CB_5H_5C\text{--}]_n\text{--}H \tag{4}$$

This indicates that the *C*-monolithium derivative undergoes facile intermolecular hydrogen abstraction that is hindered or prevented by the formation of the *C,C'*-dilithium derivative when the reaction is carried out using excess butyllithium.

The 1,6- and 1,10-dicarba-*closo*-decaborane(10), 1,6- and $1,10\text{-}C_2B_8H_{10}$, react readily with *n*-butyllithium in solution to give monolithium[24,25] or dilithium[24] derivatives:

$$1\text{-Ph-6-H-}1,6\text{-}C_2B_8H_8 + n\text{-}C_4H_9\text{-Li} \longrightarrow 1\text{-Ph-6-Li-}1,6\text{-}C_2B_8H_8 + C_4H_{10} \quad (5)$$

$$1\text{-R-}10\text{-H-}1,10\text{-}C_2B_8H_8 + n\text{-}C_4H_9\text{-Li} \longrightarrow 1\text{-R-}10\text{-Li-}1,10\text{-}C_2B_8H_8 + C_4H_{10} \quad (6)$$
$$R = CH_3, C_6H_5$$

$$1,10\text{-}C_2B_8H_{10} + 2n\text{-}C_4H_9\text{-Li} \longrightarrow 1\text{-Li-}10\text{-Li-}1,10\text{-}C_2B_8H_8 + 2C_4H_{10} \quad (7)$$

Treatment of the unsubstituted $1,6\text{-}C_2B_8H_{10}$ carborane with an equimolar amount of butyllithium produces[24] a mixture of the two possible *C*-monolithium derivative isomers:

$$1,6\text{-}C_2B_8H_{10} \xrightarrow[-C_4H_{10}]{n\text{-}C_4H_9Li} 1\text{-Li-}1,6\text{-}C_2B_8H_9 + 6\text{-Li-}1,6\text{-}C_2B_8H_9 \quad (8)$$
$$70\% \qquad\qquad 30\%$$

The dipolyhedral bis[dicarba-*closo*-decaborane(10)], $1\text{-}(1',\text{-}1',10'\text{-}C_2B_8H_9)\text{-}1,10\text{-}C_2B_8H_9$, in which two carboranyl groups are bonded through a C—C bond, has been prepared[26] by the following reactions carried out in diethyl ether:

$$2(1,10\text{-}C_2B_8H_{10}) \xrightarrow{n\text{-}C_4H_9Li} 2(1\text{-Li-}10\text{-H-}1,10\text{-}C_2B_8H_8) \xrightarrow{CuCl_2}$$
$$HCB_8H_8C\text{-}CB_8H_8CH \quad (9)$$

By treating this biscarborane with excess butyllithium the corresponding *C,C'*-dilithium biscarborane derivative is obtained.[26]

The $B_9C_2H_{11}$ carboranes survive brief exposure to moist air.[27] Attempts to prepare lithium derivatives of 1,8-dicarba-*closo*-undecaborane(11) by treatment with methyllithium produced[28] an ion formulated as $C_2B_9H_{11}CH_3^-$ that is apparently a methyl derivative of the icosahedral fragment $(3)\text{-}1,7\text{-}C_2B_9H_{12}^-$.

In common with the *closo*-dicarboranes, the *closo*-monocarborane anions[29,30] $1\text{-}CB_9H_{10}^-$ and $1\text{-}CB_{11}H_{12}^-$ readily form *C*-lithium derivatives on treatment with alkyllithium reagents in solution:

$$1\text{-}CB_9H_{10}^- + n\text{-}C_4H_9Li \longrightarrow 1\text{-Li-}1\text{-}CB_9H_9^- + C_4H_{10} \quad (10)$$

$$1\text{-}CB_{11}H_{12}^- + n\text{-}C_4H_9Li \longrightarrow 1\text{-Li-}1\text{-}CB_{11}H_{11}^- + C_4H_{10} \quad (11)$$

These *C*-lithium derivatives are useful intermediates for the synthesis of a number of *C*-substituted monocarborane anions which can be isolated[30] as tetramethylammonium salts.

3.2.2. Icosahedral Carboranes: 1,2-, 1,7-, and 1,12-$C_2B_{10}H_{12}$

The icosahedral carboranes, 1,2-dicarba-*closo*-dodecarborane(12) [*o*-carborane], 1,7-dicarba-*closo*-dodecaborane(12) [*m*-carborane], and 1,12-dicarba-*closo*-dodecaborane(12) [*p*-carborane], and their *C*-derivatives exhibit quite good thermal, hydrolytic (except in the presence of nucleophiles), and oxidative stabilities.

A very extensive organic chemistry has been developed for the icosahedral carboranes, which is based largely on metallation of one or both the carbon atoms of the carborane cage. In addition, several compounds having carboranyl groups bonded to both nontransition and transition metals through carborane carbon–metal bonds have been prepared and fully characterized.

The carbon hydrogens of $C_2B_{10}H_{12}$ can be easily replaced by the metals of certain active metallating agents such as butyllithium and phenyllithium. For synthetic purposes the *C*-monolithium and *C*, *C'*-dilithium carborane derivatives are widely used. *C*-monolithium derivatives of *o*-, *m*-, and *p*-carborane are formed[31-35] by reacting the carborane with an equivalent amount of butyllithium or phenyllithium in diethyl ether or tetrahydrofuran at room temperature or below.

$$\text{H}-\text{C}\overset{\diagdown\!\!O\!\!\diagup}{\diagup\ \ \diagdown}\text{C}-\text{H} + \text{R}-\text{Li} \longrightarrow \text{H}-\text{C}\overset{\diagdown\!\!O\!\!\diagup}{\diagup\ \ \diagdown}\text{C}-\text{Li} + \text{RH} \qquad (12a)$$
$$B_{10}H_{10} \qquad\qquad B_{10}H_{10}$$

$$\text{R}'-\text{C}\overset{\diagdown\!\!O\!\!\diagup}{\diagup\ \ \diagdown}\text{C}-\text{H} + \text{R}-\text{Li} \longrightarrow \text{R}'-\text{C}\overset{\diagdown\!\!O\!\!\diagup}{\diagup\ \ \diagdown}\text{C}-\text{Li} + \text{RH} \qquad (12b)$$
$$B_{10}H_{10} \qquad\qquad B_{10}H_{10}$$

On the other hand, by using an excess of the organometallic reagent, *C*, *C'*-dilithium carborane derivatives are readily obtained:

$$\text{H}-\text{C}\overset{\diagdown\!\!O\!\!\diagup}{\diagup\ \ \diagdown}\text{C}-\text{H} + 2\text{R}-\text{Li} \longrightarrow \text{Li}-\text{C}\overset{\diagdown\!\!O\!\!\diagup}{\diagup\ \ \diagdown}\text{C}-\text{Li} + 2\text{RH} \qquad (12c)$$
$$B_{10}H_{10} \qquad\qquad B_{10}H_{10}$$

The rate of the reaction with butyllithium to form the corresponding lithium carborane compound decreases in the order *o*-carborane > *m*-carborane > *p*-carborane, in agreement with the observed decreasing order of acidity in C–H hydrogen atoms.[7,36,37] The reactions with alkyllithium reagents can be also carried out in benzene solution,[33,38] but in this solvent the carborane metallation occurs under more severe conditions (60–80°C).

It is to be noted that the preparation of the *C*-monolithium derivative of the unsubstituted *o*-carborane is complicated by an equilibrium existing in ether or ether–benzene solution between the *C*-monolithium and *C*, *C'*-dilithium species.[39]

$$2\text{H}-\text{C}\underset{\text{B}_{10}\text{H}_{10}}{\diagup\diagdown}\text{C}-\text{Li} \;\rightleftharpoons\; \text{Li}-\text{C}\underset{\text{B}_{10}\text{H}_{10}}{\diagup\diagdown}\text{C}-\text{Li} + \text{H}-\text{C}\underset{\text{B}_{10}\text{H}_{10}}{\diagup\diagdown}\text{C}-\text{H} \tag{13}$$

In benzene solution this equilbrium appears to be shifted well to the left, and thus the *C*-monometallated species can be conveniently obtained in good yield when the metallation reaction is carried out in this medium.[40]

An analogous equilibrium between *C*-monometallated and *C, C'*-dimetallated species appears to exist also in the case of *meta*- and *para*-carborane.

However, in agreement with the observed order of acidity of the carborane C–H hydrogens, this equilibrium appears to be shifted further to the left on passing from the *ortho*- to the *meta*- to the *para*-carborane.[37,41] *C*-lithium, sodium, and potassium derivatives of dicarba-*closo*-dodecarborane(12) isomers can be also prepared by treating the carboranes with the corresponding alkali metal amides in liquid ammonia.[42,44] In this manner, both mono- and dimetal carborane derivatives are formed:

$$\text{R}-\text{C}\underset{\text{B}_{10}\text{H}_{10}}{\diagup\diagdown}\text{C}-\text{H} + \text{MNH}_2 \xrightarrow{\;\text{NH}_3\,(\text{liq})\;} \text{R}-\text{C}\underset{\text{B}_{10}\text{H}_{10}}{\diagup\diagdown}\text{C}-\text{M} \tag{14}$$

$$\text{H}-\text{C}\underset{\text{B}_{10}\text{H}_{10}}{\diagup\diagdown}\text{C}-\text{H} + 2\text{MNH}_2 \xrightarrow{\;\text{NH}_3\,(\text{liq})\;} \text{M}-\text{C}\underset{\text{B}_{10}\text{H}_{10}}{\diagup\diagdown}\text{C}-\text{M} \tag{15}$$

$$\text{M} = \text{Li, Na, K}$$

Unsubstituted and *C*-monosubstituted carboranes can also be metallated with sodium amide in boiling toluene.[45]

However, in contrast to the reactions with alkali metal amides in liquid ammonia where the unsubstituted carborane forms both mono- and dimetal derivatives, in boiling toluene the unsubstituted carborane gives only the monosodium derivative by reaction with sodium amide.[42,43]

All of the alkali metal derivatives of the carboranes are useful intermediates for the preparation of other *C*-mono- and *C, C'*-disubstituted carboranes.[7,31,37]

3.3. Magnesium and Calcium Derivatives

The mildly acidic activity of the hydrogens attached to the carbon atoms of the dicarba-*closo*-dodecarborane(12) nucleus allows facile preparation of carborane Grignard reagents through reaction with alkylmagnesium halides. Thus, ethylmagnesium bromide reacts smoothly with $1,2\text{-C}_2\text{B}_{10}\text{H}_{12}$ in tetrahydrofuran (but not in diethyl ether) to give carboranylmagnesium bromide.[32,46,47]

$$\text{H}-\text{C}\underset{\text{B}_{10}\text{H}_{10}}{\diagup\diagdown}\text{C}-\text{H} + \text{C}_2\text{H}_5\text{MgBr} \longrightarrow \text{H}-\text{C}\underset{\text{B}_{10}\text{H}_{10}}{\diagup\diagdown}\text{C}-\text{MgBr} + \text{C}_2\text{H}_6 \tag{16}$$

The 1,7–dicarba-*closo*-dodecaborane(12), *m*-carborane, forms analogous carborane Grignard compounds under similar conditions.[48]

In contrast to *ortho-* and *meta*-carborane, the *para*-carborane isomer does not react with Grignard reagents in tetrahydrofuran.[49]

Carboranylmagnesium bromides can be also obtained by reacting 1–Br–2-R-1,2-dicarba-*closo*-dodecaborane(12) with magnesium turnings in diethyl ether at reflux temperature:[31]

$$R-C{-}C-Br + Mg \longrightarrow R-C{-}C-MgBr \qquad (17)$$

$$R = n\text{-}C_3H_7,\ n\text{-}C_4H_9$$

An interesting reaction occurs in preparing a carborane Grignard compound when $1\text{-BrCH}_2\text{-}1,2\text{-}C_2B_{10}H_{11}$ and magnesium turnings are treated in tetrahydrofuran:

$$HC{-}C-CH_2Br + Mg \longrightarrow H-C{-}C-CH_2MgBr \longrightarrow BrMg-C{-}C-CH_3 \qquad (18)$$

In fact, it was found that in tetrahydrofuran (unlike ether) the first reaction product,

$$H-C{-}C-CH_2MgBr$$

isomerizes to 1-methyl-2-carboranylmagnesium bromide.[32,50] This unusual rearrangement confirms that the hydrogen atoms attached to polyhedral carbons are sufficiently labile to permit metallation.

Calcium derivatives of unsubstituted and *C*-monosubstituted dicarba-*closo*-dodecaborane(12) have been obtained by reaction of the appropriate carborane with calcium amide in liquid ammonia at low temperature.[42,43]

3.4. Derivatives of Group IIIA Elements

3.4.1. Boron

A number of compounds formed through a carborane carbon–boron bond have been reported. Bis(dimethylamino)-*o*-carboranylborane,

$$H-C{-}C-B[N(CH_3)_2]_2$$

was prepared[51] by reacting 1-Li-1,2-dicarba-*closo*-dodecaborane(12) with chlorobis(dimethalamino)borane in diethyl ether/*n*-pentane at 0°C.

$$H-C{-}{-}C-Li + ClB[N(CH_3)_2]_2 \longrightarrow H-C{-}{-}C-B[N(CH_3)_2]_2 + LiCl$$
$$\underset{B_{10}H_{10}}{\diagdown O \diagup} \qquad\qquad\qquad \underset{B_{10}H_{10}}{\diagdown O \diagup} \tag{19}$$

This carboranylborane derivative reacts with methanol giving *o*-carborane, $NH(CH_3)_2$, and $B(OCH_3)_3$, but not dimethoxy-*o*-carboranylborane.

On treating $1\text{-Li-}2\text{-}C_4H_9\text{-}1,2\text{-}C_2B_{10}H_{10}$ with *B*-trichloro-*N*-trimethylborazole, a carboranyl derivative of borazole has been prepared:[51]

$$3n\text{-}C_4H_9-C{-}{-}C-Li + (ClBNCH_3)_3 \longrightarrow (n\text{-}C_4H_9-C{-}{-}C-BNCH_3)_3 + 3LiCl$$
$$\underset{B_{10}H_{10}}{\diagdown O \diagup} \qquad\qquad\qquad\qquad \underset{B_{10}H_{10}}{\diagdown O \diagup} \tag{20}$$

From the reaction between BCl_3 and $1\text{-Li-}2\text{-R-}1,2\text{-}C_2B_{10}H_{10}$ ($R = C_2H_5$, $i\text{-}C_3H_7$) in benzene–hexane (1:1) solution at $0°C$, *C*-boron-substituted *o*-carboranes I and II have been obtained with yields of ca. 80%:[52]

$$R-C{-}{-}C-BCl_2$$
$$\underset{\underset{\mathbf{I, II}}{B_{10}H_{10}}}{\diagdown O \diagup}$$

I: $R = C_2H_5$, mp = 40–42.5°C
II: $R = i\text{-}C_3H_7$, mp = 52–57°C

The *C*-isopropyl derivative, II, undergoes substitution of the boron–halogen atoms on treatment with ROH, RSH, NH_3, NH_2R, and NHR_2. Thus, the following[52] 1,2-dicarba-*closo*-dodecaborane(12) derivatives containing a side C–B bond have been prepared:

$R\text{-}B(SC_2H_5)_2$	thick colorless liquid; bp: 183–185°C (1 mm)
$R\text{-}BCl[S(n\text{-}C_4H_9)]$:	viscous liquid; bp: 154°C (0.5 mm)
$R\text{-}B[S(n\text{-}C_4H_9)]_2$	viscous liquid; bp: 168–169°C (0.001 mm)
$R\text{-}B(OCH_3)_2$	white crystals; mp: 43–48°C; bp 123–124°C (1 mm)
$R\text{-}B(NH_2)_2$	white crystals; mp: 67–69°C
$R\text{-}B(NHCH_3)_2$	white crystals; mp: 106–107°C
$R\text{-}B[N(C_2H_5)_2]_2$	white crystals; mp: 81–82; bp: 120°C (0.03 mm)

$$R = i\text{-}C_3H_7-C{-}{-}C-$$
$$\underset{B_{10}H_{10}}{\diagdown O \diagup}$$

Compounds containing two *o*-carboranyl groups linked through carbon–boron bonds have been obtained[53] following the method of building the dicarbollide ion suggested by Hawthorne.[54] Thus, by treating a benzene suspension of the lithium salts of (3)-1,2-dicarbollide ions, $(3)\text{-}1,2\text{-}C_2B_9H_{11}^{2-}$ and $(3)\text{-}1,2\text{-}C_2B_9H_{10}(CH_3)^{2-}$, with $1\text{-}BCl_2\text{-}2\text{-H-}1,2\text{-}C_2B_{10}H_{10}$ and $1\text{-}BCl_2\text{-}2\text{-}CH_3\text{-}1,2\text{-}C_2B_{10}H_{10}$, respectively, dicarbollide insertion reactions occur to form 1-(*o*-

carboran-3-yl)-2-CH$_3$-o-carborane, III, and 3-(o-carboran-1-yl)-1-CH$_3$-o-carborane, IV.

H–C——C–H
\\O/
3–B$_{10}$H$_9$–C——C–CH$_3$
\\O/
B$_{10}$H$_{10}$

III: Mp: 208–210°C

H–C——C–CH$_3$
\\O/
3–B$_{10}$H$_9$–C——C–H
\\O/
B$_{10}$H$_{10}$

IV: mp: 191–193°C

3.4.2. Thallium

Carboranyl thallium compounds containing carbon-thallium σ-bonds have been prepared by reacting C-lithium derivatives of both 1,2- and 1,7-C$_2$B$_{10}$H$_{12}$ with anhydrous TlCl$_3$ or alkylthallium halides in ethereal solution. Thus, compounds of the type (o- and m-C$_2$B$_{10}$H$_{10}$R)$_2$TlCl and (o- and m-C$_2$B$_{10}$H$_{10}$R′) Tl(C$_4$H$_9$)Cl, where R = H, C$_6$H$_5$, CH$_2$Cl, and R′ = H, C$_6$H$_5$, have been reported.[55] In comparison with the classic thallium metallorganic compounds, the carboranyl thallium derivatives show several distinctive features due to the electronic effect of the carboranyl group. Thus, the carborane carbon-thallium bond appears not to be affected on treatment with hydrogen chloride and halogens.[55]

Thallium derivatives containing a carborane boron-thallium bond have been obtained[56,57] by direct metallation of 1,2-C$_2$B$_{10}$H$_{10}$RR′ (R = H, CH$_3$, C$_6$H$_5$; R′ = H; R = R′ = CH$_3$) and 1,7-C$_2$B$_{10}$H$_{10}$RR′ (R = R′ = CH$_3$) with Tl(OCOCF$_3$)$_3$:

$$o\text{-}, m\text{-C}_2\text{B}_{10}\text{H}_{10}\text{RR}' + \text{Tl(OCOCF}_3)_3 \longrightarrow$$
$$9\text{-[Tl(OCOCF}_3)_2]\text{-}o\text{-(or }\text{-}m\text{-)C}_2\text{B}_{10}\text{H}_9\text{RR}' + \text{CF}_3\text{COOH} \quad (21)$$

These boron-thallium bonded o- and m-carborane derivatives have also been prepared by treating the corresponding mercury derivatives (9-C$_2$B$_{10}$H$_{11}$)$_2$Hg or 9-[Hg(OCOCF$_3$)]-C$_2$B$_{10}$H$_{11}$ with Tl(OCOCF$_3$)$_3$ in diethyl ether.[56,58] Transfer of a carboranyl group from mercury to thallium occurs:

$$(9\text{-}o\text{- or }9\text{-}m\text{-C}_2\text{B}_{10}\text{H}_{11})_2\text{Hg} + \text{Tl(OCOCF}_3)_3 \longrightarrow$$
$$9\text{-[Tl(OCOCF}_3)_2]\text{-}o\text{- (or }\text{-}m\text{-)C}_2\text{B}_{10}\text{H}_{11} + 9\text{-[Hg(OCOCF}_3)]\text{-}o\text{- (or }\text{-}m\text{-)C}_2\text{B}_{10}\text{H}_{11} \quad (22)$$

Recently, a thallium derivative containing two o-carboranyl groups bonded to the metal atom through carborane carbon-thallium and carborane boron-thallium bonds, respectively, has also been prepared on treatment of 9-[Tl(OCOCF$_3$)$_2$]-o-C$_2$B$_{10}$H$_{11}$ with 1-Li-2-C$_6$H$_5$-1,2-C$_2$B$_{10}$H$_{10}$ in ether:[57]

$$\text{H–C}\underset{\displaystyle\text{O}}{—}\text{C–H} \quad + \quad \text{Li–C}\underset{\displaystyle\text{O}}{—}\text{C–C}_6\text{H}_5 \longrightarrow$$
$$9\text{–B}_{10}\text{H}_9\text{–Tl(OCOCF}_3)_2 \qquad\qquad\qquad \text{B}_{10}\text{H}_{10}$$

$$\begin{array}{c}\text{H–C}\underset{\displaystyle\text{O}}{—}\text{C–H} \qquad \text{OCOCF}_3 \\ 9\text{–B}_{10}\text{H}_9\text{–Tl} \\ \\ \text{C—C–C}_6\text{H}_5 \\ \text{B}_{10}\text{H}_{10}\end{array} \qquad + \ \text{CF}_3\text{COOLi}$$

$$(23)$$

3.5. Derivatives of Group IVA Elements

3.5.1. Silicon

Some *C*-silyl derivatives of the smallest known *closo*-carborane $1,5\text{-C}_2\text{B}_3\text{H}_5$, [1,5-dicarba-*closo*-pentaborane(5)], have been obtained[59] by flash thermolysis of 1,2-bis(trimethylsilyl)pentaborane(9) through an apparent carbon-insertion process. Thus, a mixture containing, *inter alia*, the dicarba-*closo*-pentaborane(5) derivatives $1\text{-SiH}_3\text{-}1,5\text{-C}_2\text{B}_3\text{H}_4$, $1\text{-SiH}_2\text{Me-}1,5\text{-C}_2\text{B}_3\text{H}_4$ and $1\text{-SiH}_3\text{-2-Me-}1,5\text{-C}_2\text{B}_3\text{H}_3$, and the monocarboranes CB_5H_7 and $\text{B–Me–CB}_5\text{H}_6$ (Figure 2) has been obtained from $1,2\text{-(SiMe}_3)_2\text{B}_5\text{H}_7$ by working in a hot quartz tube at 575°C. The total carborane yield was 40 mol% and the carboranes were identified and characterized by mass spectroscopy and ^{11}B nmr studies.[59] Small amounts of the silyl *closo*-carborane $\text{SiH}_3\text{-C}_2\text{B}_3\text{H}_4$ together with silyl derivatives of the *nido*-carborane $\text{C}_2\text{B}_4\text{H}_8$ have been produced by pyrolysis at 220–230°C of both $\mu\text{-SiH}_3\text{C}_2\text{B}_4\text{H}_7$ and $4\text{-SiH}_3\text{C}_2\text{B}_4\text{H}_7$.[60] *B*-silylmethyl derivatives

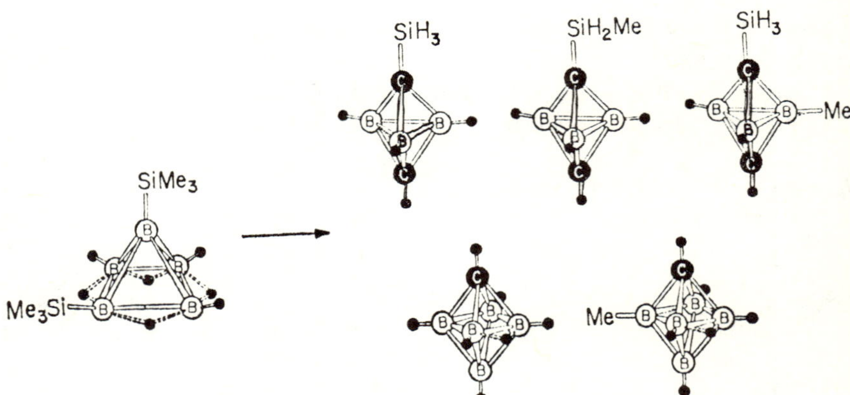

Figure 2. Carboranes formed in the thermolysis of $1,2\text{-(SiMe}_3)_2\text{B}_5\text{H}_7$, (from Reference 59).

of the *nido*-carborane $2,3$-$C_2B_4H_7^-$ have been obtained[61] by treating $2,3$-$C_2B_4H_7^-$ with $ClH_2CSiCl(CH_3)_2$ and subsequent rearrangement of the first reaction product (Figure 3). These *B*-substituted *nido*-carboranes undergo further rearrangement on heating at $690°C$ yielding *B*-substituted derivatives of both $1,5$-$C_2B_3H_5$ and $1,6$-$C_2B_4H_6$ *closo*-carboranes.

C-silyl derivatives of the *closo*-carborane anions 1-$CB_9H_{10}^-$ and 1-$CB_{11}H_{12}^-$ have been also prepared.[29,30] In fact, the anions 1-$Si(CH_3)_3$-1-$CB_9H_9^-$ and 1-$Si(CH_3)_3$-1-$CB_{11}H_{11}^-$ are the reaction products when 1-Li-1-$CB_9H_9^-$ and 1-Li-1-$CB_{11}H_{11}^-$, respectively, are treated with $(CH_3)_3SiCl$ in solution. These *C*-silyl anions were isolated as tetramethylammonium salts.

Trimethylsilyl derivatives of the *closo*-carboranes $C_2B_3H_5$ and $C_2B_5H_7$, 2-$(CH_3)_3Si$-$C_2B_3H_4$, and 2-$(CH_3)_3Si$-$2,4$-$C_2B_5H_7$, have been obtained on prolonged heating at $230°C$ of 2-$(CH_3)_3SiC_2B_4H_7$, presumably by disproportionation of the C_2B_4 cage.[60] As regards the $2,4$-$C_2B_5H_7$ *closo*-carborane, the *C*-mono(trimethylsilyl) derivative has also been prepared[21] through the reactions:

$$2,4\text{-}C_2B_5H_7 \xrightarrow{n\text{-}C_4H_9Li} Li\text{-}CB_5H_5C\text{-}Li \xrightarrow{(CH_3)_3SiCl} \tag{24}$$
$$Li\text{-}CB_5H_5C\text{-}Si(CH_3)_3 \xrightarrow{HCl} HCB_5H_5CSi(CH_3)_3$$

carried out in diethyl ether.

On the other hand, by treating the C, C'-dilithium derivative with an excess of $(CH_3)_2SiCl_2$ in diethyl ether at $0°C$, the $2,4$-bis(chlorodimethylsilyl)-$2,4$-dicarba-*closo*-heptaborane(7)

$$\begin{array}{ccc} CH_3 & & CH_3 \\ | & & | \\ ClSi\text{---}CB_5H_5C\text{---}Si\text{---}Cl \\ | & & | \\ CH_3 & & CH_3 \end{array}$$

was obtained.[22] The latter compound reacts[22] with anhydrous methanol at room temperature giving

$$\begin{array}{ccc} CH_3 & & CH_3 \\ | & & | \\ CH_3O\text{---}Si\text{---}CB_5H_5C\text{---}Si\text{---}OCH_3 \\ | & & | \\ CH_3 & & CH_3 \end{array}$$

The bis-silyl derivative of $2,4$-$C_2B_5H_7$, $2,4$-$[Si(CH_3)_2(C_5H_5)]_2$-$2,4$-$C_5B_5H_5$ has been obtained by treating $2,4$-$[Si(CH_3)_2Cl]_2$-$2,4$-$C_2B_5H_5$ with $ClMgC_5H_5$ in ether-benzene solution.[62] On refluxing $2,4$-$[Si(CH_3)_2(C_5H_5)]_2$-$2,4$-$C_2B_5H_5$ with iron pentacarbonyl in xylene, the diiron complex $2,4$-$[Si(CH_3)_2(C_5H_4)Fe(CO)_2]_2$-$2,4$-$C_2B_5H_5$ has been obtained. This complex has been isolated by column chromatography on silica gel using benzene as eluent (yield ca. 40%) and then characterized spectrally.[62] This is an example of an $[\eta^5\text{-}C_5H_5Fe(CO)_2]_2$ complex in which the cyclopentadienyl rings are linked through a bridging segment:

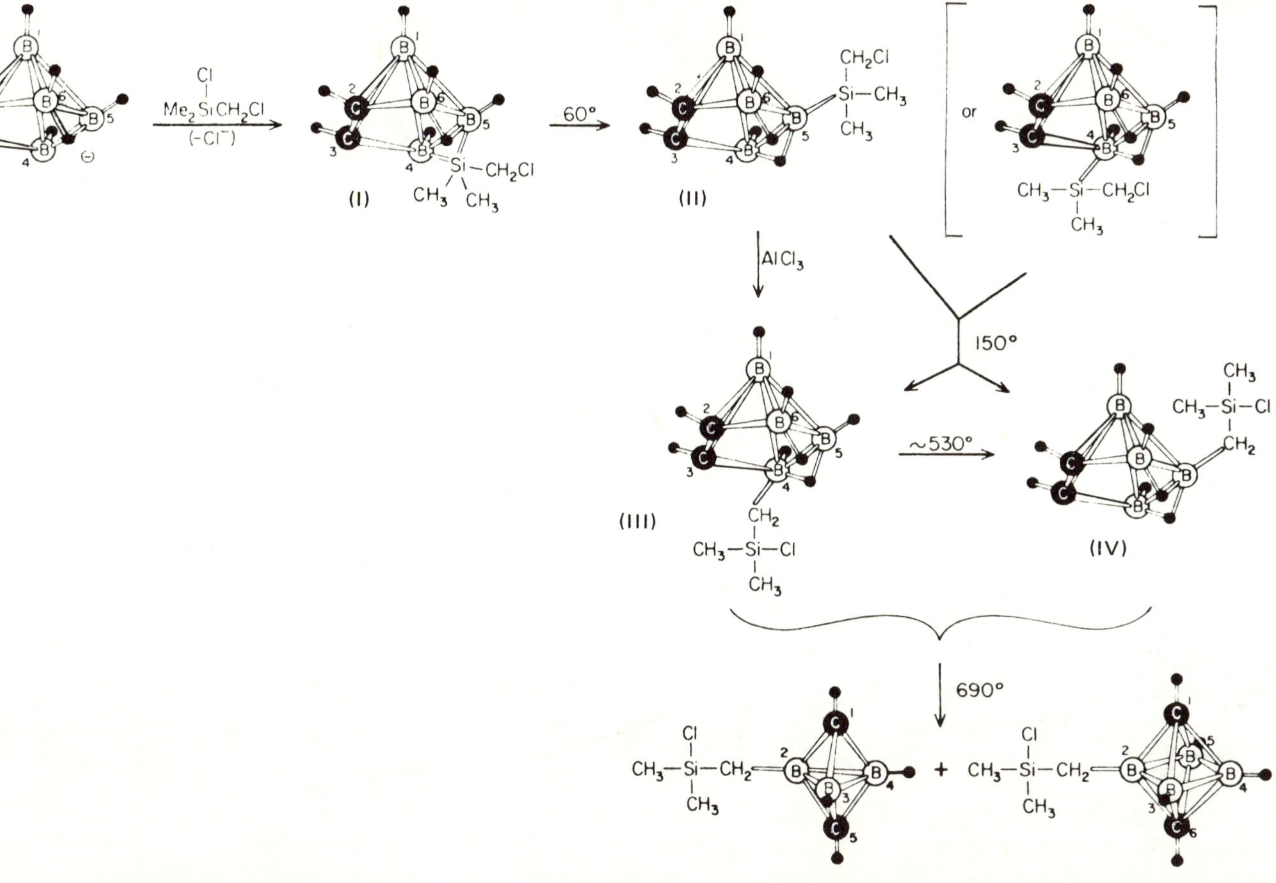

Figure 3. Reaction of $2,3\text{-}C_2B_4H_7^-$ with $(CH_3)_2SiClCH_2Cl$ and subsequent rearrangements (from Reference 61).

$$(CH_3)_2Si-CB_5H_5C-Si(CH_3)_2$$

Starting from the corresponding C,C'-dilithio derivative, C,C'-disilyl derivatives of $1,10$-$C_2B_8H_{10}$, [$1,10$-dicarba-*closo*-decaborane(10)], can be prepared and both $1,10$-bis(chlorodimethylsilyl)- and $1,10$-bis(methoxydimethylsilyl)-$1,10$-dicarba-*closo*-decaborane(10) have been reported.[22]

In addition, a C-mono (trimethylsilyl) derivative of $1,10$-$C_2B_8H_{10}$ has been obtained as white crystalline product (mp: 83-84°C) by reacting 1-C_6H_5-10-Li-$1,10$-$C_2B_8H_8$ with $(CH_3)_3SiCl$ in benzene at room temperature.[63]

A large number of $1,2$-, $1,7$-, and $1,12$-dicarba-*closo*-dodecaborane(12), $C_2B_{10}H_{12}$, derivatives containing silicon atoms bonded directly to carbon atoms of the carborane cage have been reported. Generally, mono- and bissilyl-derivatives of the $C_2B_{10}H_{12}$ isomers are obtained by treating C-lithium and C,C'-dilithium derivatives, respectively, with $SiCl_4$ or R_nSiCl_{4-n} ($n = 1$-3) in dry solvents, such as ether, tetrahydrofuran, benzene, toluene.[31,32,64-70] Thus, some silyl derivatives of o-carborane are prepared according to Equations (25)-(28).

$$\text{Li}-\text{C}\underset{B_{10}H_{10}}{\overset{O}{\diagdown\diagup}}\text{C}-\text{Li} + 2R_3SiCl \longrightarrow R_3Si-\text{C}\underset{B_{10}H_{10}}{\overset{O}{\diagdown\diagup}}\text{C}-SiR_3 + 2LiCl \qquad (25)$$

$$R'-\text{C}\underset{B_{10}H_{10}}{\overset{O}{\diagdown\diagup}}\text{C}-\text{Li} + R_3SiCl \longrightarrow R'-\text{C}\underset{B_{10}H_{10}}{\overset{O}{\diagdown\diagup}}\text{C}-SiCl_3 + LiCl \qquad (26)$$

$$\text{Li}-\text{C}\underset{B_{10}H_{10}}{\overset{O}{\diagdown\diagup}}\text{C}-\text{Li} + 2R_2SiCl_2 \longrightarrow ClR_2Si-\text{C}\underset{B_{10}H_{10}}{\overset{O}{\diagdown\diagup}}\text{C}-SiR_2Cl + 2LiCl \qquad (27)$$

$$2R-\text{C}\underset{B_{10}H_{10}}{\overset{O}{\diagdown\diagup}}\text{C}-\text{Li} + SiCl_4 \longrightarrow (R-\text{C}\underset{B_{10}H_{10}}{\overset{O}{\diagdown\diagup}}\text{C}-)_2SiCl_2 + LiCl \qquad (28)$$

The corresponding silyl derivatives of the *meta-* and *para*-carborane are obtained through analogous mathathesis reactions from lithium or dilithium derivatives of $HCB_{10}H_{10}CR$ and $H\overline{CB_{10}H_{10}C}R$, respectively (R = H, alkyl, phenyl, etc.).

The mono(trimethylsilyl) derivative of 1-C_6H_5-$1,2$-$C_2B_{10}H_{11}$ has been also obtained through the classic reaction used for the preparation of the o-

carboranes from decaborane, $B_{10}H_{14}$, and alkynes.[71] Thus, decaborane reacts[72] with 1-phenyl-2-trimethlysilyl acetylene in the presence of acetonitrile, as Lewis base, to give $1\text{-}C_6H_5\text{-}2\text{-}(CH_3)_3Si\text{-}1,2\text{-}C_2B_{10}H_{10}$.

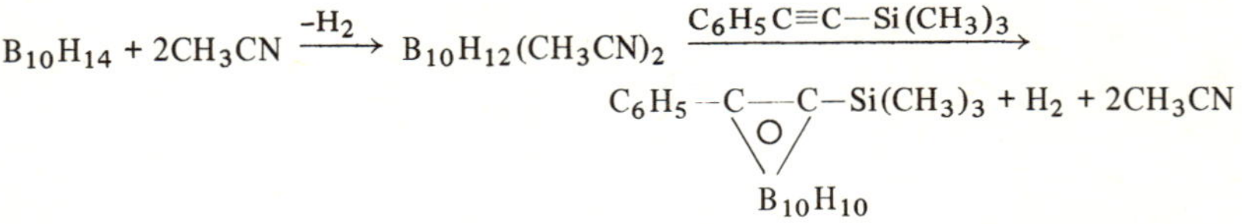

$$B_{10}H_{14} + 2CH_3CN \xrightarrow{-H_2} B_{10}H_{12}(CH_3CN)_2 \xrightarrow{C_6H_5C\equiv C-Si(CH_3)_3}$$

$$C_6H_5-\underset{\underset{B_{10}H_{10}}{\diagdown O\diagup}}{C-C}-Si(CH_3)_3 + H_2 + 2CH_3CN \tag{29}$$

On the other hand the synthesis of some 1,7–bis(silyl)–1,7–dicarba-*closo*-dodecaborane(12) derivatives was accomplished by thermal isomerization of the corresponding 1,2–carborane derivatives.

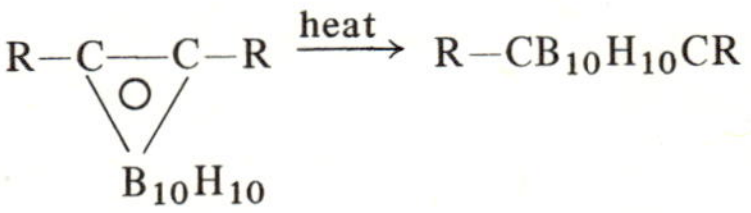

$$R-\underset{\underset{B_{10}H_{10}}{\diagdown O\diagup}}{C-C}-R \xrightarrow{heat} R-CB_{10}H_{10}CR \tag{30}$$

In this way a series of *m*-carborane derivatives in which $R = (CH_3)(C_6H_5)_2Si$, $Cl(C_6H_5)_2Si$, $(C_6H_5)(CH_3)_2Si$, and $(CH_3)_3Si$ was prepared.[73] The polyhedral rearrangement occurs at temperatures between 260° and 300°C depending on the substituents on the silicon atoms.

Treatment of $1,7\text{-}Li_2\text{-}1,7\text{-}C_2B_{10}H_{10}$ with $(Me_2SiO)_3$ yields $Li\text{-}CB_{10}H_{10}C\text{-}SiMe_2OLi$, which on subsequent reaction with $RR'MeSiCl$ ($R = H$, alkoxy groups; $R' = Me$, Ph, $CF_3CH_2CH_2$) forms $RR'MeSi\text{-}CB_{10}H_{10}C\text{-}SiMe_2OSiMeRR'$.[74] Similar *m*-carborane derivatives were also obtained[75] by reacting dilithium *m*-carborane with $(PhMeSiO)_3$ and $[(CF_3CH_2CH_2)MeSiO]_3$.

Replacement of one hydrogen atom at each silicon atom by a chlorine in both $1,2\text{-}(Me_2HSi)_2\text{-}$ and $1,7\text{-}(Me_2HSi)_2\text{-}C_2B_{10}H_{10}$ was achieved by treating the latter compounds with chlorine.[76] The bissilylchloride derivatives of 1,2–carborane react with water or amines forming five-membered disilaoxane and disilaazane rings, respectively.[77]

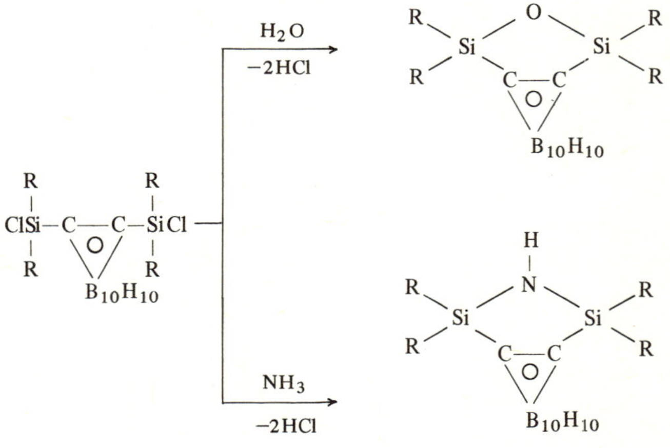

$$\tag{31}$$

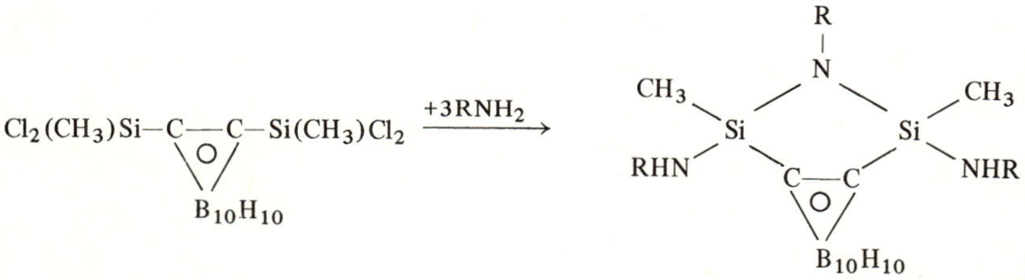

$$Cl_2(CH_3)Si-C\underset{\underset{B_{10}H_{10}}{\diagdown O\diagup}}{-}C-Si(CH_3)Cl_2 \xrightarrow{+3RNH_2}$$

(32)

The ease of formation of these five-membered rings is due to the structural peculiarities of the *ortho*-carborane molecule in which the magnitude of the carbon–carbon distance and the bond angles at the carbon atoms are particularly suited to form cycles of this geometry.

Unlike the numerous organic compounds containing nitrogen–silicon bonds which readily react with moisture, the cyclic silyl carboranes are not hydrolyzed. In addition, both the disilaoxane and disilaazane derivatives are thermally stable towards either degradation or *ortho*- to *meta*-carborane isomerization and are recovered quantitatively unchanged after heating at 500°C.[66,77]

However, cleavage of the carborane carbon–silicon bonds occurs in the presence of either inorganic or organic bases.[66]

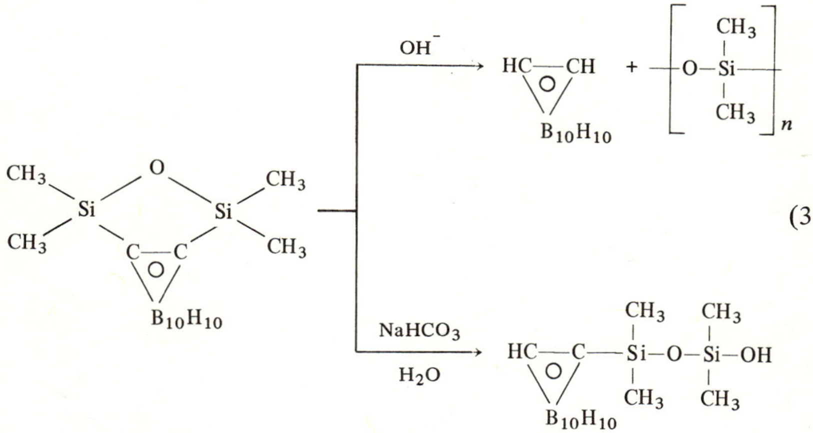

(33)

Attempts to prepare an acyclic compound by reaction of the dilithium derivative of di(*o*-carboranyl)dimethylsilane with an excess of dichlorodimethylsilane were unsuccessful and a product having a six-membered ring was obtained.[66]

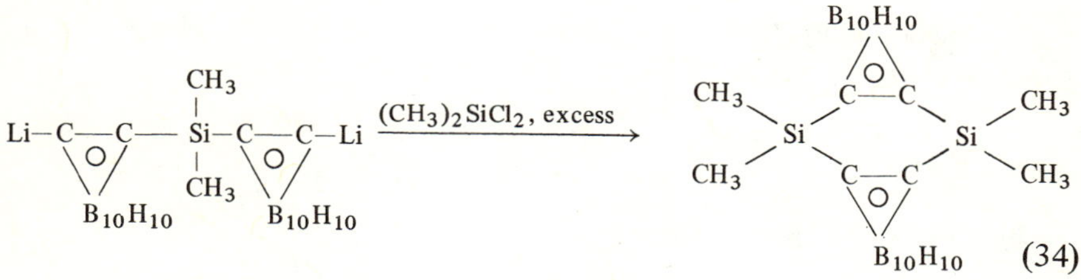

(34)

An analogous cyclic tetrachloro derivative[66] is formed when bis(trichlorosilyl)-*o*-carborane is treated with an equimolar amount of dilithio-*o*-carborane.

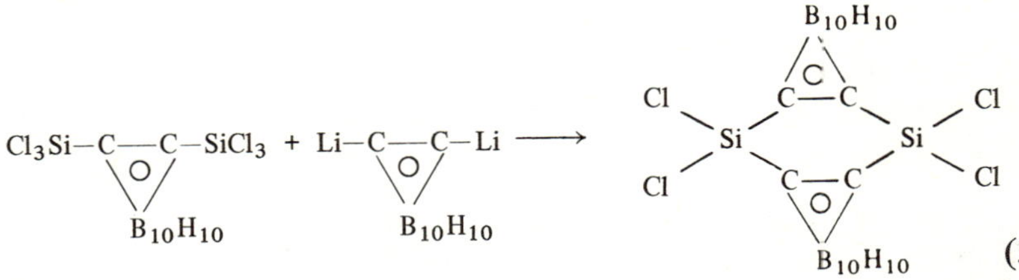

$$(35)$$

Unlike the derivatives of 1,2-dicarba-*closo*-dodecaborane(12), the C, C'-bissilyl derivatives of the 1,7-dicarba-*closo*-dodecaborane(12) isomer appear to be unable to form cyclic compounds. In fact, owing to structural reasons, the formation of small rings is not feasible in the reactions of the derivatives of the *meta*-carborane isomer. Thus, in the hydrolysis of $1,7\text{-}(ClR_2Si)_2\text{-}1,7\text{-}C_2B_{10}H_{10}$ the corresponding acyclic hydroxysilyl compounds are formed.[25,67]

$$(36)$$

$$R = CH_3, C_6H_5$$

Likewise, the reactions with ammonia or amines produces acyclic aminosilyl derivatives.[25]

$$(37)$$

The kinetics of the splitting of $HCB_{10}H_{10}C\text{-}Si(CH_3)_2R$, ($R = CH_3, C_4H_9$, CH_2Cl, $CH_2CH_2CF_3$, OCH_3, OC_2H_5), with butyl alcohol, catalyzed by LiOH, NaOH and KOH, have been investigated at $10\text{-}60°C$.[78] The process appears to be a reaction of nucleophilic splitting and can be formally described by a S_N2 mechanism.

3.5.2. Germanium, Tin, and Lead

Some tin and lead derivatives of 1,10-dicarba-*closo*-decaborane(10) formed through metal–carbon bonds have been prepared by the reactions shown in Equations (38) and (39) carried out in dry ether.[25,63]

$$2(1\text{-}C_6H_5\text{-}10\text{-}Li\text{-}1,10\text{-}C_2B_8H_8) + (CH_3)_2SnCl_2 \longrightarrow$$
$$(1\text{-}C_6H_5\text{-}1,10\text{-}C_2B_8H_8\text{-}10)_2Sn(CH_3)_2 + LiCl \qquad (38)$$

$$1\text{-}C_6H_5\text{-}10\text{-}Li\text{-}1,10\text{-}C_2B_8H_8 + (CH_3)_3PbCl \longrightarrow$$
$$1\text{-}C_6H_5\text{-}10\text{-}(CH_3)_3Pb\text{-}1,10\text{-}C_2B_8H_8 + LiCl \qquad (39)$$

When these tin and lead derivatives are boiled with an ether–benzene suspension of potassium hydroxide, cleavage of the carborane carbon–metal bond occurs.

The preparations of Ge, Sn and Pb derivatives of 1,2-, 1,7-, and 1,12-dicarba-*closo*-dodecaborane(12) are essentially analogous to those of the corresponding silicon compounds reported above.

Like silicon tetrachloride, germanium tetrachloride reacts with *C*-lithium derivatives of *o*-carborane to yield only bis(*o*-carboranyl)dichlorogermane.[64]

$$2\text{--}C_6H_5\text{--}C\text{---}C\text{--}Li + GeCl_4 \longrightarrow (C_6H_5\text{--}C\text{---}C)_2GeCl_2 + 2LiCl \tag{40}$$

In contrast, the reaction of $SnCl_4$ with *o*-phenylcarboranyl lithium gives both di- and trisubstituted compounds with a total yield of carboranyl derivatives of 62%. No tetra-substituted tin derivatives are formed.[64]

$$C_6H_5\text{--}C\text{---}C\text{--}Li + SnCl_4 \xrightarrow[\text{reflux}]{C_6H_6} (C_6H_5\text{--}C\text{---}C)_2SnCl_2 + (C_6H_5\text{--}C\text{---}C)_3SnCl \tag{41}$$

$$59\% \qquad 41\%$$

This behavior can be attributed to steric hindrance of the bulky carboranyl group. Thus, the increase in the atomic radii of the metal atom on passing from Si to Sn permits three carboranyl groups to be accommodated around the tin atom.

On reacting 1,2-dilithio-*o*-carborane with $(CH_3)_2GeCl_2$ and $(C_2H_5)_3GeCl$ in ether–benzene at 5–10°C, 1,2-bis(dimethyl-chlorogermyl)– and 1,2-bis(triethylgermyl)-*o*-carborane, respectively, are formed.[79,80] Like the corresponding chlorosilyl derivatives of *o*-carborane, 1,2-bis(dimethyl-chlorogermyl)–*o*-carborane reacts with H_2O and NH_3 to yield[79] five-membered cyclic compounds.

$$\tag{42}$$

In the reaction with the C,C'-dilithio derivative of *m*-carborane, $GeCl_4$ behaves as does $SiCl_4$[65] with formation of a hexachloro acyclic compound.

$$GeCl_4 + Li\text{-}CB_{10}H_{10}C\text{-}Li \longrightarrow Cl_3Ge\text{-}CB_{10}H_{10}C\text{-}GeCl_3 + 2LiCl \qquad (43)$$

When $(CH_3)_2GeCl_2$ is reacted with the C,C'-dilithio derivative of 1,7-dicarba-*closo*-dodecaborane(12) in a stoichiometric ratio (2:1) in diethyl ether solution, the bis(dimethyl-chlorogermyl) derivative of 1,7-carborane is formed together with a considerably greater amount of a carborane–germanium polymer.[79]

$$Li\text{-}CB_{10}H_{10}C\text{-}Li + 2(CH_3)_2GeCl_2 \xrightarrow{-LiCl} \underset{20\%}{Cl\text{-}\underset{\underset{CH_3}{|}}{\overset{\overset{CH_3}{|}}{Ge}}\text{-}CB_{10}H_{10}C\text{-}\underset{\underset{CH_3}{|}}{\overset{\overset{CH_3}{|}}{Ge}}\text{-}Cl} +$$

$$-\underset{80\%}{\left[-CB_{10}H_{10}C\text{-}\underset{\underset{CH_3}{|}}{\overset{\overset{CH_3}{|}}{Ge}}-\right]_n}- \qquad (44)$$

Several alkyltin and aryltin derivatives of both *ortho*-carborane[64,72,81] and *meta*-carborane[82,83] were prepared by reacting organotin halides with lithium derivatives of *o*- and *m*-carborane, respectively.

$$(45)$$

$$R = H, CH_3, C_6H_5; \quad R' = n\text{-}C_3H_7, C_6H_5; \quad R'' = n\text{-}C_4H_9$$

$$(46)$$

$$R = CH_3, C_6H_5; \quad R' = CH_3, C_6H_5$$

Compounds containing two tin atoms in the carborane molecule,

$$R_3Sn-C\underset{\underset{B_{10}H_{10}}{\diagup\!\diagdown}}{\overline{\quad}}C-SnR_3$$

$$(R = n-C_3H_7, C_6H_5)$$

and $R_3Sn-CB_{10}H_{10}C-SnR_3$ $(R = CH_3, C_6H_5)$, were obtained by reacting o- and m-carborane C,C'-dilithio derivatives with organotin chlorides.[64,81,82] Unlike R_2SiCl_2 and R_2GeCl_2, the dialkylchlorotin compounds form six-membered cyclic derivatives[79,81] when reacted with 1,2-dilithio-1,2-dicarba-*closo*-dodecaborane(12). However, employment of diaryl- instead of dialkylchlorotin derivatives yields[79] acyclic disubstituted o-carborane compounds.

$$(47)$$

$$R = CH_3, n-C_3H_7, n-C_4H_9$$

Recently, Russian chemists have also synthetized a series of heterocyclic compounds containing Si, Ge, and Sn as heteroatoms, directly bonded to two o-carboranyl groups:[84,85]

$$R = H, OCH_3;\ M = Si, Sn \qquad\qquad M = Si, Ge$$

In addition, a heterocyclic compound in which germanium and phosphorus atoms are both bonded to two *o*-carboranyl groups has also been reported:[86]

$$
\begin{array}{c}
B_{10}H_{10} \\
(C_6H_5)P \diamond Ge(CH_3)_2 \\
B_{10}H_{10}
\end{array}
$$

In the reaction between dilithio-*m*-carborane and $(C_6H_5)_2SnCl_2$ in ether solution at room temperature, metathesis of either one[79] or both[82] lithium atoms takes place, but the disubstituted carborane derivative is formed together with tin carborane polymers[79] when the reagents are treated in refluxing toluene.

$$
Li-CB_{10}H_{10}C-Li \xrightarrow{(C_6H_5)_2SnCl_2}
$$

ether, room temp. →

$$
\underset{\underset{C_6H_5}{|}}{\overset{\overset{C_6H_5}{|}}{HCB_{10}H_{10}C-SnCl}} + \underset{\underset{C_6H_5}{|}}{\overset{\overset{C_6H_5}{|}}{ClSnCB_{10}H_{10}CSnCl}} \underset{\underset{C_6H_5}{|}}{\overset{\overset{C_6H_5}{|}}{}}
$$

toluene, reflux →

$$
\left[\underset{\underset{C_6H_5}{|}}{\overset{\overset{C_6H_5}{|}}{-CB_{10}H_{10}CSn-}} \right]_{\approx 6} + \underset{\underset{C_6H_5}{|}}{\overset{\overset{C_6H_5}{|}}{ClSnCB_{10}H_{10}CSnCl}} \underset{\underset{C_6H_5}{|}}{\overset{\overset{C_6H_5}{|}}{}}
$$

$$(48)$$

When the C,C'-dilithium derivative of 1,12-dicarba-*closo*-dodecaborane(12), $\overline{LiCB_{10}H_{10}CLi}$, is used in this reaction in place of $LiCB_{10}H_{10}CLi$, a polymeric material appears to be the only reaction product,[79] even in diethyl ether solution. Likewise, tin carborane polymers are exclusively formed in the reaction between $Li-CB_{10}H_{10}C-Li$ and $(CH_3)_2SnCl_2$, even with the tin compound in large excess.[79]

Finally, a bis(carboranyl) derivative of tin(II) that is stable in inert atmosphere was prepared[81] by the reaction:

$$
2\ C_6H_5-\overset{B_{10}H_{10}}{\underset{}{C-C}}-Li + SnCl_2 \longrightarrow (C_6H_5-\overset{B_{10}H_{10}}{\underset{}{C-C}}-)_2Sn + 2LiCl
$$

$$(49)$$

Like the mercury carborane bond, the tin carborane bond is labile toward nucleophilic attack. Thus, an aqueous and alcoholic solution of KOH cleaves the tin carborane bond with the generation of the original carborane molecule.[87] Alkaline cleavage of *o*-, *m*-, and *p*-$HCB_{10}H_{10}C-Sn(CH_3)_3$ in CH_3OH with KOH and in CD_3OD with KOD has also been investigated.[88] The obtained rate constant values are in the order *ortho* $>$ *meta* $>$ *para*, and show an absence of any appreciable isotope effect. The overall reaction rate may be expressed by the equation: rate $= k_1 [R-Sn(CH_3)_3] [OH^-]$.

The germanium carborane and tin carborane bonds undergo cleavage even on basic alumina. Under the same conditions the silicon carborane bond is unaffected.[87]

On the other hand, the carboranyl tin derivatives are unaffected on treatment with HCl in ethanol and benzene solution.[83] The stability toward this electrophilic reagent may be due to the electron-withdrawing effect of the carboranyl groups. A study of the Mössbauer spectra of a series of carboranyl tin compounds confirmed that the carboranyl group exerts a high electron-withdrawing action on the tin atom comparable with that exhibited by the pentafluorophenyl group.[81]

3.5.3. Carborane Polymers with Group IVA Elements

Many attempts to prepare potentially useful polymers in which carborane cages are joined by Si, Ge, Sn and Pb atoms have been reported. In principle, carborane-linked polymers are of interest because of the high chemical and thermal stability of the carborane cages, and the fact that the carboranyl groups behave as energy sinks, thereby increasing the stability of neighboring bonds in the polymeric chain.

At present, a wide variety of polymers containing polyhedral carborane cages is known, and some of them exhibit truly extraordinary properties, being resistant to degradation by heat and oxidation under conditions in which conventional organic and organometallic polymers undergo deploymerization or degradation.

With respect to the *closo*-carboranes smaller than the $C_2B_{10}H_{12}$ isomers, some siloxane-type polymers have been prepared[22] starting from 2,4-dicarba-*closo*-heptaborane(7), $2,4-C_2B_5H_7$, and 1,10-dicarba-*closo*-decaborane(10), $1,10-C_2B_8H_{10}$. Thus, a polycarboranylsiloxane with molecular weight up to 12,500 containing the repeating unit

$$\left[\begin{array}{c} CH_3 \qquad\quad CH_3 \\ | \qquad\qquad\quad | \\ Si-CB_5H_5C-Si-O \\ | \qquad\qquad\quad | \\ CH_3 \qquad\quad CH_3 \end{array} \right]_n$$

was obtained[22] as a hard wax on treating equimolecular amounts of $2,4-[Si(CH_3)_2Cl]_2-2,4-C_2B_5H_5$ and $2,4-[Si(OCH_3)(CH_3)_2]_2-2,4-C_2B_5H_5$ in the presence of 2 mol% of $FeCl_3$. On the other hand, the reaction between $2,4-[Si(CH_3)_2Cl]_2-2,4-C_2B_5H_5$, $2,4-[Si(OCH_3)(CH_3)_2]_2-2,4-C_2B_5H_5$ and $1,10-[Si(OCH_3)(CH_3)_2]_2-1,10-C_2B_8H_8$ in the molar ratio 5:4:1, and in the presence of anhydrous $FeCl_3$, yielded[22] a carborane–siloxane copolymer of the type:

$$-\left[\begin{array}{c}\ \ \ CH_3\ \ \ \ \ \ \ \ \ \ \ \ \ \ CH_3\\ |\ \ \ \ \ \ \ \ \ \ \ \ \ \ \ \ \ \ |\\ -Si-CB_5H_5C-Si-O-\\ |\ \ \ \ \ \ \ \ \ \ \ \ \ \ \ \ \ \ |\\ \ \ \ CH_3\ \ \ \ \ \ \ \ \ \ \ \ \ \ CH_3\end{array}\right]_{\approx 9}\ \ \ \ \ \left[\begin{array}{c}\ \ \ CH_3\ \ \ \ \ \ \ \ \ \ \ \ \ \ CH_3\\ |\ \ \ \ \ \ \ \ \ \ \ \ \ \ \ \ \ \ |\\ -Si-CB_8H_8C-Si-O-\\ |\ \ \ \ \ \ \ \ \ \ \ \ \ \ \ \ \ \ |\\ \ \ \ CH_3\ \ \ \ \ \ \ \ \ \ \ \ \ \ CH_3\end{array}\right]_{\approx 1}-$$

An elastomeric carborane–siloxane copolymer was also obtained by copolymerization of a small amount of the larger carboranes (C_2B_8 and C_2B_{10}) with $2,4\text{-}C_2B_5H_7$ derivatives. Thus, by treating $2,4\text{-}[Si(CH_3)_2Cl]_2\text{-}2,4\text{-}C_2B_5H_5$ (50 mol%) with $2,4\text{-}[Si(OCH_3)(CH_3)_2]_2\text{-}2,4\text{-}C_2B_5H_5$ (45–35 mol%) and $1\text{-}CH_2{=}CH\text{-}2\text{-}Si(OCH_3)_2(CH_3)\text{-}1,2\text{-}C_2B_{10}H_{10}$ (5–15 mol%) ($FeCl_3$ as catalyst), a branched carborane-polymer[22] was prepared:

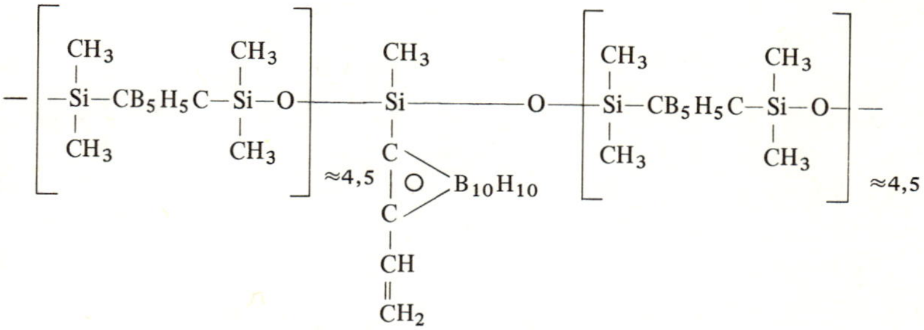

$$-\left[\begin{array}{c}CH_3\ \ \ \ \ \ \ \ \ \ CH_3\\ |\ \ \ \ \ \ \ \ \ \ \ \ \ \ |\\ -Si-CB_5H_5C-Si-O\\ |\ \ \ \ \ \ \ \ \ \ \ \ \ \ |\\ CH_3\ \ \ \ \ \ \ \ \ \ CH_3\end{array}\right]_{\approx 4,5}\!\!\!\begin{array}{c}CH_3\\ |\\ Si\\ |\\ C\\ |\ \ \ \diagdown\\ O\ \ \ \rangle B_{10}H_{10}\\ |\ \ \ \diagup\\ C\\ |\\ CH\\ \|\\ CH_2\end{array}\!\!\!-O-\left[\begin{array}{c}CH_3\ \ \ \ \ \ \ \ \ \ CH_3\\ |\ \ \ \ \ \ \ \ \ \ \ \ \ \ |\\ Si-CB_5H_5C-Si-O\\ |\ \ \ \ \ \ \ \ \ \ \ \ \ \ |\\ CH_3\ \ \ \ \ \ \ \ \ \ CH_3\end{array}\right]_{\approx 4,5}-$$

Numerous investigations have been carried out in an effort to obtain carborane polymers containing $-C_2B_{10}H_{10}$ units linked by siloxy groups or by single atoms of silicon, germanium, tin, or lead. In this manner, polymers containing the *meta*- and *para*-carborane groups were developed. Derivatives of *ortho*-carborane [1,2-dicarba-*closo*-dodecaborane(12)], however, are not suitable for formation of linear polymers. In fact, as reported above, the C,C'-difunctional *o*-carborane derivatives show a great tendency to form exopolyhedral rings, since the carbon atoms of this carborane isomer are ideally situated for participation in five- or six-membered rings.[66]

The first siloxane-linked polymers of *m*-carborane, known under the trade name Dexsil, were synthesized by a group of American chemists at the Olin Corporation.[69,89] Such polymers could not be prepared by the hydrolysis of $1,7\text{-bis(chlorodimethylsilyl)}\text{-}1,7\text{-}C_2B_{10}H_{10}$, which produces only stable diols [Equation (50)], probably owing to the influence of the strong electronegative carborane nucleus.[90]

$$\begin{array}{c}\ \ \ CH_3\ \ \ \ \ \ \ \ \ \ \ \ \ CH_3\ \ \ \ \ \ \ \ \ \ \ \ \ \ \ \ \ \ \ CH_3\ \ \ \ \ \ \ \ \ \ \ \ \ \ CH_3\\ |\ \ \ \ \ \ \ \ \ \ \ \ \ \ \ \ \ \ |\ |\ \ \ \ \ \ \ \ \ \ \ \ \ \ \ \ \ \ \ |\\ Cl-Si-CB_{10}H_{10}C-Si-Cl \xrightarrow[-HCl]{H_2O} HOSi-CB_{10}H_{10}C-SiOH\\ |\ \ \ \ \ \ \ \ \ \ \ \ \ \ \ \ \ \ |\ |\ \ \ \ \ \ \ \ \ \ \ \ \ \ \ \ \ \ \ |\\ \ \ \ CH_3\ \ \ \ \ \ \ \ \ \ \ \ \ CH_3\ \ \ \ \ \ \ \ \ \ \ \ \ \ \ \ \ \ \ CH_3\ \ \ \ \ \ \ \ \ \ \ \ \ \ CH_3\end{array}\qquad(50)$$

However, on bulk heating of equimolecular amounts of $1,7\text{-}[Si(CH_3)_2Cl]_2\text{-}1,7\text{-}C_2B_{10}H_{10}$ and $1,7\text{-}[Si(OCH_3)(CH_3)_2]_2\text{-}1,7\text{-}C_2B_{10}H_{10}$ in the presence of a catalytic quantity of $FeCl_3$ at temperatures up to $190°C$, evolution of methyl chloride occurred and a solid polymer with mol wt = 16,500 (for Dexsil 100*) having *m*-carborane units in the backbone was obtained:[69]

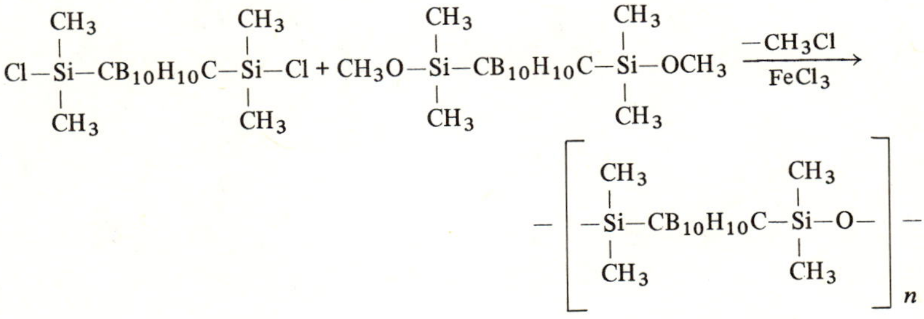

$$Cl\underset{\underset{CH_3}{|}}{\overset{\overset{CH_3}{|}}{Si}}-CB_{10}H_{10}C-\underset{\underset{CH_3}{|}}{\overset{\overset{CH_3}{|}}{Si}}-Cl + CH_3O-\underset{\underset{CH_3}{|}}{\overset{\overset{CH_3}{|}}{Si}}-CB_{10}H_{10}C-\underset{\underset{CH_3}{|}}{\overset{\overset{CH_3}{|}}{Si}}-OCH_3 \xrightarrow[FeCl_3]{-CH_3Cl}$$

$$\left[-\underset{\underset{CH_3}{|}}{\overset{\overset{CH_3}{|}}{Si}}-CB_{10}H_{10}C-\underset{\underset{CH_3}{|}}{\overset{\overset{CH_3}{|}}{Si}}-O- \right]_n - \tag{51}$$

Dexsil 100

Polymers containing two (Dexsil 200) and three (Dexsil 300) siloxy groups in the repeating units were also prepared by reacting $1,7\text{-}[Si(OCH_3)(CH_3)_2]_2\text{-}1,7\text{-}C_2B_{10}H_{10}$ with $(CH_3)_2SiCl_2$ and $Cl(CH_3)_2Si\text{-}O\text{-}Si(CH_3)_2Cl$, respectively. Iron(III) chloride was used as catalyst.[69,89,91,92]

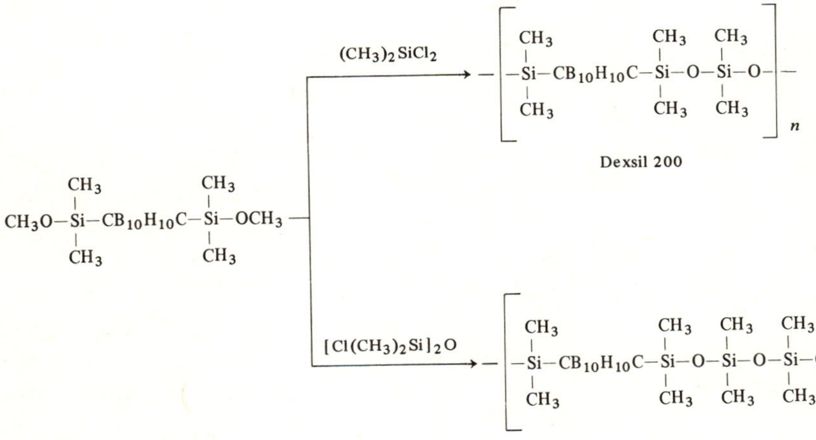

$$\tag{52}$$

with $(CH_3)_2SiCl_2$: Dexsil 200

with $[Cl(CH_3)_2Si]_2O$: Dexsil 300

Dexsil 100 is a crystalline solid and melts over the range $233\text{-}240°C$, but the higher Dexsil polymers are rubbery materials.

Elastomers with four siloxy groups per repeating units (Dexsil 400) have been also obtained:[89]

---

*The Dexsil 100 polymer has one oxygen atom per repeating unit, the Dexsil 200 has two, and so on.

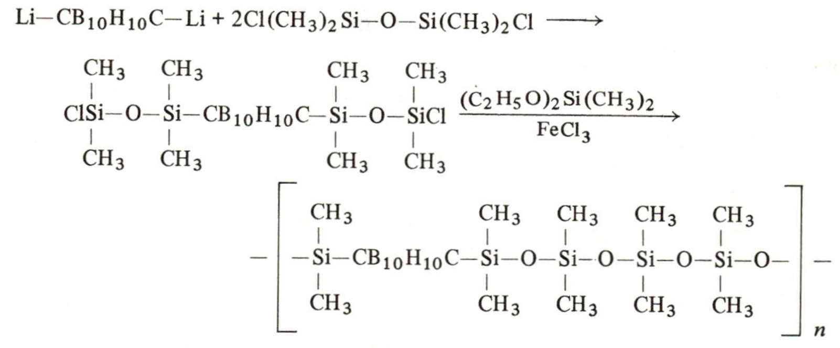

$$\text{Li}-\text{CB}_{10}\text{H}_{10}\text{C}-\text{Li} + 2\text{Cl}(\text{CH}_3)_2\text{Si}-\text{O}-\text{Si}(\text{CH}_3)_2\text{Cl} \longrightarrow$$

$$\underset{\substack{|\\ \text{CH}_3}}{\overset{\substack{\text{CH}_3\\ |}}{\text{ClSi}}}-\text{O}-\underset{\substack{|\\ \text{CH}_3}}{\overset{\substack{\text{CH}_3\\ |}}{\text{Si}}}-\text{CB}_{10}\text{H}_{10}\text{C}-\underset{\substack{|\\ \text{CH}_3}}{\overset{\substack{\text{CH}_3\\ |}}{\text{Si}}}-\text{O}-\underset{\substack{|\\ \text{CH}_3}}{\overset{\substack{\text{CH}_3\\ |}}{\text{SiCl}}} \xrightarrow[\text{FeCl}_3]{(\text{C}_2\text{H}_5\text{O})_2\text{Si}(\text{CH}_3)_2}$$

$$\left[-\underset{\substack{|\\ \text{CH}_3}}{\overset{\substack{\text{CH}_3\\ |}}{\text{Si}}}-\text{CB}_{10}\text{H}_{10}\text{C}-\underset{\substack{|\\ \text{CH}_3}}{\overset{\substack{\text{CH}_3\\ |}}{\text{Si}}}-\text{O}-\underset{\substack{|\\ \text{CH}_3}}{\overset{\substack{\text{CH}_3\\ |}}{\text{Si}}}-\text{O}-\underset{\substack{|\\ \text{CH}_3}}{\overset{\substack{\text{CH}_3\\ |}}{\text{Si}}}-\text{O}-\underset{\substack{|\\ \text{CH}_3}}{\overset{\substack{\text{CH}_3\\ |}}{\text{Si}}}-\text{O}- \right]_n$$

(53)

Dexsil 400

Branched carborane–siloxane polymers in which some of the methyl substituents on the silicon atoms of the chain are replaced by

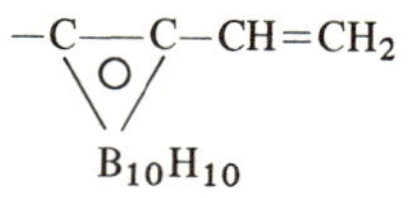

$$-\underset{\underset{\text{B}_{10}\text{H}_{10}}{\diagdown\,\text{O}\,\diagup}}{\text{C}-\!\!-\!\!\text{C}}-\text{CH}=\text{CH}_2$$

groups were obtained by adding to these reaction mixtures a small proportion of[69,93,94]

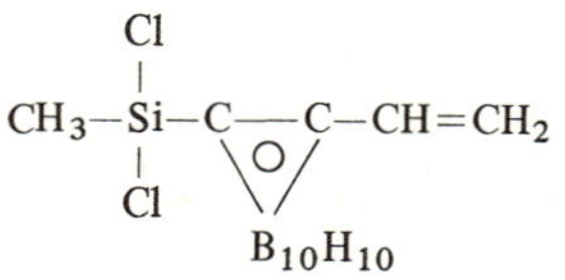

$$\text{CH}_3-\underset{\substack{|\\ \text{Cl}}}{\overset{\substack{\text{Cl}\\ |}}{\text{Si}}}-\underset{\underset{\text{B}_{10}\text{H}_{10}}{\diagdown\,\text{O}\,\diagup}}{\text{C}-\!\!-\!\!\text{C}}-\text{CH}=\text{CH}_2$$

The Dexsil-type polymers below a molecular weight of 10,000 are soluble in chlorobenzene, xylene, and decalin; above 10,000 there is little if any solubility in these solvents, but some solubility in hexamethyl phosphoramide.[69]

Differential scanning calorimetry studies both under nitrogen and in air have been carried out on the carborane–siloxane polymers.[69] The Dexsil 100, Dexsil 200 and Dexsil 300 polymers appear to be stable up to 500°C in an inert atmosphere. Upon heating in air, however, only the first, which contains one $\text{Si}(\text{CH}_3)_2$ group attached directly to the carbon atoms of the *m*-carborane moiety, remains inert. This may be due to inductive stabilization of the silyl groups bonded directly to the electron-withdrawing *m*-carborane nucleus. In fact, the thermal and oxidative stability appears to decrease gradually on increasing the number of siloxy groups not bonded directly to carboranyl units. Thus, incipient oxidative degradation has been observed at 335°C in the case of the Dexsil 200 and Dexsil 300 polymers, containing two and three $(\text{CH}_3)_2\text{SiO}$ groups per repeat unit, and at 325°C in the case of the Dexsil 400 species, which contains four siloxy groups per repeat unit.[69]

As reported above, except for the Dexsil 100 species which is a soluble crystalline product, the carborane–siloxane polymers formed through FeCl_3-catalyzed copolymerization of chloro- and methoxy-terminated monomers

[Equations (52) and (53)] are rubber-like materials which swell in organic solvents. In fact, the final stage of $FeCl_3$-catalyzed condensation involves a cross-linking step and, consequently, a network polymer is formed instead of the expected linear species.[95] Subsequently, it has been discovered that long-chain linear polymers can be obtained by simple hydrolytic condensation of $1,7$-$(ClMe_2SiOMe_2Si)_2$-$1,7$-$C_2B_{10}H_{10}$ and $1,7$-$(ClMe_2SiOMe_2SiOMe_2Si)_2$-$1,7$-$C_2B_{10}H_{10}$ at ice-bath temperature, as well as by acid-catalyzed condensation of the corresponding silanols.[95] The obtained polymers are liquids and soluble waxes with molecular weights between 16,000 and 30,000, which are potentially useful as high-temperature liquids and coatings and can be cured at room temperature to form elastomers.

Recently, chemists of the Union Carbide Corporation have developed new routes to linear, high-molecular-weight *m*-carborane–siloxane polymers.[96,97] To obviate the deleterious effects of the iron(III) chloride catalyst used in previous routes[69] to linear *m*-carborane–siloxane polymers, a process based on the condensation reaction

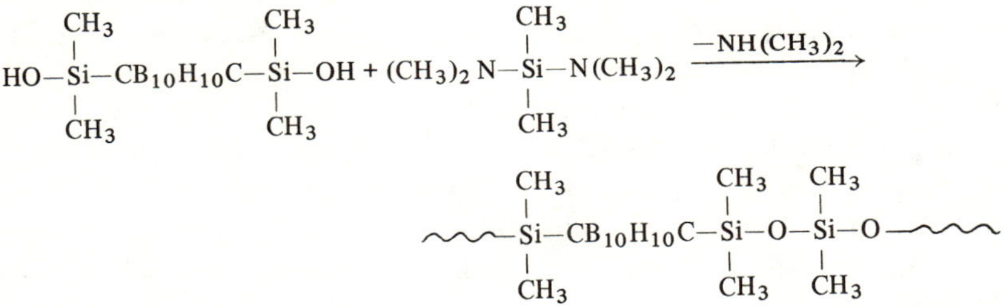

$$
\begin{array}{c}
\underset{\displaystyle \text{CH}_3}{\overset{\displaystyle \text{CH}_3}{\text{HO}-\overset{|}{\underset{|}{\text{Si}}}-\text{CB}_{10}\text{H}_{10}\text{C}}} - \underset{\displaystyle \text{CH}_3}{\overset{\displaystyle \text{CH}_3}{\overset{|}{\underset{|}{\text{Si}}}-\text{OH}}} + (\text{CH}_3)_2\,\text{N}-\underset{\displaystyle \text{CH}_3}{\overset{\displaystyle \text{CH}_3}{\overset{|}{\underset{|}{\text{Si}}}-\text{N(CH}_3)_2}} \xrightarrow{\;-\text{NH(CH}_3)_2\;}
\end{array}
$$

$$
\sim\!\!\sim\!\!-\underset{\displaystyle \text{CH}_3}{\overset{\displaystyle \text{CH}_3}{\overset{|}{\underset{|}{\text{Si}}}-\text{CB}_{10}\text{H}_{10}\text{C}}} - \underset{\displaystyle \text{CH}_3}{\overset{\displaystyle \text{CH}_3}{\overset{|}{\underset{|}{\text{Si}}}-\text{O}}} - \underset{\displaystyle \text{CH}_3}{\overset{\displaystyle \text{CH}_3}{\overset{|}{\underset{|}{\text{Si}}}-\text{O}}}\!\!\sim\!\!\sim \tag{54}
$$

has been exploited. Polymers with a degree of polymerization up to 50 (mol wt = 18,000) have been obtained under mild reaction conditions.[96] Attempts to achieve higher molecular weights were thwarted by a dimethylamine-induced cleavage of the carborane–silicon bonds in the monomer. Thus, in order to obtain products with higher molecular weight, the reaction of bisureidosilanes with *m*-carborane disilanols has also been investigated both in chlorobenzene solution at $-10°C$ and in the absence of solvents at $160°C$:[96]

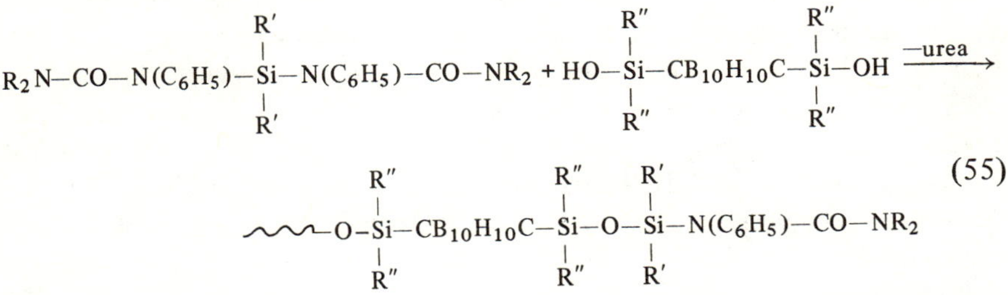

$$
\text{R}_2\text{N}-\text{CO}-\text{N(C}_6\text{H}_5)-\underset{\displaystyle \text{R}'}{\overset{\displaystyle \text{R}'}{\overset{|}{\underset{|}{\text{Si}}}}}-\text{N(C}_6\text{H}_5)-\text{CO}-\text{NR}_2 + \text{HO}-\underset{\displaystyle \text{R}''}{\overset{\displaystyle \text{R}''}{\overset{|}{\underset{|}{\text{Si}}}}}-\text{CB}_{10}\text{H}_{10}\text{C}-\underset{\displaystyle \text{R}''}{\overset{\displaystyle \text{R}''}{\overset{|}{\underset{|}{\text{Si}}}}}-\text{OH} \xrightarrow{\;-\text{urea}\;}
$$

$$
\sim\!\!\sim\!\!-\text{O}-\underset{\displaystyle \text{R}''}{\overset{\displaystyle \text{R}''}{\overset{|}{\underset{|}{\text{Si}}}}}-\text{CB}_{10}\text{H}_{10}\text{C}-\underset{\displaystyle \text{R}''}{\overset{\displaystyle \text{R}''}{\overset{|}{\underset{|}{\text{Si}}}}}-\text{O}-\underset{\displaystyle \text{R}'}{\overset{\displaystyle \text{R}'}{\overset{|}{\underset{|}{\text{Si}}}}}-\text{N(C}_6\text{H}_5)-\text{CO}-\text{NR}_2 \tag{55}
$$

The latter condensation reaction, in which the leaving group is unreactive urea, is very effective and high-molecular-weight (mol wt = 250,000) linear polymers have been obtained. These polymers are marketed under the trade name UCarSil.

Diphenyl, methylphenyl and methylvinylsiloxane groups have been incorporated into the polymers in order to improve performance and curing properties.[96] Elastomers with useful general properties after heat aging in air at 315°C for 300 hr have then been obtained.[98,99]

The linear, high-molecular-weight *m*-carborane-dimethylsiloxane polymers show a crystalline metlting point around 68°C. This crystalline phase detracted from the elastomeric properties and results in a large modulus change in the region 50–70°C. Thus, to obtain an elastomeric material, this crystallinity has been disrupted by replacing 30–50% of the *m*-carborane with *p*-carborane moieties or by incorporating phenyl substituents on the polymer backbone.[100] The latter approach yields polymers having improved thermooxidative stability. In particular, a *m*-carborane–dimethyl-methylphenylsiloxane (67/33 mol%) polymer exhibits a weight loss of only 1.5% on heating at 500°C in air. Correlations of the glass transition temperature and thermooxidative stability with the polymer structure have also been reported.[100]

Linear *m*-carborane–siloxane polymers with one (UCarSil F_1), two (UCarSil F_2) and three (UCarSil F_3) trifluoropropyl moieties per repeat unit have also been prepared by condensation reactions [see Equation (55)] of bis(*N*-pyrrolidino-*N'*-phenylureido)dialkylsilanes with bis(hydroxydialkylsilyl)-*m*-carborane, where the dialkyl groups are either methyltrifluoropropyl or dimethyl.[101] The resulting polymers, having molecular weights between 100,000 and 220,000, exhibit greater thermal and oxidative stability than fluorosilicones. In fact, TGA studies show that the fluorosilicone polymers start to lose weight at 50 to 100°C below the onset of weight loss for the trifluoropropyl-substituted UCarSil polymers.[101]

Use of the carborane–siloxane polymers in gas chromatography has also been reported. Thus, carborane–siloxane polymers of the types Dexsil 300 GC, Dexsil 400 GC, and Dexsil 410 GC have been employed as stationary phases in separations of biomedical compounds, chlorinated aromatic compounds, drugs, fatty acids, hydrocarbons, pesticides, steroids, triglycerides, vitamins, and waxes.[102] Furthermore, gas-chromatographic separation of organosilicon and organic substances has been achieved on using, as stationary phase, carborane–siloxane polymers having molecular weights of 100,000 to 1,000,000.[103]

The preparation of inorganic polymers formed through *m*- or *p*-carborane units linked by single atoms of Group IVA elements has also been developed in order to obtain thermally stable materials. Thus, by polycondensation reactions of $1,7$-Li_2-$1,7$-$C_2B_{10}H_{10}$ with R_2SiCl_2 ($R = CH_3, C_2H_5, C_6H_5$), $(C_6H_5)_2GeCl_2$, R_2SnCl_2 ($R = CH_3$, *n*-C_4H_9, C_6H_5) and $(C_6H_5)_2PbCl_2$ in tetrahydrofuran at temperatures between 0 and 60°C, hard inflexible polymers containing the repeating units

$$-CB_{10}H_{10}C-\overset{\overset{\displaystyle R}{|}}{\underset{\underset{\displaystyle R}{|}}{M}}- \qquad M = Si, Ge, Sn, Pb$$

were obtained,[79,104-106] and found to have molecular weights of up to 9,500 (osmometric method). The thermal stability of these polymers decreases in the order $Si > Ge > Sn > Pb$. The silicon carborane polymers containing $Si(CH_3)_2$ groups appear to be stable on heating in air up to $600°C$.[106]

Polymers based on *para*-carborane, 1,12-dicarba-*closo*-dodecaborane(12), have been prepared by treating $Li-CB_{10}H_{10}C-Li$ with $(CH_3)_2SnCl_2$.[79] A random copolymer with *meta*-carborane units linked by both tin and germanium atoms [Equation (56)] and a copolymer in which both *m*- and *p*-carborane moieties are connected through tin atoms [Equation (57)] have also been reported.[79]

$$2n(Li-CB_{10}H_{10}C-Li) + n(CH_3)_2GeCl_2 + n(CH_3)_2SnCl_2 \longrightarrow$$

$$\left[\left(CB_{10}H_{10}C-\underset{\underset{CH_3}{|}}{\overset{\overset{CH_3}{|}}{Ge}}\right)\left(CB_{10}H_{10}C-\underset{\underset{CH_3}{|}}{\overset{\overset{CH_3}{|}}{Sn}}\right)\right]_n + 4n(LiCl) \quad (56)$$

mp: 350-355°

$$n(Li-CB_{10}H_{10}C-Li) + n(Li-CB_{10}H_{10}C-Li) + 2n[(CH_3)_2SnCl_2] \longrightarrow$$

$$\left[\left(CB_{10}H_{10}C-\underset{\underset{CH_3}{|}}{\overset{\overset{CH_3}{|}}{Sn}}\right)\left(CB_{10}H_{10}C-\underset{\underset{CH_3}{|}}{\overset{\overset{CH_3}{|}}{Sn}}\right)\right]_n + 4n(LiCl) \quad (57)$$

mp: 370-375°

3.6. Derivatives of Group VA Elements

Numerous carborane derivatives containing nonmetallic elements of Group VA have been reported, but these are beyond the scope of this chapter.

A tris(phenyl-*o*-carboranyl)antimony[107] has been obtained by reacting $1-Li-2-C_6H_5-1,2-C_2B_{10}H_{10}$ with $SbCl_3$ in diethyl ether at $5-10°C$ [Equation (58)]:

$$3\,C_6H_5-C\underset{B_{10}H_{10}}{\diagdown_O\diagup}C-Li + SbCl_3 \longrightarrow (C_6H_5-C\underset{B_{10}H_{10}}{\diagdown_O\diagup}C-)_3Sb + 3\,LiCl \quad (58)$$

The arsenic analog can be prepared in a similar way using $AsCl_3$ in place of $SbCl_3$.[107]

B-mercurated carboranes have been shown to be convenient starting compounds for the preparation of carboranyl derivatives of Group IV-VI elements containing a carborane boron–heteroelement bond.[108] Thus, by reacting equimolecular amounts of $[9\text{-}o\text{-}$ (or $9\text{-}m\text{-}) C_2B_{10}H_{11}]_2Hg$ and MCl_3, where $M = As$ and Sb, in benzene solution at $20°C$ for a few minutes, the compounds 9–

$[AsCl_2]$-o-$C_2B_{10}H_{11}$, 9-$[AsCl_2]$-m-$C_2B_{10}H_{11}$, and 9-$[SbCl_2]$-m-$C_2B_{10}H_{11}$ have been obtained in 50% yield.[109] Likewise, the reaction of [9-o- (or 9-m-) $C_2B_{10}H_{11}]_2$Hg with $AsCl_3$ in 1,2-dichloroethane smoothly yields 9-$[AsCl_2]$-o-$C_2B_{10}H_{11}$, as shown in Equation (59), and 9-$[AsCl_2]$-m-$C_2B_{10}H_{11}$, respectively.[110]

$$(\text{H}-\text{C}\!-\!\!-\!\text{C}-\text{H})_2\,\text{Hg} + \text{AsCl}_3 \longrightarrow \text{H}-\text{C}\!-\!\!-\!\text{C}-\text{H} \qquad + \text{H}-\text{C}\!-\!\!-\!\text{C}-\text{H} \tag{59}$$

$$9-\text{B}_{10}\text{H}_9- \qquad\qquad 9-\text{B}_{10}\text{H}_9-\text{AsCl}_2 \qquad 9-\text{B}_{10}\text{H}_9-\text{HgCl}$$

These antimony and arsenic derivatives are stable in the air and resistant to heat. The group VA element in all of these compounds, as in the starting mercury derivatives, is bonded to the boron atom in position 9 of the carborane cage.

3.7. Zinc and Mercury Derivatives

3.7.1. Derivatives Formed through Carborane Carbon–Metal Bonds

A zinc derivative of 1,2-dicarba-*closo*-dodecaborane(12) has been reported.[111] By reacting 1-C_6H_5-2-H-1,2-$C_2B_{10}H_{10}$ with diethylzinc in hexamethyltriamidophosphate at elevated temperatures, bis[1-C_6H_5-1,2-dicarba-*closo*-dodecaborane(12)] zinc was obtained:

$$2\,\text{C}_6\text{H}_5-\text{C}\!-\!\!-\!\text{C}-\text{H} + \text{Et}_2\text{Zn} \longrightarrow \text{C}_6\text{H}_5-\text{C}\!-\!\!-\!\text{C}-\text{Zn}-\text{C}\!-\!\!-\!\text{C}-\text{C}_6\text{C}_5 \tag{60}$$

$$\text{B}_{10}\text{H}_{10} \qquad\qquad\qquad \text{B}_{10}\text{H}_{10} \qquad \text{B}_{10}\text{H}_{10}$$

Several mercury derivatives of 1,6– and 1,10–dicarba-*closo*-decaborane(10) and 1,2-, 1,7-, and 1,12-dicarba-*closo*-dodecaborane(12) formed through carbon–mercury bonds were reported. By the action of mercury(II) bromide on 1-C_6H_5-6-Li-1,6-$C_2B_8H_8$ and 1-C_6H_5-10-Li-1,10-$C_2B_8H_8$ in ether–benzene solution at room temperature in an argon atmosphere, high yields of symmetrical bis(carborane) mercury derivatives[63] were obtained:

$$2\text{-}C_6H_5\text{-}6\text{-}Li\text{-}1,6\text{-}C_2B_8H_8 + \text{HgBr}_2 \longrightarrow (1\text{-}C_6H_5\text{-}1,6\text{-}C_2B_8H_8\text{-}6\text{-})_2\text{Hg} + 2\text{LiBr} \tag{61}$$

$$1\text{-}C_6H_5\text{-}10\text{-}Li\text{-}1,10\text{-}C_2B_8H_8 + \text{HgBr}_2 \longrightarrow (1\text{-}C_6H_5\text{-}1,10\text{-}C_2B_8H_8\text{-}10\text{-})_2\text{Hg} + 2\text{LiBr} \tag{62}$$

Even the use of a large excess of mercury(II) bromide in these reactions does not lead to the formation of monocarborane mercury derivatives. Unsymmetrical carborane mercury derivatives were prepared by reacting the *C*-lithium derivatives of 1,6– and 1,10–$C_2B_8H_{10}$ with chloromethylmercury in ether–benzene mixtures:[63]

$$1\text{-}C_6H_5\text{-}6\text{-}Li\text{-}1,6\text{-}C_2B_8H_8 + \text{CH}_3\text{HgCl} \xrightarrow{25^\circ} 1\text{-}C_6H_5\text{-}6\text{-}CH_3Hg\text{-}1,6\text{-}C_2B_8H_8 + \text{LiCl} \tag{63}$$

$$1\text{-}C_6H_5\text{-}10\text{-}Li\text{-}1,10\text{-}C_2B_8H_8 + CH_3HgCl \xrightarrow{25^\circ} 1\text{-}C_6H_5\text{-}10\text{-}CH_3Hg\text{-}1,10\text{-}C_2B_8H_8 + LiCl$$
$$(64)$$

These carborane methylmercury derivatives are stable compounds and melt without decomposition, and no disproportionation is observed. On the other hand, when $(1\text{-}C_6H_5\text{-}1,10\text{-}C_2B_8H_8\text{-}10\text{-})_2Hg$ is reacted with $HgBr_2$ in boiling chlorobenzene for 25 hr, disproportionation occurs:

$$(1\text{-}C_6H_5\text{-}1,10\text{-}C_2B_8H_8\text{-}10\text{-})_2Hg + HgBr_2 \longrightarrow 2(1\text{-}C_6H_5\text{-}10\text{-}BrHg\text{-}1,10\text{-}C_2B_8H_8)$$
$$(65)$$

Carborane methylmercury undergoes cleavage of the CH_3-Hg bond on treatment with bromine in benzene at room temperature:

$$1\text{-}C_6H_5\text{-}6(10)\text{-}CH_3Hg\text{-}1,6(1,10)\text{-}C_2B_8H_8 + Br_2 \longrightarrow$$
$$1\text{-}C_6H_5\text{-}6(10)BrHg\text{-}1,6(1,10)\text{-}C_2B_8H_8 + CH_3Br$$
$$(66)$$

However, the carborane $C-Hg$ bond is cleaved when 1-phenyl-10-methylmercury-1,10-dicarba-*closo*-decaborane(10) is treated with KOH in a benzene-water suspension or with HCl in ethanol at reflux temperatures:

$$1\text{-}C_6H_5\text{-}10\text{-}CH_3Hg\text{-}1,10\text{-}C_2B_8H_8 \left\{ \begin{array}{l} \xrightarrow[10\ h]{KOH} 1\text{-}C_6H_5\text{-}10\text{-}H\text{-}1,10\text{-}C_2B_8H_8 \\ \\ \xrightarrow[9\ h]{HCl} 1\text{-}C_6H_5\text{-}10\text{-}H\text{-}1,10\text{-}C_2B_8H_8 + CH_3HgCl \end{array} \right.$$
$$(67)$$

A large number of mercury derivatives of 1,2-, 1,7-, and 1,12-dicarba-*closo*-dodecaborane(12) have been prepared and their unusual chemical behavior investigated. In general, these compounds are obtained directly by treating lithiocarboranes or carboranyl Grignard reagents with a mercury(II) halide in benzene or ether–benzene solution at 45–50°C.[64,72,112–114]

The use of mercury(II) dihalides leads only to the formation of symmetrical bis(*o*-carboranyl)mercury derivatives [Equation (68)] even when the mercury halide reagent is present in large excess.

$$2\ R-C\underset{B_{10}H_{10}}{\overset{}{\diagup\!\!\diagdown}}C-Li + HgX_2 \longrightarrow R-C\underset{B_{10}H_{10}}{\overset{}{\diagup\!\!\diagdown}}C-Hg-C\underset{B_{10}H_{10}}{\overset{}{\diagup\!\!\diagdown}}C-R + 2LiX$$
$$(68)$$

$$R = H,\ CH_3,\ CH_2{=}CH,\ C_6H_5\ ;\ X = Cl,\ Br$$

This behavior is an unusual feature in the chemistry of organomercury compounds.

Monocarboranyl mercury derivatives are formed by treating the lithiocarborane with an alkyl- or arylmercury halide:

$$R-C{\diagdown}\!\!\diagup\!\!C-Li + R'HgX \longrightarrow R-C{\diagdown}\!\!\diagup\!\!C-HgR' + LiX \qquad (69)$$

$$B_{10}H_{10} \qquad\qquad\qquad B_{10}H_{10}$$

R - H, CH_3, $CH_2{=}CH$, C_6H_5; R' = CH_3, C_6H_5, ferrocenyl

The $1\text{-}(CH_3Hg)\text{-}2\text{-}(ClCH_2)\text{-}1,2\text{-}C_2B_{10}H_{10}$ derivative has also been prepared and its structure determined by X-ray diffraction.[115] The crystals are monoclinic and the molecular conformation appears to be distorted so that the Cl atom is near the Hg atom. Thus the Hg···Cl coordination is attained with a distance of 3.27(3) Å.

Unlike $1,7\text{-}C_2B_{10}H_{12}$, $1,2\text{-}C_2B_{10}H_{12}$ and its derivatives can be directly mercuriated by the action of alkyl- or arylmercury hydroxides:[116]

$$R-C{\diagdown}\!\!\diagup\!\!C-H + R'HgOH \longrightarrow R-C{\diagdown}\!\!\diagup\!\!C-HgR' + H_2O \qquad (70)$$

$$B_{10}H_{10-n}X_n \qquad\qquad\qquad B_{10}H_{10-n}X_n$$

R = H, CH_3, C_6H_5; R' = CH_3, C_6H_5; X = Cl, Br; n = 0, 2, 4

This reaction takes place on treating the reagents in benzene or toluene solution and occurs more readily in the case of the *B*-halo-derivatives of *o*-carborane and *C*(1)-substituted *o*-carboranes with electron-attracting substituents.

Both symmetrical and unsymmetrical carboranyl mercury derivatives in general are exceptionally stable. Thus, the symmetrical bis(carboranyl)mercury compounds are unchanged even on prolonged heating at 300–350°C and the unsymmetrical derivatives, RHgR' (R = carboranyl), exhibit no tendency to disproportionate to R_2Hg and $R_2'Hg$ at the melting point.[112] In spite of the fact that the cleavage of the carbon–mercury bond in dialkyl- and diarylmercury compounds is easily achieved by the action of electrophilic reagents such as HCl, Br_2 and $HgCl_2$, the corresponding reactions with *o*-carboranyl mercury derivatives, especially the symmetrical ones, occur either with difficulty under severe conditions or not at all. Thus, symmetrical bis(*o*-carboranyl)mercury derivatives are unaffected by prolonged boiling (20 hr) with HCl in ethanol or by the action of bromine in boiling dichloromethane and benzene.

Unlike conventional organometallic R_2Hg compounds for which the reaction $R_2Hg + HgCl_2 \longrightarrow 2RHgCl$ usually takes place very easily, bis(phenyl-*o*-carboranyl)mercury is stable on continuous refluxing with $HgCl_2$ in ethanol. However, phenylcarboranylmercury chloride is formed on treating bis(phenyl-*o*-carboranyl) mercury with $HgCl_2$ over a long period in nitrobenzene at 210°C [Equation (71)] but even then the yield of the chloride is only 20%.

$$C_6H_5-C{\diagdown}\!\!\diagup\!\!C-Hg-C{\diagdown}\!\!\diagup\!\!C-C_6H_5 + HgCl_2 \xrightarrow[\text{8 hr}]{210^\circ} 2\,C_6H_5-C{\diagdown}\!\!\diagup\!\!C-HgCl \qquad (71)$$

$$B_{10}H_{10} \qquad B_{10}H_{10} \qquad\qquad\qquad B_{10}H_{10}$$

Moreover, dimethylmercury reacts with bis(phenyl-*o*-carboranyl)mercury under severe conditions (210°C for 9 hr) giving methyl(phenyl-*o*-carboranyl)mercury in 70% yield.[117] On treating methyl(phenyl-*o*-carboranyl)mercury with bromine in benzene at room temperature, cleavage of the Hg—CH$_3$ bond occurs:

$$C_6H_5-C\underset{\underset{B_{10}H_{10}}{}}{-}C-Hg-CH_3 + Br_2 \longrightarrow C_6H_5-C\underset{\underset{B_{10}H_{10}}{}}{-}C-HgBr + CH_3Br \qquad (72)$$

Mercury(II) chloride reacts[113] in the same manner, the *o*-carboranyl–mercury bond being unaffected:

$$C_6H_5-C\underset{\underset{B_{10}H_{10}}{}}{-}C-Hg-CH_3 + HgCl_2 \xrightarrow[78^\circ C,\ 12\ hr]{C_2H_5OH} \left[C_6H_5-C\underset{\underset{B_{10}H_{10}}{}}{-}C-Hg \overset{CH_3}{\underset{Cl}{\cdots}} HgCl \right] \longrightarrow$$

$$C_6H_5-C\underset{\underset{B_{10}H_{10}}{}}{-}C-HgCl + CH_3HgCl \qquad (73)$$

On the other hand, the action of hydrogen chloride on methyl(phenyl-*o*-carboranyl) mercury in boiling ethanol breaks the carborane carbon–mercury bond, forming phenyl-*o*-carborane quantitatively:

$$C_6H_5-C\underset{\underset{B_{10}H_{10}}{}}{-}C-Hg-CH_3 + HCl \xrightarrow[78^\circ C,\ 8\ hr]{C_2H_5OH} C_6H_5-C\underset{\underset{B_{10}H_{10}}{}}{-}CH + CH_3HgCl \qquad (74)$$

Considering that the carboranyl radical is a strong electron acceptor, the stability of the carborane carbon–mercury bond in the bis(*o*-carboranyl)mercury derivatives toward the action of HCl in ethanol may be due to the reduction in electron density on the carborane carbon atom.[113] Vice versa, in methyl(phenyl-*o*-carboranyl)mercury, an electron donation from the methyl group through the mercury atom compensates this carboranyl electron absorption and thus the carborane carbon–mercury bond can be cleaved by HCl.[66] On replacing the methyl group with phenyl, this compensation occurs to a minor extent, so that the reaction with dry HCl in benzene–ether (3:1) solution at reflux temperature proceeds slowly:

$$C_6H_5-C\underset{\underset{B_{10}H_{10}}{}}{-}C-Hg-C_6H_5 + HCl \xrightarrow{20\ hr} \begin{cases} C_6H_5-C\underset{\underset{B_{10}H_{10}}{}}{-}C-H + C_6H_5HgCl \\[2em] C_6H_5-C\underset{\underset{B_{10}H_{10}}{}}{-}C-HgCl + C_6H_6 \end{cases} \qquad (75)$$

Unlike most "ordinary" organomercury derivatives, and in agreement with the anomalous properties of the carborane C—Hg bonds, the unsymmetrical *o*-carboranyl mercury compounds, such as phenyl-*o*-carboranylmercury chloride and bromide, appear to be unaffected by strong symmetrizing agents such as potassium cyanide and iodide.

Nucleophilic agents react with symmetrical bis(*o*-carboranyl)mercury derivatives, usually giving products of cleavage of the C—Hg bonds.[117] Thus, they are reduced by $LiAlH_4$ to carborane [Equation (76)] or converted by n-C_4H_9Li to lithiocarborane [Equation (77)]:

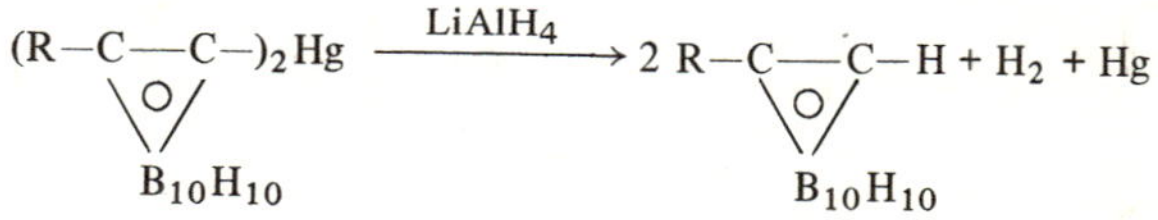

$$(R-C{-}C{-})_2Hg \xrightarrow{\ LiAlH_4\ } 2\,R-C{-}C{-}H + H_2 + Hg \qquad (76)$$

with $B_{10}H_{10}$

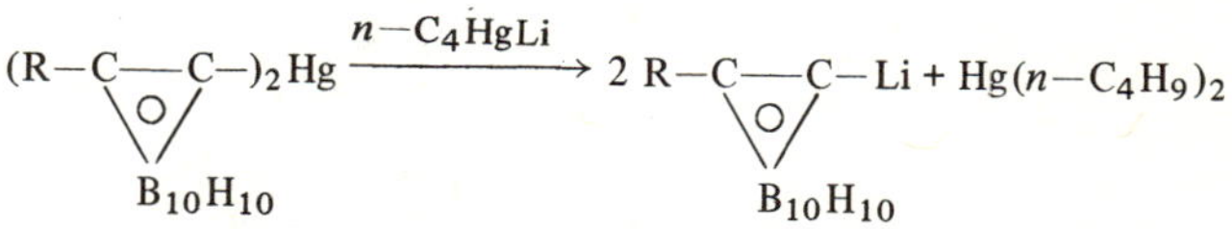

$$(R-C{-}C{-})_2Hg \xrightarrow{\ n-C_4H_9Li\ } 2\,R-C{-}C{-}Li + Hg(n-C_4H_9)_2 \qquad (77)$$

with $B_{10}H_{10}$

Both bis(phenyl-*o*-carboranyl) and bis(methyl-*o*-carboranyl)mercury react with metallic lithium in 1,2-dimethoxyethane at 20°C to give carborane; however, they are inert toward lithium and aluminum when heated in an inert medium such as toluene, even at 150°C in a sealed tube.[117] It has been suggested that the rate of the reaction with lithium in dimethoxyethane probably is determined by an electron-transfer process from the metal to the organomercury molecule.[117]

Symmetrical *o*-carboranyl derivatives of mercury readily form complexes[117] when treated with α,α'-bipyridyl in alcohol or benzene solution:

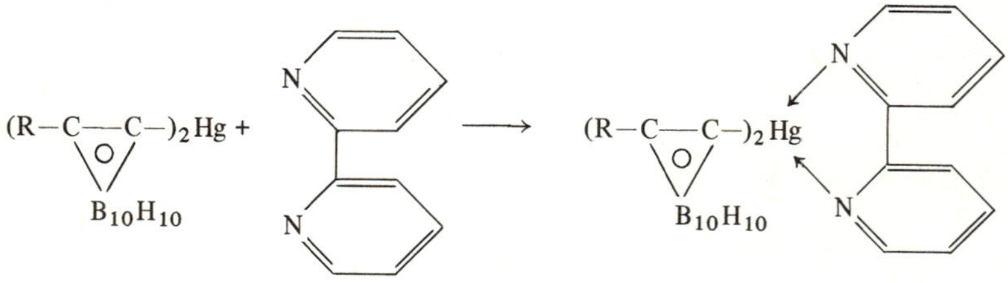

$$(R-C{-}C{-})_2Hg + \text{(bipyridyl)} \longrightarrow (R-C{-}C{-})_2Hg\cdot\text{(bipyridyl)} \qquad (78)$$

$$R = H,\ CH_3,\ CH_2{=}CH,\ C_6H_5$$

Both the symmetrical bis(*o*-carboranyl)$_2$Hg and the unsymmetrical *o*-carboranyl mercury compounds also form stable 1:1 adducts with *o*-phenanthroline.[118] The unsymmetrical complexes do not undergo symmetrization on being heated in benzene. All of these adducts melt with decomposition.[117,118] *Ortho*-carboranyl mercury hydroxides, substances that are unstable in air, were obtained[117] by reacting *o*-carboranyl mercury bromide with alcoholic potassium hydroxide:

$$R-\overset{\displaystyle C\!-\!\!-\!\!C}{\underset{B_{10}H_{10}}{\diagdown\!\!O\!\!\diagup}}-HgBr + KOH \xrightarrow{\;C_2H_5OH\;} R-\overset{\displaystyle C\!-\!\!-\!\!C}{\underset{B_{10}H_{10}}{\diagdown\!\!O\!\!\diagup}}-HgOH + KBr \tag{79}$$

The carborane carbon–mercury bonds of the bis(*o*-carboranyl) mercury derivatives are slowly cleaved by an alcoholic KOH solution, even at room temperature. However, this reaction is apparently complicated by the formation of the corresponding dicarbaundecaborate anion.[119]

Like the *C*-lithium derivatives of $1,2\text{-}C_2B_{10}H_{12}$ (*o*-carborane), the *C*-lithium derivatives of both 1,7- and $1,12\text{-}C_2B_{10}H_{12}$ (*m*- and *p*-carborane, respectively) react with mercury halides giving the corresponding *m*- and *p*-carboranyl mercury derivatives.[120,121] Thus, symmetrical *m*-carboranyl-mercury compounds can be easily prepared[120] by reacting lithium derivatives of $1,7\text{-}C_2B_{10}H_{12}$ with $HgCl_2$ in benzene–tetrahydrofuran mixtures at reflux for 2 hr:

$$R-CB_{10}H_{10}C-Li + HgCl_2 \longrightarrow (R-CB_{10}H_{10}C)_2Hg + 2LiCl \tag{80}$$

$$R = CH_3,\ C_6H_5$$

Methyl(*m*-carboranyl)mercury compounds. $R-HgCH_3$ ($R = R'CB_{10}H_{10}C-$, in which $R' = CH_3,\ C_6H_5$), are readily obtained by the reaction between *C*-lithium *m*-carborane derivatives and CH_3HgCl in benzene at reflux.

By reacting $1\text{-}Li\text{-}7\text{-}C_6H_5\text{-}1,7\text{-}C_2B_{10}H_{10}$ with C_6H_5HgCl in benzene solution, phenyl(phenyl-*m*-carboranyl)mercury (56% yield) is obtained together with bis(phenyl-*m*-carboranyl) mercury and diphenyl mercury. The amount of the latter species appears to increase on addition of tetrahydrofuran to the reaction mixture, and their formation may be due to the following exchange reaction [Equation (81)] :

$$C_6H_5\text{-}CB_{10}H_{10}C\text{-}Li + C_6H_5HgCl \longrightarrow C_6H_5\text{-}CB_{10}H_{10}C\text{-}HgCl + C_6H_5Li \tag{81}$$

Phenyl(phenyl-*m*-carboranyl)mercury is obtained in good yield from phenyl-*m*-carboranylmercury chloride or bromide and phenylmagnesium bromide. Mixed mercury derivatives containing both *o*-carboranyl and *m*-carboranyl radicals have also been obtained:[120]

$$C_6H_5\text{-}CB_{10}H_{10}C\text{-}HgCl + C_6H_5-\overset{\displaystyle C\!-\!\!-\!\!C}{\underset{B_{10}H_{10}}{\diagdown\!\!O\!\!\diagup}}-Li \longrightarrow C_6H_5\text{-}CB_{10}H_{10}C\text{-}Hg-\overset{\displaystyle C\!-\!\!-\!\!C}{\underset{B_{10}H_{10}}{\diagdown\!\!O\!\!\diagup}}-C_6H_5 + LiCl \tag{82}$$

On the other hand, when phenyl-*o*-carboranylmercury bromide is reacted with $CH_3\text{-}CB_{10}H_{10}C\text{-}Li$ a mixture of bis(methyl-*m*-carboranyl)mercury, bis(phenyl-*o*-carboranyl)mercury and methyl-*m*-carboranyl(phenyl-*o*-carboranyl)mercury is formed, probably as a result of an exchange reaction of the type:

$$CH_3\text{-}CB_{10}H_{10}C\text{-}Li + C_6H_5-\overset{\displaystyle C\!-\!\!-\!\!C}{\underset{B_{10}H_{10}}{\diagdown\!\!O\!\!\diagup}}-HgBr \rightleftharpoons CH_3\text{-}CB_{10}H_{10}C\text{-}HgBr + C_6H_5-\overset{\displaystyle C\!-\!\!-\!\!C}{\underset{B_{10}H_{10}}{\diagdown\!\!O\!\!\diagup}}-Li \tag{83}$$

Unlike the bis(*o*-carboranyl)mercury compounds, the bis(*m*-carboranyl)mercury compounds undergo[120] cleavage of the C—Hg bonds on prolonged boiling with HCl in ethanol.

$$(R-CB_{10}H_{10}C)_2Hg + HCl \xrightarrow[78°]{C_2H_5OH} R-CB_{10}H_{10}CHgCl + R-CB_{10}H_{10}CH \quad (84)$$

The reactions of some mixed *m*-carboranyl mercury derivatives with alcoholic hydrogen chloride under reflux have also been investigated.

$$R-CB_{10}H_{10}C-Hg-CH_3 \xrightarrow{HCl} \begin{cases} R-CB_{10}H_{10}C-HgCl + CH_4 \\ \\ R-CB_{10}H_{10}CH + CH_3HgCl \end{cases} \quad (85)$$

$$R = CH_3, C_6H_5$$

$$C_6H_5-CB_{10}H_{10}C-Hg-C_6H_5 + HCl \longrightarrow C_6H_5-CB_{10}H_{10}C-HgCl + C_6H_6 \quad (86)$$

$$C_6H_5-CB_{10}H_{10}C-Hg-\overset{\displaystyle C----C}{\underset{B_{10}H_{10}}{\diagdown O \diagup}}-C_6H_5 + HCl \longrightarrow C_6H_5-CB_{10}H_{10}C-HgCl + C_6H_5-\overset{\displaystyle C----CH}{\underset{B_{10}H_{10}}{\diagdown O \diagup}}$$

$$(87)$$

Symmetrical bis(*m*-carboranyl)mercury compounds react very slowly with bromine in benzene solution but the reaction is complicated by side reactions with evolution of hydrogen bromide. On the other hand, the methyl(*m*-carboranyl)mercury derivatives readily react with bromine in benzene undergoing cleavage of the Hg—CH$_3$ bond:[120]

$$R-CB_{10}H_{10}C-Hg-CH_3 + Br_2 \xrightarrow{20°} R-CB_{10}H_{10}C-Hg-Br + CH_3Br \quad (88)$$

$$R = CH_3, C_6H_5$$

Symmetrical and unsymmetrical *m*-carboranylmercury compounds react with mercury(II) chloride more easily than the corresponding *o*-carboranyl derivatives.[120]

The *m*-carboranylmercury halides, $R-CB_{10}H_{10}C-HgX$, appear to be unaffected when treated with symmetrizing agents, such as KCN and KI, in an ethanol solution at the reflux temperature.

As with the corresponding *o*-carboranyl derivatives, both symmetrical and unsymmetrical *m*-carboranyl mercury compounds form 1:1 adducts by treatment with *o*-phenanthroline;[118] e.g., $(R-CB_{10}H_{10}C)_2Hg \cdot C_{12}H_8N_2$ and $(R-CB_{10}H_{10}C)HgX \cdot C_{12}H_8N_2$. The latter unsymmetrical complexes do not undergo symmetrization on being treated in benzene.

On comparing with the chemical behavior of the *o*-carboranyl derivatives, the *m*-carboranylmercury compounds appear to be more reactive towards electrophilic reagents. This may be due to the lower electron-withdrawing power exhibited by the *m*-carboranyl group.[122]

The strong electron-acceptor effect of the carboranyl groups is demonstrated by the half-wave potentials, $E_{1/2}$, which were found[114,120] for the polargraphic reduction in dimethylformamide of some bis(carboranyl)mercury derivatives. The data in Table 2 indicate that the polarity of the C—Hg bond in the *m*-carboranyl derivatives is smaller than the polarity of the C—Hg bond in the *o*-carboranyl derivatives.[123]

Further evidence that *o*-carborane is more electronegative than *m*-carborane, which in turn is more electronegative than *p*-carborane, was obtained[121] by investigation of the ^{1}H nmr spectra of a number of methylmercury derivatives of *ortho*-, *meta*-, and *para*-carborane, carb—Hg—CH$_3$. In particular, the spin-spin coupling constants $J(^{199}$Hg—C—^{1}H) obtained for a series of unsymmetrical (carboranyl) Hg—CH$_3$ componds were used as a measure of the polarity of the Hg—CH$_3$ bond and thus the electron-acceptor power of the *o*-, *m*-, and *p*-carboranyl radicals could be evaluated.[121]

Triethylgermyl- and triethylsilylmercury derivatives of both 1,2- and 1,7-dicarba-*closo*-dodecaborane(12) have been synthesized[80] through exchange

Table 2. Polarity of the C—Hg Bond
in *m*–Carboranyl and *o*–Carboranyl Derivatives

Compound	$E_{1/2}$, V
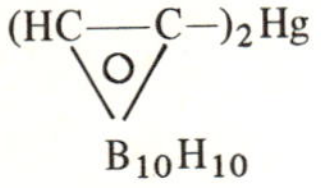 $(HC{-}C{-})_2Hg$, $B_{10}H_{10}$	−0.78
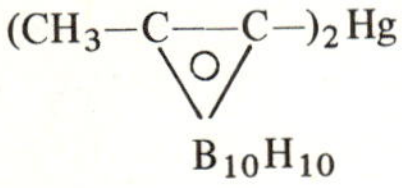 $(CH_3{-}C{-}C{-})_2Hg$, $B_{10}H_{10}$	−0.80
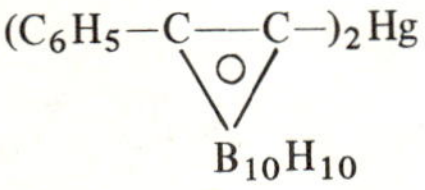 $(C_6H_5{-}C{-}C{-})_2Hg$, $B_{10}H_{10}$	−0.89
$(CH_3{-}CB_{10}H_{10}C{-})_2Hg$	−1.36
$(C_6H_5)_2Hg$	−2.60

reactions of carboranylmercury derivatives with bis(triethylgermyl)mercury and bis(triethylsilyl)mercury respectively, in boiling benzene:

$$(\text{Et}_3\text{M})_2\text{Hg} + (\text{Ph}-\underset{\underset{\text{B}_{10}\text{H}_{10}}{}}{\text{C}-\overset{\text{O}}{\diagdown\diagup}\text{C}}-)_2\text{Hg} \longrightarrow 2\,\text{Et}_3\text{M}-\text{Hg}-\underset{\underset{\text{B}_{10}\text{H}_{10}}{}}{\text{C}-\overset{\text{O}}{\diagdown\diagup}\text{C}}-\text{Ph} \qquad (89)$$

$$\text{M} = \text{Si, Ge}$$

$$(\text{Et}_3\text{Ge})_2\text{Hg} + \text{XHg}-\underset{\underset{\text{B}_{10}\text{H}_{10}}{}}{\text{C}-\overset{\text{O}}{\diagdown\diagup}\text{C}}-\text{Ph} \longrightarrow \text{Et}_3\text{Ge}-\text{Hg}-\underset{\underset{\text{B}_{10}\text{H}_{10}}{}}{\text{C}-\overset{\text{O}}{\diagdown\diagup}\text{C}}-\text{Ph} + \text{Et}_3\text{GeX} + \text{Hg} \qquad (90)$$

Linear polymers containing the repeating $-\text{CB}_{10}\text{H}_{10}\text{C}-\text{Hg}-$ unit were obtained[124] by reacting $1,7\text{–Li}_2\text{–}1,7\text{–C}_2\text{B}_{10}\text{H}_{10}$ with an equimolar amount of HgCl_2 in diethyl ether or tetrahydrofuran at $30°\text{C}$.

$$n\text{Li}-\text{CB}_{10}\text{H}_{10}\text{C}-\text{Li} + n\text{HgCl}_2 \xrightarrow{\;-\text{LiCl}\;} {\ +\!\!-}\!\text{CB}_{10}\text{H}_{10}\text{C}-\text{Hg}{+}\!\!-\!\big]_n \qquad (91)$$

Polymers with a molecular weight of 10,000 (osmometric) were obtained when the reaction was carried out in diethyl ether solution; however, when conducted in tetrahydrofuran, polymers of lower molecular weight (≈ 3000) were obtained. These carborane–mercury polymers are soluble in tetrahydrofuran but not in hydrocarbons or CH_2Cl_2. They are not decomposed on heating up to $300°\text{C}$.

3.7.2. Derivatives Formed through Carborane Boron–Mercury Bonds

The mercuration of 1,2-, 1,7-, and $1,12\text{–C}_2\text{B}_{10}\text{H}_{10}\text{R}_2$ at one boron atom of the carborane cage has been carried out with mercury(II) trifluoroacetate in trifluoroacetic acid:[125,126]

$$o\text{-, }m\text{-, }p\text{-B}_{10}\text{H}_{10}\text{C}_2\text{R}_2 \xrightarrow{\;\text{Hg(OCOCF}_3)_2\;} (o\text{-, }m\text{-, }p\text{-B}_{10}\text{H}_9\text{C}_2\text{R}_2)\text{Hg(OCOCF}_3) \qquad (92)$$

The mercuration of *para*-carborane occurs under more drastic conditions than that of the *ortho*- and *meta* isomers. In fact, *ortho*- and *meta*-carborane react with mercury(II) trifluoroacetate at room temperature, but the mercuration of *para*-carborane takes place at the reflux temperature.[126] The unsubstituted and C,C'-disubstituted 1,2- and $1,7\text{–C}_2\text{B}_{10}\text{H}_{12}$ derivatives are mercurated at the boron atom in position 9 of the carborane cage. The mercuration of *C*-monosubstituted carboranes occurs either in position 9 or 12, and thus a mixture of isomers is formed. On the other hand, $1,12\text{–C}_2\text{B}_{10}\text{H}_{12}$ undergoes mercuration at the boron atom in position 2.

If a large excess (up to 10 times) of $\text{Hg(OCOCF}_3)_2$ is used in the reaction seen in Equation (92), polymercuration takes place and a C,C'-unsubstituted *ortho*-carborane derivative containing four $-\text{HgOCOCF}_3$ groups is formed.[126] A pentamercury-*o*-carborane derivative has also been obtained by reacting $1,2\text{–C}_2\text{B}_{10}\text{H}_{12}$ with a large excess of HgO in CF_3COOH at the reflux temperature for 20 hr.[127] After treatment of the reaction product with NaCl in acetone–water, penta(chloro-mercury)–$1,2\text{–B}_{10}\text{H}_5\text{C}_2\text{H}_2$ has been isolated.[127]

The *B*-chloromercury carboranes are obtained according to the reaction

$$CF_3CO_2Hg-B_{10}H_9C_2R_2 + NaCl \longrightarrow ClHg-B_{10}H_9C_2R_2 + NaOCOCF_3 \qquad (93)$$

carried out in acetone–water solution.[126] The *B*-carboranyl mercury trifluoro-acetates react with Grignard reagents in ether to give the corresponding alkyl-mercury derivatives:[128]

$$(o\text{-}, m\text{-}p\text{-}C_2B_{10}H_{11}\text{-}9)Hg(OCOCF_3) \xrightarrow{EtMgBr} (o\text{-}, m\text{-}p\text{-}C_2B_{10}H_{11}\text{-}9)HgEt \qquad (94)$$

The *B*-mercurated carboranes differ from the corresponding *C*-mercurated carboranes in their reactivity and properties. In fact, the behavior of the *B*-derivatives appears to be more similar to that shown by the aliphatic organo-mercury compounds. Thus, unlike the *C*-mercurated carboranes, the *B*-mercurated derivatives easily undergo cleavage of the B–Hg bond on treatment with Br_2 in CCl_4 or HCl in acetone–water solution at room temperature.[126] However, when $(o\text{-}C_2B_{10}H_{11}\text{-}9)HgCH_3$ and $(m\text{-}C_2B_{10}H_{11}\text{-}9)HgCH_3$ are treated with HCl in ethanol at $20°C$ cleavage of the $Hg-CH_3$ bond occurs, but not of the Hg–B bond:[126,127]

$$(o\text{-}, m\text{-}C_2B_{10}H_{11}\text{-}9)HgCH_3 + HCl \xrightarrow[\text{ethanol}]{20°} (o\text{-} m\text{-}C_2B_{10}H_{11}\text{-}9)HgCl + CH_4 \qquad (95)$$

The unsymmetrical *B*-mercurated carboranes are much less stable than the corresponding *C*-mercurated derivatives toward symmetrization.[128] Thus, on heating at $100°C$ *in vacuo*, they give symmetrical products:

$$2(o\text{-}, m\text{-}C_2B_{10}H_{11}\text{-}9)HgEt \xrightarrow[\text{in vacuo}]{100°C} (o\text{-} m\text{-}C_2B_{10}H_{11}\text{-}9)_2Hg + HgEt_2 \qquad (96)$$

A symmetrical compound containing two *o*-carboranyl groups linked through a B–Hg–B bridge was obtained in the reaction of $(o\text{-}C_2B_{10}H_{11}\text{-}3)HgCl$ with bis(triethylgermyl)mercury.[80]

$$\begin{array}{ccc} HC\!\!-\!\!\!-\!\!CH & & HC\!\!-\!\!\!-\!\!CH \\ \diagdown\!\!\!\bigcirc\!\!\!\diagup & & \diagdown\!\!\!\bigcirc\!\!\!\diagup \\ 3-B_{10}H_9\!\!-\!\!Hg\!\!-\!\!(3-B_{10}H_9) \end{array}$$

In order to estimate the relative stabilities of the borane ions of the carborane series, polarographic reduction potentials have been measured for several unsymmetrical and symmetrical *B*-mercurated *o*-, *m*-, and *p*-carborane compounds in dimethylformamide at $25°C$.[129] The potential required for reduction of the B–Hg bond is much higher than that required for reduction of the C–Hg bond in the corresponding carborane derivatives; this is further evidence that the *B*-mercurated carboranes resemble the alkyl- and arylmercury derivatives in their properties.[130] The B–Hg bond undergoes reduction more readily in the *p*-carborane than in either *o*- or *m*-carborane derivatives. This is due to the effect of the neighboring electron-deficient carbon atom in the *para*-carborane, where the mercurated boron atom is in position 2 of the carborane cage. Moreover, the electrochemical data show that the 2-borane ion of *p*-carborane is more stable than the 9-borane ions of both *o*- and *m*-carborane.[129]

Carboranyl mercury(II) derivatives containing B—Hg and C—Hg bonds are the products of thermal and initiated (benzoyl peroxide, UV, etc.) decarboxylation reactions of mercury(II) salts of *o*-carboranyl-9(12)-, *m*-carboranyl-9(10)- and *o*-carboranyl-1-carboxylic acids.[131]

4. TRANSITION METAL CLOSO-CARBORANE DERIVATIVES

The chemistry of organometallic derivatives of transition metals incorporating *closo*-carboranyl groups through carborane carbon–metal σ-bonds is a very young area of interest. The first examples[132] of compounds of this class involved platinum(II) complexes and were reported in 1968. Subsequently, several other interesting carborane transition metal derivatives were prepared and characterized and this chemistry appears to be a promising field for further growth.

The peculiar electronic and steric properties of the carboranyl anionic ligands tend to promote the formation of transition metal complexes having unusual coordination numbers and geometries, which are characterized by highly stable carborane carbon–metal σ-bonds.

4.1. Derivatives Formed through Carborane Carbon–Metal σ-Bonds

4.1.1. Manganese and Iron Derivatives

A series of stable manganese(I) and iron(II) complexes containing *closo*-carboranyl groups bonded through metal–carbon σ-bonds has been reported.[133,134] Thus, the complexes $1\text{-}[Mn(CO)_5]\text{-}10\text{-}CH_3\text{-}1,10\text{-}(\sigma\text{-}C_2B_8H_8)$ and $1\text{-}[(\eta^5\text{-}C_5H_5)Fe(CO)_2]\text{-}10\text{-}CH_3\text{-}1,10\text{-}(\sigma\text{-}C_2B_8H_8)$ were prepared by treating $1\text{-}Li\text{-}10\text{-}CH_3\text{-}1,10\text{-}C_2B_8H_8$ with $[MnBr(CO)_5]$ and $[(\eta^5\text{-}C_5H_5)FeI(CO)_2]$, respectively.

A diiron derivative of the 1,10–dicarba-*closo*-decaborane(10) was obtained when the C,C'-dilithium carborane was reacted with $[(\eta^5\text{-}C_5H_5)FeI(CO)_2]$ in the molar ratio 1:2. The suggested structure is pictured in Figure 4. On treating the latter complex with excess triphenylphosphine, substitution of one carbonyl group took place and the complex $1\text{-}[(\eta^5\text{-}C_5H_5)Fe(CO)_2]\text{-}10\text{-}\{(\eta^5\text{-}C_5H_5)Fe(CO)[P(C_6H_5)_3]\}\text{-}1,10\text{-}(\sigma\text{-}C_2B_8H_8)$ was isolated and characterized.[133]

With respect to the higher carboranes, manganese(I) and iron(II) complexes containing the $\text{-}2\text{-}R\text{-}1\,2\text{-}C_2B_{10}H_{10}$ ligands have been reported. Thus, on treatment[133] of $1\text{-}Li\text{-}2\text{-}R\text{-}1,2\text{-}C_2B_{10}H_{10}$ (R = CH_3, C_6H_5) with $[MnBr(CO)_5]$ and $[(\eta^5\text{-}C_5H_5)FeI(CO)_2]$, the complexes $1\text{-}[Mn(CO)_5]\text{-}2\text{-}R\text{-}1,2\text{-}(\sigma\text{-}C_2B_{10}H_{10})$ and $1\text{-}[(\eta^5\text{-}C_5H_5)Fe(CO)_2]\text{-}2\text{-}R\text{-}1,2\text{-}(\sigma\text{-}C_2B_{10}H_{10})$, respectively, were obtained (Figure 5).

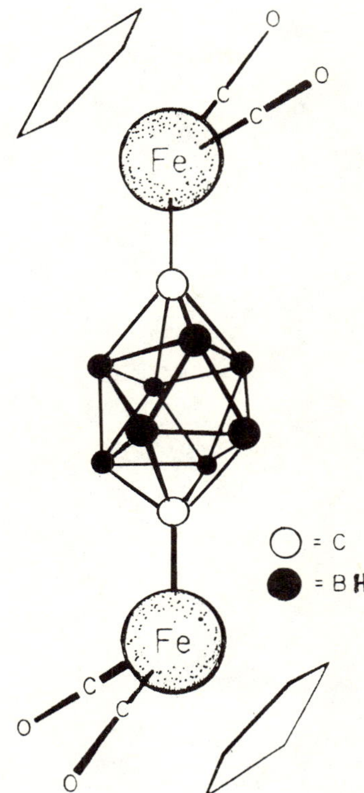

Figure 4. Proposed structure of $1,10\text{-}[(\eta^5\text{-}C_5H_5)$ $Fe(CO)_2]_2\text{-}1,10\text{-}(\sigma\text{-}C_2B_8H_8)$ (from Reference 134).

An analogous iron(II) complex containing a *m*-carboranyl group in place of the *o*-carboranyl group[135] was synthesized by a different preparative route:

$$HCB_{10}H_{10}CCOCl + Na[Fe(\eta^5\text{-}C_5H_5)(CO)_2] \xrightarrow{-NaCl}$$

$$HCB_{10}H_{10}CCOFe(CO)_2(\eta^5\text{-}C_5H_5) \xrightarrow{\Delta} \tag{97}$$

$$1\text{-}[(\eta^5\text{-}C_5H_5)Fe(CO)_2]\text{-}7\text{-}H\text{-}1,7\text{-}(\sigma\text{-}C_2B_{10}H_{10}) + CO$$

Initially, the *m*-carborane carboxylic acid chloride reacts with $Na[Fe(\eta^5\text{-}C_5H_5)(CO)_2]$ giving a σ-acyl iron derivative which, on heating at reflux temperature in *p*-xylene, loses CO to give as final product the σ-*m*-carboranyl iron(II) complex. In the same manner a C,C'-*m*-carboranyldiiron σ-complex was obtained by using *m*-carboranyldicarboxylic acid dichloride.[135] Unlike *m*-carborane carboxylic acid chlorides, *o*-carborane carboxylic acid chlorides react with $Na[Fe(\eta^5\text{-}C_5H_5)(CO)_2]$ to yield complexes containing *o*-carboranyl–cyclopentadiene bonds.[136] In fact, in the latter reaction, a rearrangement of the initially formed σ-acyl com-

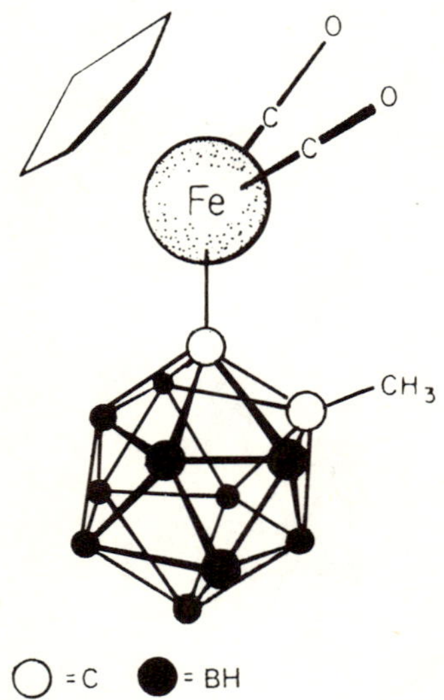

O = C   ● = BH

Figure 5. Proposed structure of $1\text{-}[(\eta^5\text{-}C_5H_5)Fe(CO)_2]\text{-}2\text{-}CH_3\text{-}1,2\text{-}(\sigma\text{-}C_2B_{10}H_{10})$ (from Reference 133).

plex occurs, leading to [*exo*-(substituted-*o*-carboranyl)-cyclopentadiene] iron tricarbonyl complexes:

$$R-C{-}C-COCl + Na[Fe(\eta^5\text{-}C_5H_5)(CO)_2] \xrightarrow[\text{THF}]{20^\circ} R-C{-}C{-} \qquad Fe(CO)_3 \qquad (98)$$

$$B_{10}H_{10} \qquad\qquad\qquad\qquad\qquad\qquad B_{10}H_{10}$$

A study of the chemistry of a series of these iron complexes containing both *o*- and *m*-carboranyl groups σ-bonded was carried out by Zakharkin *et al.*[135] Thus, it was found that on treatment of these complexes with excess Br_2, bromination of the carborane cage occurs, the carborane carbon–iron bond being unaffected. Conversely, when the corresponding σ-alkyl and σ-aryl derivatives, $(\eta^5\text{-}C_5H_5)FeR(CO)_2$, are treated with halogens under the same conditions, cleavage of the C–Fe bond is observed.

Iron(II) complexes containing σ-bonded *o*- and *m*-carboranyl groups, $1\text{-}[(\eta^5\text{-}C_5H_5)(CO)_2Fe]\text{-}2\text{-}H\text{-}1,2\text{-}(\sigma\text{-}C_2B_{10}H_{10})$, $1\text{-}[(\eta^5\text{-}C_5H_5)(CO)_2Fe]\text{-}7\text{-}H\text{-}1,7\text{-}(\sigma\text{-}C_2B_{10}H_{10})$, $1,2\text{-}[(\eta^5\text{-}C_5H_5)(CO)_2Fe]_2\text{-}1,2\text{-}$ and $1,7\text{-}[(\eta^5\text{-}C_5H_5)(CO)_2Fe]_2\text{-}1,7\text{-}(\sigma\text{-}C_2B_{10}H_{10})$, have also been prepared by reacting $(\eta^5\text{-}C_5H_5)Fe(CO)_2Br$ with the corresponding carboranyl copper(I) derivatives in benzene-tetrahydrofuran at 20°C for 3 hr and then at 50–60°C for 2 hr.[137]

4.1.2. Cobalt, Rhodium, and Iridium Derivatives

Reaction of 2,2′-dilithiobiscarborane

$$Li-C{-}C-C{-}C-Li$$

$$B_{10}H_{10} \quad B_{10}H_{10}$$

with 0.5 mol equivalent of anhydrous $CoCl_2$ as an ether slurry provided an oily purple solid, from which bright violet air-stable needles of $[(C_2H_5)_4N]_2Co[(C_2B_{10}H_{10})_2]_2$ were obtained by treating the crude product with $(C_2H_5)_4NBr$ in CH_2Cl_2.[138,139]

Reaction of the biscarborane derivative of cobalt(II) with lithium amalgam in acetone yielded a red-brown solution which presumably contained a formal cobalt(I) derivative. From this solution, the cobalt(II) biscarborane derivative can be restored upon exposure to air. A diamagnetic derivative of cobalt(III) was isolated as $[(C_2H_5)_4N]Co[(C_2B_{10}H_{10})_2]_2$ by treating the corresponding cobalt(II) derivative with a large excess of anhydrous $CuCl_2$ in dry CH_2Cl_2. The crystal structure of the biscarborane cobalt(III) derivative is shown in Figure 6a.[140] It is found that the four carborane icosahedra are tetrahedrally arranged about the cobalt atom (Figure 6b). However, the four bonding carbon atoms show severe distortion from tetrahedral symmetry. In addition, one of the boron atoms is rather close (2.29 Å) to the cobalt atom and this suggests the possible existence of a B—H—Co bridge. Thus, the coordination about the cobalt atom appears to consist of one B—H—Co bridge bond and four carborane carbon–cobalt σ-bonds.

Earlier, several bis(carboranyl)cyclopentadienyl cobalt(III) derivatives of

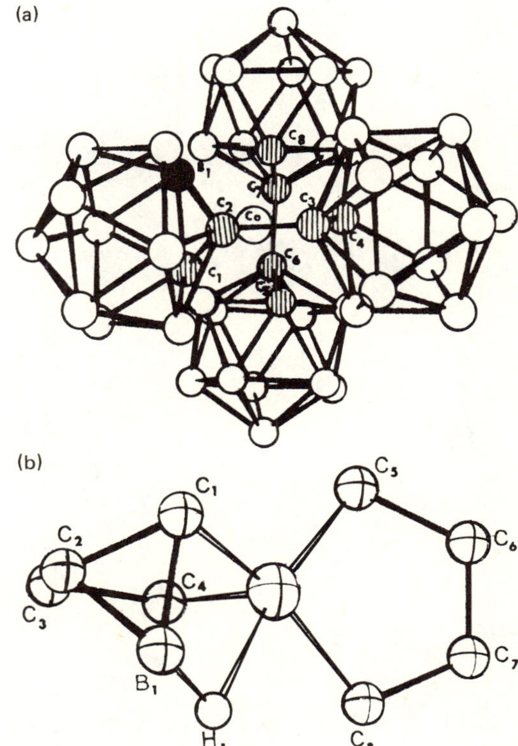

Figure 6. (a) The structure of the $Co[(C_2B_{10}H_{10})_2]_2^-$ anion (from Reference 140). (b) The coordination about the cobalt atom in the $Co[(C_2B_{10}H_{10})_2]_2^-$ anion (from Reference 140).

the general formula $(\eta^5\text{-}C_5H_5)Co(PPh_3)(\sigma\text{-carb})_2$ in which carb = -2-CH$_3$-1,2- and -7-C$_6$H$_5$-1,7-C$_2$B$_{10}$H$_{10}$, were reported,[141] but these complexes were not fully characterized owing to their poor stability in solution.

The reaction of *C*-monolithio derivatives of 1,2- and 1,7-dicarba-*closo*-dodecaborane(12) with RhCl(PPh$_3$)$_3$ and *trans*-RhCl(CO)(PPh$_3$)$_2$ in ethereal solvents gave rise to a novel series of neutral rhodium(I)-carborane complexes containing a metal–carbon σ-bond.[142] Thus, by treating RhCl(PPh$_3$)$_3$ with an excess of 1-Li-2-R-1,2- and 1-Li-7-R-1,7-C$_2$B$_{10}$H$_{10}$ (R = H, CH$_3$, C$_6$H$_5$), the complexes 1-[Rh(PPh$_3$)$_2$]-2-R-1,2-(σ-C$_2$B$_{10}$H$_{10}$) and 1-[Rh(PPh$_3$)$_2$]-7-R-1,7-(σ-C$_2$B$_{10}$H$_{10}$), respectively, were obtained and fully characterized. In the solid state these complexes are rather insensitive toward air and moisture, but are highly air-sensitive when placed in solution. These compounds are the first examples of neutral three-coordinate species of rhodium(I), as confirmed by X-ray analysis.[143] In particular, the crystal structure of 1-[Rh(PPh$_3$)$_2$]-2-C$_6$H$_5$-1,2-(σ-C$_2$B$_{10}$H$_{10}$) shows that the rhodium atom is surrounded by three ligands, namely two PPh$_3$ groups and a 2-phenyl-1,2-dicarba-*closo*-dodecaborane(12) radical, the metal and the coordinated atoms being in a substantially planar arrangement (Figure 7). In addition, one of the boron atoms of the car-

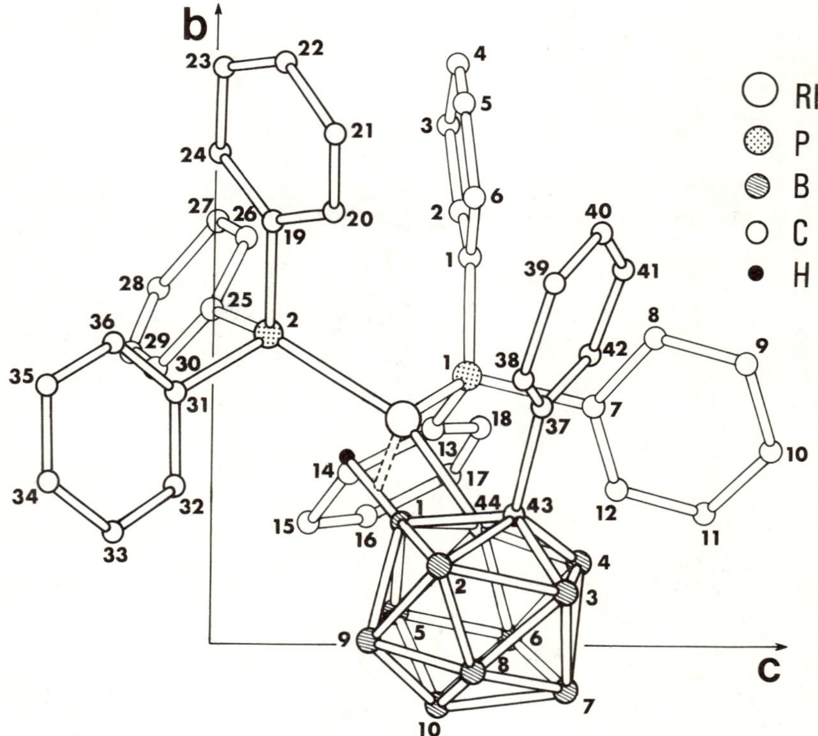

Figure 7. Molecular structure of 1-[Rh(PPh$_3$)$_2$]-2-C$_6$H$_5$-1,2-(σ-C$_2$B$_{10}$H$_{10}$) (from Reference 143).

borane cage is rather close (2.35 Å) to the metal atom, suggesting that the coordination of carborane to rhodium(I) is supplied by a Rh—C σ-bond and possibly also by a Rh—H—C bridge bond. The crystal structure indicates that the presence of a large carboranyl ligand does not allow the coordination of more than two bulky triphenylphosphine ligands to the metal. These formally three-coordinated complexes react with neutral ligands having small strain-free cone angles, such as C_6H_5CN and CO to give four-coordinated species.[142] Thus, on treatment with C_6H_5CN the four-coordinated nitrile complex $1-[Rh(PPh_3)_2(C_6H_5CN]-2-C_6H_5-1,2-(\sigma-C_2B_{10}H_{10})$ was prepared. The reaction with carbon monoxide afforded four-coordinate bis(carbonyl)derivatives:

$$\text{Rh}(\sigma\text{-carb})(\text{PPh}_3)_2 \xrightleftharpoons[\ ]{+CO,\ -\text{PPh}_3} \text{Rh}(\sigma\text{-carb})(\text{CO})_2(\text{PPh}_3) \qquad (99)$$

$$\text{carb} = -2-C_6H_5-1,2-(\sigma-C_2B_{10}H_{10})$$

This carbonylation is reversed on flushing with dry nitrogen. In contrast, the four-coordinate monocarbonyl-rhodium(I) derivative $1-[Rh(CO)(PPh_3)_2]-2-C_6H_5-1,2-(\sigma-C_2B_{10}H_{10})$ was prepared from the reaction between *trans*$[RhCl(CO)(PPH_3)_2]$ and $1-Li-2-C_6H_5-1,2-C_2B_{10}H_{10}$. On treatment under a CO atmosphere, the monocarbonyl complex undergoes displacement of one phosphine ligand yielding the corresponding bis(carbonyl) complex mentioned above.

The three-coordinate complex $Rh(\sigma\text{-carb})(PPh_3)_2$ undergoes facile ligand exchange on treatment with free trisubstituted phosphines having less bulk than the triphenylphosphine; in such cases, three and/or four coordinated species are obtained, depending on the steric requirements of the entering ligand.[144] On treatment with $PMePh_2$ the complexes $Rh(\sigma\text{-carb})(PMePPh_2)_2$ and $Rh(\sigma\text{-carb})(PMePh_2)_3$ were isolated, but the reaction with less hindered entering ligands, such as PMe_2Ph and $P(OPh)_3$, afforded only the four-coordinated complexes $Rh(\sigma\text{-carb})(PR_3)_3$, where carb = $-2-R-1,2-$ and $-7-C_6H_5-1,7-C_2B_{10}H_{10}$ (R = H, CH_3). On the other hand, the complex $1-[Rh(PPh_3)_2]-2-C_6H_5-1,2(\sigma-C_2B_{10}H_{10})$ behaves in a different way in these ligand exchange reactions, a stable hydridocarboranylrhodium(III) derivative being formed in the reaction with PMe_2Ph (Figure 8). In this case the substitution reaction is followed by a

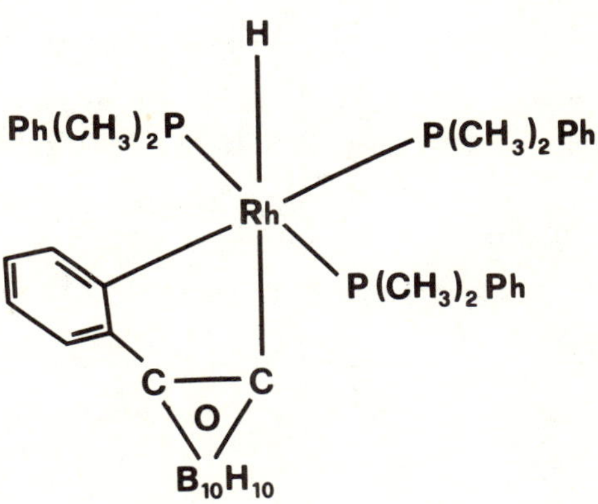

Figure 8. Proposed structure of $1-[Rh(H)(PMe_2Ph)_3]-2-C_6H_4-1,2-(\sigma-C_2B_{10}H_{10})$.

fast intramolecular oxidative addition involving an *ortho* C—H bond of the C(2)-carborane phenyl substituent.

Ortho- and *meta*-carboranyl derivatives of iridium(I) and iridium(III) have been reported.[145] Substitution reactions of *trans*[IrCl(CO)(PPh$_3$)$_2$] with the lithium derivatives 1-Li-2-R-1,2-C$_2$B$_{10}$H$_{10}$ (R = H, CH$_3$) and 1-Li-7-R'-1,7-C$_2$B$_{10}$H$_{10}$ (R' = H, CH$_3$, C$_6$H$_5$) in benzene–diethyl ether afforded a series of iridium(I) complexes of general formula *trans*[Ir(σ-carb)(CO)(PPh$_3$)$_2$]. An analogous carborane complex containing PMePh$_2$ has also been prepared, namely, 1-[Ir(CO)(PMePh$_2$)$_2$]-7-C$_6$H$_5$-1,7-(σ-C$_2$B$_{10}$H$_{10}$), whose ^{1}H nmr spectrum confirms a *trans* arrangement of the phosphine ligands. When the lithium derivative of the phenyl-*o*-carborane, 1-Li-2-C$_6$H$_5$-1,2-C$_2$B$_{10}$H$_{10}$ was employed, the hydridocarboranyliridium(III) complex 1-[Ir(H)(CO)(PPh$_3$)$_2$]-2-C$_6$H$_4$-1,2-(σ-C$_2$B$_{10}$H$_{10}$), was obtained as a result of an intramolecular oxidative addition reaction involving one *ortho* C—H bond of the *C*(2)-carborane phenyl substituent.

Intramolecular oxidative addition was also found to occur in the reaction between *trans*[IrCl(CO)(PMePh$_2$)$_2$] and 1-Li-2-C$_6$H$_5$-1,2-C$_2$B$_{10}$H$_{10}$.[146] But in this case, as shown by ^{1}H and ^{31}P nmr spectra, internal metallation occurs at an *ortho* C—H bond of one phenyl substituent of the phosphine ligands to give the complex:

$$1-\{Ir(H)(CO)[P(C_6H_4)(MePh)](PMePh_2)\}-2-C_6H_5-1,2-(\sigma-C_2B_{10}H_{10}).$$

The carborane iridium(I) complexes react with hydrogen giving three different isomers of *cis*-addition as a result of solvent dependence on the hydrogen uptake.[145] Figure 9 depicts the reaction scheme found for the hydrogen addition to the carborane iridium(I) complexes in solution at room temperature. The solvent dependence of the stereochemical course of the oxidative addition of H$_2$ to *trans*[Ir(σ-carb)(CO)(PPh$_3$)$_2$] was confirmed by further investigations on the complexes 1-[Ir(CO)(PPh$_3$)(RCN)]-7-C$_6$H$_5$-1,7-C$_2$B$_{10}$H$_{10}$ (RCN = C$_6$H$_5$CN and CH$_3$CN), which can be obtained[147] by the following ligand exchange reaction:

$$trans[Ir(\sigma\text{-carb})(CO)(PPh_3)_2] + RCN \rightleftharpoons trans[Ir(\sigma\text{-carb})(CO)(RCN)(PPh_3)] + PPh_3$$

$$carb = -7\text{-}C_6H_5\text{-}1,7\text{-}C_2B_{10}H_{10} \tag{100}$$

These nitrile complexes react with hydrogen to give the corresponding dihydrido-carborane iridium(III) derivatives having the structure:

$$\begin{array}{ccc} & H & \\ Ph_3P & | & H \\ & \diagdown \!\! | \!\! \diagup & \\ & Ir & \\ OC & | & carb \\ & NCR & \end{array}$$

The nitrile ligand in turn can be substituted by PPh$_3$ giving the analogous di-

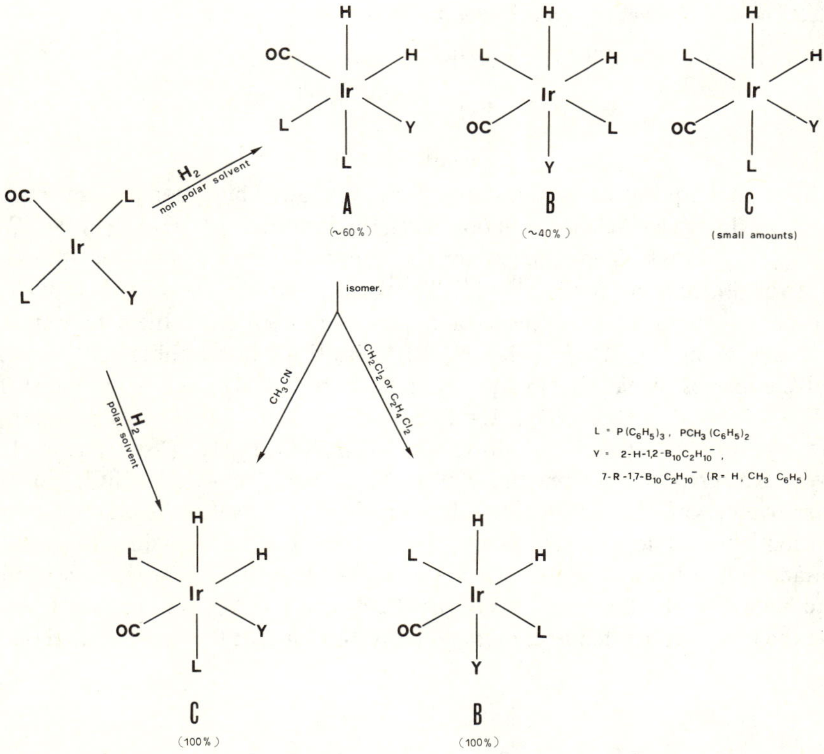

Figure 9. Reaction scheme for hydrogen addition to *trans*$[Ir(\sigma\text{-carb})(CO)(PR_3)_2]$ (from Reference 145).

hydrido isomer, which was directly obtained by the addition reaction of H_2 to complexes *trans*$[Ir(\sigma\text{-carb})(CO)(PPh_3)_2]$ carried out in nitrile solutions.

The carboranyl complexes, $[Ir(\sigma\text{-carb})(CO)(RCN)(PPh_3)]$, containing a weakly bonded nitrile ligand, appear to be catalysts for the selective homogeneous hydrogenation of 1-alkenes and alkynes under very mild conditions.[148] From ^1H nmr evidence, the hydrogenation appears to proceed through formation of the dihydrido carboranyl adducts which interact with the organic substrate forming monohydrido σ-alkyl or σ-alkenyl intermediates. The latter species in turn undergo reductive elimination of the alkane and alkene molecule, respectively, and this stage represents the reaction determining step.

Finally, it is to be noted that the reaction of $[IrCl(PPh_3)_3]$ with *C*-monolithium derivatives of *o*- and *m*-carborane does not give iridium carborane complexes; instead the products were internally metallated derivatives of iridium(I), $Ir[P(C_6H_4)Ph_2](PPh_3)_2$, or iridium(III), $\{Ir(H)Cl[P(C_6H_4)Ph_2](PPh_3)_2\}$, depending on the type of the lithium carborane reactant.[149,150]

4.1.3. Nickel, Palladium, and Platinum Derivatives

Reaction of the dilithio derivative of bis(*o*-carborane)

$$\text{Li}-\text{C}\overset{}{\underset{B_{10}H_{10}}{\diagdown\diagup}}\text{C}-\text{C}\overset{}{\underset{B_{10}H_{10}}{\diagdown\diagup}}\text{C}-\text{Li}$$

with 0.5 mol equivalent of anhydrous $NiBr_2$, followed by cation exchange with $(C_2H_5)_4NBr$, gave the diamagnetic nickel(II) derivative $[(C_2H_5)_4N]_2 Ni[(C_2B_{10}H_{10})_2]_2$, in which the nickel atom is σ-bonded to the four carbon atoms of the two biscarborane units.[138,139] When this complex was stirred in dry dichloroethane with a large excess of anhydrous cupric chloride, the paramagnetic derivative of nickel(III), $[(C_2H_5)_4N]Ni[(C_2B_{10}H_{10})_2]_2$ was obtained. Carboranyl complexes of nickel(II) of the type $[Ni(\sigma\text{-carb})_2L_2]$, carb = –2–$CH_3$–1,2– and –7–$C_6H_5$–1,7–$C_2B_{10}H_{10}$, L = PPh_3, PEt_3, and $[(\eta^5\text{-}C_5H_5)Ni(\sigma\text{-carb})\text{-}(PPh_3)]$, carb = –2–R–1,2–, –7–R–1,7–, –7–H–1,7–$C_2B_{10}H_{10}$ (R = CH_3, C_6H_5) have been prepared by reacting $(PR_3)_2NiCl_2$ and $[(\eta^5\text{-}C_5H_5)NiCl(PPh_3)]$, respectively, with the appropriate *C*-lithio carborane derivatives in ether benzene solution.[151] While the $[Ni(\sigma\text{-carb})_2L_2]$ complexes are unstable, the cyclopentadienyl carboranyl derivatives are quite stable both in solution and in the solid state. The reaction of excess of $(\eta^5\text{-}C_5H_5)NiCl(PPh_3)$ with $LiCB_{10}H_{10}CLi$ gave the monosubstituted *m*-carboranyl derivative but not the disubstituted one.

A nickel(II) complex containing a chelate carboranyl group was obtained[152] in the reaction of *C, C′*-dilithio-*o*-carborane with $[NiCl_2(PPh_3)_2]$. The product, $[Ni(C_2B_{10}H_{10})(PPh_3)_2]$, contains a chelate three-membered ring formed through two carbon–metal σ-bonds whose structure, as shown by X-ray analysis, is essentially planar about the nickel atom, but severely distorted from a square (Figure 10).

A few palladium derivatives containing σ-bonded carboranyl groups have been reported. A cyclometallated σ-carboranyl palladium(II) complex with the following suggested structure[153] has been obtained by reacting dipyridyl palladium(II) dichloride with 1–Li–2–C_6H_5–1,2–$C_2B_{10}H_{10}$. The substitution reaction is accompanied by an intramolecular metallation of an *ortho* C–H bond of the phenyl substituent of the carboranyl ligand, with the formation of two metal–carbon σ-bonds.

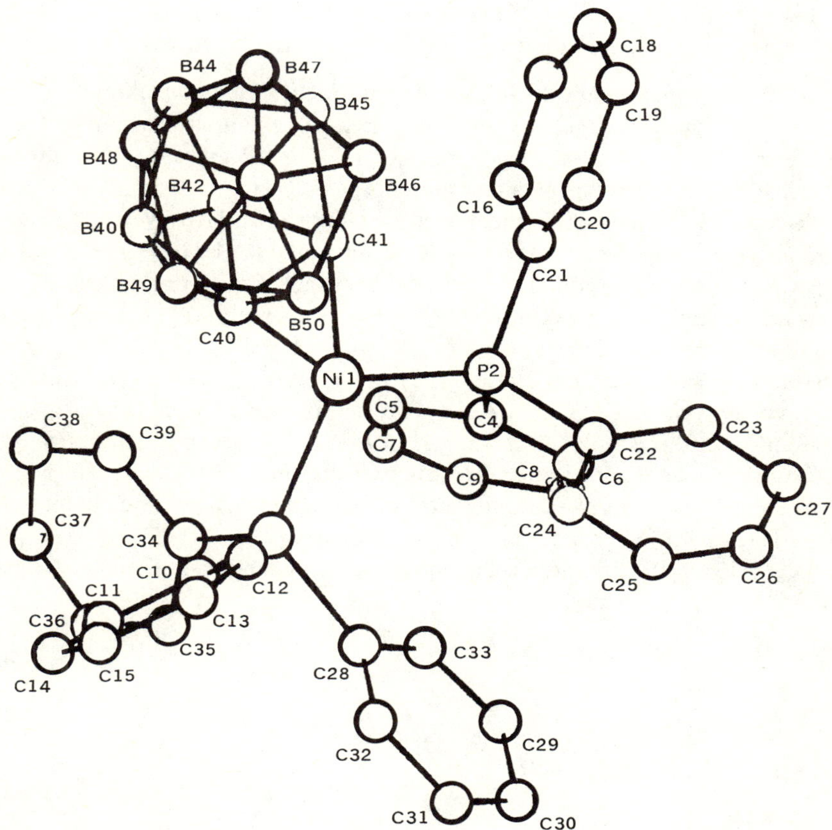

Figure 10. Molecular structure of $[(C_6H_5)_3P]_2NiC_2B_{10}H_{10}$ (from Reference 152).

Two cyclometallated carboranyl palladium(II) derivatives in which the metallation occurs at one carbon atom of an alkyl substituent of the coordinated phosphines, $1-[PdPEt_2(C_2H_4)(PEt_3)]-2-R-1,2-(\sigma-C_2B_{10}H_{10})$, R = CH$_3$ and C$_6$H$_5$, were obtained by treating *trans*$[PdCl_2(PEt_3)_2]$ with the appropriate *C*-lithio carborane in an ethereal solvent.[154] These complexes, together with some analogous platinum(II) derivatives were the first examples of transition metal compounds containing σ-bonded *o*- and *m*-carboranyl groups, and were reported in 1968.[132] As with the palladium derivatives, the carboranyl platinum(II) complexes were obtained by reacting either *trans*$[PtCl_2L_2]$, L = PEt$_3$, P(*n*-Pr)$_3$, with $1-Li-2-R-1,2-$ and $1-Li-7-R-1,7-C_2B_{10}H_{10}$ (R = H, CH$_3$, C$_6$H$_5$)[132,154] or *cis*$[PtCl_2L_2]$, L = PMePh$_2$, P(CH$_2$C$_6$H$_5$)$_3$ with $1-Li-2-C_6H_5-1,2-C_2B_{10}H_{10}$ and $1-Li-2-CH_3-1,2-C_2B_{10}H_{10}$, respectively,[155,156] in ether, THF, or benzene solution at room temperature or below. These reactions occur by substitution of one Cl with a carboranyl group followed by loss of hydrogen from one alkyl group of the coordinated phosphines and concomitant formation of a second Pt—C σ-bond to give an exocycle P—Pt—C. It is to be noted that identical cyclometallated monocarboranyl platinum(II) derivatives are formed by using either an excess or equimolecular amount of the lithium reagents. At first, on the basis of ^{1}H and ^{31}P nmr data, the authors suggested[132,154] that the metallation occurs at the β-carbon atom of one phosphine alkyl group, forming a four-membered ring. Subsequently, however, X-ray analysis of the complexes containing the phosphines $P(C_2H_5)_3$, $P(n\text{-}C_3H_7)_3$, and $P(CH_2C_6H_5)_3$ showed[156-158] that the metallation invariably occurs at the α-carbon atom of an alkyl group of the phosphines with formation of an unusual three-membered chelate ring Pt—P—C (Figure 11 and Figure 12). The molecular structures show that the platinum(II) atom is coordinated by the carboranyl group through its $C(1)$ atom and by two mutually *cis* phosphines through its P atoms, the four-coordination being achieved by internal metallation involving a C atom bonded to P of one alkyl group of the phosphine ligands. The platinum and the four-coordinated atoms are substantially coplanar and the carborane C—Pt bond is a "pure" σ-bond, as shown by the Pt—C bond lengths of 2.09–2.13 Å. On considering the bond lengths and angles, the geometry of these

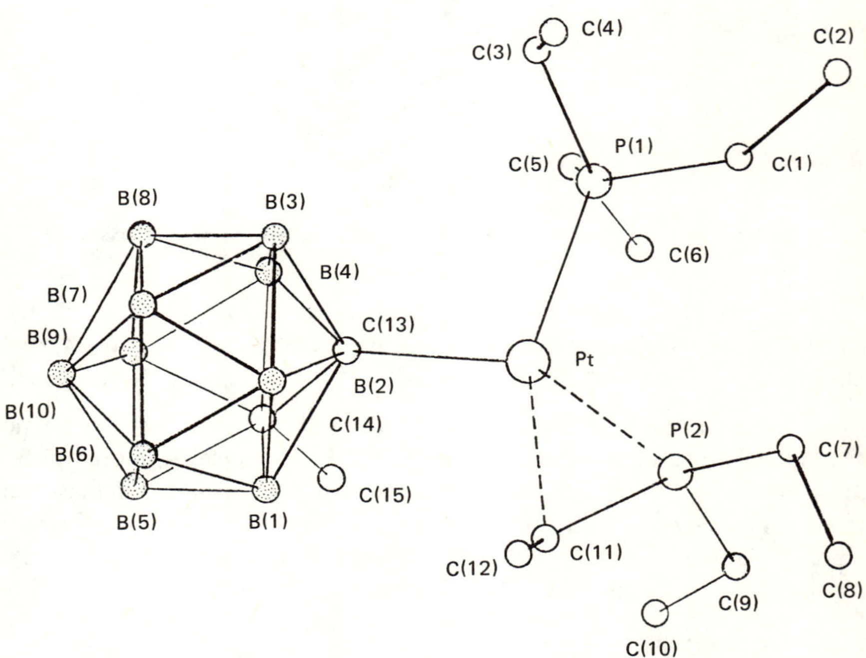

Figure 11. Molecular structure of 1-[(Et$_3$P)Pt(PEt$_2$CHCH$_3$)]-2-CH$_3$-1,2-(σ-C$_2$B$_{10}$H$_{10}$) (from Reference 157).

three-membered rings $\underset{\llcorner\!\_\!\_\!\_\!\lrcorner}{Pt-P-C}$ is better described in terms of a π-olefin-like bond between Pt and a $P=C$ double bond, rather than two $Pt-C$ and $Pt-P$ σ-bonds.[156] Such results further demonstrate that the bulky carboranyl ligands σ-bonded to a transition metal atom are able to promote intramolecular cyclo-metallation reactions[144] and that the formation of a three-membered $Pt-P-C$ ring takes place even when rings of sterically more favored size could be formed. On treating the cyclometallated complexes with dry HCl in benzene,[154] cleavage of the platinum alkyl carbon bond was observed

$$R = C_2H_5,\ R' = CH_3;\ R = n\text{-}C_3H_7,\ R' = C_2H_5 \tag{101}$$

but no cleavage of the carborane carbon–platinum bond took place even when these complexes were treated with excess HCl at temperatures up to 80°C. On the other hand, ring opening and cleavage of the carboranyl group occurred sequentially when these complexes reacted with dry HCN in benzene solution at room temperature:

$$[Pt(CN)_2(PR_3)_2] \tag{102}$$

The reaction between $cis[PtCl_2(PR_3)_2]$, $(R = C_2H_5,\ n\text{-}C_3H_7,\ C_6H_5)$ and 1-Li-2-R-1,2-$C_2B_{10}H_{10}$ $(R = CH_3,\ C_6H_5)$ produced[154] only monocarboranyl chloroplatinum(II) complexes:

$$cis[PtCl_2(PR_3)_2] + \text{Li-carb} \longrightarrow cis[PtCl(\sigma\text{-carb})(PR_3)_2] + \text{LiCl}$$

$$\text{carb} = -2\text{-}R\text{-}1,2\text{-}C_2B_{10}H_{10}\ (R = CH_3,\ C_6H_5) \tag{103}$$

No dicarboranyl derivatives were obtained, owing to the high steric requirements of the carborane moiety. On heating a solution of the $cis[PtCl(\sigma\text{-carb})(PR_3)_2]$ complexes in 1,2-dimethoxyethane, at the reflux temperature, in the presence of a catalytic amount of the appropriate free phosphine, a very fast cyclometalla-tion reaction via HCl elimination occurred to quantitatively give the corresponding cyclometallated compounds reported above.

However, it has been reported[159] that the palladium(II) and platinum(II) complexes $[MCl_2(PPh_3)_2]$, $M = Pd$ and Pt, react with 1,2-Li_2-1,2-$C_2B_{10}H_{10}$ in

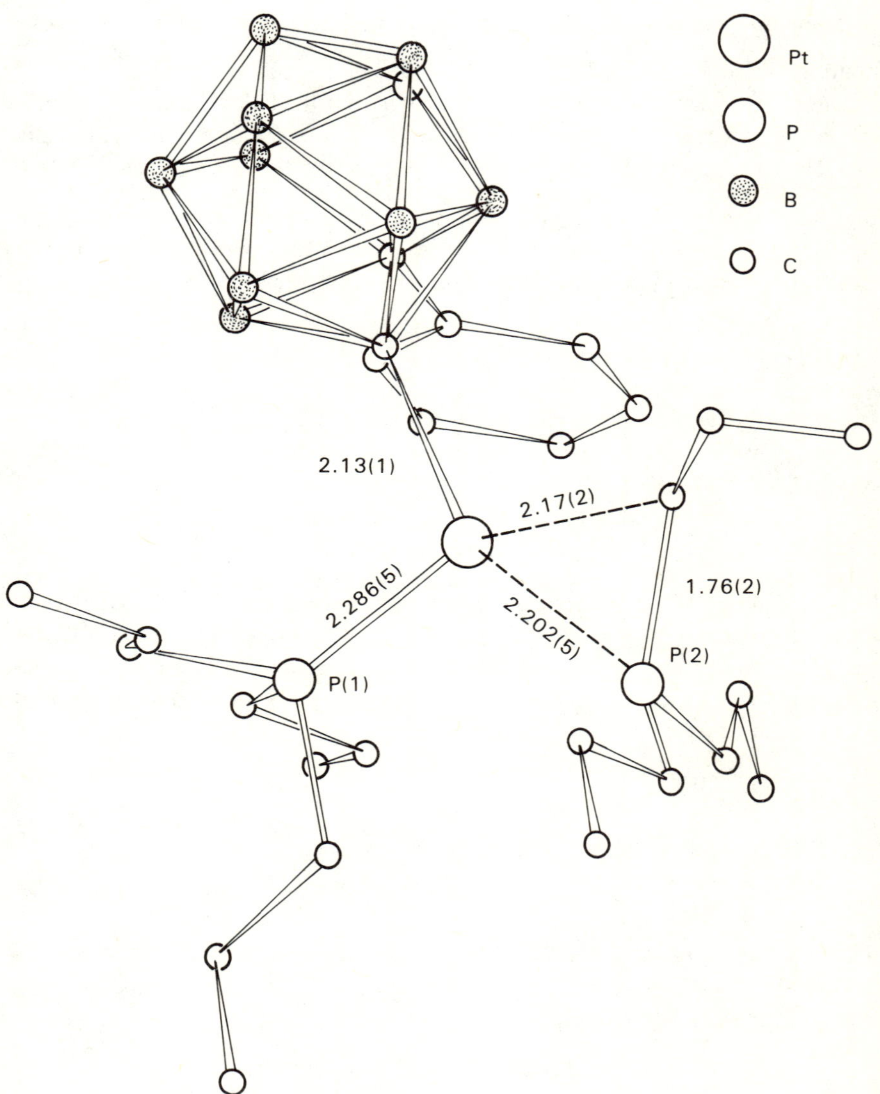

Figure 12. A view with some bond lengths and angles of the molecule of 1-[(n-Pr$_3$P) Pt(P-n-Pr$_2$CHCH$_2$CH$_3$)]-2-C$_6$H$_5$-1,2-(σ-C$_2$B$_{10}$H$_{10}$) (from Reference 158).

ether–benzene at 0°C to give compounds for which the structure containing a chelate *o*-carboranyl group

$$(Ph_3P)_2M \diagdown \diagup \begin{array}{c} C \\ | \; O \\ C \end{array} \diagdown \diagup B_{10}H_{10}$$

has been suggested on the basis of analytical data.

Neutral platinum (II) monohydridocarboranyl complexes of general formula *cis*- and *trans*[PtH(σ-carb)L$_2$], where L = PEt$_3$, PPh$_3$, PMePh$_2$, PMe$_2$Ph, and carb = -2-R-1,2- or -7-R-1,7-C$_2$B$_{10}$H$_{10}$ (R = H, CH$_3$, C$_6$H$_5$), have been prepared[160,161] from the reaction of *trans*[Pt(H)ClL$_2$] with excess Li-carb in diethyl ether solution or suspension at room temperature:

$$trans[Pt(H)ClL_2] + \text{Li-carb} \longrightarrow cis\text{- or }trans[Pt(H)(\sigma\text{-carb})L_2] \qquad (104)$$

The stereochemical course of this substitution reaction is dependent both on the coordinated phosphines and on the entering carboranyl ligand. Thus, in the case of bis(triethylphosphine) derivatives, the reaction in Equation (104) occurs with retention of configuration when the entering carboranyl ligands are those having the highest steric hindrance, namely the *C*(2)-substituted derivatives of the 1,2-carborane. Conversely, complexes having the *cis* configuration are the only reaction products obtained independently of the carboranyl ligands, when the substitution reaction involves platinum(II) complexes in which L = PPh$_3$, PMePh$_2$, or PMe$_2$Ph. While the *trans*-complexes are stable as solids and in solution, the *cis*-complexes, except those containing PPh$_3$ as neutral ligands, slowly decompose in solution yielding uncharacterized products.

Palladium(II) and platinum(II) complexes containing chelate *o*-carboranyl groups have also been reported.[162] In fact, a mixture of *cis*- and *trans*- bischelate *o*-carboranyl complexes of Pd(II), is formed when 1-Li-2-Et$_2$NCH$_2$-1,2-C$_2$B$_{10}$H$_{10}$ reacts with [PdCl$_2$(C$_6$H$_5$CN)$_2$] in a benzene–ether solution at −3 to 0°C. These palladium compounds are not hydrolyzed by water and are stable when stored.[162]

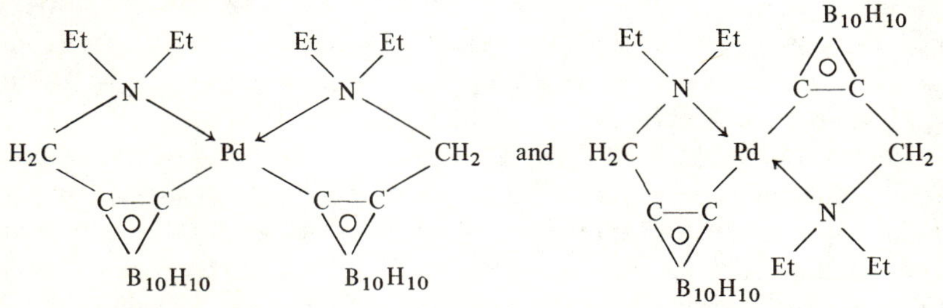

By reacting 1-Li-2-Et$_2$NCH$_2$-1,2-C$_2$B$_{10}$H$_{10}$ with *trans*[PtCl$_2$(SEt$_2$)$_2$] in benzene at 5 to 8°C, the following Pt(II) derivative is obtained as stable colorless crystals.[162] On the other hand, the five-coordinated platinum(II) complex

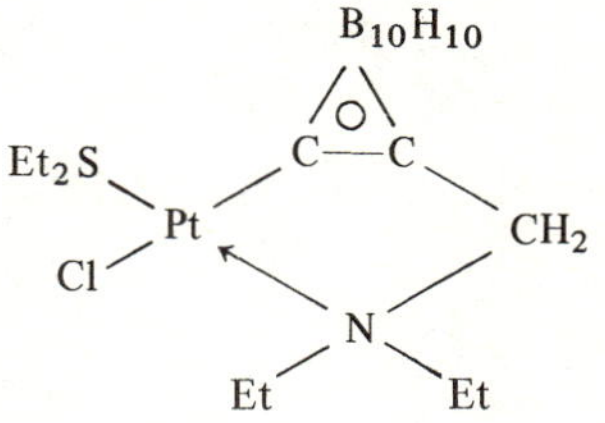

$$\text{Et–N(Et)–CH}_2\text{–C–C(B}_{10}\text{H}_{10}\text{)}\cdots\text{Pt(PhCN)}\cdots\text{C–C(B}_{10}\text{H}_{10}\text{)–CH}_2\text{–N(Et)–Et}$$

is formed when $[\text{PtCl}_2(\text{PhCN})_2]$ is treated with $1\text{-Li-2-Et}_2\text{NCH}_2\text{-1,2-C}_2\text{B}_{10}\text{H}_{10}$ in benzene solution at 5 to 8°C. The latter complex is a pale yellow crystalline solid, unstable in air.[162]

4.1.4. Copper and Gold Derivatives

A copper(III) compound containing two chelate biscarborane groups, {1-[1′,-1′,2′-dicarba-*closo*-dodecaborane(12)]-1,2-dicarba-*closo*-dodecaborane-(12)}, formed through four metal–carbon σ-bonds was reported.[138,139] Treatment of an ether slurry of

$$\text{Li–C–C–C–C–Li}\quad(\text{B}_{10}\text{H}_{10}\ \text{B}_{10}\text{H}_{10})$$

with 1 equiv of anhydrous CuCl_2 at reflux for 3 hr followed by cation exchange with $(\text{C}_2\text{H}_5)_4\text{NBr}$ afforded the diamagnetic $[(\text{C}_2\text{H}_5)_4\text{N}]\,\text{Cu}[(\text{C}_2\text{B}_{10}\text{H}_{10})_2]_2$ compound. Reduction of this anionic complex with lithium in acetone or with zinc metal in dichloromethane containing $(\text{C}_2\text{H}_5)_4\text{NBr}$ gave the salt of the corresponding derivative of copper(II), $\text{Cu}[(\text{C}_2\text{B}_{10}\text{H}_{10})_2]_2{}^{2-}$. For these copper anionic complexes structures were proposed analogous to the ones attributed to the biscarborane complexes of nickel and cobalt.[139]

C-copper(I) derivatives of 1,2-, 1,7-, and 1,12-dicarba-*closo*-dodecaborane (12) are easily prepared from the corresponding C-lithio carboranes and CuCl in tetrahydrofuran.[163] Treatment of o-, m-, and p-$\text{C}_2\text{B}_{10}\text{H}_{11}\text{Cu}$ with phenyllithium in dimethylformamide at 100°C affords the corresponding C-phenyl substituted carboranes in yields of 30–35%.[163] C-copper(I) derivatives of 1,2-, 1,7-, and 1,12-$\text{C}_2\text{B}_{10}\text{H}_{12}$ have also been reported as reaction intermediates in the preparation of C-ethynyl- and C-vinylcarborane(12) derivatives through the reaction sequences:

$$o\text{-, }m\text{-, }p\text{-HCB}_{10}\text{H}_{10}\text{C-Li}\xrightarrow{\text{CuCl}}o\text{-, }m\text{-, }p\text{-HCB}_{10}\text{H}_{10}\text{C-Cu}\xrightarrow{\text{BrC}\equiv\text{CR}}$$

$$o\text{-, }m\text{-, }p\text{-HCB}_{10}\text{H}_{10}\text{C–C}\equiv\text{CR}$$

$$(\text{R} = \text{Si(CH}_3)_3,\ \text{alkyl, aryl})\tag{105}$$

$$o\text{-, }m\text{-HCB}_{10}\text{H}_{10}\text{C-Li}\xrightarrow{\text{CuCl}}o\text{-, }m\text{-HCB}_{10}\text{H}_{10}\text{C-Cu}\xrightarrow{\text{ICH}=\text{CHX}}$$

$$o\text{-, }m\text{-HCB}_{10}\text{H}_{10}\text{CCH}=\text{CHX}$$

$$\text{X} = \text{H, Cl, I}\tag{106}$$

carried out in tetrahydrofuran solution.[164,165] A *C,C'*-bis(halovinyl)-*m*-carborane derivative has also been prepared through the reaction[165]

$$Cu-CB_{10}H_{10}C-Cu + 2IHC{=}CHI \longrightarrow 1,7-(IHC{=}CH)_2-1,7-C_2B_{10}H_{10} + 2CuI \qquad (107)$$

As reported above, carboranyl copper derivatives react with $(\eta^5\text{-}C_5H_5)Fe(CO)_2Br$ in benzene–tetrahydrofuran to give σ-carboranyl iron complexes.[137]

The reaction of $1\text{-Li-2-Et}_2NCH_2\text{-}1,2\text{-}C_2B_{10}H_{10}$ with CuCl in diethyl ether at temperatures between -3 and $0°C$ yields the chelate o-carboranyl copper(I) derivative[162]

The compound is a light brown solid, unstable in air. On the other hand, a copper(II) derivative containing two chelate o-carboranyl groups is formed in the reaction

$$2Li-C{-}C-CH_2NMe_2 + 2CuCl \longrightarrow \qquad + Cu + 2LiCl \qquad (108)$$

carried out in ether–benzene solution at -2 to $0°C$. This copper(II) derivative is a dark-green crystalline solid, stable in the air for a short time.[162]

A series of triphenylphosphine gold(I)–carborane compounds

$$(R = H, CH_3, C_6H_5)$$

has been prepared by treating $[AuCl(PPh_3)]$ with *C*-monolithio-*o*-carborane derivatives.[166] The carborane–gold σ-bond shows unusual stability. Thus, these complexes undergo no insertion reactions with tetrafluroethylene and on treatment with bromine give carboranyl–gold(III) derivatives by oxidative addition reaction. The gold(III) complexes are stable towards reductive elimination reactions. This observed stability may be related to the electron-withdrawing effect of the carboranyl ligand.[166]

4.2. Derivatives Formed through Carborane Boron-Transition Metal σ-Bonds

Only few compounds of this type have been reported in the literature. Systematic investigations appear not to have been carried out in this area and the reported complexes in most cases were fortuitously obtained.

4.2.1. Rhenium Derivatives

A compound containing a carborane boron–rhenium σ-bond has been reported.[167] By reacting sodium pentacarboranylrhenate, $Na[Re(CO)_5]$, with the chloride of the 3-carborane carboxylic acid in tetrahydrofuran at $-70°C$, an acyl derivative of rhenium is obtained. When this compound is heated in octane solution at $120°C$, decarbonylation occurs and the complex

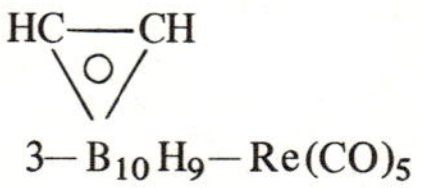

$$HC\!-\!CH$$
$$3\!-\!B_{10}H_9\!-\!Re(CO)_5$$

is formed in quantitative yield.

4.2.2. Iron Derivatives

On treatment of $Na[Fe(\eta^5\text{-}C_5H_5)(CO)_2]$ with o-carborane-3-carbonyl-chloride in tetrahydrofuran at $-70°C$, an acyl iron compound is formed.[168] On heating to $170°C$, the latter compound undergoes decarbonylation giving an iron complex containing a boron–metal σ-bond:

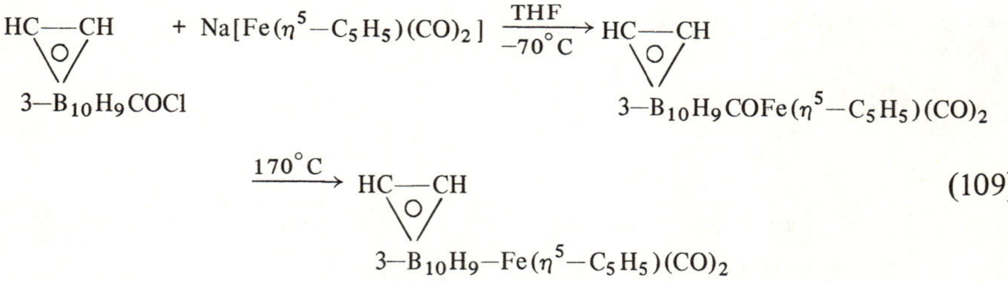

$$HC\!-\!CH \quad + \; Na[Fe(\eta^5\!-\!C_5H_5)(CO)_2] \xrightarrow[-70°C]{THF} HC\!-\!CH$$
$$3\!-\!B_{10}H_9COCl \qquad\qquad\qquad\qquad 3\!-\!B_{10}H_9COFe(\eta^5\!-\!C_5H_5)(CO)_2$$

$$\xrightarrow{170°C} HC\!-\!CH \tag{109}$$
$$3\!-\!B_{10}H_9\!-\!Fe(\eta^5\!-\!C_5H_5)(CO)_2$$

In the same manner, 1,2–dimethyl-3-o-carborane carboxylic acid chloride reacts with $Na[Fe(\eta^5\text{-}C_5H_5)(CO)_2]$ leading to σ–(1,2–dimethyl-o-carboran-3-yl)-η^5-cyclopentadienyl iron dicarbonyl,[169]

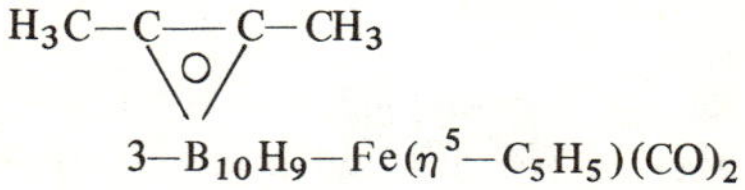

$$H_3C\!-\!C\!-\!C\!-\!CH_3$$
$$3\!-\!B_{10}H_9\!-\!Fe(\eta^5\!-\!C_5H_5)(CO)_2$$

The reactions of the complex containing the C-unsubstituted carborane with bromine and chlorine in CCl_4 [169,170] lead to an almost quantitative yield of

η^5-(*o*-carboran-3-yl)cyclopentadienyl dicarbonyliron bromide and chloride, respectively.

$$3-B_{10}H_9-Fe(CO)_2(C_5H_5) + X_2 \xrightarrow[20°]{CCl_4} X-Fe(CO)_2(C_5H_4-B_{10}H_9) \qquad (110)$$

Thus, the reactions with Br_2 and Cl_2 occur through the cleavage of the B—Fe σ-bond with the migration of the 3-*o*-carboranyl group from iron to the cyclopentadienyl group.

Unlike bromine and chlorine, iodine does not react under the same conditions even upon refluxing in CCl_4. The complex

$$3-B_{10}H_9-Fe(\eta^5-C_5H_5)(CO)_2 \quad [H_3C-C-C-CH_3]$$

reacts with chlorine and bromine in a somewhat different manner.[169] The action of 2 mol of chlorine upon 1 mol of the latter complex in CCl_4 over the temperature range -10-$+5°C$ followed by hydrolysis results in cleavage of the carborane boron–iron σ-bond with formation of 1,2-dimethyl-3-*o*-carboranecarboxylic acid [Equation (111)].

$$3-B_{10}H_9-Fe(\eta^5-C_5H_5)(CO)_2 \xrightarrow{Cl_2,\ H_2O} 3-B_{10}H_9-COOH \qquad (111)$$

On the other hand, the reaction with excess Br_2 in CCl_4 carried out at $20°C$ gives pentabromocyclopentane, 1,2-dimethyl-3-*o*-carborane- and 1,2-dimethyl-*B*-bromo-*o*-carborane carboxylic acids:

$$3-B_{10}H_9-Fe(\eta^5-C_5H_5)(CO)_2 \xrightarrow{Br_2,\ H_2O} C_5H_5Br_5 + 3-B_{10}H_9-CO_2H + B_{10}H_8BrCO_2H \qquad (112)$$

The reactions of both carborane iron complexes with PPh_3 and $HgCl_2$ were also studied.[169] Thus, on treatment of these complexes with PPh_3, an exchange reaction of one coordinated carbonyl group occurs, yielding the complex:

$$3-B_{10}H_9-Fe(\eta^5-C_5H_5)(CO)(PPh_3) \quad [RC-CR]$$

$$(R = H,\ CH_3)$$

The reaction with $HgCl_2$ proceeds, yielding a mercury-carborane compound formed through a B—Hg σ-bond. The latter compound undergoes cleavage of the B—Hg bond by reaction with bromine:

$$R-C\underset{\underset{3-B_{10}H_9-Fe(\eta^5-C_5H_5)(CO)_2}{\diagdown O \diagup}}{\!-\!\!-\!}C-R \quad + HgCl_2 \xrightarrow[\text{toluene}]{110^\circ} R-C\underset{\underset{3-B_{10}H_9-HgCl}{\diagdown O \diagup}}{\!-\!\!-\!}C-R \xrightarrow{Br_2} R-C\underset{\underset{3-B_{10}H_9Br}{\diagdown O \diagup}}{\!-\!\!-\!}C-R \qquad (113)$$

4.2.3. *Iridium Derivatives*

A number of iridium complexes containing a carborane boron–iridium σ-bond have been prepared by both intra- and intermolecular oxidative addition of terminal boron–hydrogen bonds to iridium(I) species.[171-173] Thus, when an excess of the carboranyl phosphines $1\text{-}PMe_2\text{-}1,2\text{-}C_2B_{10}H_{11}$, L, was added to a suspension of $[Ir(C_8H_{14})_2Cl]_2$ in cyclohexane and then heated at reflux temperature, a complex formulated as $[Ir(1\text{-}PMe_2\text{-}1,2\text{-}C_2B_{10}H_{11})_2(1\text{-}PMe_2\text{-}1,2\text{-}C_2B_{10}H_{10})Cl]$ was obtained:

$$\tfrac{1}{2}[Ir(C_8H_{14})_2Cl]_2 + 3L \xrightarrow[-C_8H_{14}]{\text{cyclohexane}} [IrClL_3] \xrightarrow{\Delta} \qquad (114)$$

$$L = 1\text{-}PMe_2\text{-}1,2\text{-}C_2B_{10}H_{11}$$

This hydrido iridium(III) complex is the product of an intramolecular oxidative addition to an intermediate $IrClL_2$ species that was not itself isolated in a pure state. The metallation reaction involves a carborane B—H bond as confirmed by synthesizing the above reported complex using carboranylphosphine ligands labeled with deuterium.

Reaction of $[IrCl(PPh_3)_2]$, formed *in situ* from $[IrCl(C_8H_{14})_2]_2$ and two equivalents of PPh_3 in various solvents, with $1,2\text{-}C_2B_{10}H_{12}$ affords[173] the complex $3\text{-}[Ir(H)(Cl)(PPh_3)_2]\text{-}1,2\text{-}C_2B_{10}H_{11}$ formed through oxidative addition of a terminal B—H bond to an iridium(I) species:

$$\tfrac{1}{2}[Ir(C_8H_{14})_2Cl]_2 + 2PPh_3 + 1,2\text{-}C_2B_{10}H_{12} \xrightarrow{-C_8H_{14}}$$

$$3\text{-}[Ir(H)(Cl)(PPh_3)_2]\text{-}1,2\text{-}C_2B_{10}H_{11} \qquad (115)$$

The complex appears to be formed through a carborane boron–iridium σ-bond. The position of substitution on the carborane cage of this complex was determined by deuterium labeling experiments and ^1H and ^{11}B nmr studies. Similar complexes have been obtained from the reaction between $[IrClL_2]$, $(L = PPh_3$

or $AsPh_3$), and each of the three icosahedral *o-, m-, p-*carborane isomers. Unlike the synthesis with $1,2\text{-}C_2B_{10}H_{12}$, which gives nearly analytically pure crude products with a stoichiometric quantity of carborane, synthesis with 1,7- or $1,12\text{-}C_2B_{10}H_{12}$ requires a large excess of carborane. The proposed structures of the complexes obtained are shown in Figure 13.

A fast reductive elimination of the carborane moiety[173] was observed on reacting the complex $3\text{-}[Ir(H)(Cl)(PPh_3)_2]\text{-}1,2\text{-}C_2B_{10}H_{11}$ with excess CO in benzene solution at room temperature, generating Vaska's complex, $[IrCl(CO)(PPh_3)_2]$:

$$3\text{-}[Ir(H)(Cl)(PPh_3)_2]\text{-}1,2\text{-}C_2B_{10}H_{11} + CO \longrightarrow \tag{116}$$
$$trans\,[IrCl(CO)(PPh_3)_2] \quad + \quad 1,2\text{-}C_2B_{10}H_{12}$$

Upon heating $3\text{-}[Ir(H)(Cl)(PPh_3)_2]\text{-}1,2\text{-}C_2B_{10}H_{11}$ with an excess of PPh_3 in toluene, a similar reaction involving reductive elimination of carborane was observed.[173]

However, as this reaction with PPh_3 is carried out under vigorous conditions, the carborane elimination is followed by an intramolecular oxidative addition occurring at an *ortho* C—H bond of a phenyl substituent on a coordinated phosphine. Thus, the $[IrCl(PPh_3)_3]$ compound first produced is partially converted to the *ortho*-metallated derivative of iridium(III):

$$3\text{-}[Ir(H)(Cl)(PPh_3)_2]\text{-}1,2\text{-}C_2B_{10}H_{11} + PPh_3 \xrightarrow{-1,2\text{-}C_2B_{10}H_{12}}$$
$$[IrCl(PPh_3)_3] \xrightarrow{\Delta} [Ir(H)(Cl)(PPh_3)_2\,PPh_2(C_6H_4)] \tag{117}$$

The *B-σ*-carboranyl–iridium complexes can be models for intermediate species postulated for transition-metal catalyzed deuterium exchange at terminal boron–hydrogen bonds in a wide variety of B—H-containing species and are themselves the most active agents thus far discovered for such exchange reactions.[174]

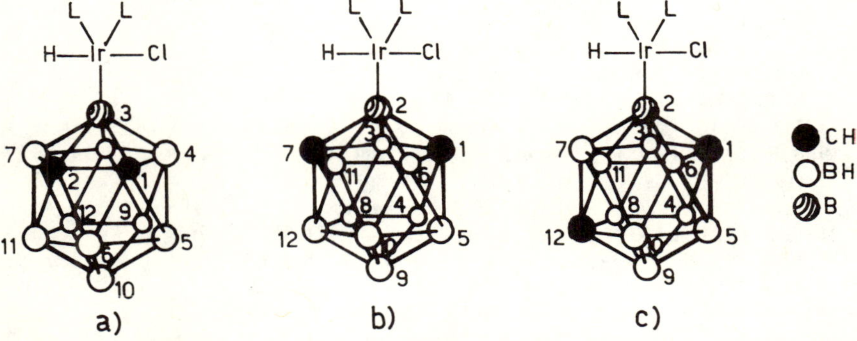

Figure 13. The proposed structures (from Reference 173) of: (a) $3\text{-}[Ir(H)(Cl)L_2]\text{-}1,2\text{-}C_2B_{10}H_{11}$; (b) $2\text{-}[Ir(H)(Cl)L_2]\text{-}1.7\text{-}C_2B_{10}H_{11}$; (c) $2\text{-}[Ir(H)(Cl)L_2]\text{-}1,12\text{-}C_2B_{10}H_{11}$ (*Note:* L = PPh_3, $AsPh_3$).

Attempts to oxidatively add $1,2\text{-}C_2B_{10}H_{12}$ to rhodium(I) complexes such as $[Rh(Cl)PPh_3)_3]$ or $[Rh(Cl)(PPh_3)_2]_2$ were unsuccessful.[173]

4.2.4. Platinum Derivatives

Stable *o*- and *m*-carborane derivatives containing a carborane boron–platinum σ-bond have been obtained[175] on reacting $Pt(PPh_3)_2$ with $(1,2\text{-}C_2B_{10}H_{11}\text{-}9)HgCl$ or $(1.7\text{-}C_2B_{10}H_{11}\text{-}9)HgCl$ in benzene solution at $20°C$:

$$(o\text{-},\ m\text{-}C_2B_{10}H_{11}\text{-}9)HgCl + Pt(PPh_3)_3 \longrightarrow (o\text{-},\ m\text{-}C_2B_{10}H_{11}\text{-}9)PtCl(PPh_3)_2 + Hg + PPh_3$$

$$(118)$$

ACKNOWLEDGMENTS: The author is grateful to Dr. Bruno Longato and Dr. Franco Morandini for helpful support during the writing of this contribution.

REFERENCES

1. W. N. Lipscomb *Boron Hydrides*, W. A. Benjamin, Inc., New York (1963).
2. M. F. Hawthorne, *Endeavour*, **25**, 146 (1966).
3. E. L. Muetterties and W. H. Knoth, *Polyhedral Boranes*, Marcel Dekker, New York (1968).
4. F. N. Tebbe, P. M. Garrett, and M. F. Hawthorne, *J. Am. Chem. Soc.*, **90**, 869 (1968).
5. R. E. Williams, *Progress in Boron Chemistry*, Vol. 2, p. 51, Pergamon Press, Oxford (1970).
6. R. N. Grimes, *Carboranes*, Academic Press, New York (1970).
7. E. L. Muetterties, *Boron Hydride Chemistry*, Academic Press, New York (1975); and references therein.
8. R. Hoffmann and W. N. Lipscomb, *J. Chem. Phys.*, **36**, 3489 (1962).
9. The nomenclature of boron compounds, *Inorg. Chem.*, 7, 1945 (1968).
10. R. N. Grimes, *J. Am. Chem. Soc.*, **88**, 1070 (1966).
11. R. N. Grimes, *J. Am. Chem. Soc.*, **88**, 1895 (1966).
12. R. E. Williams and F. S. Gerhart, *J. Am. Chem. Soc.*, **87**, 3513 (1965).
13. F. N. Tebbe, P. M. Garrett and M. F. Hawthorne, *J. Am. Chem. Soc.*, **90**, 869 (1968).
14. E. L. Muetterties (Ed.), *Inorganic Synthesis*, Vol. 10, p. 91, McGraw-Hill, New York (1967).
15. D. Grafstein and J. Dvorah, *Inorg. Chem.*, **2**, 1228 (1963).
16. S. Papetti and T. L. Heying, *J. Am. Chem. Soc.*, **86**, 2295 (1964).
17. R. R. Olsen and R. N. Grimes, *Inorg. Chem.*, **10**, 1103 (1971).
18. T. Onak, F. S. Gerhart and R. E. Williams, *J. Am. Chem. Soc.*, **85**, 3378 (1963).
19. S. F. Ditter, E. B. Klusmann, J. D. Oakes, and R. E. Williams, *Inorg. Chem.*, **9**, 889 (1970).
20. R. A. Beaudet and R. L. Poynter, *J. Am. Chem. Soc.*, **86**, 1258 (1964).
21. R. R. Olsen and R. N. Grimes, *J. Am. Chem. Soc.*, **92**, 5072 (1970).
22. R. E. Kesting, K. F. Jackson, E. B. Klusmann, and F. J. Gerhart, *J. Appl. Polym. Sci.*, **14**, 2525 (1970).
23. R. Hoffmann and W. N. Lipscomb, *J. Chem. Phys.*, **36**, 3484 (1962).
24. P. M. Garrett, J. C. Smart, and M. F. Hawthorne, *J. Am. Chem. Soc.*, **91**, 4707 (1969).
25. L. I. Zakharkin, V. M. Kalinin and E. G. Rys, *Zh. Obshch. Khim.*, **42**, 477 (1972).
26. L. I. Zakharkin and A. I. Kovredov, *Zh. Obshch. Khim.*, **44**, 1840 (1974).
27. F. Tebbe, P. M. Garrett and M. F. Hawthorne, *J. Am. Chem. Soc.*, **86**, 4222 (1964).

28. D. A. Owen and M. F. Hawthorne, *J. Am. Chem. Soc.,* **90**, 5912 (1968).

29. W. H. Knoth, *J. Am. Chem. Soc.,* **89**, 1274 (1967).

30. W. H. Knoth, *Inorg. Chem.,* **10**, 598 (1971).

31. T. L. Heying, S. W. Ager, S. L. Clark, R. P. Alexander, S. Papetti, S. A. Reid, and S. J. Trotz, *Inorg. Chem.,* **2**, 1097 (1963).

32. D. Grafstein, S. Bobinski, J. Dvorak, H. Smith, N. Schwartz, M. S. Cohen, and M. M. Fein, *Inorg. Chem.,* **2**, 1120 (1963).

33. L. I. Zakharkin, V. I. Stanko, A. I. Klimova and Yu. A. Chapovskii, *Izv. Akad. Nauk SSSR, Ser. Khim.,* 2236 (1963).

34. E. W. Cox and T. L. Heying, U. S. Patent 3,137,734 (1964); *Chem. Abstr.,* **61**, 4392 (1964).

35. E. L. Muetterties, *Boron Hydride Chemistry,* Academic Press, New York, p. 316 (1975).

36. V. I. Bregadze and O. Yu. Okhlobystin, *Russ. Chem. Rev.,* **37**, 173 (1968).

37. L. I. Zakharkin, V. N. Kalinin and L. S. Podvisotskaya, *Izv. Akad. Nauk SSSR, Ser. Khim.,* 2661 (1968).

38. L. I. Zakharkin, A. I. L'Vov and L. S. Podvisotskaya, *Izv. Akad. Nauk. SSSR, Ser. Khim.,* 1905 (1965).

39. R. N. Grimes, *Carboranes,* Academic Press, New York (1970); and references therein.

40. L. I. Zakharkin, V. I. Stanko, V. A. Brattsev, Yu. A. Chapovskii, A. I. Klimova, O. Yu. Okhlobystin and A. A. Ponomarenko, *Dokl. Akad. Nauk SSSR,* **155**, 1119 (1964).

41. L. I. Zakharkin, A. V. Grebennikov and A. V. Karantsev, *Izv. Akad. Nauk SSSR, Ser. Khim.,* 2077 (1967).

42. L. I. Zakharkin, *Tetrahedron Lett.,* **33**, 2255 (1964).

43. L. I. Zakharkin, *Izv. Akad. Nauk SSSR, Ser. Khim.,* 158 (1965).

44. L. I. Zakharkin and A. V. Karantsev, *Zh. Obshch. Khim.,* **35**, 1123 (1965); 37, 1211 (1967).

45. L. I. Zakharkin, V. I. Stanko, and Yu. A. Chapovsky, *Izv. Akad. Nauk SSSR, Ser. Khim.,* 582 (1964).

46. V. I. Stanko and G. A. Anorova, *Zh. Obshch. Khim.,* **36**, 946 (1966).

47. V. I. Stanko, G. A. Anorova, and T. V. Klimova, *Zh. Obshch. Khim.,* **36**, 1174 (1966).

48. V. I. Stanko and A. I. Klimova, *Zh. Obshch. Khim.,* **36**, 159 (1966).

49. V. I. Stanko, G. A. Anorova, and T. V. Klimova, *Zh. Obshch. Khim.,* **39**, 2143 (1969).

50. V. I. Stanko, *Zh. Obshch. Khim.,* **35**, 1139 (1965).

51. J. L. Boone, R. J. Brotherton, and L. L. Patterson, *Inorg. Chem.,* **4**, 910 (1965).

52. B. H. Mikailov, E. A. Shagova, and T. P. Potapova, *Izv. Akad. Nauk SSSR, Ser. Khim.,* 2048 (1970).

53. L. I. Zakharkin and V. S. Korlova, *Zh. Obshch. Khim.,* **42**, 476 (1972).

54. M. F. Hawthorne and P. A. Wegner, *J. Am. Chem. Soc.,* **90**, 896 (1968).

55. V. I. Bregadze, A. Ya. Usyatinskii, and N. N. Godovikov, *Dokl. Akad. Nauk SSSR,* **241**, 364 (1978).

56. V. I. Bregadze, V. Ts. Kampel, A. Ya. Usyatinskii, and N. N. Godovikov, *J. Organomet. Chem.,* **154**, C1 (1978).

57. V. I. Bregadze, A. Ya. Usyatinskii, and N. N. Godovikov, *Izv. Akad. Nauk SSSR, Ser. Khim.,* 1876 (1979).

58. V. I. Bregadze, V. Ts. Kampel, A. Ya. Usyatinskii and N. N. Godovikov, *Izv. Akad. Nauk SSSR, Ser. Khim.,* 1467 (1978).

59. J. B. Leake, G. Oates, S. Tang, and T. Onak, *J. Chem. Soc., Dalton Trans.,* 1018 (1975).

60. M. L. Thompson and R. N. Grimes, *Inorg. Chem.,* **11**, 1925 (1972).

61. C. B. Ungermann and T. Onak, *Inorg. Chem.,* **16**, 1428 (1977).

62. P. A. Wegner, V. A. Uski, R. P. Kiester, S. Dabestani, and V. W. Day, *J. Am. Chem. Soc.,* **99**, 4846 (1977).

63. L. I. Zakharkin, V. N. Kalinin, and E. G. Rys, *Zh. Obshch. Khim.,* **43**, 847 (1973).

64. L. I. Zakharkin, V. I. Bregadze, and O. Yu. Okhlobystin *J. Organomet. Chem.*, **4**, 211 (1965).
65. S. Papetti and T. L. Heying, *Inorg. Chem.*, **2**, 1105 (1963).
66. S. Papetti, B. B. Schaeffer, M. J. Troscianiec, and T. L. Heying, *Inorg. Chem.*, **3**, 1444 (1964).
67. S. Papetti and T. L. Heying, *Inorg. Chem.*, **3**, 1448 (1964).
68. N. N. Schwartz, E. O'Brien, S. Karlan, and N. M. Fein, *Inorg. Chem.*, **4**, 661 (1965).
69. S. Papetti, B. B. Schaeffer, A. P. Gray, and T. L. Heying, *J. Polymer Sci.*, A–1, **4**, 1623 (1966).
70. H. Schroeder, O. G. Shaffling, T. B. Larchar, F. F. Frulla, and T. L. Heying, *Rubber Chem. Technol.*, **39**, 1189 (1966).
71. T. L. Heying, J. W. Ager, Jr., S. L. Clark, D. J. Mangold, H. L. Goldstein, M. Hillman, R. J. Polak, and J. W. Szymanski, *Inorg. Chem.*, **2**, 1089 (1963).
72. L. I. Zakharkin, V. I. Bregadze and O. Yu. Okhlobystin, *Izv. Akad. Nauk SSSR, Ser. Khim.*, 1539 (1964).
73. R. M. Salinger and C. L. Frye, *Inorg. Chem.*, **4**, 1815 (1965).
74. V. V. Korshak, S. V. Vinogradova, A. I. Solomantina, N. I. Bekasova, M. A. Andrieva, and E. G. Lagutina, USSR Patent 497300 (1975); *Chem. Abstr.*, **84**, 105747m (1976).
75. V. V. Korol'ko, E. G. Kagan, G. V. Dotsenko, T. I. Saratovkina, and G. A. Ivanova, USSR Patent 491638; *Chem. Abstr.*, **84**, 44340z (1976).
76. Yu. A. Yuzhelevski, E. B. Dmokhovskaya, E. G. Kagan, T. I. Lezhen, and O. N. Larionova, USSR Patent 485119 (1975); *Chem. Abstr.*, **84**, 59723a (1976).
77. S. Papetti and T. L. Heying, *Inorg. Chem.*, **2**, 1105 (1963).
78. V. V. Korol'ko, V. I. Vecherskaya, V. P. Mileshkevic, and N. D. Il'ina, *Zh. Obshch. Khim.*, **48**, 596 (1978).
79. H. Schroeder, S. Papetti, R. P. Alexander, J. F. Sieckhaus, and T. L. Heying, *Inorg. Chem.*, **8**, 2444 (1969).
80. O. A. Kruglaya, L. I. Zakharkin, B. I. Petrov, G. S. Kalinina, and N. S. Vyarankin, *Synth. Inorg. Met.-Org. Chem.*, **3**, 63 (1973).
81. A. Yu. Aleksandrov, V. I. Bregadze, V. I. Gol'danskii, L. I. Zakharking, O. Yu. Oklobystin, and V. V. Krapov, *Dokl. Akad. Nauk SSSR*, **165**, 593 (1965).
82. S. Bresadola, F. Rossetto, and G. Tagliavini, *Ann. Chim. (Rome)*, **58**, 598 (1968).
83. S. Bresadola, G. Plazzogna, G. Cecchin, and G. Tagliavini, *Gazz. Chim. Ital.*, **100**, 175 (1970).
84. L. I. Zakharkin and N. F. Shemyalkin, *Izv. Akad. Nauk SSSR, Ser. Khim.*, 2350 (1977).
85. L. I. Zakharkin and N. F. Shemyalkin, *Izv. Akad. Nauk SSSR, Ser. Khim.*, 1450 (1978).
86. N. G. Bokii, A. I. Yanovskii, Yu. T. Struchnov, N. F. Shemyalkin, and L. I. Zakharkin, *Izv. Akad. Nauk SSSR, Ser. Khim.*, 380 (1978).
87. V. I. Bregadze and O. Yu. Oklobystin, *Izv. Akad. Nauk SSSR, Ser. Khim.*, 2084 (1967).
88. V. I. Stanko, T. V. Klimova, and I. P. Beletskaya, *J. Organomet. Chem.*, **61**, 191 (1973).
89. T. L. Heying, S. Papetti, and O. G. Shaffling, U. S. Patents 3,388,090 (1968); 3,388,091 (1968); 3,388,092 (1968); *Chem. Abstr.*, **69**, 28297a; 28298b; 36897m (1968).
90. M. F. Hawthorne, T. E. Berry, and P. A. Wegner, *J. Am. Chem. Soc.*, **87**, 4746 (1965).
91. H. J. Dietrich, R. P. Alexander, T. L. Heying, H. Kwasnik, C. O. Obenland, and H. A. Shroeder, *Makromol. Chem.*, **175**, 425 (1974).
92. L. P. Dorofeenko, V. F. Gridina, A. L. Klebanskii, L. E. Krupnova, N. I. Shkambarnaya, I. F. Ermachkova, L. M. Vasil'eva, A. F. Zhigach, V. N. Siryatskaya, and V. V. Korol'ko, USSR Patent 319622 (1971); *Chem. Abstr.*, **76**, 100353e (1972).
93. H. Schroeder, *Inorg. Macromol. Rev.*, **1**, 45 (1970).
94. J. F. Sieckhaus, R. N. Scott, and T. B. Larchar, U. S. Patent 3,899,456 (1975); *Chem. Abstr.*, **83**, 194897e (1975).

95. K. O. Knollmueller, R. N. Scott, H. Kwasnik, and J. F. Sieckhaus, *J. Polymer Sci.,* A-1, **9**, 1071 (1971); and references therein.

96. E. Hedaya, J. H. Kawakami, P. W. Kopf, G. T. Kwiatkowski, D. W. McNeil, D. A. Owen, E. N. Peters, and R. W. Tulis, *J. Polym. Sci., Polym. Chem. Ed.,* **15**, 2229 (1977); and references therein.

97. E. Hedaya, J. H. Kawakami and G. T. Kwiatkowski, German Patent 2,435,385 (1975); *Chem. Abstr.,* **82**, 171723h (1975).

98. E. N. Peters, E. Hedaya, J. H. Kawakami, G. T. Kwiatkowski, D. W. McNeil, and R. W. Tulis, *Rubber Chem. Technol.,* **48**, 14 (1975).

99. E. N. Peters, D. D. Stewart, J. J. Bokan, G. T. Kwiatkowski, C. D. Beard, R. Moffitt, and E. Hedaya, *J. Elastomers Plast.,* **9**, 177 (1977).

100. E. N. Peters, J. H. Kawakami, G. T. Kwiatkowski, E. Hedaya, B. L. Joesten, D. W. McNeil, and D. A. Owen, *J. Polym. Sci., Polym. Phys. Ed.,* **15**, 723 (1977).

101. E. N. Peters, D. D. Stewart, J. J. Bokan, R. Moffitt, C. D. Beard, G. T. Kwiatkowski, and E. Hedaya, *J. Polym. Sci., Polym. Chem. Ed.,* **15**, 973 (1977); and references therein.

102. J. A. Yancey and T. R. Lynn, *Analabs Res. Notes,* **14**, 1 (1974); and references therein.

103. I. P. Yudina, Yu. A. Yuzhelevskii, E. B. Dmokhovskaya, L. N. Karmanova, and K. I. Sakodynskii, USSR Patent 654,894 (1976); *Chem. Abstr.,* **90**, 197198h (1979).

104. S. Bresadola, F. Rossetto, and G. Tagliavini, *Chem. Commun.,* 623 (1966).

105. S. Bresadola, F. Rossetto, and G. Tagliavini, *Chim. Ind. (Milan),* **49**, 531 (1967).

106. S. Bresadola, F. Rossetto, and G. Tagliavini, *Eur. Polym. J.,* **4**, 75 (1968).

107. L. I. Zakharkin, V. I. Bregadze, and O. Yu. Okhlobystin, *J. Organomet. Chem.,* **4**, 211 (1965).

108. V. I. Bregadze, V. Ts. Kampel, and N. N. Godovikov, *J. Organomet. Chem.,* **157**, C1 (1978).

109. V. I. Bregadze, V. Ts. Kampel, and N. N. Godovikov, *Izv. Akad. Nauk SSSR, Ser. Khim.,* 1951 (1978).

110. L. I. Zakharkin and I. V. Pisareva, *Izv. Akad. Nauk SSSR, Ser. Khim.,* 1226 (1978).

111. O. Yu. Oklobystin and L. I. Zakharkin, *J. Organomet. Chem.,* **3**, 257 (1965).

112. L. I. Zakharkin, V. I. Bregadze, and O. Yu. Okhlobystin, *Zh. Obshch. Khim.,* **36**, 761 (1966).

113. L. I. Zakharkin, V. I. Bregadze, and O. Yu. Okhlobystin, *J. Organomet. Chem.,* **6**, 228 (1966).

114. V. I. Pakhomov, A. V. Medvedev, V. I. Bregadze, and O. Yu. Okhlobystin, *J. Organomet. Chem.,* **29**, 15 (1971).

115. N. G. Bokii, Yu. T. Struchkow, V. N. Kalinin, and L. I. Zakharkin, *Zh. Strukt. Khim.,* **19**, 380 (1978).

116. L. I. Zakharkin, V. N. Kalinin, and L. S. Podvisotskaya, *Izv. Akad. Nauk SSSR, Ser. Khim.,* 688 (1968).

117. V. I. Bregadze and O. Yu. Okhlobystin, *Dokl. Akad. Nauk SSSR,* **177**, 347 (1967).

118. L. I. Zakharkin and L. S. Podvisotskaya, *Izv. Akad. Nauk SSSR, Ser. Khim.,* 681 (1968).

119. R. A. Wiesboeck and M. F. Hawthorne, *J. Am. Chem. Soc.,* **86**, 1642 (1964).

120. L. I. Zakharkin and L. S. Podvisotskaya, *J. Organomet. Chem.,* **7**, 385 (1967).

121. L. A. Fedorov, V. N. Kalinin, E. I. Fedin, and L. I. Zakharkin, *Izv. Akad. Nauk SSSR, Ser. Khim.,* 849 (1970).

122. R. E. Dessy and J. Y. Kim, *J. Am. Chem. Soc.,* **83**, 1167 (1961).

123. L. I. Zakharkin, V. N. Kalinin, and L. S. Podvisotskaya, *Izv. Akad. Nauk SSSR, Ser. Khim.,* 679 (1968).

124. S. Bresadola, F. Rossetto, and G. Tagliavini, *Chim. Ind. (Milan).,* **50**, 452 (1968).

125. V. I. Bregadze, V. Ts. Kampel, and N. N. Godovikov, *J. Organomet. Chem.,* **112**, 249 (1976).

126. V. I. Bregadze, V. Ts. Kampel, and N. N. Godovikov, *J. Organomet. Chem.*, **136**, 281 (1977).

127. L. I. Zakharkin and I. V. Pisareva, *Izv. Akad. Nauk SSSR, Ser. Khim.*, 1885 (1977).

128. V. I. Bregadze, V. Ts. Kampel, and N. N. Godovikov, *Izv. Akad. Nauk SSSR, Ser. Khim.*, 1630 (1977).

129. V. Ts. Kampel, K. P. Butin, V. I. Bregadze, and N. N. Godovikov, *Izv. Akad. Nauk SSSR, Ser. Khim.*, 1508 (1978).

130. K. P. Butin, A. N. Kashin, I. P. Beletskaya, and O. A. Reutov, *J. Organomet. Chem.*, **10**, 197 (1967); I. P. Beletskaya, K. P. Butin, A. N. Ryabtsev, and O. A. Reutov, *J. Organomet. Chem.*, **59**, 1 (1973).

131. Yu. A. Ol'dekop, N. A. Maier, A. A. Erdman, and V. P. Prokopovich, *Dokl. Akad. Nauk SSSR*, **243**, 933 (1978).

132. S. Bresadola, P. Rigo, and A. Turco, *Chem. Commun.*, 1205 (1968).

133. D. A. Owen, J. C. Smart, P. M. Garrett, and M. F. Hawthorne, *J. Am. Chem. Soc.*, **93**, 1362 (1971).

134. J. C. Smart, P. M. Garrett, and M. F. Hawthorne, *J. Am. Chem. Soc.*, **91**, 1031 (1969).

135. L. I. Zakharkin, L. V. Orlova, and L. I. Denisovich, *Zh. Obshch. Khim.*, **42**, 2217 (1972).

136. L. I. Zakharkin, L. V. Orlova, A. I. Kovredov, L. A. Fedorov, and B. V. Lokshin, *J. Organomet. Chem.*, **27**, 95 (1971).

137. L. I. Zakharkin, A. I. Kovredov, M. G. Meiramov, and A. V. Kazantsev, *Izv. Akad. Nauk SSSR, Ser. Khim.*, 1673 (1977).

138. D. A. Owen and M. F. Hawthorne, *J. Am. Chem. Soc.*, **92**, 3194 (1970).

139. D. A. Owen and M. F. Hawthorne, *J. Am. Chem. Soc.*, **93**, 873 (1971).

140. R. A. Love and R. Bau. *J. Am. Chem. Soc.*, **94**, 8274 (1972).

141. S. Bresadola, G. Cecchin, and A. Turco, *Proceedings of the Thirteenth International Conference on Coordination Chemistry, Cracow–Zakopane*, Polish Academy of Sciences, Warsaw, p. 160 (1970).

143. G. Allegra, M. Calligaris, R. Furlanetto, G. Nardin, and L. Randaccio, *Cryst. Struct. Commun.*, **3**, 69 (1974).

144. S. Bresadola, B. Longato, and F. Morandini, *Coord. Chem. Rev.*, **16**, 19 (1975).

145. B. Longato, F. Morandini, and S. Bresadola, *Inorg. Chem.*, **15**, 650 (1976).

146. S. Bresadola and F. Morandini, unpublished results.

147. B. Longato and S. Bresadola, *Inorg. Chim. Acta*, **33**, 189 (1979).

148. S. Bresadola, B. Longato, and F. Morandini, *Abstracts of the Ninth International Conference on Organometallic Chemistry, Dijon, France*, Les Presses de l'Imprimerie Universitaire, Dijon, p. C4 (1979).

150. S. Bresadola, B. Longato, and F. Morandini, *Inorg. Chim. Acta*, **25**, L135 (1977).

151. S. Bresadola, G. Cecchin, and A. Turco, *Gazz. Chim. Ital.*, **100**, 682 (1970).

152. A. A. Sayler, H. Beall, and J. F. Sieckhaus, *J. Am. Chem. Soc.*, **95**, 5790 (1973).

153. L. I. Zakharkin and A. I. Kovredov, *Zh. Obshch. Khim.*, **44**, 1832 (1974).

154. S. Bresadola, A. Frigo, B. Longato, and G. Rigatti, *Inorg. Chem.*, **12**, 2788 (1973).

155. S. Bresadola, B. Longato, and F. Morandini, *J. Organomet. Chem.*, **128**, C5 (1977).

156. S. Bresadola, N. Bresciani-Pahor, and B. Longato, *J. Organomet. Chem.*, **179**, 73 (1979).

157. N. Bresciani-Pahor, *Acta Crystallogr.*, **B33**, 3214 (1977).

158. N. Bresciani, M. Calligaris, P. Delise, G. Nardin, and L. Randaccio, *J. Am. Chem. Soc.*, **96**, 5642 (1974).

159. L. I. Zakharkin and A. I. Kovredov, *Izv. Akad. Nauk SSSR, Ser. Khim.*, 2619 (1975).

160. S. Bresadola, B. Longato, and F. Morandini, *Chem. Commun.*, 510 (1974).

161. B. Longato, F. Morandini, and S. Bresadola, *J. Organomet. Chem.*, **121**, 113 (1976).

162. L. I. Zakharkin and I. S. Savel'eva, *Izv. Akad. Nauk SSSR, Ser. Khim.*, 1381 (1979).

163. L. I. Zakharkin and A. I. Kovredov, *Izv. Akad. Nauk SSSR, Ser. Khim.*, 740 (1974).

164. L. I. Zakharkin and A. I. Kovredov, *Izv. Akad. Nauk SSSR, Ser. Khim.*, 1593 (1976).

165. L. I. Zakharkin, A. I. Kovredov, V. A. Ol'shevskaya and V. V. Kobak, *Zh. Obshch. Khim.*, 48, 2132 (1978).

166. C. M. Mitchell and F. G. A. Stone, *Chem. Commun.*, 1263 (1970).

167. L. I. Zakharkin and L. V. Orlova, *Izv. Akad. Nauk SSSR, Ser. Khim.*, 1847 (1971).

168. L. I. Zakharkin and L. V. Orlova, *Izv. Akad. Nauk SSSR, Ser. Khim.*, 2417 (1970).

169. L. I. Zakharkin, L. V. Orlova, B. V. Lokshin, and L. A. Fedorov, *J. Organomet. Chem.*, 40, 15 (1972).

170. L. I. Zakharkin and L. V. Orlova, *Izv. Akad. Nauk SSSR, Ser. Khim.*, 209 (1972).

171. E. L. Hoel and M. F. Hawthorne, *J. Am. Chem. Soc.*, 95, 2712 (1973).

172. E. L. Hoel and M. F. Hawthorne, *J. Am. Chem. Soc.*, 96, 6770 (1974).

173. E. L. Hoel and M. F. Hawthorne, *J. Am. Chem. Soc.*, 97, 6388 (1974).

174. E. L. Hoel, M. Talebinasab-Savari and M. F. Hawthorne, *J. Am. Chem. Soc.*, 99, 4356 (1977).

175. L. I. Zakharkin and I. V. Pisareva, *Izv. Akad. Nauk SSSR, Ser. Khim.*, 252 (1978).

6

Electrochemistry of Metallaboron Cage Compounds

William E. Geiger Jr.

1. INTRODUCTION AND SCOPE

The first mention of voltammetric measurements on transition metal-carboranes or metal–boranes seems to be that found in the 1965 note[1] describing the preparation of the first metal dicarbollide compound, $(C_2B_9H_{11})_2Fe^-$. Although no data were given, it was stated that the polarographic reduction of the monoanion gave the corresponding dianion. Since that time, Hawthorne *et al.* have reported electrochemical data, usually via cyclic voltammetry measurements, in surveying the physical properties of most of the metallacarboranes they have synthesized. Taken together with our data and data from other groups, we now have a moderately large body of electrochemical data on this class of compounds, and it seems timely to review this area to see what we have learned about the redox properties of metal-containing boron clusters.

The small number of in-depth electrochemical studies has made it difficult to provide a wide-ranging critical review of this area. For the most part, only the potentials of redox couples, perhaps with an assignment to metal oxidation state changes, and a notation of "reversible" or "irreversible" is reported in the original literature. These notations must be passed along to the reader as such, unless there are other indications that the electrochemical mechanisms might be more complicated than reported. Where cyclic voltammetry measurements have been employed, it may in general be assumed that systems reported to be reversible show at least moderate stability of the oxidized and reduced forms of the

---

William E. Geiger Jr. ● Department of Chemistry, University of Vermont, Burlington, Vermont.

redox couple, and that the electron-transfer rate leading to the redox change is reasonably rapid. The latter factor is important because in this case (low electrodic overpotential) the reported $E_{1/2}$ (or E°) values do have thermodynamic significance, regardless of whether or not relatively slow chemical reactions follow the electron-transfer step itself.

Undoubtedly, many of the metallacarborane redox processes are more complicated than reported. Cyclic voltammetry experiments, using sweep rates appropriate for pen-and-ink recorders, imply a lifetime of the order of 20 sec or so for products of "reversible" electron-transfer reactions. But a myriad of electrochemical techniques exist for investigating either shorter-lived species, or those which may undergo interesting chemical reactions over longer time scales. It should be remembered that "irreversible" merely means that a chemical reaction or structure change has followed (or been concomitant with) the electron-transfer step. One of my goals is to bring attention to some redox processes which might, in fact, warrant closer scrutiny and broader investigations. For example, few attempts have been made to generate, through controlled potential electrolytic methods, metallacarborane ions which are highly reduced or oxidized. It is probable that careful work of this type could yield interesting new metallacarborane chemistry and lead to better knowledge of electronic structures of metallacarboranes through electron spin resonance (or other) studies on the electrolytically generated ions.

In the tables contained in this chapter, all potentials are referred to the aqueous saturated calomel electrode (sce). Unless otherwise noted, the electrolyte medium may be assumed to be acetonitrile as solvent, and a tetraalkylammonium salt of perchlorate or hexafluorophosphate as supporting electrolyte (usually $0.1M$). Electrode materials have been exclusively either platinum or mercury, and in our experience metallacarboranes seldom are subject to adsorption or other phenomena which would make the choice of electrode material crucial. In the tables, redox processes are referred to as reversible (R), irreversible (IR), or in a few cases, quasi-reversible (QR). In the latter case, use of this term is somewhat ambiguous in the inorganic literature. It should properly refer to a couple in which the electron-transfer rate is slow enough to be measured by the experimental technique chosen. Thus it should refer to the electron-transfer step and not to the presence of a chemical reaction following the electron transfer. Instances in which chemical follow-up reactions were found after a reversible charge-transfer are designated (R,F). Put most simply, quasi-reversible systems are those with moderately slow electron-transfer steps (k_s in scheme below), and follow-up reactions refer to the rate, k_c, of decomposition of the primary electrode product. Purely irreversible systems can arise either when k_s is very low, or k_c is very large.

$$O + ne^{-} \underset{k_s}{\rightleftharpoons} R \xrightarrow{k_c} Y$$

I shall not cover in this review the electrochemistry of boron clusters which do not contain transition metals, but the reader is referred to the literature for this important work.[2-6] Also, I shall only cover compounds for which voltammetric redox potentials have been reported, although the recent success of Cooksey *et al.*[7] and Morris[8] in isolation of metallaboranes using high voltages and sacrificial metal anodes suggests that electrochemical approaches to formation of metal–boron compounds may prove fruitful in future work.

2. METAL DICARBOLLIDE COMPLEXES: COMPARISON WITH METALLOCENES

In 1965 Hawthorne produced the first examples of carborane clusters into which transition metals had been inserted.[1] In a formal sense, the metallacarborane cluster could be described as the interaction of a metal-containing fragment, e.g., $CpFe^{2+}$ ($Cp = \eta^5\text{-}C_5H_5$), with the anionic carborane ligand, e.g., $C_2B_9H_{11}^{2-}$, to give a π-complex, e.g., $CpFe(C_2B_9H_{11})$. This was a particularly useful formalism, since the analogy with metal sandwich compounds, such as the metallocenes, allowed a focus for understanding the chemical and physical properties of the metallacarboranes. Indeed, much has been written about the analogy[9] between the "dicarbollide" ion, $C_2B_9H_{11}^{2-}$, and the isoelectronic cyclopentadienide ion, $C_5H_5^-$, and their metal complexes. Various experiments which established that the metallocenes and their metal dicarbollide analogs have similar electronic structures[10-15] prompted a calculational (MO) investigation of these compounds, which supported the qualitative analogy.[16] However, in the last few years, following the theoretical concepts of Wade,[17] Mingos,[18] and Rudolph,[19] it has been more popular to view metallacarboranes as covalently-bonded clusters, and the metal-ligand distinction has been deemphasized. The cluster model has the attraction of being more generally applicable, since the $C_2B_9H_{11}^{2-}$ ion is one of the few carborane "ligands" shown to have existence free of its metal complexes. However, the π-complex model remains very useful in discussing the redox behavior of the metal dicarbollide complexes. As will be seen below, there is a remarkable analogy in the electrochemical behavior of metal dicarbollide compounds and the metallocenes.

Redox properties of the metallocenes and of metal dicarbollide "sandwich" compounds are collected in Table 1. Besides the metallocenes themselves, we include mixed-sandwich (Cp-dicarbollide) and biscarboranyl compounds of the *ortho*-dicarbollide ligand $(1,2\text{-}C_2B_9H_{11})^{2-}$, certain carbon- or boron-substituted analogs, and complexes of the *meta*-dicarbollide ligand, $(1,7\text{-}C_2B_9H_{11})^{2-}$. Typical structures are shown in Figure 1. With few exceptions, these compounds give one or more reversible redox processes. In fact, some compounds undergo several reversible reductions and/or oxidations, giving fairly extensive electron-transfer series. For example, some cobaltacarboranes can transverse four different

Table 1. E° Values of Metal Dicarbollide Compounds and Metallocenes[a]

Entry number	Metal	Compound[b]	Charge on M(III)	$E^\circ(V)^{a,c,d}$ M(4/3)	M(3/2)	M(2/1)[e]	Comments[f]	References
E1	Ti	$(\eta^8-C_8H_8)Ti(1,2-C_2B_9H_{11})^-$	1−	−0.91			Irreversible oxidation of neutral compound at +1.70 V	31,32
E2	Cr	$(1,2-C_2B_9H_{11})_2Cr^-$	1−	(No reversible reduction)				21
E3	Cr	Cp_2Cr	1+		−0.67	−2.30		24
E4	Fe	$(1,2-C_2B_9H_{11})_2Fe^{2-}$	1−		−0.42		Acetone/H_2O/LiClO$_4$	27,33
E5	Fe	$[1,2-(CH_3)_2C_2B_9H_9]_2Fe^{2-}$	1−		−0.54		Acetone/H_2O/LiClO$_4$	27
E6	Fe	$[1,2-(C_6H_5)_2C_2B_9H_9]_2Fe^{2-}$	1−		−0.46		Acetone/H_2O/LiClO$_4$	27
E7	Fe	$3,1,2-CpFeC_2B_9H_{11}$	0		−0.08		Acetone/H_2O/LiClO$_4$	27
E8	Fe	Cp_2Fe	1+		+0.31			35
E9	Co	$(1,2-C_2B_9H_{11})_2Co^-$	1−	+1.63	−1.36	−2.24		25,27, 36,37
E10	Co	$(1,7-C_2B_9H_{11})_2Co^-$	1−	+1.68	−1.14	−2.52		25,27
E11	Co	$(1,2-C_2B_9H_8Br_3)_2Co^-$	1−		−0.48	−1.52		27,38
E12	Co	$[1,2-(CH_3)_2C_2B_9H_9]_2Co^-$	1−		−1.13			27
E13	Co	$[1,2-(C_6H_5)_2C_2B_9H_9]_2Co^-$	1−		−1.28			27
E14	Co	$(1,2-C_4H_4-1,2-C_2B_9H_9)_2Co^-$	1−	(Single reduction: −1.14 V)			dc polarography, 2e$^-$ wave	39
E15	Co	$3,1,2-CpCoC_2B_9H_8Br_3$	0		−0.82	−1.80		34
E16	Co	$[\mu(1,2)-C_3H_6-1,2-C_2B_9H_9]_2Co^-$	1−		−1.00			40
E17	Co	$3,1,2-CpCoC_2B_9H_{11}$	0	+2.07 (R,F)	−1.21	−2.11	Co(I) unstable in CH_3CN, stable in THF	25,27, 34,38

E18	Co	$3,1,7\text{-}CpCoC_2B_9H_{11}$	0		−1.03		Other isomers also reported	41
E19	Co	$1,2\text{-}(CH_3)_2\text{-}3,1,2\text{-}CpCoC_2B_9H_9$	0		−1.20		Other isomers also reported	42
E20	Co	$\mu(1,2)\text{-}C_3H_6\text{-}3,1,2\text{-}CpCoC_2B_9H_9$	0		−1.14		Other isomers also reported	42
E21	Co	Cp_2Co^+	1+		−0.94	−1.88		43
E22	Rh	$(1,2\text{-}C_2B_9H_{11})_2Rh^-$	1−		−1.82 (R,F)			44
E23	Rh	Cp_2Rh^+	1+		−1.41 (R,F)	−2.18 (R,F)	Dimer forms after first reduction	44
E24	Ni	$(1,2\text{-}C_2B_9H_{11})_2Ni$	1−	+0.25	−0.57	−2.10	Ni(I) unstable in CH_3CN, stable in THF	20,25, 27,45
E25	Ni	$[1,2\text{-}(CH_3)_2\text{-}C_2B_9H_9]_2Ni$	1−	+0.22	−0.52	−1.98		38,46
E26	Ni	$[1,2\text{-}(C_6H_5)_2\text{-}C_2B_9H_9]_2Ni$	1−	+0.35	−0.52			46
E27	Ni	$[\mu(1,2)\text{-}C_3H_6\text{-}1,2\text{-}C_2B_9H_9]_2Ni^-$	1−	+0.60	−0.45			46
E28	Ni	$(1,7\text{-}C_2B_9H_{11})_2Ni^-$	1−	+0.55	−0.92	−2.09 (R,F)	Ni(I) unstable in CH_3CN, stable in glyme (−2.25 V)	25,27, 45
E29	Ni	$(1,2\text{-}C_4H_4\text{-}1,2\text{-}C_2B_9H_9)_2Ni^-$	1−		−0.72	−1.14	Direct current polarography; no test for reversibility	39
E30	Ni	$3,1,2\text{-}CpNiC_2B_9H_{11}$	0	+0.55	−0.52			20
E31	Ni	Cp_2Ni	1+	+0.77	−0.09	−1.66 (R,F)		23,24
E32	Pd	$(1,2\text{-}C_2B_9H_{11})_2Pd$	1−	−0.14	−0.56			27,46
E33	Pd	$[1,2\text{-}(CH_3)_2\text{-}C_2B_9H_9]_2Pd^{2-}$	1−	ca. 0 V (IR)	−0.44			46
E34	Cu	$(1,2\text{-}C_2B_9H_{11})_2Cu^-$	1−		−0.35	ca. −1 V (QR)		27

(continued overleaf)

Table 1 (*cont.*)

Entry number	Metal	Compound[b]	Charge on M(III)	E° (V)[a,c,d] M(4/3)	M(3/2)	M(2/1)[e]	Comments[f]	References
E35	Au	$(1,2\text{-}C_2B_9H_{11})_2Au$	1^-	+1.16(IR)	−0.63	−0.81	Oxidation of M(III) is an irreversible, $2e^-$ process	27,38

[a] Potentials are reported vs the aqueous saturated calomel electrode (sce).

[b] $Cp = (\eta^5\text{-}C_5H_5)$.

[c] In this table, redox processes are reversible (R) unless otherwise designated as irreversible (IR), quasi-reversible (QR), or reversible with a follow-up reaction (R,F).

[d] Data obtained by cyclic voltammetry, unless otherwise designated.

[e] Symbol M(4/3) represents the M(IV)/M(III) couple, M(3/2) the M(III)/M(II) couple, etc.

[f] Solvent-supporting electrolyte system was CH_3CN with R_4NClO_4 or R_4NPF_6, unless otherwise designated.

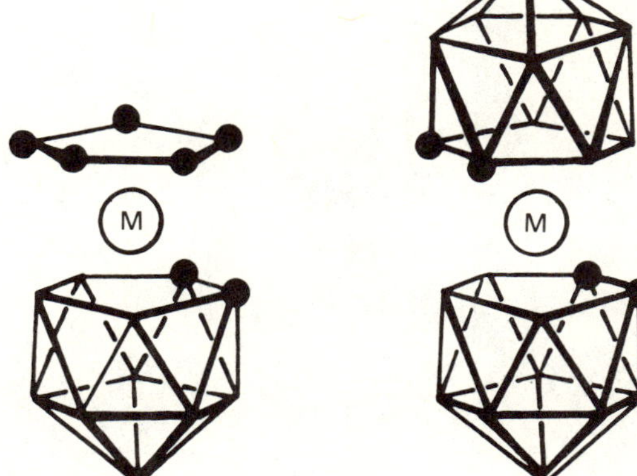

Figure 1. Structures of metal complexes of the dicarbollide ion: (left) $3,1,2$-$CpMC_2B_9H_{11}$; (right) $(1,2$-$C_2B_9H_{11})_2M$. Note: ●, CH.

metal oxidation states, as in the bis(*ortho*-dicarbollyl) cobalt compound [E9]; (line entry numbers for the tables are in brackets for easy location of the data being discussed).

$$(1,2\text{-}C_2B_9H_{11})_2Co^{IV} \underset{\rightleftharpoons}{\overset{e^-}{}} (1,2\text{-}C_2B_9H_{11})_2Co^{III} \underset{\rightleftharpoons}{\overset{e^-}{}} (1,2\text{-}C_2B_9H_{11})_2Co^{II2-}$$

$$\underset{\rightleftharpoons}{\overset{e^-}{}} (1,2\text{-}C_2B_9H_{11})_2Co^{I3-}$$

It may appear inconsistent, when perusing Table 1, that some metal carboranes show 3 electron-transfer reactions linking 4 oxidation states, but others, similar in structure, may show only 1 redox process. The conditions used to investigate these compounds are the source of these differences. Many of the compounds showing more extensive electron-transfer series have reductions or oxidations at very low (negative) or high (positive) potentials, respectively. Observation of these waves requires carefully purified solvents and properly prepared electrodes, and a blank entry in the table does not mean that the redox couple in question is inaccessible. Rather, in many cases, it could be measured with proper attention to electrochemical details. A cyclic voltammogram showing the three reversible couples of $(1,2\text{-}C_2B_9H_{11})_2Ni$ is shown in Figure 2.

There is a remarkable consistency between the redox potentials of bis(*ortho*-dicarbollyl)metallates and the metallocenes, for a process involving the same metal oxidation state changes. These data are plotted in Figure 3. A straight line of unity slope and an intercept of +0.5 V in favor of the metallacarborane is found. That is, given a change involving the same metal oxidation states, a bis (*ortho*-dicarbollyl) compound will be 0.4–0.5 V easier to oxidize (harder to reduce) than its metallocene analog. This is the source of the well-known tendency of the dicarbollide ligand to stabilize high formal metal oxidation states.[20] Although a more thorough discussion of this tendency appears in Section 6, we can point out here that this stabilization undoubtedly occurs because of the dif-

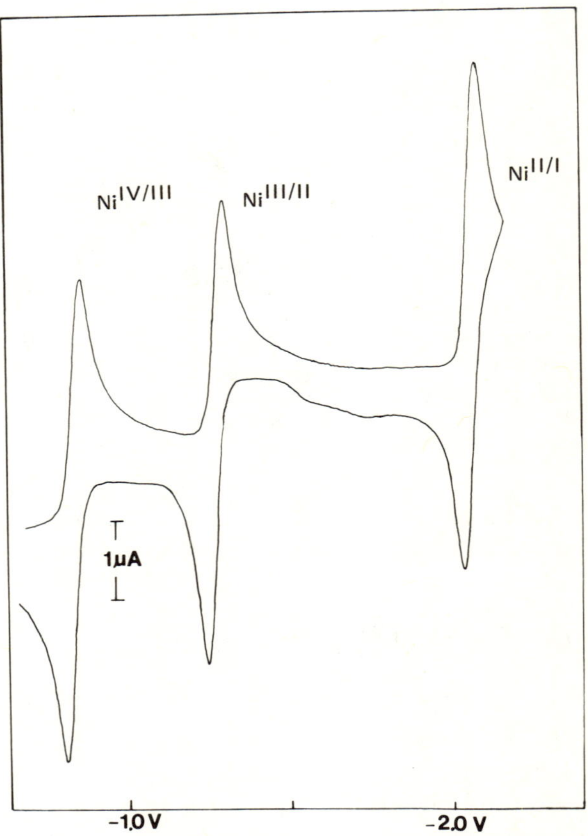

Figure 2. Cyclic voltammogram of $(1,2\text{-}C_2B_9H_{11})_2Ni$ at hanging mercury drop electrode in dimethoxyethane/0.1 M Bu_4NPF_6. Scan rate 100 mV/sec. Three reversible reductions are indicated.

ference in charge on the ligand, the dinegative $(C_2B_9H_{11})^{2-}$ versus the mononegative $(C_5H_5)^-$. If two species with different ligands have similar electronic structures, it will be easier to oxidize $(L_2M^{II})^{2-}$ than $L_2'M^{II}$. Data on the mixed sandwich compounds [E7, E17, E30] give further credence to the importance of overall charge, since their redox potentials approximately split the difference between the metallocenes and the bisdicarbollides. The charge effect can be followed in Table 1 most easily by noting the column in which the overall charge on the compound in the +3 metal oxidation state is given. Compared to the metallocenes, there is an important enhancement in stability of electrochemically generated metal dicarbollide ions which seems to parallel the well-documented air- and moisture-stability of metal dicarbollides.[21] Many examples could be cited, but perhaps the best example comes in the comparison of the redox behavior of nickelocene [E31] with that of *bis*(dicarbollyl) nickel [E24]. Each

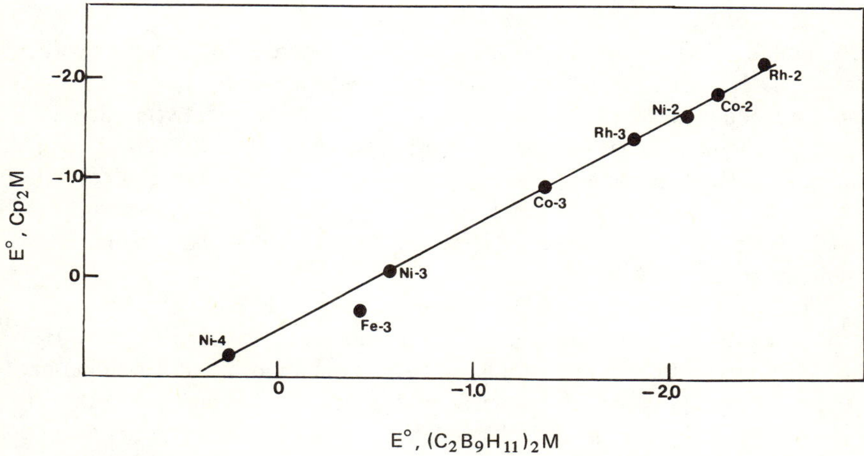

Figure 3. Plot of E° potentials (vs. sce) of bisdicarbollyl metal compounds (abcissa) vs. E° potentials of metallocenes (ordinate). Points are labelled by metal and by the highest formal metal oxidation state of the redox couple. Thus Ni-4 relates E° for $Cp_2 Ni^{2+/+}$ [E31] and $(1,2-C_2 B_9 H_{11})_2 Ni^{0/-}$ [E24]. Other entry numbers (from Table 1) are: Fe-3, [E4 and E8]; Co-3 and Co-2, [E9 and E21]; Rh-3 and Rh-2, [E22 and E23].

compound exhibits a four-membered electron-transfer series, encompassing the formal nickel oxidation states of +4, +3, +2 and +1. The Ni^{IV} dication of

$$L_2 Ni^{IV} \overset{e^-}{\rightleftharpoons} L_2 Ni^{III} \overset{e^-}{\rightleftharpoons} L_2 Ni^{II} \overset{e^-}{\rightleftharpoons} L_2 Ni^{I}$$

nickelocene is very susceptible to nucleophilic attack and the $Cp_2 Ni^{+/2+}$ oxidation is irreversible except at low temperatures[20] or under rigorously dry high-vacuum conditions.[22] However, the neutral Ni(IV) dicarbollide compound, $(1,2-C_2 B_9 H_{11})_2 Ni$, is an air-stable, sublimable compound. At the other end of the electron-transfer series, the Ni(I) species show similar differences in stability. The Ni(I) nicleocene monoanion is highly reactive, and low temperatures (below 220 K) are needed to observe a reversible $Cp_2 Ni$ reduction. But again, the $Ni^{(I)}$ *trianion*, $(1,2-C_2 B_9 H_{11})_2 Ni^{3-}$, forms in a reversible reduction at room temperature in some solvents.[25] These metallacarboranes in outlying metal oxidation states would most likely display some interesting physical properties or chemical reactivities, but little has been done with them up to this point.

Those metal dicarbollide compounds for which there is no reported metallocene analog deserve special mention. These are the compounds containing Pd [E32], Cu [E34], and Au [E35]. $(1,2-C_2 B_9 H_{11})_2 Pd$ undergoes reversible reductions first to the Pd(III) monoanion and then the Pd(II) dianion at relatively mild potentials (-0.56 V for the second wave). By analogy with $(1,2-C_2 B_9 H_{11})_2 Ni$, more negative scans would probably reveal reduction to a formal

Pd(I) trianion, which might prove amenable to esr study (^{105}Pd, $I = 5/2$, 22.2% nat. abund.). The formal M(II) dianions have different magnetic properties, $(1,2-C_2B_9H_{11})_2Pd^{2-}$ being diamagnetic,[26] and $(1,2-C_2B_9H_{11})_2Ni^{2-}$ having two unpaired electrons.[27] Interestingly, the difference between successive redox potentials is much smaller for the Pd compound (0.4 V) than for the Ni analog (0.8 V). A similar trend is also observed in comparing the cobalt and rhodium compounds (0.9 V vs. 0.6 V) and the copper and gold compounds (0.7 V vs. 0.2 V). In each case, the redox steps for the later row-transition metal compound are more closely spaced.

The copper(II) and gold(II) compounds, $(1,2-C_2B_9H_{11})_2M^{2-}$, can be reversibly oxidized to the formal M(III) monoanions, but Warren and Hawthorne[26] report that the reductions of these compounds are not well-defined. We have investigated the voltammetry of the gold system and found that the Au(II)/Au(I) reduction is indeed reversible, but subject to electrode fouling, which can be avoided by cleaning the electrode surface before each scan. Hence it appears that the gold and copper systems undergo the redox steps M(III)/ M(II)/M(I). The most reduced of these is a formal *22-electron* metal system! The structural and electronic reasons why metal dicarbollides can support very electron-rich systems have been widely discussed,[29] but few details of the electron-transfer mechanisms of these electron-rich compounds (e.g., the influence of structural rearrangements or electron-transfer rates) have been reported. This subject is currently under investigation in our laboratory.

Figure 3 may also be used to comment on predicted values of as yet unobserved redox couples of metallocenes. We shall restrict ourselves to two examples, that of Cu and Co. The data on $(1,2-C_2B_9H_{11})_2Cu^{2-}$, combined with the relationship in Figure 3, predict a value of *ca.* +0.1 V for the Cu(III/II) couple in the hypothetical cupricene molecule. This relatively positive value suggests a high tendency for Cp_2Cu^+ to be reduced to the 21-electron Cu(II) species, which probably has very poor stability. It will take moderately strong oxidizing conditions for formation of the cupricene cation.

These relationships also shed light on the possibility of generating the dication of cobaltocene, Cp_2Co^{2+}, which would be a d^5 system. Again, using the relationship between metallocene and metal dicarbollide redox potentials, and noting that the Co(IV/III) couple for $(1,2-C_2B_9H_{11})_2Co$ occurs at +1.63 V [E9], one can predict a value of about +2.1 V ± 0.2 V for the process $Cp_2Co^{2+/+}$. If the E° value lies at the high end of this range, detection in normal electrolyte media may be difficult. A better system in which to look for a d^5 cobalt metallocene might be $[\eta^5-C_5(CH_3)_5]_2Co$, which is reported by Robbins[28] to have an E° value for the Co(III/II) couple shifted negative by several tenths of a volt from that of unsubstituted cobaltocene.

Returning to the dicarbollide compounds, a brief comment on substituent effects is in order. Substitution of Br for H at cage *boron* atoms results, as expected, in a considerable positive shift of the reduction potentials of the metallacarborane (ca. 0.9 V comparing the bis(tribromo) cobalt compound [E11] with its unsubstituted analog [E9]). But, unexpectedly (to me), substitution of CH_3 for H at cage *carbons* also leads to a positive shift for all but the *bis*-dicarbollyl iron derivatives [E4 and E5]. The reported shifts are as large as 360 mV ([E9 and E16]; see also [E9 and E12]; [E17 and E20]; [E24, E25, and E27]). Since methyl substitution generally results in negative shifts in reduction potentials for metal π-compounds (see, for example, the data on substituted cobalticinium salts),[30] the reason for the different behavior of metallacarboranes is not apparent to us.

3. METALLAHETEROBORANES AND -CARBORANES

In this category we place metallaboranes and -carboranes which also have a group V (P, As, Sb) or group VI (S) heteroatom in the pentagonal face of the borane which is bonded to the metal. Relatively few compounds of this general type have been subjected to electrochemical measurements. In Table 2, we have collected data extracted from two papers published over ten years ago, and have supplemented with data very recently obtained by David E. Brennan, working at Vermont. The published work is difficult to evaluate, because the electrochemical experiments were not a major focus of the papers, and few experimental details are reported. For the purpose of discussion, we have assumed that the gross features of the redox processes in Table 2 are as reported, but an appropriate amount of skepticism should be injected into the reading. For example, the E° values reported in Reference 48 were obtained using the rotated platinum electrode, and there are no inherent problems in that, but it is hard to know what claims of "reversible" or "irreversible" mean under these experimental conditions. Actually, if the reduction of $(SB_{10}H_{10})_2Co^-$ were irreversible, as reported,[48] it would be quite unusual, and perhaps unique, among reported Co(III/II) cobaltaborane or cobaltacarborane redox processes. These redox couples are invariably reversible.

One may also anticipate that, at least for the cobalt compounds, further reductions [to Co(I)] would have been observed if conditions allowing scans to more negative potentials had been employed. This was borne out by our reinvestigation of [E43], a cobaltaphosphacarborane, in which a Co(II/I) couple at -1.93 V was found. With the single exception of [E43], all of the entries in Table 2 refer to acetonitrile solutions containing tetraalkylammonium salts as electrolyte media.

The general features of the redox behavior of the heteroatom-containing compounds are similar to those discussed previously for the metal dicarbollides. Potentials for a particular redox couple of a metal depend mostly on overall charge on the complex. Take, for example, the Fe(III)/Fe(II) reduction. For cationic Fe(III) compounds this occurs at about $+1.5$ V [E39 and E40] but the reduction potential shifts negative by about 0.7 V for a neutral Fe(III) compound [E38] and by a similar amount for anionic compounds [E36, E37, and

Table 2. Voltammetric Behavior of Metallaheteroboranes and Metallaheterocarboranes

Entry number	Metal	Compound	Charge on $M(III)$	Results[a,b]	Comments[c,d]	References
E36	Fe	$(1,2-PCB_9H_{10})_2Fe^{2-}$	1−	$Fe(III/II) = +0.08$	CV/Pt	47
E37	Fe	$(1,7-PCB_9H_{10})_2Fe^{2-}$	1−	$Fe(III/II) = +0.05$	CV/Pt	47
E38	Fe	$(1,7-PCB_9H_{10})Fe(1-CH_3-1,7-PCB_9H_{10})$	0	$Fe(III/II) = +0.74$	CV/Pt	47
E39	Fe	$(1-CH_3-1,2-PCB_9H_{10})_2Fe$	1+	$Fe(III/II) = +1.60$	CV/Pt	47
E40	Fe	$(1-CH_3-1,7-PCB_9H_{10})_2Fe$	1+	$Fe(III/II) = +1.46$	CV/Pt	47
E41	Fe	$(SB_{10}H_{10})_2Fe^{2-}$	1−	$Fe(III/II) = +0.21$	RPE; further irreversible oxidation at +1.35 V	48
E42	Co	$(1-CH_3-1,2-PCB_9H_{10})_2Co$	1+	$Co(II/I) = -0.78$ $Co(III/II) = +0.43$	CV/Pt; unsuccessful attempt to isolate cation	47
E43	Co	$(1-CH_3-1,7-CB_9H_{10})_2Co$	1+	$Co(I/0) = -1.93(R,F)$ $Co(II/I) = -0.62$ $Co(III/II) = +0.41$	Dimethoxyethane/ Bu_4NPF_6; CV/hmde CV/hmde CV/Pt	38,47
E44	Co	$(SB_{10}H_{10})_2Co^-$	1−	$Co(III/II) = -0.77$ (IR)	RPE; reported irreversible	48
E45	Co	$CpCo(PB_{10}H_{10})^-$	1−	$Co(II/I) = -2.55$ (R,F) $Co(III/II) = -2.00$ $Co(III)$ oxidation $+1.14$ (IR)	CV/hmde CV/hmde CV/Pt	49
E46	Co	$(AsB_{10}H_{10})_2Co^{3-}$	3−	$Co(IV/III) = +0.04$	CV/Pt hmde; further, irreversible oxidation at +1.22 V	49
E47	Co	$(1,2-$ or $1,7-PCB_9H_{10})_2Co^-$	1−	(No waves reported between +1.5 and −1.0V)		47

E48	Co	$CpCo(AsB_{10}H_{10})^-$	1−	$Co(III/II) = -0.02$	CV/hmde	49
				$Co(IV/III) = +1.18 (R,F)$	CV/Pt	
E49	Co	$(As_2B_9H_9)_2Co^-$	1−	Co(I) reduction -2.57 (IR)	CV/hmde	49
				$Co(II/I) = -1.87$	CV/hmde	
				$Co(III/II) = -1.00$	CV/hmde	
				Co(III) oxidation $+1.79$ (IR)	CV/Pt	
E50	Co	$CpCo(As_2B_9H_9)$	0	$Co(II/I) = -1.79$	CV/hmde	49
				$Co(III/II) = -1.00$	CV/hmde	
E51	Ni	$(AsB_{10}H_{10})_2Ni^{2-}$	3−	$Ni(IV/III) = -1.56$	CV/hmde	38
				$Ni(III/II) = -2.39 (R,F)$		

[a] Potentials in volts vs. aqueous saturated calomel electrode (sce).

[b] Reversible, unless otherwise stated: (IR) = irreversible; (R,F) = reversible with follow-up reaction.

[c] In CH_3CN/Bu_4ClO_4, unless otherwise stated.

[d] CV = cyclic voltammetry; RPE = rotating platinum electrode; hmde = hanging mercury drop electrode.

E41]. Comparison with metal dicarbollides shows that, for compounds with the same overall charges, reduction to Fe(II) is markedly easier for carboranes containing either phosphorus or sulfur in the cage.

Trends are less definite for the cobalt compounds which have been investigated. Co(III)/Co(II) reductions for monoanionic complexes vary from −0.8 V to −2.0 V, quite a large range. We have investigated a few of these cobalt heteroboranes, and some do support the electron-transfer series Co(IV)/Co(III)/Co(II)/Co(I) [E43 and E45].

It might prove possible to take advantage of the considerable variations in formal charge on the ligands to attempt actual isolation of cobaltacarboranes in unusual oxidation states. For example, neutral $[P(CH_3)CB_9H_{10}]_2Co$ [E43] is a formal Co(II) compound which can be reduced to the formal Co(I) monoanion at the relatively mild potential of −0.78 V, giving promise that the Co(I) species might be isolable. Conversely, the cobaltaarsaborane [E46], $(AsB_{10}H_{10})_2Co^{3-}$, can be oxidized to the formal Co(IV) *dianion* at +0.04 V, again indicating the possibility of isolation of an unusual species, this time in a high metal oxidation state.

Unlike the cobalt compounds containing the dicarbollide ion [E9, E10, E11, E15, and E17], electrogenerated Co(I) phosphaboranes are very reactive, and for [E43] and [E45], moderately rapid cyclic voltammetry sweep rates (ca. 1 V/sec) must be employed to outrun the rate of chemical reactions following the electron-transfer step. The electrochemical data helps little in speculation about the nature of the reactions these compounds are undergoing, and isolation of electrolysis products is sorely needed if interest in the electrochemistry of heteroatom-substituted metallaboranes is to grow.

4. METALLABORANES

As mentioned in Section 1, some metallaboranes have been synthesized by electrolysis of solutions of borane anions at sacrificial metal anodes.[7] However, very few metallaboranes have been studied using voltammetric methods. Table 3 collects the available data on voltammetry of this class of compounds. Except for the as yet unpublished data on $CpCoB_4H_8$ [E52], only data on nickelaboranes has been reported, most of it in a paper by Leyden *et al.*[51] Seven of the compounds studied are *closo*-clusters [E54, E55, and E57–E61]. Each of these compounds undergoes reversible reduction over a wide range of potentials, but all oxidations were reported to be irreversible. Hence it appears that these systems are better able to accommodate an excess, rather than a deficiency, of electrons in the cluster. A rationale for this observation may be found in the fact that the formal nickel oxidation state in the *closo*-compounds listed is +4, and reduction involves the Ni(IV/III) couple, so these processes should be reasonably facile. In fact, it would not be surprising if these compounds were found to undergo further reversible reductions at more negative potentials.

However, it is not true that *nido*-nickelaboranes can be reversibly *oxidized* [E53, E56, and E62]. For example, $(B_{10}H_{12})_2Ni^{2-}$ is reported to undergo an irreversible oxidation [E53]. This raises a point worth considering. Rules for skeletal electron-counting of clusters predict that loss of two electrons from a *nido*-structure should result in formation of a *closo*-structure.[17-19] Structure changes which accompany electron-transfer reactions lead to irreversibility (or quasi-reversibility if the changes are small), and one is left wondering what the nature of the follow-up reactions is for these nickelaboranes. It may be that the most interesting electrochemistry of metallaboranes is that associated with the irreversible, rather than reversible, processes. Work is needed on the mechanism of irreversible metallaborane redox processes.

5. MULTIMETAL BORANES AND CARBORANES

A fairly large number of boron cage compounds containing two or more metals is known, and a number of these have been investigated electrochemically. The pertinent data are collected in Table 4. Not surprisingly, these compounds exhibit rather rich electrochemical behavior, generally undergoing one or more reversible reductions or oxidations.

Possibly the most important general observation one can make from the published data is that all reported reversible waves are *1-electron* processes. The significance of this was first pointed out by Francis and Hawthorne[36] and discussed in more detail later on by Dustin and Hawthorne.[55] If the metals in these cage compounds were electronically isolated from one another, one would expect to see multielectron processes in compounds having chemically equivalent metals. But, at least with respect to the number of electron transfer processes per compound, and the potential spacing between successive redox steps, multimetal carboranes are very similar to monometallic carboranes. This implies that charge can be efficiently delocalized in these clusters.

The most extensive electron-transfer series yet reported for this class of compounds are found for the dicobalt compounds with a formal "triple-decker sandwich" structure. These compounds, with the structures shown in Figure 4, were prepared by Grimes *et al.*[63] and by Siebert *et al.*[64] These compounds, [E71, E72, and E74], undergo as many as four reversible reductions or oxidations, making a 5-membered electron-transfer series of the type

$$Co-Co^{2+} \underset{\rightleftharpoons}{\overset{-e^-}{}} Co-Co^+ \underset{\rightleftharpoons}{\overset{-e^-}{}} Co-Co \underset{\rightleftharpoons}{\overset{e^-}{}} Co-Co^- \underset{\rightleftharpoons}{\overset{e^-}{}} Co-Co^{2-}$$

Radical ions of multimetallic carboranes could provide rich ground for detailed investigations of metal–metal interactions in these clusters (i.e., mixed-valence chemistry). To date, there do not appear to be any reports of esr studies of radicals produced by electrochemical reduction or oxidation of these clusters, but likely candidates for such studies are readily apparent from a casual perusal

Table 3. Voltammetric Behavior of Metallaboranes

Entry number	Compound	Results[a]			Conditions[b]	References
E52	$2\text{-CpCoB}_4\text{H}_8$	Oxidation;	Co(III/IV)	$+1.14$ V (IR)	CH_3CN/0.1 M $B\mu_4NPF_6$, Pt or Hg electrode	38
		Reduction;	Co(III/II)	-1.39 V (R)		
		Reduction;	Co(II/I)	-2.55 V (IR)		
E53	$(B_{10}H_{12})_2\,Ni^{2-}$	Oxidation		$+1.03$ V (IR)	CH_3CN	50
E54	$CpNi(B_{11}H_{11})$	Reduction		-1.50 V (R)	CH_3CN/Pt electrode	51
		Oxidation		$+1.70$ V (IR)		
E55	$(CpNi)_2(B_{10}H_{10})$	Reductions		-0.27 V (R); -1.35 V (R)	CH_2Cl_2	51
		Oxidation		$+1.73$ V (IR)	CH_3CN/0.1 M Et_4NClO_4	
E56	$CpNi(B_{10}H_{12})^-$	Oxidation		$+0.61$ V (QR)	CH_3CN/Et_4NClO_4	51
E57	$Cp_2NiCo(B_{10}H_{10})^{2-}$	Reduction		-1.55 V (QR)	Hg electrode	51
		Oxidation		$+0.45$ V (R)	CH_3CN/Et_4NClO_4	
E58	$2\text{-CpNi}(B_9H_9)^-$	Reduction		-1.42 V	CH_3CN/Et_4NClO_4	51
		Oxidation		$+0.87$ V (IR)	Pt electrode	
E59	$1\text{-CpNi}(B_9H_9)^-$	Reduction		-1.52 V	CH_3CN/Et_4NClO_4	51
		Oxidation		$+0.77$ V (IR)	Pt electrode	
E60	$2\text{-CpNi}(B_9Cl_9)$	Reduction		-0.40 V (R)	CH_3CN/Et_4NClO_4	51
E61	$1\text{-CpNi}(B_9Cl_9)$	Reduction		-0.35 V (R)	CH_3CN/Et_4NClO_4	51
E62	$CpNi(B_{10}H_{13})$	Oxidation		$+0.79$ V (IR)	CH_3CN/Et_4NClO_4	51

[a]Volt vs. sce.; (IR) = irreversible, (R) = reversible, (QR) = quasi-reversible.

[b]Solvent supporting electrolyte (where given), electrode employed.

Table 4. Electrochemistry of Multimetallic Boranes and Carboranes

Entry number	Compound	Results[a]	References
Fe–Fe compounds			
E63	$(1,2-C_2B_9H_{11})_2Fe_2(CO)_4^{2-}$	Reductions -1.62 V (IR), -2.16 V (IR)[b,e]	38,52
E64	$1,6-Cp_2-1,6,2,3-Fe_2(C_2B_6H_8)$	Reduction -1.17 V IR)[c,d] Oxidation $+0.80$ V (R)	53
E65	$4,5-Cp_2-4,5-Fe_2-(2,3-C_2B_9H_{11})$	Reduction -0.59 V (R)[c,d] Oxidation $+1.36$ V (IR)	54
E66	$[4-Cp-5-(C_2B_9H_{11})-4,5-Fe_2(2,3-C_2B_9H_{11})]^-$	Reduction -0.70 V (R)[c,d] Oxidations $+1.07$ V (IR), $+1.76$ V (IR)	54
E67	$Cp_2Fe_2(C_2B_8H_{10})$	Reductions -0.56 V (R), -1.40 V (QR)[c,d]	54
Fe–Co compounds			
E68	$4,5-Cp_2-4-Co-5-Fe-1,8-[C_2B_9H_9(CH_3)_2]$	Reduction -0.47 V (R)[c,d] Oxidation $+0.92$ V (R)	55
E69	$4,5-Cp_2-4-Co-5-Fe-1,8-(C_2B_9H_{11})$	Reduction -0.42 V (R)[c,d] Oxidation $+0.90$ V (R)	55
E70	$Cp_2FeCo[C_3B_2(C_2H_5)_4(CH_3)]$	Reduction -1.77 V (R)[d,f] Oxidation -0.06 V (R)[e,f]	49
Multi-Co Compounds			
Co–Co Compounds			
E71	$1,7,2,3-Cp_2Co_2(C_2B_3H_5)$	Reductions -1.44 V (R), -2.23 V (R)[e,f] Oxidations $+0.51$ V (R), $+1.72$ V (IR)[d,f]	56
E72	$1,7,2,4-Cp_2Co_2(C_2B_3H_5)$	Reductions -1.35 V (R), -2.30 V (R)[d,f] Oxidation $+0.89$ V (R)[d,f]	56
E73	$1,2,3,5-Cp_2Co_2(C_2B_3H_5)$	Reductions -1.60 V (R), -2.48 V (IR)[e,f] Oxidation $+0.68$ V (IR)[d,f]	49

(continued overleaf)

Table 4. (cont.)

Entry number	Compound	Results[a]	References
E74	$Cp_2Co_2[C_3B_2(C_2H_5)_4(CH_3)]$	Reductions $-1.53\,V$ (R), $-2.56\,V$ (R,F)[e,f] Oxidation $-0.53\,V$ (R), $+1.74\,V$ (IR)[d,f]	49
E75	$Cp_2Co_2(C_2B_5H_7)$	Reduction $-1.45\,V$ (R), $-2.58\,V$ (R)[e,f] Oxidation $+0.82\,V$ (IR)[d,f]	49
E76	Isomers of $Cp_2Co_2(C_2B_5H_7)$	One reduction (R), one oxidation (R)[c,d]	57
E77	$2,7,1,10$–$Cp_2Co_2(C_2B_6H_8)$	Reduction $-1.50\,V$[g] Oxidation $+1.37\,V$[g]	58
E78	Isomers of $Cp_2Co_2(C_2B_7H_9)$	One reduction between -0.8 and $-1.4\,V$[c,d] One oxidation between $+0.8$ and $+1.2\,V$	57,58
E79	Isomers of $Cp_2Co_2(C_2B_8H_{10})$	One reduction between -1.2 and $-1.7\,V$[g] One oxidation $ca.$ $+1.4\,V$	58,59
E80	$2,11$–Cp_2–$2,11$–Co_2–1–$(CB_9H_{10})^-$	Oxidation $+0.20\,V$ (R)[d,f]	60
E81	$[(1,2$–$C_2B_9H_{11})_2Co_2(C_2B_8H_{10})]^{2-}$	Reductions $-1.48\,V$ (R), $-2.36\,V$ (R)[e,f] Oxidations $+0.97\,V$ (R), $+1.60\,V$ (IR)[d,f]	25,36
E82	Isomers of $Cp_2Co_2(C_2B_9H_{11})$ or $Cp_2Co_2[C_2B_9H_9(CH_3)_2]$	One reduction between -0.4 and $-1.3\,V$[c,d] One oxidation between $+0.5$ and $+1.0\,V$	55
E83	$1,14,2,10$–$Cp_2Co_2(C_2B_{10}H_{12})$	Reduction $-1.10\,V$ (R) Oxidation $+1.17\,V$ (R)	61
Tricobalt Compounds			
E84	$[(1,2$–$C_2B_9H_{11})Co(C_2B_8H_{10})]_2Co^{3-}$	Reduction $-1.53\,V$ (R)[c,d] Oxidation $+0.70\,V$ (R)	36
E85	$[CpCo(C_2B_8H_{10})]_2Co^-$	Reduction $-1.42\,V$ (R)[c,d] Oxidation $+0.75\,V$ (R)	59
E86	$[CpCo(C_2B_9H_{11})]_2Co^-$	Reduction $-0.84\,V$ (R)[c,d] Oxidation $+0.56\,V$ (R)	55

E87	$Cp_3Co_3(C_2B_7H_9)$	Reduction -0.87 V (R)^{c,d} Oxidation $+0.65$ V (R)	57
Co–Ni Compounds			
E88	$Cp_2CoNi[C_3B_2(C_2H_5)_4(CH_3)]$	Reductions -1.65 V (R,F)^{e,f}; -1.58 V (R)^{e,h} Oxidations $+0.02$ V (R)^{e,f}, $+0.93$ V (IR)^{d,f}	49
E89	$Cp_2CoNi(CB_7H_{8-n}Br_n)$ and isomers (n = 0.1.2)	One reduction between -0.6 and -1.0 V^{d,f} One oxidation, no potentials given	62
E90	$Cp_2CoNi(B_{10}H_{10})^{2-}$	See Table 3 [(E57)]	51
Multi-Nickel Compounds			
E91	$Cp_2Ni_2(B_{10}H_{10})$	See Table 3 [(E55)]	51
E92	$Cp_3Ni_3[CB_5H_5(CH_3)]$	Reduction -0.95 V (R) Oxidations (all IR) $+0.49$ V, $+0.95$ V, $+1.25$ V^i	60
E93	$Cp_2Ni_2[C_3B_2(C_2H_5)_4(CH_3)]$	Reduction -1.30 V (R)^{d,f} Oxidations -0.13 V (R), $+1.26$ V (IR)^{d,f}	49

[a]Volt vs. sce.; (R) = irreversible, (QR) = quasi-reversible, (R) = reversible, (R,F) = follow-up reaction. [b]Dimethoxyethane/0.1 M Bu$_4$NPF$_6$. [c]CH$_3$CN/0.1 M Et$_4$NClO$_4$. [d]Pt electrode. [e]Hg electrode. [f]CH$_3$CN/0.1 M Bu$_4$NPF$_6$. [g]Electrolyte and reversibility unspecified. [h]tetrahydrofuran/0.1 M Bu$_4$NPF$_6$. [i]Oxidations at $+0.95$ V and $+1.25$ V are due to products of irreversible oxidation at $+0.49$ V.

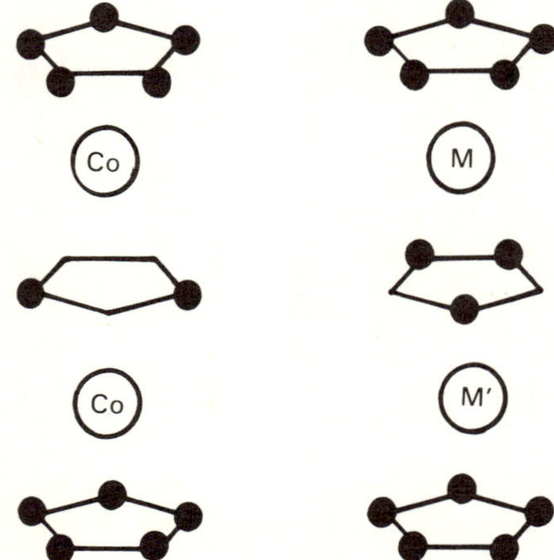

Figure 4. Metallacarboranes described as triple-decker sandwich compounds. Left, $Cp_2Co_2C_2B_3H_5$ [E72] and right, Cp_2M_2 $(C_3B_2Et_4Me)$ [E70, E74, E88, and E93]. A second isomer of the $Cp_2Co_2C_2B_3H_5$ system, in which the central ring carbon atoms are adjacent, has also been extensively studied: Ref. 63; also Grimes, R.N., *Acc. Chem. Res.*, **11**, 420 and references therein. Note: ●, CH.

of the literature. For example, Salentine *et al.*[60] have reported the synthesis of the icosahedral dicobalt anion $Cp_2Co_2(CB_9H_{10})^-$. Reversible oxidation to the formally Co(III)–Co(IV) neutral compound occurred at a mild potential (+0.20 V) [E80] and the paramagnetic species might even prove isolable. Salentine[62] has reported that three isomers of the mixed metal species Cp_2CoNi (CB_7H_8) [E89] can be reversibly reduced electrochemically to give air-sensitive green solutions of the monoanions, but no physical studies of the anions were reported. We believe that a comprehensive study interfacing electrochemistry with spectroscopy (esr, optical spectroscopy) would be a powerful approach to the question of metal–metal interactions in metallaborane cages, and it is hoped that in time such studies will appear.

6. STABILIZATION OF UNUSUAL OXIDATION STATES

The ability of carborane ligands to stabilize metals in unusually high oxidation states is well established. Co(IV) complexes of carboranes have been isolated first by Knoth[65] and later by Dustin and Hawthorne.[66] Other metallaboranes and metallacarboranes which fit into this class are those containing Ni(IV),[27,51,65,67,68] Pd(IV),[27] and Mn(IV).[69].

As recognized very early[20] and developed in greater detail in Section 2, the high negative charge of the ligand seems to be a major factor in this stabilization. For example, three of the cages which stablize either Co(IV) or Ni(IV) are formally trinegatively charged:

$$(CB_7H_8)^{3-},[66] \ (CB_{10}H_{11})^{2-},[65,68,69] \ and \ (B_{11}H_{11})^{3-}[51]$$

The potential of the M(IV/III) couple may be used as a guide for attempts to synthesize Co(IV) carboranes from the normally more accessible Co(III) species.

For example, cyclic voltammetry data shows that a neutral, formally Co(IV) species can be electrogenerated by 1-electron oxidation of $(C_2B_9H_{11})_2Co^-$ [E9 and E10]. But the potentials of these oxidations are rather high (ca. +1.6 V) and it would be expected that isolation of the Co(IV) species would be very difficult, since the Co(III) compound would be easily reformed on contact with even weakly nucleophilic solvents or impurities. However, the related anion $CpCo(CB_7H_8)^-$ is oxidized to the neutral Co(IV) compound at +0.08 V[66] [E112]. The latter can be isolated, after oxidation of a dichloromethane solution of the anion with $FeCl_3$, as a green solid which is stable if stored at cool temperatures.

Some carboranes can also be said to stabilize *low* metal oxidation states. For example, the dicarbollide ion, which stabilizes Ni(IV) in $(1,2\text{-}C_2B_9H_{11})_2Ni$, also stabilizes the Ni(I) oxidation state. Evidence for this is found in cyclic voltammetry studies of $(1,2\text{-}C_2B_9H_{11})_2Ni$, which in some nonaqueous solvents (e.g., THF, DMF) shows three reversible reductions to an eventual Ni(I) trianion, as we pointed out in Section 2 (Figure 2). Obviously, it is not true that stabilization in the thermodynamic sense (E° values) occurs for both low and high metal-oxidation states for compounds of the *same* ligand. To be sure, the Ni(II/I) reduction [E24] for $(1,2\text{-}C_2B_9H_{11})_2Ni^{2-}$ is more negative than that of nickelocene [E31], but Ni(I) produced from the carborane is kinetically much more stable than that generated from the metal–hydrocarbon. The same points can be made for Co(I) carboranes, which often are produced at very negative potentials, but, at least in larger cage systems, seem to have appreciable stability. Reductions to Co(I) are quite common for cobaltacarboranes which have been investigated with electrolytes having sufficiently negative potential ranges. Probably the first evidence for a Co(I) carborane was found in the cyclic voltammetry data on $(1,2\text{-}C_2B_9H_8Br_3)_2Co^-$ [E11], in which the Co(II/I) wave was pushed positive enough by the bromine substituent effect to allow detection of the reduction to the trianion.[27] Cobaltacarboranes formed from uninegative ligands can be reduced to Co(I) at less negative potentials. For example, reduction occurs for $[PCB_9H_{10}(CH_3)]_2Co^{II}$ at ca. -0.7 V [E42], and even formally Co(0) is accessible in this system [E43]. The variety of borane and carborane ligands known, with the range of charges they carry, allows routes to either high or low oxidation state compounds.

7. REDOX-INDUCED STRUCTURAL REARRANGEMENTS

One of the very interesting properties of carborane clusters is the ability of some of these compounds to undergo thermally-induced isomerizations. Paxson *et al.*[40] and Warren and Hawthorne[46] have shown that for nickel or palladium bisdicarbollides, these rearrangements are much more facile in higher oxidation state compounds, and they were able to effect structural changes after electrolytic oxidation of a Ni(III) carborane anion to the corresponding Ni(IV) species.

Specifically, there is a tendency for the carbons lying in the pentagonal face bonded to the nickel to move away from the metal, into the 7 position, in the next five-membered ring. Three isomers of the dimethyl derivatives, which for steric reasons undergo these rearrangements at lower temperatures,[46] have been studied and are depicted in Figure 5. In the original work, the 1,2-1′,2′ isomer was designated isomer **A**, the 1,2-1′,7′ isomer **B**, and the 1,7-1′,7′ isomer **C**, so we will retain these designations for ease of discussion.

The Ni(III) monoanion rearranges from **A** to **B** if heated to 200°, but the isomerization proceeds at room temperature (or below) if oxidation to Ni(IV) is effected, as in treatment of Ni(III)-**A** solutions with ferric ion.

Paxson *et al.*,[40] have studied the redox reactions of the dimethyl compounds and the corresponding μ-1,2-trimethylene compounds, in which the two carbon atoms are tied together by a trimethylene bridge, in more quantitative electrochemical detail. The latter were found to undergo the same sort of facile ligand isomerizations as found for the dimethylcarboranyl species. Figure 6 reproduces a cyclic voltammogram reported for the oxidation of the **A** isomer of the Ni(III)-μ-trimethylene anion. Note that in the initial sweep positive (bottom half of Figure 6), starting from 0 V, no wave is seen at +0.3 V, the potential of the Ni(IV/III) couple of isomer **B** (top half of Figure 6), but after the oxidation to Ni(IV)-**A**, the couple for isomer **B** appears as small waves in the reverse scan. Note also that the cathodic reverse wave in the Ni(IV/III) couple for isomer **A** is diminished from what would be expected for a completely reversible couple. Thus, the scan shows that even at −8° C, measureable amounts of **B** are formed from Ni(IV)-**A** over the time period of the cyclic scan. Paxson *et al*,[40] used reverse current chronopotentiometry to measure the kinetics of these processes and, for example, found a first order rate constant of 2.60 sec^{-1} for the **A** → **B** isomerization of the Ni(IV) trimethylene compound at 15°C.

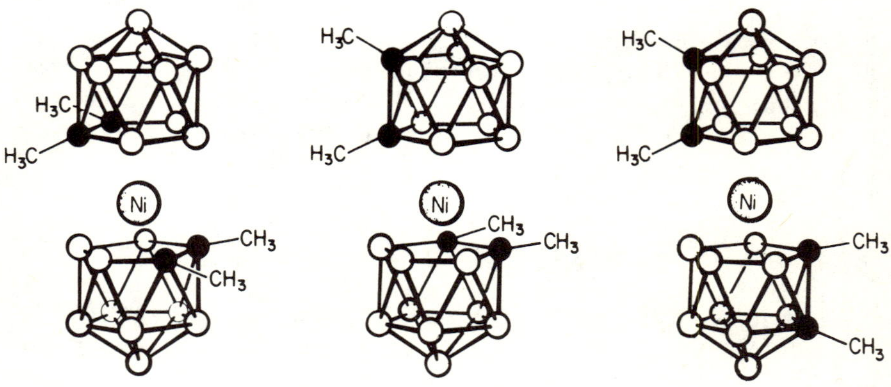

Figure 5. Three positional isomers of $[(CH_3)_2 C_2 B_9 H_9]_2 Ni$. Left, isomer A; center, isomer B; right, isomer C [from *Advances in Organometallic Chemistry*, **14**, 145 (1976)]. Note: ●, CH; ○, BH.

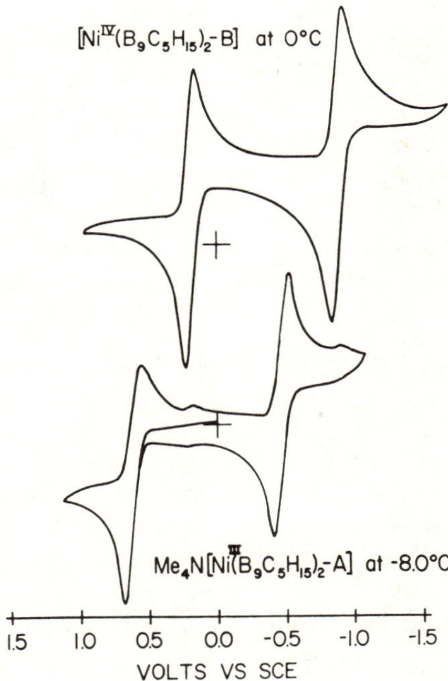

Figure 6. Cyclic voltammograms of *bis*-(1,2–trimethylene–1,2–dicarbollyl) nickel compounds in CH_3CN/0.1 *M* Et_4NClO_4. Scan rate 100 mv/sec. Top, isomer **B**; bottom isomer **A**. [*J. Am. Chem. Soc.*, **94**, 4882 (1972)].

Other than the thoroughly studied Ni(IV) rearrangements, only the isomerization of the analogous Pd(IV) species has been reported.[46] There seems to be no inherent reason why similar rearrangements could not also occur in other metallacarboranes in high oxidation states. Certainly, Co(IV) dicarbollide complexes are accessible electrochemically (Table 1), and the Co(III/IV) oxidation is generally subject to slow follow-up reactions. It is possible that these species are also undergoing isomerizations, and a thorough electrochemical study of this possibility might be worthwhile.

A topic that must be covered as a possibility, rather than a demonstrated fact, is the use of electrochemical methods to effect opening up of clusters, or in other ways to control the gross structures of metallacarborane clusters. Reduction of either *closo*-carboranes or *closo*-metallacarboranes is the key step in the polyhedral expansion reactions, which have proven to be an elegant route whereby cluster size may be increased. In this procedure,[70] two or more equivalents of a strong reducing agent, such as sodium napthalenide, are reacted with a carborane or metallacarborane in the presence of a metal (e.g., $CoCl_2$ and sodium cyclopentadienide) in order to incorporate the metal moiety (usually CpCo) into the cluster, thereby expanding by one the number of vertices in the cluster. It is assumed that this process occurs by a mechanism first involving formation of a *nido*-dianion by reduction of the *closo*-cage by two electrons.[71]

One might expect that an electrochemical version of this process might be straightforward, in which the *closo*-cluster could be opened by a 2-electron reduction, and the dianion thereby produced could be reacted with a suitable metal reagent which would be incorporated into the cluster. However, to our knowledge, no successes along these lines have been reported. This is probably due to the fact that, as pointed out in the preceding section, little work has been done on metallacarboranes in low oxidation states. We believe that the question of whether or not structure changes can be induced by production of highly electron-rich clusters is worthy of systematic study. Some of the "irreversible" reductions of metallacarboranes may well be due to *closo* to *nido* or other structural changes that accompany the electron-transfer step.

8. ELECTRON-TRANSFER RATES

Very little has been published on the electron-transfer rates of metallaboron compounds. The available data shows that these rates are very rapid. As part of a study of ferrocene–ferricinium and related electron-transfer rates, Pladziewicz and Espenson[72] investigated the oxidation of $CpFe(C_2B_9H_{11})^-$ by Fe^{3+} using stopped-flow spectrophotometry and obtained a rate constant of about $5 \times 10^6 \ M^{-1} \ sec^{-1}$. The oxidation of the analogous bisdicarbollide complex, $Fe(C_2B_9H_{11})_2^{2-}$, was too fast to measure, and a lower limit of $10^8 \ M^{-1} \ sec^{-1}$ was reported.

Electrochemical, heterogeneous, electron-transfer rates are also rapid. The present author, working with D. E. Smith, investigated seven redox couples involving metallacarboranes by phase selective alternating current polarography,[45] in acetonitrile at the dropping mercury electrode. The measured standard heterogeneous electron-transfer rates fall in the range 0.3 to 1.6 cm sec^{-1}. These values are typical of rapid, reversible, electron-transfer processes. In at least one of these steps, namely the reduction $Ni(1,2-C_2B_9H_{11})_2^- \rightleftharpoons Ni(1,2-C_2B_9H_{11})_2^{2-}$, the metal–ligand bonding is expected to change due to the well-documented "slippage" of the carborane away from the normal sandwich structure in electron-rich dicarbollide complexes. However, this step was not abnormally slow ($k_s = 0.66$ cm sec^{-1}), implying that the activation barrier to interconversion of the sandwich and slipped-sandwich forms is quite low. Put another way, even in cases in which the metallacarborane is expected to undergo modest structural changes during electron-transfer, very low overpotentials (high reversibility) are observed.

9. OTHER METALLABORON COMPOUNDS

Electrochemical data have appeared on a variety of other metallaboron compounds, and these data are collected in Tables 5 and 6. The results on compounds of the $(C_2B_{10}H_{12})^{2-}$ ligand (Table 5) are particularly noteworthy, because there is otherwise very little data on metallacarobranes containing early transition metals.

Table 5. Electrochemistry of Compounds Containing the $(C_2B_{10}H_{12})^{2-}$ Ligand

Entry number	Metal	Compound	Results[a]		References
E94	Ti	$4-(\eta^5-C_5H_5)-4-Ti(1,6-C_2B_{10}H_{12})^-$	Oxidations $+0.51$ V, $+0.85$ V (IR) Reduction -1.78 V (R) Formal	M(II)/M(I)	31,32
E95	Ti	$4-(\eta^8-C_8H_8)-4-Ti(1,6-C_2B_{10}H_{12})$	Oxidation $+1.77$ V (IR) Reduction -0.55 V (R)	M(IV)/M(III)	31,32
E96	Ti	$[4,4'-Ti(1,6-C_2B_{10}H_{12})_2]^{2-}$	Oxidation -0.24 V (R) Reduction -2.03 V (R)	M(III)/M(II) M(II)/M(I)	32
E97	Ti	$\{4,4'-Ti[1,6-C_2B_{10}H_{10}(CH_3)_2]_2\}^{2-}$	Oxidation -0.34 V (IR) Reduction -2.10 V (R)	M(II)/M(I)	32
E98	Ti	$[4-(\eta^5-C_5H_5)-4-Ti(1,6-C_2B_{10}H_{10}(CH_3)_2]^-$	Oxidation $+0.51$ V (IR) Reduction -1.82 V (R)	M(II)/M(I)	32
E99	Hf	$\{4,4'-Hf[1,6-C_2B_{10}H_{10}(CH_3)_2]_2\}^{2-}$	Oxidation -0.52 V (IR)		32
E100	V	$[4,4'-V(1,6-C_2B_{10}H_{12})_2]^{2-}$	Oxidation $+0.04$ V (R) Reduction -1.79 V (R)	M(III)/M(II) M(II)/M(I)	31,32
E101	V	$\{4,4'-V[1,6-C_2B_{10}H_{10}(CH_3)_2]_2\}^{2-}$	Oxidation -0.12 V (R) Reduction -1.80 V (R)	M(III)/M(II) M(II)/M(I)	32
E102	Cr	$[4,4'-Cr(1,6-C_2B_{10}H_{12})_2]^{2-}$	Oxidation -0.11 V (R) Reduction -1.90 V (R)		32
E103	Mn	$[4,4'-Mn(1,2-C_2B_{10}H_{12})_2]^{4-}$	Oxidation -1.11 V (R,F) Reduction -1.52 V (R)		32

[a]Potentials reported vs aqueous sce. Electrolyte was acetonitrile/0.1 M $Et_4NPF_6 \cdot$Pt electrode.

Table 6. Other Metallacarborane Electrochemistry

Entry number	Compound	Results[a]		Comments	References
7 Vertices					
E104	$1,2,3\text{–}CpCo(C_2B_4H_6)$	Reductions	$-1.62\,V$ (R), $-2.50\,V$ (R)[b]		56
		Oxidation	$+1.52\,V$ (IR)[b]	Rev. in CH_2Cl_2	
E105	$1,2,3\text{–}CpCo[C_2B_4(CH_3)_2]$	Reduction	$-1.69\,V$ (R)[b]		38
E106	$1,2,4\text{–}CpCo(C_2B_4H_6)$	Reduction	$-1.43\,V$ (R)[b]		56
		Oxidation	$+1.70\,V$ (IR)	Rev. in CH_2Cl_2	
9 Vertices					
E107	$4,1,8\text{–}CpFe(C_2B_6H_8)$	Reduction	$-0.58\,V$		74
E108	$1,4,6\text{–}CpFe(C_2B_6H_8)$	Reduction	$-0.17\,V^c$		74
E109	$1,4,5\text{–}CpCo(C_2B_6H_8)$	Reduction	$-1.09\,V^c$		74
E110	$1,4,6\text{–}CpCo(C_2B_6H_8)$	Reduction	$-0.84\,V^c$		74
E111	$(4,5\text{–}C_2B_6H_8)_2Co^-$	Reduction	$-1.75\,V$ (IR)^c		74
		Oxidation	$+0.69\,V$ (IR)^c		
E112	$1,5\text{–}CpCo(CB_7H_8)^-$	Oxidation	$+0.08\,V$ (R)^c		66
E113	$5\text{–}CH_3\text{–}1,5\text{–}CpCo(CB_7H_7)^-$	Oxidation	$+0.00\,V$ (R)^c		66
10 Vertices					
E114	$(1,2\text{–}C_2B_9H_{11})Co(1',6'\text{–}C_2B_7H_9)^-$	Reduction	$-1.29\,V$ (R)^c		73
E115	$(1,2\text{–}C_2B_9H_{11})Co(1',10'\text{–}C_2B_7H_9)^-$	Reduction	$-1.33\,V$ (R)^c		73
E116	$8,6,7\text{–}CpCo(C_2B_7H_{11})$	Reduction	$-0.87\,V$ (R)^c		73
E117	$CpCo(C_2B_7H_9)$ (isomer unknown)	Reduction	$-1.33\,V$ (R)^c		73
E118	$2,6,9\text{–}CpCo(C_2B_7H_9)$	Reduction	$-1.17\,V$ (R)^c		73
E119	$2,1,10\text{–}CpCo(C_2B_7H_9)$	Reduction	$-1.10\,V$ (R)^c		73
E120	$2,1,6\text{–}CpCo(C_2B_7H_9)$	Reduction	$-1.03\,V$ (R)^c		73
E121	$2,3,10\text{–}CpCo(C_2B_7H_9)$	Reduction	$-1.15\,V^c$		74

11 Vertices

E122	$1,2,3$–$CpFe(C_2B_8H_{10})$	Reduction	$-0.17\,V^c$	74
E123	$1,2,4$–$CpCo(C_2B_8H_{10})$	Reduction	$-0.81\,V\,(R)^c$	73
E124	$(1,2$–$C_2B_9H_{11})Co(2',4'$–$C_2B_8H_{10})^-$	Reductions	$-0.93\,V\,(R),\,-1.70\,V\,(R)^c$	73
E125	$1,2,3$–$CpCo(C_2B_8H_{10})$	Reduction	$-0.77\,V^c$	74
E126	$(2,3$–$C_2B_8H_{10})_2Co^-$	Reduction	$-0.28\,V^c$	74
		Oxidation	$+1.27\,V^c$	
E127	2–$[1'$–$(1',10'$–$C_2B_8H_9)]$–$1,2,3$–$CpCo$–$(C_2B_8H_9)$	Reduction	$-0.75\,V^c$	74
E128	$2,3$–$(CH_3)_2$–$10,2,3$–$CpCo(C_2B_8H_9)$	Reductions	$-1.23\,V\,(R),\,-1.57\,V\,(R)^g$	79
E129	$2,4$–$(CH_3)_2$–$1,2,4$–$CpCo(C_2B_8H_8)$	Reductions	$-0.83\,V\,(R),\,-1.64\,V\,(QR)$	79
E130	$2,3$–$(CH_3)_2$–$1,2,3$–$CpCo(C_2B_8H_8)$	Reductions	$-0.75\,V\,(R),\,-1.66\,V\,(IR)$	79

12 Vertices

E131	$(1,2$–$C_2B_9H_{11})_2FeH^-$	Oxidation	$-0.44\,V\,(QR)^d$	Analog of Cp_2FeH	33
E132	$[8$–$(C_2H_5)_2S$–$1,2$–$C_2B_9H_{10}]_2Fe$	Oxidation	$+0.48\,V\,(R)^e$		33
E133	$[8$–$(C_2H_5)_2S$–$1,2$–$C_2B_9H_{10}]Fe(1,2$–$C_2B_9H_{11})$	Reduction	$+0.13\,V\,(R)^e$		33
E134	$(CB_{10}H_{11})Fe^{3-}$	Reduction	$-0.73\,V$	Fe(III/II)	75
E135	$[2$–(C_8H_8)–2–$Ti(1,7$–$C_2B_9H_{11})]^-$	Oxidation	$-0.91\,V\,(R)^f$		32
E136	$(CB_{10}H_{11})_2Co^{3-}$	Oxidation	$+0.37\,V$	Co(IV/III)	75
E137	$(CB_{10}H_{11})_2Ni^{2-}$	Reduction	$-1.12\,V$	Ni(IV/III)	75
E138	$[C(NH_3)B_{10}H_{10}]_2Ni$	Reductions	$-0.49\,V,\,-0.60\,V$	Ni(IV/III/II)	75
E139	$(C_{10}H_8)Co(CB_{10}H_{11})$	Reductions	$-0.62\,V\,(R),\,-1.31\,V\,(QR)^f$		62

13 Vertices

E140	$[(C_2B_{10}H_{10})_2]_2Cu^{2-}$	Reduction	$-1.36\,V\,(IR)^c$	[E131–E134] are sigma-(carbon) bonded compounds	76,77
		Oxidation	$+0.15\,V\,(R)^c$		
E141	$[(C_2B_{10}H_{10})_2]_2Ni^-$	Reductions	$+0.86\,V\,(R),\,-2.07\,V\,(R)^{c,h}$		76,77
E142	$[(C_2B_{10}H_{10})_2]_2Co^-$	Reductions	$+0.62\,V\,(R),\,-2.35\,V\,(R)^{c,h}$	Make solutions of Co(I) trianion by reduction with Li(Hg)	76,77

(continued overleaf)

Table 6. (*cont.*)

Entry number	Compound	Results[a]		Comments	References
E143	$[(C_2B_{10}H_{10})_2]_2Zn^{2-}$	Reduction	$-2.36\,V$ (R)[c,h]		76,77
E144	$CpFe(C_2B_{10}H_{10})$	Reduction	$-0.16\,V$ (R)[c]		78
E145	$(C_2B_{10}H_{10})_2Fe^{2-}$	Oxidation	$-0.02\,V$ (R)[c]		78
E146	$CpCo(C_2B_{10}H_{10})$	Reduction between -0.7 and $-1.2\,V$ (R)[c]		3 isomers	78
E147	$(C_2B_{10}H_{10})_2Co^-$	Reduction	$-0.33\,V$ (R)[c]		78
E148	$(C_2B_{10}H_{10})_2Ni^{2-}$	Oxidation	$-0.03\,V$ (R)[c]	Reported 2-electron Oxidation—no details given	78

[a]Potentials vs aqueous sce. Unless otherwise noted, cyclic voltammetry data reported. [b]$CH_3CN/0.1\,M\ Bu_4NPF_6$/Hg or Pt electrode. [c]$CH_3CN/0.1\,M\ Et_4ClO_4$/Pt electrode. [d]10% $HClO_4$/90% CH_3OH/Hg electrode. [e]$CH_2Cl_2/0.1\,M\ Bu_4NClO_4$. [f]$CH_3CN/0.1\,M\ Bu_4NPF_6$/Pt electrode. [g]$CH_3CN/0.1\,M\ Bu_4NBr$/Pt or Hg electrode/$0.2\,V\ sec^{-1}$ scan rate. [h]Direct current polarography data.

ACKNOWLEDGMENTS: I am indebted to David E. Brennan, who helped with the literature search for this chapter, and to the National Science Foundation, for support during the time in which it was written.

REFERENCES

1. M. F. Hawthorne, D. C. Young, and P. A. Wegner, *J. Am. Chem. Soc.*, 87, 1818 (1965).
2. E. B. Rupp, D. E. Smith, and D. F. Shriver, *J. Am. Chem. Soc.*, 89, 5562 (1967).
3. D. E. Smith, E. B. Rupp, and D. F. Shriver, *J. Am. Chem. Soc.*, 89, 5568 (1967).
4. R. L. Middaugh and F. Farha, Jr., *J. Am. Chem. Soc.*, 88, 4147 (1966).
5. R. J. Wiersema and R. L. Middaugh, *Inorg. Chem.*, 8, 2074 (1969).
6. R. J. Wiersema and R. L. Middaugh, *J. Am. Chem. Soc.*, 92, 223 (1970).
7. B. G. Cooksey, J. D. Gorham, J. H. Morris, and L. Kane, *J. Chem. Soc. Dalton Trans.* 141 (1978).
8. J. H. Morris, *Proc. 1, 19th Internat. Conf. Coordination Chem., Prague*, p. 84 (1978).
9. M. F. Hawthorne and P. A. Wegner, *J. Am. Chem. Soc.*, 87, 4392 (1965).
10. C. B. Harris, *Inorg. Chem.*, 7, 1517 (1968).
11. A. H. Maki and T. E. Berry, *J. Am. Chem. Soc.*, 87, 4437 (1965).
12. R. H. Herber, *Inorg. Chem.*, 8, 174 (1969).
13. R. Birchall and I. Drummond, *Inorg. Chem.*, 10, 399 (1971).
14. D. N. Hendrickson, Y. S. Sohn, and H. B. Gray, *Inorg. Chem.*, 10, 1559 (1971).
15. R. J. Wiersema and M. F. Hawthorne, *J. Am. Chem. Soc.*, 96, 761 (1974).
16. D. A. Brown, Chambers, M. O. Fanning, and N. J. Fitzpatrick, *Inorg. Chem.*, 17, 1620 (1979).
17. K. Wade, *Adv. Inorg. Chem. Radiochem Trans.*, 8, 1, (1976), and references therein.
18. D. M. P. Mingos, *J. Chem. Soc. Dalton Trans.* 602 (1977) and references therein.
19. R. W. Rudolph, and W. R. Pretzer, *Inorg. Chem.*, 11, 1974 (1972).
20. R. J. Wilson, L. F. Warren, Jr., and M. F. Hawthorne, *J. Am. Chem. Soc.*, 91, 758 (1969).
21. H. W. Rhule and M. F. Hawthorne, *Inorg. Chem.*, 7, 2279 (1968).
22. R. P. Van Duyne and C. N. Reilly, *Anal. Chem.*, 44, 158 (1972).
23. J. D. L. Holloway, W. L. Bowden, and W. E. Geiger, Jr., *J. Am. Chem. Soc.*, 99, 7089 (1977).
24. J. D. L. Holloway and W. E. Geiger, Jr., *J. Am. Chem. Soc.*, 101, 2038 (1979).
25. W. E. Geiger, Jr. and D. E. Smith, *Chem. Commun.*, 8 (1971).
26. L. F. Warren, Jr. and M. F. Hawthorne, *J. Am. Chem. Soc.*, 90, 4823 (1968).
27. M. F. Hawthorne, D. C. Young, T. D. Andrews, D. V. Howe, R. L. Pilling, A. D. Pitts, M. Reintjes, L. F. Warren, Jr., and P. A. Wegner, *J. Am. Chem. Soc.*, 90, 879 (1968).
28. J. H. Robbins, Ph.D. Dissertation, University of California, Berkeley (1979).
29. (For recent discussions and references leading to earlier work) D. M. P. Mingos, M. I. Forsyth, and A. J. Welch, *J. Chem. Soc. Chem. Commun.*, 605 (1977); and P. A. Wegner, *Inorg. Chem.*, 14, 212 (1975).
30. N. El Murr and E. Laviron, *Can. J. Chem.*, 54, 3350 (1976).
31. C. G. Salentine and M. F. Hawthorne, *Chem. Commun.*, 848 (1975).
32. C. G. Salentine and M. F. Hawthorne, *Inorg. Chem.*, 15, 2872 (1976).
33. M. F. Hawthorne, L. F. Warren, Jr., K. P. Callahan, and N. F. Travers, *J. Am. Chem. Soc.*, 93, 2407 (1971).
34. W. E. Geiger, Jr., W. L. Bowden, and N. El Murr, *Inorg. Chem.*, 18, 2358 (1979).
35. T. Kuwana, D. Bublitz, and G. Hoh, *J. Am. Chem. Soc.*, 82, 5811 (1960).
36. J. N. Francis and M. F. Hawthorne, *Inorg. Chem.*, 10, 863 (1971).
37. J. Koryta, P. Vanysek, G. Janchenova, and M. Brezina, *Elektrokhimiya*, 13, 706, (1977); *Sov. Electrochem.*, 13, 605 (1977).

38. W. E. Geiger, Jr., unpublished results.
39. D. S. Matteson and R. E. Grunzinger, Jr., *Inorg. Chem.*, **13**, 671 (1974).
40. T. E. Paxson, M. K. Kaloustian, G. M. Tom, R. J. Wiersema, and M. F. Hawthorne, *J. Am. Chem. Soc.*, **94**, 4882 (1971).
41. M. Kaloustian, R. J. Wiersema, and M. F. Hawthorne, *J. Am. Chem. Soc.*, **93**, 4912 (1971).
42. M. Kaloustian, R. J. Wiersema, and M. F. Hawthorne, *J. Am. Chem. Soc.*, **94**, 6679 (1972).
43. W. E. Geiger, Jr., *J. Am. Chem. Soc.*, **96**, 2632 (1974).
44. N. El Murr, J. E. Sheats, W. E. Geiger, Jr., and J. d. L. Holloway, *Inorg. Chem.*, **18**, 1443 (1979).
45. W. E. Geiger, Jr. and D. E. Smith, *J. Electroanal. Chem.*, **50**, 31 (1974).
46. L. F. Warren, Jr. and M. F. Hawthorne, *J. Am. Chem. Soc.*, **92**, 1157 (1970).
47. J. L. Little, P. S. Welcker, N. J. Loy, and L. J. Todd, *Inorg. Chem.*, **9**, 63 (1970).
48. W. R. Hertler, F. Klanberg, and E. L. Muetterties, *Inorg. Chem.*, **6**, 1696 (1967).
49. D. E. Brennan, Ph.D. Thesis, University of Vermont (1980).
50. A. R. Siedle and L. J. Todd, *Inorg. Chem.*, **15**, 2938 (1976).
51. R. N. Leyden, B. P. Sullivan, R. T. Baker, and M. F. Hawthorne, *J. Am. Chem. Soc.*, **100**, 3758 (1978).
52. M. F. Hawthorne and H. W. Ruhle, *Inorg. Chem.*, **8**, 176 (1969).
53. K. P. Callahan, W. J. Evans, F. Y. Lo, C. E. Strouse, and M. F. Hawthorne, *J. Am. Chem. Soc.*, **97**, 296 (1975).
54. C. G. Salentine and M. F. Hawthorne, *Inorg. Chem.*, **17**, 1498 (1978).
55. D. F. Dustin and M. F. Hawthorne, *J. Am. Chem. Soc.*, **96**, 3462 (1974).
56. D. E. Brennan and W. E. Geiger, Jr., *J. Am. Chem. Soc.*, **101**, 3399 (1979).
57. W. J. Evans and M. F. Hawthorne, *Inorg. Chem.*, **13**, 869 (1974).
58. W. J. Evans, C. J. Jones, B. Stibr, R. A. Grey, and M. F. Hawthorne, *J. Am. Chem. Soc.*, **96**, 7405 (1974).
59. C. J. Jones and M. F. Hawthorne, *Inorg. Chem.*, **12**, 608 (1973).
60. C. G. Salentine, C. E. Strouse, and M. F. Hawthorne, *Inorg. Chem.*, **15**, 1832 (1976).
61. W. J. Evans and M. F. Hawthorne, *Chem. Commun.*, 38 (1974).
62. C. G. Salentine and M. F. Hawthorne, *J. Am. Chem. Soc.*, **97**, 6382 (1975).
63. R. N. Grimes, D. C. Beer, L. G. Sneddon, V. R. Miller, and R. Weiss, *Inorg. Chem.*, **13**, 1138 and references therein (1974).
64. W. Siebert, J. Edwin, and M. Bochmann, *Angew. Chem. Int. Ed. Engl.*, **17**, 868 (1979).
65. W. H. Knoth, *J. Am. Chem. Soc.*, **89**, 3342 (1967).
66. D. F. Dustin and M. F. Hawthorne, *Inorg. Chem.*, **12**, 1380 (1973).
67. D. E. Hyatt, J. L. Little, J. R. Moran, F. R. Scholer, and L. J. Todd, *J. Am. Chem. Soc.*, **89**, 3342 (1967).
68. R. R. Rietz, D. F. Dustin, and M. F. Hawthorne, *Inorg. Chem.*, **13**, 1580 (1974).
69. W. H. Knoth, *Inorg. Chem.*, **10**, 598 (1971).
70. G. B. Dunks and M. F. Hawthorne, *J. Am. Chem. Soc.*, **92**, 7213 (1970).
71. K. P. Callahan and M. F. Hawthorne, *Pure Appl. Chem.*, **39**, 475 and references therein (1974).
72. J. R. Pladziewicz, and J. H. Espenson, *J. Am. Chem. Soc.*, **95**, 56 (1973).
73. C. J. Jones, J. N. Francis, and M. F. Hawthorne, *J. Am. Chem. Soc.*, **94**, 8391 (1972).
74. W. J. Evans, G. B. Dunks, and M. F. Hawthorne, *J. Am. Chem. Soc.*, **95**, 4565 (1973).
75. L. J. Todd, *Adv. Organomet. Chem.*, **8**, 87 (1970).
76. D. A. Owen and M. F. Hawthorne, *J. Am. Chem. Soc.*, **93**, 873 (1971).
77. D. A. Owen and M. F. Hawthorne, *J. Am. Chem. Soc.*, **92**, 3195 (1970).
78. D. F. Dustin, G. B. Dunks, and M. F. Hawthorne, *J. Am. Chem. Soc.*, **95**, 1109 (1973).
79. G. D. Mercer, M. Tribo, and F. R. Scholer, *Inorg. Chem.*, **14**, 764 (1975).

7

Boron Clusters with Transition Metal–Hydrogen Bonds

Russell N. Grimes

1. INTRODUCTION

Transition metal hydride chemistry has experienced enormous growth and impact in recent decades,[1] accelerated by the widespread interest in these species as homogeneous and heterogeneous catalysts. This work has paralleled the rapid development of metallaborane and metallacarborane chemistry,[2] and it was inevitable that the two fields would overlap and that metal–boron clusters having metal–hydrogen bonds would be discovered. What was not really foreseen, however, was the vast scope of structures and chemistry that has emerged in this area. Indeed, it is the case that in terms of isolable monomeric species, the variety of stable transition metal–hydrogen bonding modes in known metal–boron cluster compounds greatly exceeds that found in "pure" (nonboron) metal cluster hydrides. The reasons would seem to be, first, that boron frameworks in general provide inherently stable matrices in which metal atoms with hydrogen ligands can reside; second, that borane and metallaborane cages tend to be hydrogen-rich, setting the stage for metal–hydrogen interactions; and third, that hydrogen bridging is a common feature in electron-deficient boron clusters, and the involvement of metals in such systems can be seen (at least in retrospect) as a natural extension of this phenomenon.

Figure 1 illustrates schematically several types of metal–hydrogen interactions that are possible in metallaboron clusters. Surprising though it may seem, examples of all of these types have been identified in *isolable* compounds and structurally characterized, most of them since 1970. Moreover, many of the

Russell N. Grimes ● Department of Chemistry, University of Virginia, Charlottesville, Virginia.

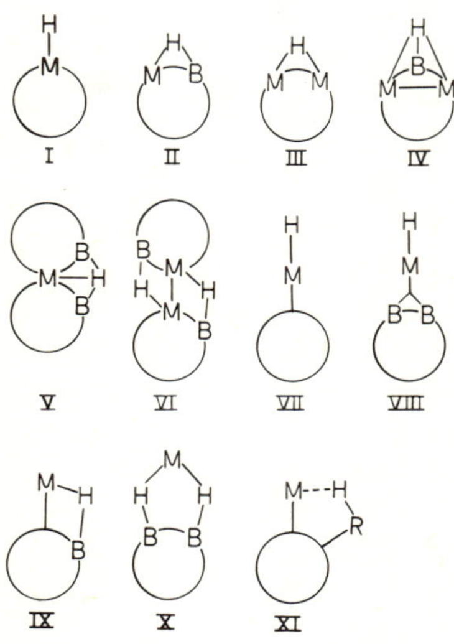

Figure 1. Schematic representation of established modes of transition metal-hydrogen bonding interactions in boron cluster compounds. Circles represent open or closed polyhedral boron cages of four or more vertices, and may contain atoms of other elements in addition to boron. Metal atoms (M) may have other ligands in addition to hydrogen (not shown). I, terminal M–H bond, metal in cage. II, M–H–B bridge, metal in cage. III, M–H–M bridge, both metals in cage. IV, hydrogen capping M–M–B face on cluster. V, hydrogen linked to metal common to two polyhedra. VI, cages linked by M–H–B bridges and by direct M–M bond (metals in cages). VII, terminal M–H bond, metal σ-bonded to cage. VIII, terminal M–H bond, metal μ-bonded to cage. IX, M–H–B bridge, metal σ-bonded to cage. X, metal linked to cage by M–H–B bonds only. XI, M–H–R bridge (R = C or other atom), M and R σ-bonded to cage.

arrangements depicted have occurred frequently, with a variety of cage geometries and different transition metals. Still other bonding modes are conceivable for metallaboron cluster hydrides: for example, two polyhedra linked solely by M–H–M bridges.

This review is intended to deal primarily with true polyhedral clusters, which for present purposes are defined as cages of four or more vertices; chelate complexes (such as those of BH_4^-) are excluded, as are those of $B_3H_8^-$, which are dealt with separately in Chapter 3. In general, the discussion is limited to species actually isolated and characterized, as opposed to unstable intermediates in which M–H bonding has been postulated; however, selected instances of the latter type of situation will be mentioned. The evidence for M–H bonding in most cases is derived from X-ray crystallographic, infrared, and/or proton nmr data. Hydrogen atoms in the vicinity of transition metals are often impossible to resolve in X-ray studies, and even when they are found, there is sometimes ambiguity as to the extent of metal-hydrogen interaction; consequently, the most reliable diagnostic test for M–H bonding is, in general, the presence of a high-field proton NMR signal (usually 5 ppm or more to the high field of tetramethylsilane).[3] Since no other commonly encountered functionalities exhibit proton shifts in this region (e.g., B–H–B and B–H$_{terminal}$ resonances are nearly always to low field of $\delta-5$), the M–H signal can, in most cases, be unambiguously interpreted.

Most of the compounds described herein have not been extensively examined with respect to their chemical behavior, and hence this review is centered mainly on structures and bonding descriptions. To the extent it is possible to do so, an attempt has been made to indicate the relationships of various cluster types to each other and to nonboron cluster compounds. This task is greatly aided by the skeletal electron-count theory for polyhedral molecules,[4] which provides a powerful holistic framework for dealing with clusters of all types.

2. CLUSTERS CONTAINING TERMINAL M—H GROUPS

The stereochemically simplest type of metal-hydrogen interaction, at least in principle, is that in which a hydrogen ligand (usually designated formally as hydride, H^-) occupies a specific coordination site on the metal. In such compounds the hydrogen ligand is bound to the metal atom by an electron-pair bond and does not significantly interact directly with any other atoms. Such a situation is designated terminal M—H bonding. In this general category there are different compound types based on the location and role of the *metal* atom: (1) those in which the metal is external to the cage and is linked to it by a single electron-pair bond, (2) those in which the metal occupies a vertex in the cage, and (3) intermediate cases in which the metal atom is bound to the edge of the cage by a B—M—B three-center, two-electron bond. A number of examples of the first two types are known, but there is only one reported species of the third variety, mentioned at the end of this section.

Compounds containing one or more terminal M—H bonds, to the extent that the hydrogen ligands are free of bridging interactions with boron or other atoms, would be expected to exhibit a chemistry typical of metal hydrides having conventional ligands such as phosphines or amines. In such a case, the boron cage affects the reactivitiy of the M—H group only indirectly via steric or long-range electronic effects. However, a different situation arises if the boron ligand directly participates in reactions involving the M—H bond; in that event, there is the distinct possibility that the boron cage-metal hydride complex may function in ways entirely different from other metal hydride complexes, perhaps including new modes of catalytic hydrogenation. There is reason to believe that this may be the case in at least some metallaboron cluster hydride species, as will be shown in the discussion to follow.

2.1. Clusters in Which the Metal is Part of the Cage Framework

2.1.1. Icosahedral Metallacarboranes

Relatively few boron clusters containing true terminal metal hydride ligands are known, but several of these are potentially important because of their demonstrated catalytic activity. Most of such complexes have an icosahedral structure with the general formula $LH_2MC_2B_9H_{11}$ or $L_2HMC_2B_9H_{11}$ where L is a trialkyl- or triarylphosphino ligand.

Direct attachment of hydrogen to a metal in an existing metallacarborane is difficult to achieve, but, in general, metallacarborane hydrides can be prepared by the reaction of *nido*-carborane anions with metal halide reagents. Thus, treatment of the 7,8- or 7,9-$C_2B_9H_{12}^-$ ion with triphenylrhodium(I) cation generates, respectively, the 3,1,2 (adjacent-carbon) and 2,1,7 (nonadjacent-carbon) isomers of $[(C_6H_5)_2P]_2Rh(H)C_2B_9H_{11}$:[5]

$$[(C_6H_5)_3P]_3Rh^+ + (CH_3)_3NH^+C_2B_9H_{12}^- \xrightarrow[60°]{-CH_3OH} [(C_6H_5)_3P]_2Rh(H)C_2B_9H_{11} \quad >80\%$$

The metal-bound hydride ligand presumably is derived from the "extra" proton which occupies a $B-H-B$ bridging location on the $C_2B_9H_{12}^-$ ion. In this sense, the reaction is similar to the preparation of small metallacarborane metal hydrides from the 2,3-$C_2B_4H_7^-$ ion and its derivatives (Section 4),[6,7] although the products in the latter case are of a different structural class. The analogous treatment of the 1-C_6H_5-$C_2B_9H_{11}^-$ ion gave an optically active (90/5%) mixture of enantiomers of $[(C_6H_5)_3P]_2Rh(H)(C_6H_5)C_2B_9H_{10}$.[8]

The structure of 3,1,2-$[(C_6H_5)_3P]_2Rh(H)C_2B_9H_{11}$, a yellow crystalline solid, has been established in an X-ray investigation[9] (Figure 2), which revealed that the hydride ligand occupies a pseudo-octahedral coordination site on rhodium with an $Rh-H$ distance of 1.54(9) Å. Other relevant data are given in Table 1.

The iridium congeners of these complexes, 3,1,2- and 2,1,7-$[(C_6H_5)_3P]_2$-$Ir(H)C_2B_9H_{11}$, have been prepared by analogous reactions,[10] and the closely related ruthenium species 3,1,2- and 2,1,7-$[(C_6H_5)_3P]_2RuH_2C_2B_9H_{11}$ have been obtained from the reaction of the 7,8- or 7,8-$C_2B_9H_{12}^-$ ion with $[(C_6H_5)_3P]_3RuHCl$ in refluxing ethanol.[11,12] From spectroscopic and X-ray data, the light blue 2,1,7 complex has been formulated as a formally seven-coordinate Ru(IV) complex (Figure 3), although the hydrogen ligands on the metal were not directly located in the crystallographic study.[11] Elimination of H_2 from the 2,1,7 isomer occurs at 160°C under vacuum, yielding white

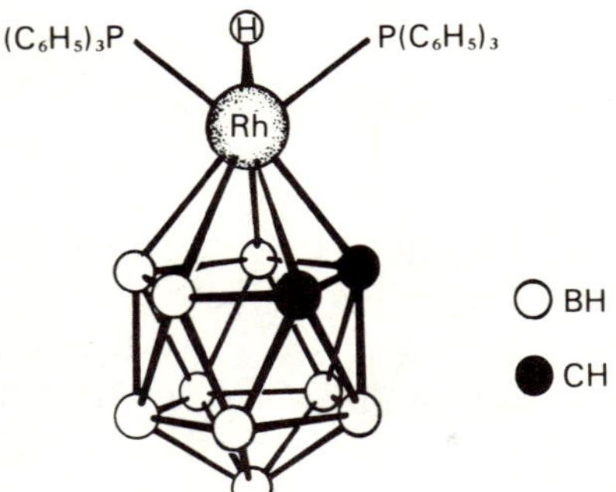

Figure 2. Structure of 3,1,2-$[(C_6H_5)_3P]_2$-$Rh(H)C_2B_9H_{11}$ (Ref. 9).

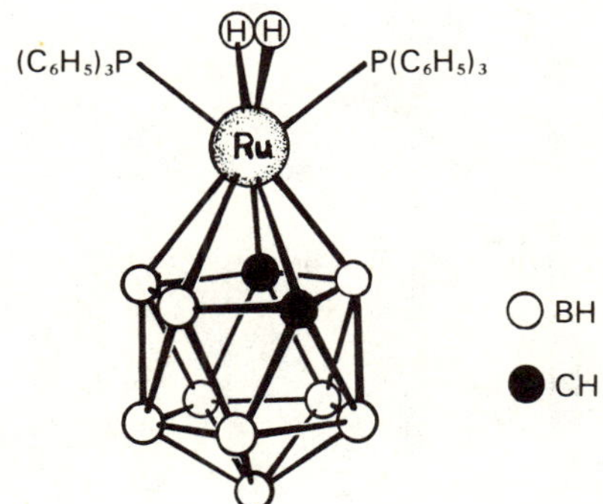

Figure 3. Structure of $2,1,7$-$[(C_6H_5)_3P]_2$-$RuH_2C_2B_9H_{11}$. Hydrogen ligand positions are calculated (Ref. 11).

$[(C_6H_5)_3P]_2RuC_2B_9H_{11}$; the latter species takes up H_2 at room temperature to regenerate the dihydride:

$$[(C_6H_5)_3P]_2Ru(H_2)C_2B_9H_{11} \rightleftharpoons [(C_6H_5)_3P]_2RuC_2B_9H_{11} + H_2$$

The 3,1,2 (adjacent-carbon) isomer of the dihydride also was observed to lose H_2 at 160°C, but no characterizable products were obtained.[12]

Both isomers react with CO and HCl to displace H_2 and generate the corresponding substituted complexes:

$$[(C_6H_5)_3P]_2Ru(H_2)C_2B_9H_{11} \begin{cases} \xrightarrow{\text{CO}} [(C_6H_5)_3P]_2Ru(CO)C_2B_9H_{11} \\ \xrightarrow{\text{HCl}} [(C_6H_5)_3P]_2Ru(H)(Cl)C_2B_9H_{11} \end{cases}$$

An alternative synthesis of $3,1,2$-$[(C_6H_5)_3P]_2Ru(CO)C_2B_9H_{11}$ utilizes the reaction of $C_2B_9H_{11}{}^{2-}$ ion with $[(C_6H_5)_3P]_2Ru(CO)_2Cl_2$ in refluxing benzene.[13] A pyridyl-B-substituted derivative of the 3,1,2 dihydride isomer has been prepared by allowing $[(C_6H_5)_3P]_2RuHCl$ to react with a B-substituted carborane anion:[12]

$$C_5H_5N{-}C_2B_9H_{10}{}^- + [(C_6H_5)_3P]_2RuHCl \cdot C_6H_5CH_3 \longrightarrow$$
$$4{-}C_5H_5N{-}3,1,2{-}[(C_6H_5)_3P]_2Ru(H)C_2B_9H_{10}$$

Interest in the metallacarborane hydrides has centered primarily on their function as homogeneous (and, more recently, heterogenerous) catalysts in hydrogenation, isomerization, and other processes. The performance of some of these catalysts is impressive in terms of conversion rate and selectivity; however, development of this field is in an early stage and few mechanistic details of the catalytic processes have been established. Moreover, many of the published reports in this area are short and fragmentary with a minimum of detail, making it difficult to present a broad picture. However, a summary of the main findings will be given. Both the 3,1,2 and 2,1,7 isomers of $[(C_6H_5)_3P]_2Rh(H)C_2B_9H_{11}$

Table 1. Structural and Spectroscopic Data on Clusters Containing Terminal M—H Groups

Compound[a]	M—H distance (Å)	^{1}H nmr δ_{M-H}(J, Hz)[b]	Solvent[c]	ν_{M-H} (cm^{-1})	Solvent[d]	Ref.
Metallacarboranes with metal in cage framework						
Nickel						
$8-Ph_3P-3,1,2-(Ph_3P)Ni(H)C_2B_9H_{10}$	—	−18.4	A	1984	—	23
Ruthenium						
$3,1,2-(Ph_3P)_2Ru(H)_2C_2B_9H_{11}$	—	−6.3(28)	B	2060, 2040	N	12
$2,1,7-(Ph_3P)_2Ru(H)_2C_2B_9H_{11}$	—	−6.9(29)	B	2060, 2000	N	12
$3,1,2-(Ph_3P)_2Ru(H)(Cl)C_2B_9H_{11}$	—	—	—	2160	N	12
$2,1,7-(Ph_3P)_2Ru(H)(Cl)C_2B_9H_{11}$	—	—	—	2160	N	12
$4-C_5H_5N-3,1,2-(Ph_3P)_2Ru(H)C_2B_9H_{10}$	—	−9.6 −21.0(31)	B	2100	N	12
Rhodium						
$6,2,3-(Ph_3P)_2Rh(H)C_2B_7H_9$	—	−13.45(16)	A	2081	N	24
$6,2,3-[(p-MeC_6H_4)_3P]_2Rh(H)C_2B_7H_9$	—	−12.24(16)	C	—	—	24
$[3,1,2-(Ph_3P)_2Rh(H)C_2B_9H_{11}]_2$	1.78(6) 1.77(6)	−8.0, −9.0	A	—	—	18
$3,1,2-(Ph_3P)Rh(H)C_2B_9H_{11}$	—	−10.5(20)	D	2220	K	13a
$3,1,2-(Ph_3P)_2Rh(H)C_2B_9H_{11}$	—	−8.4(17 ± 2)	A	2080, 2120	N	5
$3,1,2-(Ph_3P)_2Rh(H)C_2B_9H_{11}$	1.54(9)	—	—	—	—	9
$1,3-[\mu-(CH_2=CHCH_2CH_2)]-3-H-3-(Ph_3P)-3,1,2-RhC_2B_9H_{10}$	1.65(3)	−7.42(20)	A	2060	—	13a
$2,1,7-(Ph_3P)_2Rh(H)C_2B_9H_{11}$	—	−11.1(14 ± 2)	A	2070, 2110	N	5
$1-C_6H_5-3,1,2-(Ph_3P)_2Rh(H)C_2B_9H_{10}$	—	—	—	2110	—	8
Iridium						
$3,1,2-(Ph_3P)_2Ir(H)C_2B_9H_{11}$	—	—	—	—	—	10
$2,1,7-(Ph_3P)_2Ir(H)C_2B_9H_{11}$	—	—	—	—	—	10
$6,2,3-(Ph_3P)_2Ir(H)C_2B_7H_9$	—	−14.47(25)	A	2201	N	24
$1,2,4-(Ph_3P)_2Ir(H)C_2B_8H_{10}$	—	−6.30(14, 32)	A	2133	N	24

Metallaboranes with metal in cage framework

$nido$-$(Ph_3P)_2Pt(H)SB_9H_{10}$	1.66	–	–	2214	–	25
$(Ph_3P)_2Rh(H)SB_{10}H_{10}$	–	–7.2(18)	E	2080	–	26

Metallacarboranes with metal σ-bonded to cage carbon atom

Platinum

cis-1-$(Et_3P)_2Pt(H)$-1,2-$C_2B_{10}H_{11}$	–	–6.40(980)	C	2116	G	30
cis-1-$(Et_3P)_2Pt(H)$-1,7-$C_2B_{10}H_{11}$	–	–5.78(952)	F	2130	F	30
cis-1-$(Et_3P)_2Pt(H)$-7-Me-1,7-$C_2B_{10}H_{10}$	–	–5.66(953)	F	2130	G	30
cis-1-$(Et_3P)_2Pt(H)$-7-Ph-1,7-$C_2B_{10}H_{10}$	–	–5.74(951)	F	2128	G	30
$trans$-1-$(Et_3P)_2Pt(H)$-2-Me-1,2-$C_2B_{10}H_{10}$	–	–14.04(680)	F	2133	G	30
$trans$-1-$(Et_3P)_2Pt(H)$-2-Ph-1,2-$C_2B_{10}H_{10}$	–	–14.71(691)	F	2130	G	30
$trans$-1-$(Et_3P)_2Pt(H)$-7-Ph-1,7-$C_2B_{10}H_{10}$	–	–13.92(664)	F	2118	F	29
cis-1-$(Ph_3P)_2Pt(H)$-2-Ph-1,2-$C_2B_{10}H_{10}$	–	–7.32(977)	E	2127	C	30
cis-1-$(Ph_3P)_2Pt(H)$-7-Ph-1,7-$C_2B_{10}H_{10}$	–	–6.59(1015)	C	2106	N	30
cis-1-$(Ph_2MeP)_2Pt(H)$-1,2-$C_2B_{10}H_{11}$	–	–6.14(1039)	C	2090	N	30
cis-1-$(Ph_2MeP)_2Pt(H)$-2-Me-1,2-$C_2B_{10}H_{10}$	–	–6.59(981)	C	2127	J	30
cis-1-$(Ph_2MeP)_2Pt(H)$-2-Ph-1,2-$C_2B_{10}H_{10}$	–	–6.73(968)	C	2126	J	30
cis-1-$(Ph_2MeP)_2Pt(H)$-7-Ph-1,7-$C_2B_{10}H_{10}$	–	–5.75(1026)	K	2116	J	30
cis-2-$(PhMe_2P)_2Pt(H)$-2-Ph-1,2-$C_2B_{10}H_{10}$	–	–6.79(978)	E	2132	C	30

Iridium

1-$(Ph_3P)_2(CO)IrH_2$-1,2-$C_2B_{10}H_{11}$						
Phosphines cis; CO and carborane $trans$	–	–9.97	C	2162, 2144	N	31a
Phosphines $trans$; CO and carborane cis	–	–15.08, –8.59	C	2170, 2096	N	31a
1-$(Ph_3P)_2(CO)IrH_2$-2-Me-1,2-$C_2B_{10}H_{10}$						
Phosphines cis; CO and carborane cis	–	–11.48, –8.44	C	2211, 2150	N	31a
1-$(Ph_3P)_2(CO)IrH_2$-1,7-$C_2B_{10}H_{11}$						
Phosphines cis; CO and carborane $trans$	–	–10.10	C	–	–	31a
Phosphines $trans$; CO and carborane cis	–	–14.74, –8.47	C	2177, 2122	N	31a
Phosphines cis; CO and carborane cis	–	–10.74, –8.09	C	2211, 2138	N	31a
1-$(Ph_3P)_2(CO)IrH_2$-7-Me-1,7-$C_2B_{10}H_{10}$						
Phosphines cis; CO and carborane $trans$	–	–10.20	C	2140	N	31a

(continued overleaf)

Table 1 *(cont.)*

Compound[a]	M–H distance (Å)	^{1}H nmr δ_{M-H} (J, Hz)[b]	Solvent[c]	ν_{M-H} (cm^{-1})	Solvent[d]	Ref.
Phosphines *trans*; CO and carborane *cis*	–	–14.70, –8.50	C	2195, 2114, 2078	N	31a
Phosphines *cis*; CO and carborane *cis*	–	–10.84, –8.07	C	2204, 2142	N	31a
$1-(Ph_3P)_2(CO)IrH_2-7-Ph-1,7-C_2B_{10}H_{10}$						
Phosphines *cis*; CO and carborane *trans*	–	–9.52	C	2161	N	31a
Phosphines *trans*; CO and carborane *cis*	–	–14.80, –8.46	C	2210, 2149, 2064	N	31a
Phosphines *cis*; CO and carborane *cis*	–	–10.77, –7.94	C	2196, 2138	N	31a
$1-(Ph_3P)_2(CO)IrH_2-7-Ph_2MeP-1,7-C_2B_{10}H_{10}$						
Phosphines *cis*; CO and carborane *trans*	–	–9.52	C	–	–	31a
Phosphines *trans*; CO and carborane *cis*	–	–15.24, –8.65	C	2156, 2112	N	31a
$1-[(Ph_2PCH_2)_2(CO)IrH_2]-7-Ph-1,7-C_2B_{10}H_{10}$	–	–9.21, –9.06	E	2122, 2084	N	31b
$1-[(PhCN)(CO)(PPh_3)IrH_2]-7-Ph-1,7-C_2B_{10}H_{10}$	–	–7.04, –17.65	E	2114, 2110	N	31b
$1-[(MeCN)(CO)(PPh_3)IrH_2]-7-Ph-1,7-C_2B_{10}H_{10}$	–	–7.15, –18.20	E	2235, 2119	N	31b
$1-[(CO)_2(PPh_3)IrH_2]-7-Ph-1,7-C_2B_{10}H_{10}$	–	–8.81	E	2154, 2132	N	31b
Rhodium						
$1-(Ph_3P)_2Rh-2-Ph-1,2-C_2B_{10}H_{12}$	2.1	–	–	–	–	76
Cobalt						
$[Co(1,2-C_2B_{10}H_{10})_2]_2^-$	$1,82(7)^e$	–	–	–	–	75
Metallacarboranes and metallaboranes with metal σ-bonded to cage boron atom						
$3-(Ph_3P)_2IrHCl-1,2-C_2B_{10}H_{11}$	–	–18.08	A	2208, 2198, 2203	N	33
$3-(Ph_3As)_2IrHCl-1,2-C_2B_{10}H_{11}$	–	–20.17	C	2195, 2179, 2201	N	33
$2-(Ph_3P)_2IrHCl-1,7-C_2B_{10}H_{11}$	–	–18.96	C	2214	N	33
$2-(Ph_3As)_2IrHCl-1,7-C_2B_{10}H_{11}$	–	–21.17	C	2203	N	33
$2-(Ph_3P)_2IrHCl-1,12-C_2B_{10}H_{11}$	–	–19.08	C	2200	N	33
$2-(Ph_3As)_2IrHCl-1,12-C_2B_{10}H_{11}$	–	–21.23	C	2190	N	33
$Ir[1-PMe_2-1,2-C_2B_{10}H_{11}]_2[1-PMe_2-1,2-C_2B_{10}H_{10}]HCl$	–	–20.9	C	2220	N	33

$2\text{-}(Ph_3P)_2\,IrHCl\text{-}1\text{-}SB_9H_8$	—	−18.3	E	2199	—	26
$2\text{-}(Ph_3P)_2\,IrHCl\text{-}1\text{-}SB_{11}H_{10}$	—	−17.3	E	2191	—	26
Metallacarboranes and metallaboranes with metal μ-bonded to cage boron atoms						
$nido\text{-}\mu\text{-}(4,5)\text{-}trans\text{-}(Ph_3P)_2\,Pt(H)\text{-}2,3\text{-}C_2B_4H_7$	—	Xray; Pt−H not located	—	2055	—	39
$nido\text{-}\mu\text{-}(4,5)\text{-}trans\text{-}(Ph_3P)_2\,Pt(H)\text{-}2,3\text{-}Me_2\text{-}2,3\text{-}C_2B_4H_5$	—	—	—	2040	—	39
$1\text{-}Cl\text{-}2\text{-}IrClH(CO)[PMe_3]_2\text{-}B_5H_7$	—	−7.79	H	1985	N	38
$2\text{-}IrClH(CO)(PMe_3)_2\text{-}B_5H_8$	—	−6.10	A	1978	N	38

[a] Abbreviations: Ph = C_6H_5; Me = CH_3; Et = C_2H_5.

[b] J value given for proton−metal spin coupling.

[c] Code: A = CD_2Cl_2, B = tetrahydrofuran, C = CH_2Cl_2, D = C_6D_6, E = $CDCl_3$, F = C_6H_6, G = CCl_4, H = toluene-d_8, I = $(C_2D_5)_2O$, J = $CHCl_3$, and K = $1,2\text{-}C_2H_4Cl_2$.

[d] Code: N = Nujol mull, K = KBr pellet, and G = CCl_4.

[e] B−H$_{bridge}$ distance is 1.11 (7) Å.

rapidly catalyze the hydrogenation of 1-hexene in benzene solution at room temperature under 1 atm of hydrogen; other unsaturated substrates including olefinic esters, acids, and vinylcycloalkenes, have also been hydrogenated under mild conditions.[5] Very recently $1,3\text{-}[\mu\text{-}(3\text{-}CH_2{=}CHCH_2CH_2)]\text{-}3\text{-}H\text{-}3\text{-}P(C_6H_5)_3\text{-}3,1,2\text{-}RhC_2B_9H_{10}$, a metallacarborane designed specifically for the purpose, has been reported to catalyze the hydrogenation of alkenes and alkynes extremely rapidly.[13b] In this complex the butenyl side chain, anchored to a framework carbon atom, is also η^2-coordinated to rhodium via the C=C double bond; under conditions employed for hydrogenation of unsaturated hydrocarbons (low H_2 pressure, 25°C), the butenyl group is converted to butyl, leaving a vacant coordination site on rhodium which functions effectively as a site for catalytic hydrogenation of organic substrates.

The same metallacarboranes catalyze the isomerization of 1-hexene, although the process is not highly selective; a mixture of 1- and 2-*cis*-hexene, *trans*-2-hexene, and *cis*- and *trans*-3-hexene is obtained, the major product being *trans*-2-hexene. The versatility of these rhodacarborane catalysts is further demonstrated by their use in the hydrosilylation of ketones[5,14] and alkenes and alkynes,[15] and in promoting the isomerization of unsaturated alcohols to aldehydes and ketones.[16] Deuterium exchange in a variety of boranes, carboranes, and metallacarboranes is also catalyzed by the rhodium complexes and by their iridium counterparts.[5,10] The mechanism proposed[10] for the exchange involves the oxidative addition of B–H bonds to the metallacarborane catalyst via the equilibrium:

$$[(C_6H_5)_3P]_2M(H)C_2B_9H_{11} \rightleftharpoons [(C_6H_5)_3P]M(H)C_2B_9H_{11} + P(C_6H_5)_3$$

If the coordinatively unsaturated mono-triphenylphosphine complex is the active agent, one would expect hydrogen exchange to be inhibited in the presence of excess $P(C_6H_5)_3$, and this is found to be the case;[10] on the other hand, the rate of exchange is tremendously *increased* in the presence of oxygen [evidently due to the formation of $(C_6H_5)_3PO$], thereby invalidating early rate measurements which had been conducted with oxygen-contaminated D_2 gas.[10,17]

In prior independent work Siedle isolated a monoligand complex of the type shown in the previous equilibrium, when orange $[(C_6H_5)_3P]Rh(H)C_2B_9H_{11}$ was obtained in low yield from the treatment of $[(C_6H_5)_3P]_3RhCl$ with $C_2B_9H_{11}^{2-}$ ion.[13a] However, the major product of this reaction was a purple complex which was later shown[18] to be a dibenzene solvate of a dimer $\{[(C_6H_5)_3P]RhC_2B_9H_{11}\}_2$. This dimer was subsequently obtained by Hawthorne *et al.*, via oxidation of the *bis*-(triphenylphosphine) complex, $3,1,2\text{-}[(C_6H_5)_3P]_2Rh(H)C_2B_9H_{11}$, and its molecular structure (Figure 4) was elucidated by X-ray crystallography.[18] The dimer, like the monomer, is an active catalyst for the homogeneous hydrogenation of olefins, and it has been reasonably suggested that the dimer represents an intermediate in the oxidative addition of terminal B–H bonds to the metal via Rh–H–B bridging.[18]

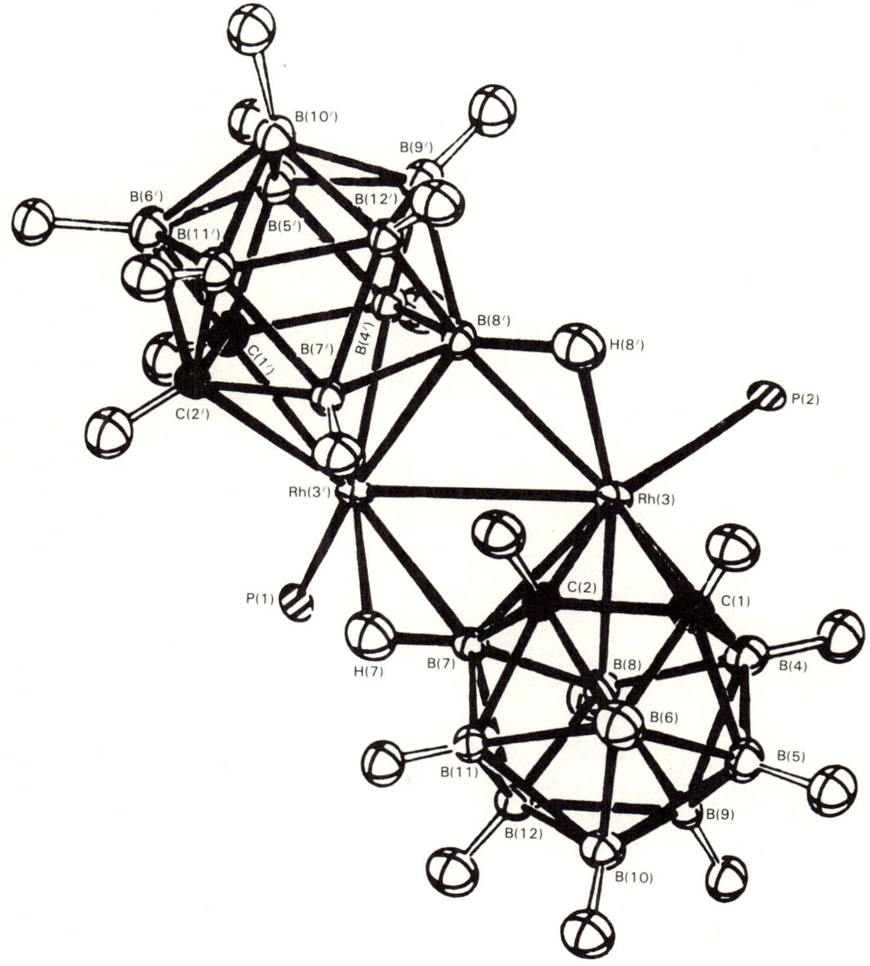

Figure 4. Structure of $[(C_6H_5)_3PRhC_2B_9H_{11}]_2$ (Ref. 18).

In studies of the catalytic effectiveness of 3,1,2- and 2,1,7-$[(C_6H_5)_3P]_2$-$M(H)C_2B_9H_{11}$ (M = Rh or Ir) in promoting deuterium exchange of B—H bonds with D_2, it was shown that (1) all four metallacarboranes are much more active than $[(C_6H_5)_2P]_3RuHCl$; (2) the 2,1,7 isomers are more active than the 3,1,2; and (3) the identity of the metal (Rh or Ir) makes little difference. In contrast to $[(C_6H_5)_3P]_3RuHCl$, the metallacarborane catalysts exhibited very low stereoselectivity.[10]

The metal-bound hydride ligand in 3,1,2-$[(C_6H_5)_3P]_2Rh(H)C_2B_9H_{11}$ is readily displaced by sulfuric acid, generating a complex in which a monodentate HSO_4^- ligand is bound to rhodium by an Rh—O link; the process is reversible, and the hydrosulfate species can be treated with H_2 gas to regenerate the original hydride complex:[19]

$$[(C_6H_5)_3P]_2Rh(H)C_2B_9H_{11} + H_2SO_4 \rightleftharpoons [(C_6H_5)_3P]_2(HSO_4)RhC_2B_9H_{11} + H_2$$

From nmr data it was concluded that the isolable HSO_4^--containing complex exists in solution in equilibrium with a mono(triphenylphosphino) species in which the HSO_4^- ligand is attached to the metal in bidentate fashion:

$$[(C_6H_5)_3P]_2RhC_2B_9H_{11} \rightleftharpoons (C_6H_5)_3P + [(C_6H_5)_3P]RhC_2B_9H_{11}$$

Similarly, $3,1,2\text{-}[(C_6H_5)_3P]_2Rh(H)C_2B_9H_{11}$ reacts with nitric acid to produce $3,1,2\text{-}[(C_6H_5)_3P](NO_3)RhC_2B_9H_{11}$, a species containing a bidentate NO_3^- ligand. This complex reacts with CO and aqueous HCl to generate air-stable $3,1,2\text{-}[(C_6H_5)_3P](CO)ClRhC_2B_9H_{11}$, or with $(C_6H_5)_3P$ to yield $[(C_6H_5)_3P]_2(NO_3)RhC_2B_9H_{11}$, which evidently undergoes partial dissociation in solution in the manner of the HSO_4^- complex described above.[20]

A recent extension of the work on rhodacarborane hydride catalysts has been the attachment of the $3,1,2\text{-}[(C_6H_5)_3P]_2Rh(H)C_2B_9H_{10}$ moiety to a solid polymeric backbone via a boron–carbon bond. Thus, $4\text{-}(polystyrylmethyl)\text{-}3,1,2\text{-}[(C_6H_5)_3P]_2Rh(H)C_2B_9H_{10}$ has been prepared by reaction of the $7,8\text{-}C_2B_9H_{11}^{2-}$ ion with chloromethylated polystyrene polymer to give a carborane-substituted polymer, into which rhodium is inserted by treatment with $[(C_6H_5)_3P]_3RhCl$.[21] The product, in which the rhodium atoms evidently have a chemical environment similar to those in the monomer, is an effective heterogeneous catalyst for the isomerization of 1-octene and for the reduction of sterically hindered alkenes to alkanes with H_2 (1 atm) at $40°C$.[21] An analogous, but different, polymer catalyst has been prepared by a rather different route, in which C-lithio-*o*-carborane $(LiC_2B_{10}H_{11})$ was allowed to react with chloromethylated polystyrene-divinylbenzene copolymer; the attached icosahedral $C_2B_{10}H_{11}$ units were then degraded to $C_2B_9H_{10}^-$ with piperidine and the rhodium inserted with $[(C_6H_5)_3P]_3RhCl$ as before.[22] This heterogeneous catalyst effects hydrogenation of 1-octene at $25°$ under 100 atm pressure of H_2; at 1 atm no hydrogenation was observed although isomerization to internal olefins occurred to a slight extent.

An entirely different route to *closo*-metallacarborane hydrides involves the intramolecular exchange of triphenylphosphine and hydrogen ligands between metal and boron atoms in the complex $3,1,2\text{-}[(C_6H_5)_3P]_2NiC_2B_9H_{11}$ (Figure 5).[23] Under reflux in dry benzene, the nickelacarborane rearranges to the hydride complex shown, the net effect of which is to interchange the original

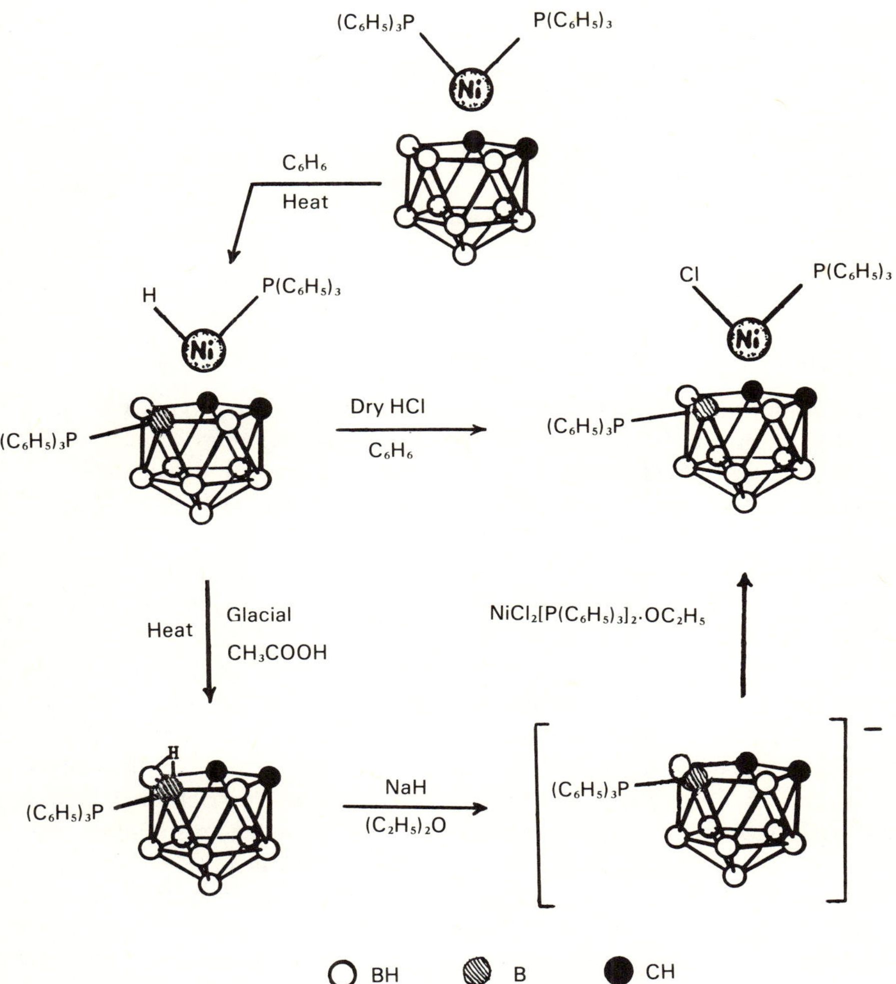

Figure 5. Rearrangement of ligands in 3,1,2-[(C$_6$H$_5$)$_3$P]$_2$NiC$_2$B$_9$H$_{11}$ (Ref. 23).

hydrogen on B(8) with one of the P(C$_6$H$_5$)$_3$ groups on nickel. The same species can also be prepared directly from Tl$_2$C$_2$B$_9$H$_{11}$ by treatment with [(C$_6$H$_5$)$_3$P]$_2$-NiCl$_2$ in refluxing benzene. Other aspects of the chemistry of these complexes are indicated in the figure.[23]

The hydride complex 3,1,2-[(C$_6$H$_5$)$_3$P]Ni(H)C$_2$B$_9$H$_{10}$-8-P(C$_6$H$_5$)$_3$ (Figure 5), a red–orange solid, is the only reported icosahedral metallacarborane of the *terminal* M—H type in which M is a first-row transition metal. Other metalla-boron clusters containing first-row metal–hydrogen bonds are known (see the following), but involve more complex interactions such as M—H—B bridges.

2.1.2. Nonicosahedral Metallacarboranes

Closo-metallacarboranes containing terminal metal–hydrogen bonds mostly involve the isomeric $C_2B_9H_{11}^{2-}$ ion or substituted derivatives. However, insertion of metals into the $C_2B_7H_{12}^-$ and $C_2B_8H_{11}^-$ ions generates, respectively, *closo*-10- and 11-vertex metallacarborane hydrides:[24]

$$NaC_2B_7H_{12} + MCl[P(C_6H_5)_3]_3 \xrightarrow{\text{M=Rh,Ir}} 6,2,3\text{-}[(C_6H_5)_3P]_2M(H)C_2B_7H_9$$

$$\xrightarrow{RuHCl[P(C_6H_5)_3]_3} 6,2,3\text{-}[(C_6H_5)_3P]_2RuC_2B_7H_9$$

$$NaC_2B_8H_{11} + IrCl[P(C_6H_5)_3]_3 \longrightarrow 1,2,4\text{-}[P(C_6H_5)_3]_2Ir(H)C_2B_8H_{10}$$

The complex $6,2,3\text{-}[(C_6H_5)_3P]_2Rh(H)C_2B_7H_9$ catalyzes the hydrogenation of alkenes, as do the $[(C_6H_5)_3P]_2Rh(H)C_2B_9H_{11}$ isomers described previously. However, the ruthenium species $[(C_6H_5)_3P]_2RuC_2B_7H_9$, in contrast to its analog $2,1,7\text{-}[(C_6H_5)_3]_2RuC_2B_9H_{11}$, shows almost no catalytic activity and does not add H_2 at atmospheric pressure; it does, however, react with CO to form $[(C_6H_5)_3P]Ru(CO)_2C_2B_7H_9$.[24]

2.1.3. Metallaboranes

Boron clusters in which metal atoms *in the absence of carbon* are incorporated into the cage framework are called metallaboranes. A large, structurally very diverse family of such species now exists, and many of these compounds contain metal–hydrogen bonds; in most cases, the hydrogens are of the bridging type (see the following), and few examples of metallaboranes with *terminal* M—H bonds exist. The few that have been prepared contain cage sulfur atoms and are described as metallathiaboranes. The first established species of this class was a *nido* 11-vertex complex, $[(C_2H_5)_3P]_2Pt(H)SB_9H_{10}$, prepared by the reaction of $Cs^+SB_9H_{12}^-$ with *trans*-$[(C_2H_5)_3P]_2Pt(H)Cl$, which generated H_2 gas and CsCl together with the metallathiaborane.[25] An X-ray crystal structure analysis revealed the structure shown in Figure 6, in which the platinum and sulfur atoms are located in adjacent positions on the open face of an 11-vertex icosahedral fragment. Remarkably, the hydride ligand was observed and refined to give a Pt—H bond distance of 1.66 Å; the "extra" hydrogen atom was not located, but presumably occupies a B—H—B bridging position on the open face of the cage[25] [a few platinacarboranes with that feature have been prepared (see the following)]. No chemistry of the hydride ligand in this compound has been reported.

A rhodathiaborane containing a hydride ligand terminally bound to the metal has been synthesized by treatment of the $SB_{10}H_{11}^-$ ion, an 11-vertex *nido* (icosahedral fragment) cage, with tris(triphenylphosphine) rhodium(I) chloride in refluxing ethanol:[26]

$$[(C_6H_5)_3P]_3RhCl + SB_{10}H_{11}^- \longrightarrow [(C_6H_5)_3P]_2Rh(H)SB_{10}H_{10}$$

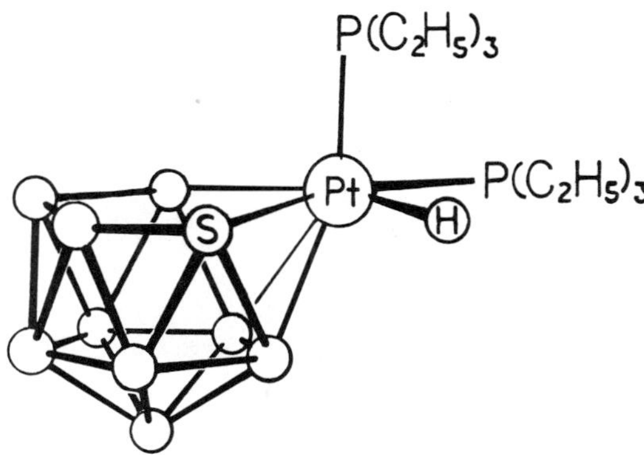

Figure 6. Structure of $[(C_2H_5)_3-P]_2Pt(H)SB_9H_{10}$ (Ref. 25).

The product, a yellow air-stable solid obtained in $>90\%$ yield, is believed to have *closo*-icosahedral geometry with Rh and S atoms occupying adjacent vertices, as in its isoelectronic analog $3,1,2-[(C_6H_5)_3P]_2Rh(H)C_2B_9H_{11}$ described earlier.

Like the latter complex, the rhodathiaborane exhibits catalytic activity; 1-octene is isomerized and hydrogenated at 25°C, but only under high (100 atm) hydrogen pressures. A dissociative mechanism involving loss of a $(C_6H_5)_3P$ ligand is indicated by the observation that catalysis is inhibited by the addition of triphenylphosphine.[26] Meneghelli and Rudolph[27] have put forward the intriguing suggestion that the thiaborane ligand $(SB_{10}H_{10}{}^{2-})$ actively participates in the hydrogenation process by effecting hydroboration of the alkene. By extension of this idea, it is proposed[26] that carborane ligands such as $C_2B_9H_{11}{}^{2-}$ play a similar role in catalysis; thus, catalysis of olefin hydrogenation by $[(C_6H_5)_3P]_2Rh(H)C_2B_9H_{11}$ may also involve a hydroboration step. As yet, however, direct evidence for this hypothesis has not been presented.

No other polyhedral metallaboranes containing terminal M—H bonds have been isolated and characterized, although iridium complexes of the type $[(C_6H_5)_3P](CO)Ir(h_2)B_{10}H_{12}$ (h = H,D) have been postulated (although not detected) in the reversible exchange of terminal B—H hydrogen atoms with deuterium in the treatment of *nido*-$[(C_6H_5)_3P]_2Ir(CO)B_{10}H_{12}$ with D_2 gas.[28] Implied in this suggested mechanism is the reversible transfer of hydrogen (deuterium) between the metal and selected boron atoms.

2.2. Clusters in Which the Metal is σ-Bonded to the Cage

2.2.1. Metal–Carbon σ-Bonds

An extensive class of *o*- and *m*-carborane $(1,2-$ and $1,7-C_2B_{10}H_{12})$ derivatives containing a metal substituent on a cage carbon atom is known, primarily through the work of Bresadola and his co-workers (see Chapter 5 in this volume). In a few of these complexes the metal group has a terminal hydride ligand, as in the following platinum(II) species:[29,30]

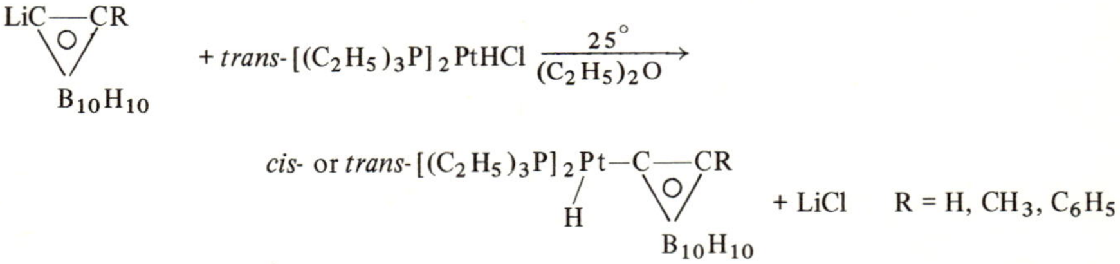

$$\text{LiC—CR (B}_{10}\text{H}_{10}) + \textit{trans-}[(C_2H_5)_3P]_2PtHCl \xrightarrow[\text{(C}_2\text{H}_5)_2\text{O}]{25^\circ}$$

$$\textit{cis-} \text{ or } \textit{trans-}[(C_2H_5)_3P]_2Pt\text{—C—CR (B}_{10}\text{H}_{10}) + LiCl \qquad R = H, CH_3, C_6H_5$$

Analogous complexes have been prepared from $1,7\text{-}C_2B_{10}H_{12}$ and from other *cis-* and *trans-*L_2PtH reagents, where L is $(C_2H_5)_3P$, $(C_6H_5)_3P$, $(C_6H_5)_2(CH_3)P$, or $(C_6H_5)(CH_3)_2P$.[29,30] In all cases, the compounds are obtained in 60–80% yields as crystalline air-stable solids. The *cis* isomers are somewhat unusual, since prior to their synthesis, known monohydrido platinum(II) derivatives containing two monodentate neutral ligands were exclusively *trans.*. It will be noted that the proton nmr chemical shifts of the *trans* complexes are substantially to higher field of the *cis* species (Table 1).

A series of iridium-carborane derivatives related to Vaska's complex, *trans-*$[(C_6H_5)_3P]_2Ir(CO)Cl$, has been synthesized and shown to exhibit some highly unusual stereochemical behavior. Treatment of a benzene suspension of Vaska's complex with excess $Li^+(1,2\text{-}RC_2B_{10}H_{10})^-$ or $Li^+(1,7\text{-}RC_2B_{10}H_{10})^-$, where R is H, CH_3, or C_6H_5, generates complexes of the type $1\text{-}\{[(C_6H_5)_3P]_2Ir(CO)\}\text{-}2\text{-}R\text{-}C_2B_{10}H_{10}$. These products are stable under an inert atmosphere in solution or in the solid state, but are highly reactive toward oxygen. When the carborane substrate is $1\text{-}C_6H_5C_2B_{10}H_{10}^-$, the initial iridium-substituted product evidently undergoes, on standing, intramolecular oxidative addition of a phenyl hydrogen atom to the metal:[31]

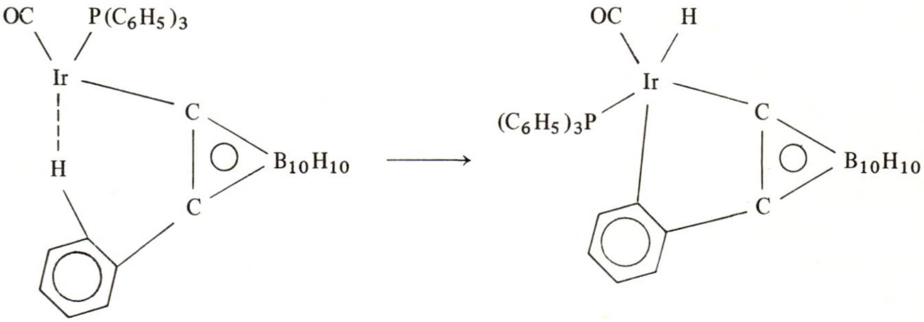

orange   white

The white final product was characterized primarily from infrared spectral data since its low solubility precluded nmr measurements.

The species derived from 1-methyl-*o*-carborane also exhibited unusual properties, including an anomalously low ν_{CO} stretching frequency (1950 cm^{-1}), very low solubility in organic solvents, and exceptionally high reactivity toward addition of H_2 in the solid state. A direct interaction between the methyl group and the iridium has been suggested:[31]

Complexes of the above type [*trans*-$(R_2 R'P)_2$ Ir(CO)(σ-carb)] where R = C_6H_5, R$'$ = C_6H_5 or CH_3, and σ-carb is 1,2- or 1,7-$RC_2B_{10}H_{11}$, undergo rapid oxidative addition with H_2 or D_2, giving dihydrides or dideuterides of the formula $(R_2 R'P)_2$ Irh$_2$(CO)(σ-carb) where h = H or D. The dihydrido products are stable with respect to thermal dissociation of carborane or H_2, but are sensitive to light. They are isoelectronic and isostructural with Vaska's complex, but differ from it in two respects: the addition of H_2 is irreversible, and it is solvent dependent, giving rise to three distinguishable *cis* isomers:

carb = $RC_2B_{10}H_{10}$; L = $(C_6H_5)_3$P, $(C_6H_5)_2(CH_3)$P

Species of type A are unstable in solution and slowly convert to B (in nonpolar solvents) or C (in polar solvents). The isomer interconversions have been shown to be kinetically controlled and apparently are strongly influenced by steric crowding effects of the bulky phosphine and carborane ligands.[31]

2.2.2. Metal–Boron σ-Bonds

Attachment of σ-ligand groups at boron positions on polyhedral boranes is not as straightforward as C-substitution, since boron-bound terminal protons are far less acidic than carboranyl C$-$H protons and, hence, are more difficult to replace with other groups. However, Hawthorne and co-workers have achieved B-σ-metallation via oxidative addition of terminal B$-$H groups in carboranes to low-valent metal reagents. Thus, iridium(I) complexes have been found to undergo both intra- and intermolecular oxidative addition as shown:[33-35]

$$[Ir(C_8H_{14})_2Cl]_2 + 6 \quad \underset{B_{10}H_{10}}{H-C{-}C-P\begin{smallmatrix}CH_3\\\\CH_3\end{smallmatrix}} \quad \xrightarrow[C_6H_{12}]{\Delta}$$

$$2[(CH_3)_2PC_2B_{10}H_{11}]_3IrCl \longrightarrow 2$$

$$L = -C{-}C-P(CH_3)_2$$
$$B_{10}H_{10}$$

$$\tfrac{1}{2}[Ir(C_8H_{14})_2Cl]_2 + 2(C_6H_5)_3P + \underset{B_{10}H_{10}}{HC{-}CH} \longrightarrow$$

These reactions are patterned after the intramolecular orthometallations of the complex $[(C_6H_5)_3P]_3IrCl$, and experiments with B-deuterated carboranes have shown that the process involves oxidative addition at a boron vertex (most probably 3,6) of $1,2\text{-}C_2B_{10}H_{12}$. Direct reaction of $[(C_6H_5)_3P]_3IrCl$ with $1,2\text{-}C_2B_{10}H_{12}$ also generates the intermolecular addition product (Figure 7) in low yield; however, the major product of this reaction is $[(C_6H_5)_3P]_2Ir[(o\text{-}C_6H_4)P(C_6H_5)_2]HCl$, a consequence of intramolecular ortho-metallation in $(C_6H_5)_3PIrCl$.[36] The corresponding treatment of 1,7- and $1,12\text{-}C_2B_{10}H_{12}$ is reported to give no identifiable B-o-carboranyl complexes.[33]

Analogs of the phosphino complexes containing $(C_6H_5)_3As$ together with a 1,7- or $1,12\text{-}C_2B_{10}H_{11}$ group on iridium have also been characterized, and a few reactions of the phosphino compounds have been examined; thus, treatment with CO at room temperature generates $trans\text{-}[(C_6H_5)_3P]_2(CO)IrCl$ and free $1,2\text{-}C_2B_{10}H_{12}$. The reaction with $(C_6H_5)_3P$ proceeds in a similar manner.[33]

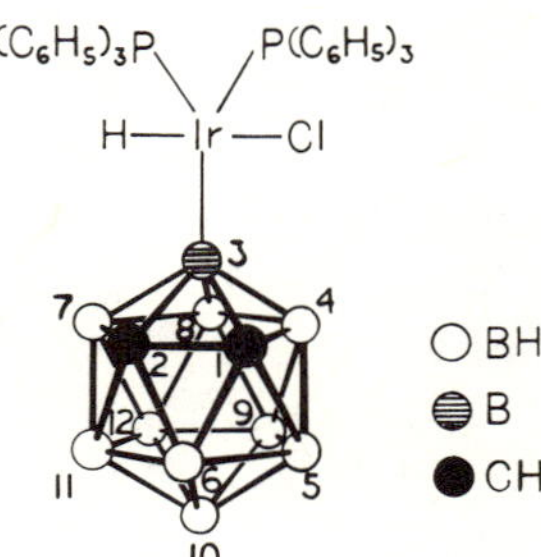

Figure 7.   Structure of 3-[(C₆H₅)₃P]₂Ir(H)(Cl)-1,2-
C₂B₁₀H₁₁ (Ref. 33).

The complex 2-{[(C₆H₅)₃P]₂IrHCl}-1,7-C₂B₁₀H₁₁ (identical to the species in Figure 7 except that the cage carbon atoms occupy the 1,7 vertices) has been shown to be very active in catalyzing exchange of deuterium between 1,7-C₂B₁₀H₁₂ and D₂ gas; the replacement of B—H with B—D is not highly selective, but occurs at all sites within 24 hr at 65°C.[10] The corresponding arsenic compound is slightly less active. In both cases, the actual catalytic agents are believed to be the L₂IrCl species {L = [(C₆H₅)₃P or [(C₆H₅)₃As] } or their hydrogen adducts, rather than the B-σ-carboranyl iridium complexes per se.[10]

A related ruthenium species [(C₆H₅)₃P][1-P(CH₃)₂-1,2-C₂B₁₀H₁₁]₂-RuHCl {formed by replacement of two triphenylphosphine ligands in trigonal bipyramidal [(C₆H₅)₃P]₃RuHCl with two dimethylphosphino-carboranyl units}, effectively catalyzes deuterium exchange of 1-(CH₃)₂P-1,2-C₂B₁₀H₁₁ with D₂.[10] A possible mechanism involves *ortho*-metallated intermediates as was previously proposed[37] for *ortho*-deuteration of triphenylphosphine by [(C₆H₅)₃P]₃RuHCl, but such a process would effect deuteration only at the four positions *ortho* to the carbons; since up to six deuterium atoms are incorporated in the present case, this mechanism cannot be the only one operative in this system.

Thiaborane derivatives containing σ-IrL₂HCl groups [L = P(C₆H₅)₃ or As(C₆H₅)₃] have been prepared by direct reactions of the L₂IrCl reagent with *closo*-1-SB₉H₉ and *closo*-1-SB₁₁H₁₁, which generate respectively 2-(L₂IrHCl)-1-SB₉H₈ (Figure 8) and its higher homologue, 2-(L₂IrHCl)-1-SB₁₁H₁₀.[26] In both cases, infrared and proton nmr data suggest that the metal retains tri-

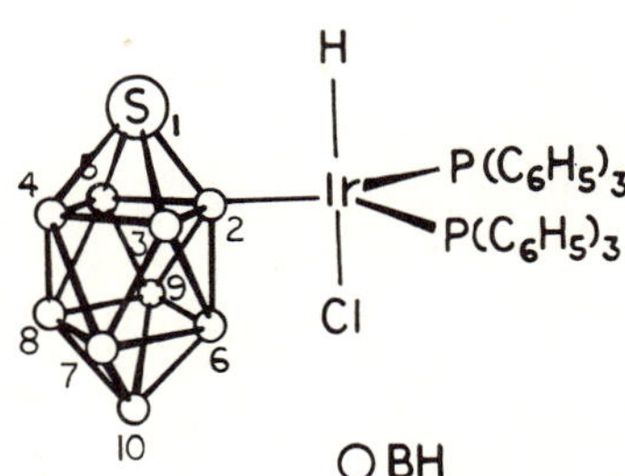

Figure 8. Structure of 2-Ir(H)(Cl)[P(C₆H₅)₃]₂-
1-SB₉H₈ (Ref. 26).

gonal bipyramidal coordination with the H and Cl ligands occupying *trans* positions.

Only one example currently exists of a small borane σ-bonded to a metal containing a terminal hydride group. Reaction of *trans*-IrCl(CO)[P(CH$_3$)$_3$]$_2$ with 1- or 2-XB$_5$H$_8$ (X = Br or Cl) results in metal insertion into a basal B—H bond to give 1-X-2-{[IrHX(CO)[P(CH$_3$)$_3$]$_2$}B$_5$H$_7$ which subsequently rearranges, interchanging the boron-bound halogen with the metal hydride ligand to give 2-{IrX$_2$(CO)[P(CH$_3$)$_3$]$_2$}B$_5$H$_8$ (Figure 9).[38] The conversion is more rapid in the bromo system, but in both cases it occurs via oxidative addition of a B—H moiety to the pentavalent iridium complex. The fact that the product is 2-substituted (on the pentaborane cage) in all cases is in accord with the theoretical prediction that nucleophilic substitution should occur at a basal position on the framework.[38]

2.3. Clusters with Metal-to-Cage Bridge Bonds

One example is known of a boron cluster to which a metal, bound to a hydride ligand, is attached via a three-center, two-electron B—M—B bridge. The reaction of the pentagonal pyramidal *nido*-carborane 2,3-(CH$_3$)$_2$C$_2$B$_4$H$_6$ with Pt[P(C$_2$H$_5$)$_3$]$_2$ produced pale yellow μ(4,5)-[(C$_2$H$_5$)$_3$P]$_2$PtH-(CH$_3$)$_2$C$_2$B$_4$H$_5$, whose structure was established by X-ray crystallography.[39] The hydrogen atoms were not located, but from the *trans* arrangement of the phosphine ligands, the hydride ligand was assumed to lie between them with the Pt—H link bisecting the P—Pt—P angle (Figure 10). The corresponding species μ(4,5)-[(C$_2$H$_5$)$_3$P]$_2$PtH-C$_2$B$_4$H$_7$ was also prepared. Proton nmr data were not given for either compound, so that direct evidence of the Pt—H interaction has not been presented.

On pyrolysis at 100°C both species converted to 7-vertex PtC$_2$B$_4$ *closo* cages with loss of the Pt—H and B—H—B hydrogens as molecular H$_2$.[39]

3. CLUSTERS CONTAINING M—H—M or M—H—B BRIDGES

It has been known from the earliest X-ray investigations of boron hydride structures that hydrogen can adopt bridging, as well as terminal, locations on borane frameworks. This is a principal distinguishing feature between the boranes and the hydrocarbons, and is one means by which the boron hydrides adapt to their so-called "electron deficient" condition in which there are insufficient electrons to bind the molecule together entirely via two-center, two-electron bonds. In a B—H—B bridge, two electrons bind the three atoms together in stable fashion, and such bridges are a standard feature in many open-faced (e.g., *nido* and *arachno*) boranes and carboranes.

The nature of the boron-hydrogen bridge has been extensively investigated both experimentally and theoretically. In general, it is treated as involving an overlap of a 1s orbital on H with hybrid (e.g., sp^2 or sp^3) orbitals on two boron

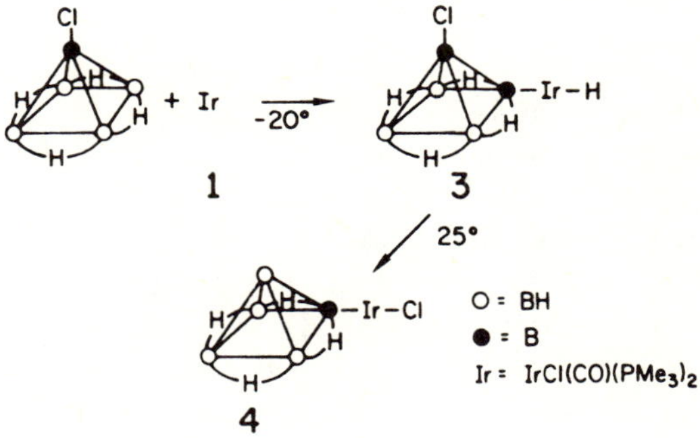

Figure 9. Synthesis of 2–IrCl₂–(CO)[P(CH₃)₃]₂–B₅H₈ (Ref. 38).

atoms, generating three molecular orbitals of which only one is bonding. Normally, B—H—B interactions are weaker than B—H$_\text{terminal}$ bonds, and are acidic toward hydride ion and other strong bases. Thus, *nido* species such as B₅H₉, B₁₀H₁₄, B₆H₁₀, and 2,3-C₂B₄H₈ react easily with sodium hydride in tetrahydrofuran to release H₂ and generate the corresponding anion,[40] e.g.,

$$B_5H_9 + H^- \longrightarrow B_5H_8^- + H_2$$

In each instance, the proton extracted is exclusively from a bridging location; the terminal hydrogen atoms exhibit no significant acidity, as shown by labeling experiments.

A straightforward extension of the B—H—B bonding concept is the metal-hydrogen-boron (M—H—B) bond, and bridges of this type have been known for many years in the form of metal chelates of borane anions such as the BH₄⁻ and B₃H₈⁻ ions (see Chapter 3 of this volume). However, M—H—B three-center bonds in *polyhedral metallaboron clusters* are of relatively recent vintage, the first established examples appearing only within the last 10 years. It is clear now that M—H—B bonding is as ubiquitous a feature as B—H—B; for example, each of the four species mentioned above (B₅H₉, B₁₀H₁₄, B₆H₁₀, and C₂B₄H₈) has metallaborane or metallacarborane structural analogs containing well-defined M—H—B bridges, and there are many other examples. In the present discussion we shall restrict ourselves to clusters containing direct metal-boron links, excluding the chelate-type species in which the metal is not actually part of a boron cage.

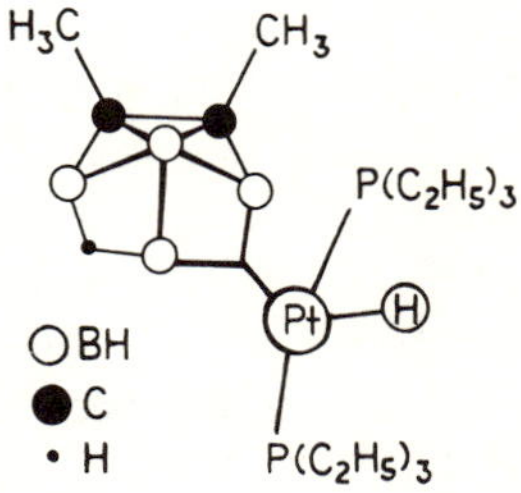

Figure 10. Structure of μ–Pt(H)[P(C₂H₅)₃]–2,3–(CH₃)₂–C₂B₄H₅ (Ref. 39).

3.1. NIDO-METALLABORANES

Few broad generalizations can be offered on the synthesis or structures of metal–boron cages, so huge and diverse is this area, but it can be noted that nearly all metallaboranes to this point have been obtained by insertion of metal groups into boron hydride substrates. When the inserted moiety is a two-electron donor group such as $Co(\eta^5\text{-}C_5H_5)$ or $Fe(CO)_3$ (formally analogous to BH),[4] the metallaborane product is often a direct structural counterpart of a neutral boron hydride such as B_5H_9 or $B_{10}H_{14}$ with the metal(s) occupying vertices on the open face; in such cases there are usually nonterminal hydrogen atoms which tend to adopt B—H—M bridging roles. As will be seen, however, exceptions to all of these observations are frequent.

3.1.1. Pyramidal Metallaborane Clusters

The $B_5H_8^-$ ion, mentioned previously, is a fertile source of metallaboranes in reactions with transition metal reagents, and many of the products contain M—H—B bridges. Thus, treatment of $Na^+B_5H_8^-$ with $CoCl_2$ and $Na^+C_5H_5^-$ in cold THF generates $2\text{-}(\eta^5\text{-}C_5H_5)CoB_4H_8$ (Figure 11), a red air-stable solid whose structure was deduced from nmr data[41,42] and later proved by X-ray crystallography.[43] The molecule is analogous to B_5H_9 both structurally and (with respect to formal assignment[4] of framework bonding electrons) electronically, and can be regarded as derived from B_5H_9 by replacement of a basal BH unit with $Co(\eta^5\text{-}C_5H_5)$. As shown, the cluster contains two B—H—B and two B—H—Co bridges, of which the latter contain the less strongly bound and more reactive protons. Boron–boron and boron–terminal hydrogen bond lengths in this cobaltaborane and B_5H_9 are closely similar, but distances involving the hydrogen bridges are significantly different. In B_5H_9 the uniform $B-H_{bridge}$ length[44] is 1.352(4) Å, whereas in $2\text{-}(\eta^5\text{-}C_5H_5)CoB_4H_8$ the B—H—B hydrogen is 1.30(6) Å from B(3), but only 1.09(6) Å from B(4); the B—H—Co hydrogen, in contrast, is symmetrically placed at 1.43(5) Å from both boron and cobalt.[43] These distances imply that the B—H—Co hydrogen is weakly bonded relative to the B—H—B, and the known chemistry of the molecule bears this out; like B_5H_9, $2\text{-}(\eta^5\text{-}C_5H_5)CoB_4H_8$ reacts vigorously with sodium hydride in THF, liberating H_2 and forming the $(C_5H_5)CoB_4H_7^-$ anion.[45] The proton removed is exclusively at the B—H—Co position, as shown by deuterium-labeling and nmr experiments.

Other aspects of the chemistry of $2\text{-}(\eta^5\text{-}C_5H_5)CoB_4H_8$ also resemble that of B_5H_9: reaction with alkynes (RC_2R') effects insertion of the acetylenic moiety into the cage, and treatment with metal reagents leads to metal incorporation.[45] For example, $2\text{-}(\eta^5\text{-}C_5H_5)CoB_4H_7^-$ reacts with $CoCl_2$ and $Na^+C_5H_5^-$ in THF to produce dicobalt, tricobalt, and tetracobalt metallaboron clusters. However, the role of the B—H—Co bridging hydrogens in these processes has not been determined. On heating at 200°C the 2-isomer rearranges to $1\text{-}(\eta^5\text{-}C_5H_5)CoB_4H_8$, in which the cobalt atom occupies the apex of the

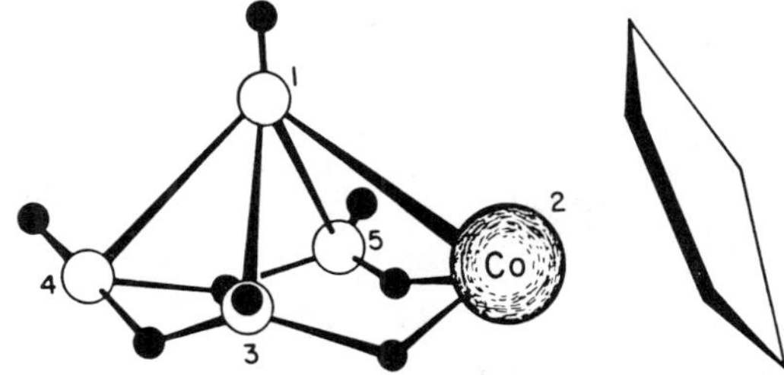

Figure 11.   Structure of 2-(η^5-C$_5$H$_5$)CoB$_4$H$_8$ (Ref. 42).

square pyramid; this species, the thermodynamically favored geometry, has four B—H—B but no B—H—Co bridges.[41,42] The formal B$_4$H$_8$$^{2-}$ ligand is isoelectronic with cyclobutadienide, C$_4$H$_4$$^{2-}$; hence, the complex is analogous to the known sandwich species (η^5-C$_5$H$_5$)Co(η^4-C$_4$H$_4$).

A related ferraborane is 1-(CO)$_3$FeB$_4$H$_8$,[46] which similarly has no metal-hydrogen bonds in that the iron adopts the apex position (the corresponding 2-isomer, which would have two Fe—H—B bridges, is not yet known). However, formal replacement of one of the basal BH groups with Fe(CO)$_3$ generates 1,2-(CO)$_6$Fe$_2$B$_3$H$_7$, a dimetallic B$_5$H$_9$ analog which does have two B—H—Fe groups (Figure 12). This molecule has been prepared by reaction of B$_5$H$_9$ with Fe(CO)$_5$ in the presence of LiAlH$_4$, and the structure shown has been established crystallographically.[47] Bond distances involving the bridging hydrogens can be compared to those cited above for 2-(η^5-C$_5$H$_5$)CoB$_4$H$_8$: B—H$_{bridge}$, 1.19(5) and 1.33(4) Å, and Fe—H$_{bridge}$, 1.57(4) and 1.58(5) Å; the Fe—Fe distance is 2.559(2) Å. No chemistry has been reported for this cluster, but nmr evidence suggests that the two B—H—B bridging protons participate in a rapid intramolecular exchange with the unique terminal proton, H(4).[47]

The reaction of B$_5$H$_9$, Fe(CO)$_5$, and LiAlH$_4$ when conducted under somewhat different conditions from those of the (CO)$_6$Fe$_2$B$_3$H$_7$ synthesis yields a different diiron cluster, (CO)$_6$Fe$_2$B$_2$H$_6$ (Figure 13).[48] The structure depicted was proposed from spectroscopic data and consists of a distorted Fe$_2$B$_2$ tetrahedron with the four equivalent Fe—B edges bridged by hydrogen atoms; the ^{1}H nmr resonance at δ-10.3 is strong evidence for Fe—H—B interactions. Like the majority of 4-vertex metallo clusters, this molecule has 12 skeletal electrons[4,49] and is isoelectronic with (CO)$_6$Co$_2$C$_2$H$_2$, (CO)$_{12}$Ru$_4$H$_4$, and many

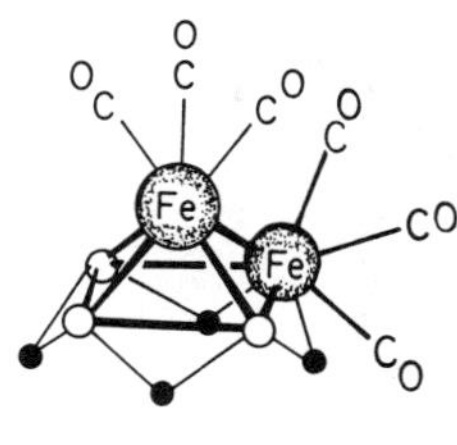

Figure 12.   Structure of (CO)$_6$Fe$_2$B$_3$H$_7$ (Ref. 47).

○ BH

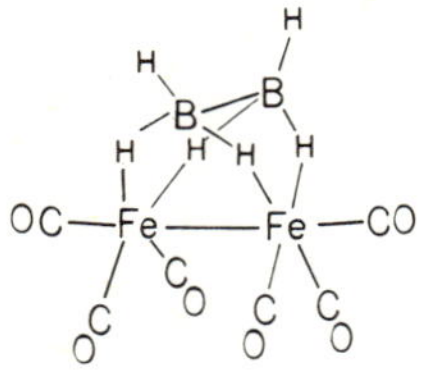

Figure 13. Proposed structure of $(CO)_6Fe_2B_2H_6$ (Ref. 48).

other species. It has a close structural relationship to the cluster $(CO)_{10}Mn_3B_2H_7$ (Figure 14), which has not only four B–H–Mn groups, but an Mn–H–Mn bridge as well.[50] This latter molecule can be regarded as a 4-vertex Mn_2B_2 cluster to which an "extra" $Mn(CO)_4$ unit is attached via two Mn–H–B bridge bonds; viewed in this sense the basic cluster is isoelectronic with $(CO)_6Fe_2B_2H_6$. The manganese species has been fully characterized crystallographically with all hydrogens located,[50] but the full structural details (including B–H and Mn–H distances) have not been published. The molecule is also related to complexes of BH_4^- which feature bidentate or tridentate coordination to the metal via B–H–M links; however, BH_4^- complexes lack B–M and B–B bonds and, hence, for purposes of this review are not regarded as clusters. Metal complexes of $B_3H_8^-$ do contain B–B bonds and are treated elsewhere in this volume.

Still other novel ferraboranes containing Fe–H–B bridges have been obtained from the reaction of B_5H_9 and $Fe(CO)_5$ at 230–260°C in a glass hot-cold reactor. The species $2\text{-}(CO)_5FeB_5H_9$ was isolated as a deep red air-sensitive liquid and assigned the structure in Figure 15.[51,52] The presence of an Fe–H–B proton is clearly indicated by a high-field (δ-16.8 ppm) ^{1}H nmr resonance, which is split into a doublet by H–H spin coupling between the bridging proton and neighboring terminal hydrogens. From variable temperature nmr studies between −80 and +25°C, it appears that the Fe–H–B proton is fluxional at room

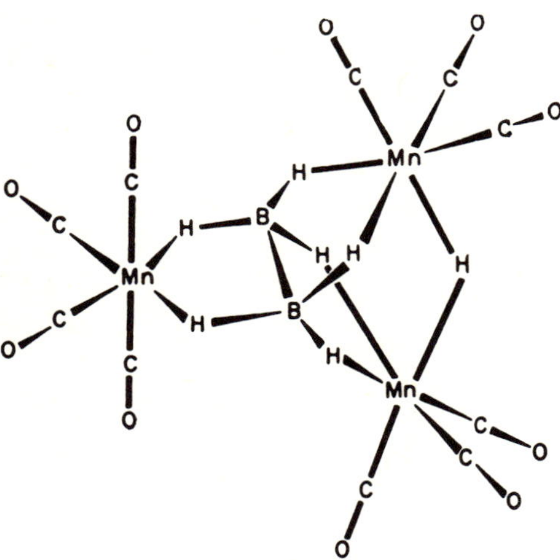

Figure 14. Structure of $(CO)_{10}Mn_3$-B_2H_7 (Ref. 50).

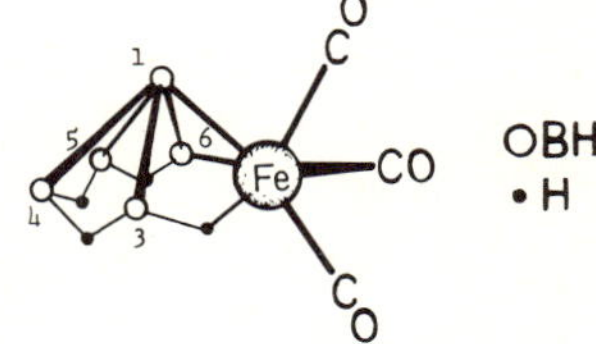

Figure 15.   Proposed structure of $2\text{-}(CO)_3FeB_5H_9$
(Ref. 51).

temperature, alternating between the two available Fe—B sites; at $-80°$, the proton is locked into a static asymmetric structure as shown in Figure 15. The movement of the bridging proton is remarkable in that it evidently remains bound to iron at all times, swinging back and forth between the Fe—B(3) and Fe—B(6) locations; there is no evidence of involvement of other hydrogens in the molecule.[51-52]

The neutral species $2\text{-}(CO)_3FeB_5H_9$ is deprotonated on treatment with KH at $-78°$ in $(CH_3)_2O$, generating the $(CO)_3FeB_5H_8^-$ ion, and curiously it is the B(3)—H—B(4) rather than the B—H—Fe proton that is removed.[52] X-ray structure determination on the anion confirmed the skeletal structure indicated for the neutral species (Figure 15)[52] although hydrogen atoms were not located.

The $2\text{-}(CO)_3FeB_5H_8^-$ ion contains a vacant (i.e., nonprotonated) B—B site on its base, and as is typical[7,53] of pyramidal boron cluster anions of this class it is susceptible to attack by cationic metal groups that insert to form a B—M—B bridged neutral species. Accordingly, *tris*-(triphenylphosphine)copper was introduced,[54] although not in the location of the B—B bond in the $(CO)_3FeB_5H_8^-$ substrate:

$$K^+(CO)_3FeB_5H_8^- + [(C_6H_5)_3P]_3CuCl \xrightarrow[-78°]{(CH_3)_2O}$$

$$(CO)_3FeB_5H_8\text{-}\mu\text{-}Cu[P(C_6H_5)_3]_2 + KCl + P(C_6H_5)_3$$

The copper-bridged product, an air-stable yellow solid, has been structurally characterized by X-ray analysis[54] and has a most unusual geometry (Figure 16) in which the copper atom bonds significantly with the terminal hydrogen on a basal boron atom, forming a Cu—H—B bridge. A consequence of this interaction is that the copper atom is nearly coplanar with the base of the FeB_5 pentagonal pyramid, in sharp contrast to earlier structurally characterized *nido*-boron clusters containing B—M—B bridging groups, in which the metal invariably lies well below the plane of the open face of the cage. In those species for which X-ray data are available, which include $\mu\text{-}[(C_6H_5)_3P]_2Cu\text{-}B_5H_8$,[53g] $\mu\text{-}(CO)_4\text{-}Fe\text{-}B_7H_{12}^-$[53h] $\mu\text{-}(CH_3)_3Si\text{-}B_5H_7Br$,[53i] $\mu\text{-}[(\eta^5\text{-}C_5H_5)Be]\text{-}B_5H_8$,[53j] *trans*-$\mu\text{-}PtCl_2\text{-}(B_6H_{10})_2$,[53k] $\mu\text{-}HgCl\text{-}1,2,3\text{-}[\eta^5\text{-}C_5(CH_3)_5Co](CH_3)_2C_2B_3H_4$,[53m] and $\mu\text{-}Hg[(\eta^5\text{-}C_5H_5)Co(CH_3)_2C_2B_3H_4]_2$,[53m] there is no indication of metal-*terminal* hydrogen bonding (although in the last-mentioned complex there is evidence of weak Hg-*bridge* hydrogen interaction[53m]). Thus, in none of these latter compounds is there any inducement for the metal (or metalloid) atom to

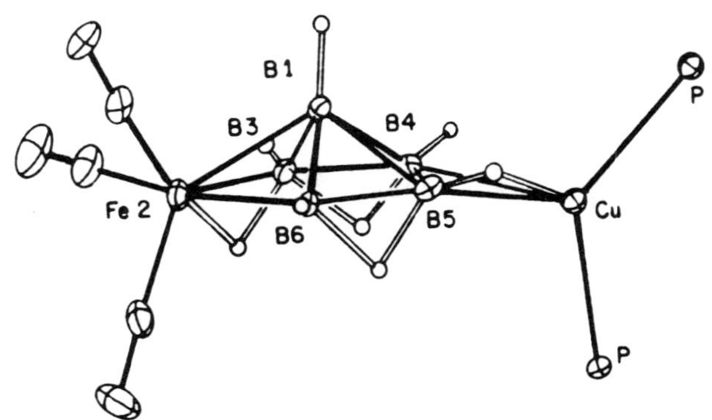

Figure 16. Structure of $\mu\text{-}[(C_6H_5)_3P]_2Cu\text{-}(CO)_3FeB_5H_8$ (Ref. 54).

be drawn into near-coplanarity with the pyramidal base. Particularly striking is the difference between $(CO)_3FeB_5H_8\text{-}\mu\text{-}Cu[P(C_6H_5)_3]_2$ and $\mu\text{-}[(C_6H_5)_3P]_2\text{-}Cu\text{-}B_5H_8,$[53g,55] which may reflect the fact that in the former complex the pentagonal pyramidal shape allows valence bond angles for the basal hydrogens that favor interaction with the bridging metal; in the latter species, the square pyramidal shape probably discourages $Cu\text{--}H\text{--}B$ bridge formation.[54]

A ferraborane isoelectronic with $2\text{-}(CO)_3FeB_5H_8$ is the violet air-sensitive compound $2\text{-}(\eta^5\text{-}C_5H_5)FeB_5H_{10}$, which was obtained from $FeCl_2$, NaC_5H_5, and $Na^+B_5H_8{}^-$ in THF and characterized from ^1H and ^{11}B nmr spectra as a pentagonal pyramid with iron occupying a basal position [Figure 17(A)] .[56,57] The presence of equivalent $Fe\text{--}H\text{--}B$ bridging protons is clearly revealed by the high-field ($\delta\text{-}16$) signal in the ^1H nmr spectrum; there is coupling between these protons and ^{11}B nuclei, since the $Fe\text{--}H\text{--}B$ resonance appears as an approximate 1:3:3:1 quartet ($J = 70$ Hz) which collapses to a sharp singlet on ^{11}B decoupling. The chemistry of this complex has not been extensively studied, but it rearranges at $175°C$ to the 1-isomer [Figure 17(B)] which is electronically analogous to ferrocene $[(\eta^5\text{-}C_5H_5)_2Fe]$ and contains a cyclic $B_5H_{10}{}^-$ ligand, a counterpart of $C_5H_5{}^-$.

From a skeletal electron-counting viewpoint, 1- and $2\text{-}(\eta^5\text{-}C_5H_5)FeB_5H_{10}$ are isoelectronic and isostructural with $2\text{-}(CO)_3FeB_5H_9$ (mentioned earlier), $1\text{-}(\eta^5\text{-}C_5H_5)CoB_5H_9,$[58] and many pentagonal pyramidal carboranes and metallacarboranes (e.g., $2,3\text{-}C_2B_4H_8$ and $1,2,3\text{-}(\eta^5\text{-}C_5H_5)CoC_2B_3H_7$) as well

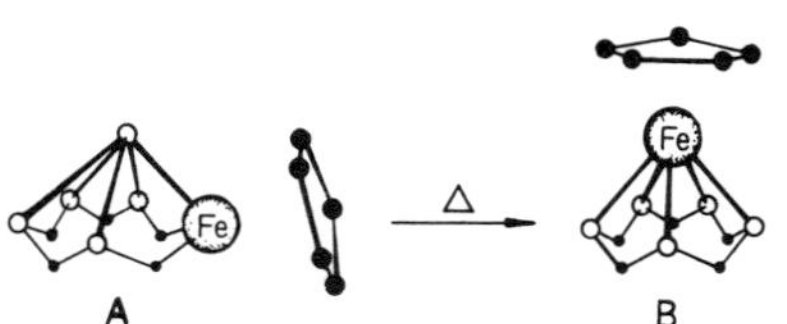

Figure 17. Rearrangement of (A) $2\text{-}(\eta^5\text{-}C_5H_5)\text{-}FeB_5H_{10}$ to (B) $1\text{-}(\eta^5\text{-}C_5H_5)FeB_5H_{10}$ (Ref. 57).

as the parent borane B_6H_{10}. A recent addition to this series is $2\text{-}(CO)_3MnB_5H_{10}$ (Figure 18) a yellow, moderately air-stable liquid which forms on reaction of B_5H_9 with $HMn(CO)_5$ (or with H_2 and $Mn_2(CO)_{10}$) at elevated temperatures.[59] From ^1H and ^{11}B nmr observations it has been deduced that the manganese occupies a basal position and participates in two equivalent Mn$-$H$-$B bridging interactions. Treatment with sodium hydride in diethyl ether abstracts a proton and generates the $(CO)_3MnB_5H_9^-$ anion which retains both Mn$-$H$-$B bridges; the proton removed is that bridging B(4) and B(5) (furthest from manganese) in the neutral complex. In this respect the molecule differs from both $2\text{-}(CO)_3\text{-}FeB_5H_9$, which on attack of H$^-$ loses its B(3)$-$H$-$B(4) proton,[52] and from $2\text{-}(\eta^5\text{-}C_5H_5)CoB_4H_8$, which selectively loses a proton bridging Co and B.[45] The acidity of other types of metal–hydrogen groups will be discussed in later sections.

3.1.2. Large Metallaborane Clusters

A number of metallaboranes are known that are analogous to $B_{10}H_{14}$ and have a similar basket-like icosahedral fragment geometry incorporating one or two metal atoms. In many of these complexes the metal(s) occupy positions on the open face and form M$-$H$-$B bridges with adjacent boron atoms. Among these are several species obtained by reaction of the $B_9H_{14}^-$ ion with $Mn(CO)_5Br$ or $Re(CO)_5Br$:[60-62]

$$B_9H_{14}^- + Mn(CO)_5Br \xrightarrow{\ R_2O\ } 6\text{-}(CO)_3MnB_9H_{13}^- + 2CO + HBr$$

Neutral THF derivatives, 2- and $5\text{-}(CH_2)_4O\text{-}6(CO)_3MnB_9H_{12}$, have also been prepared and crystallographically characterized, and on treatment of the 2-THF derivative with triethylamine a triethylaminebutoxy-substituted complex was isolated.[63] All of these derivatives have the cage structure shown in Figure 19, having manganese in the 6 position and participating in two Mn$-$H$-$B bridges. The cage frameworks in the three structurally characterized species (Table 2) all closely resemble $B_{10}H_{14}$.

Several cobaltaboranes analogous to $(CO)_3MnB_9H_{13}^-$ have been isolated and structurally characterized by X-ray studies, and all of these have Co$-$H$-$B interactions. The species $5\text{-}(\eta^5\text{-}C_5H_5)CoB_9H_{13}$ [Figure 20(A)][64] differs from the manganese complexes in metal atom location, and has only one Co$-$H$-$B bridge; however, the borane skeleton is closely similar to that of $B_{10}H_{14}$. This compound was isolated together with numerous other cobalta-

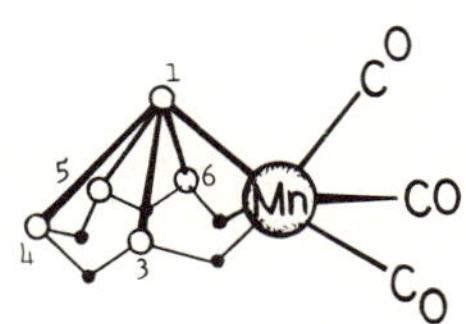

Figure 18. Proposed structure of $2\text{-}(CO)_3MnB_5H_{10}$ (Ref. 59).

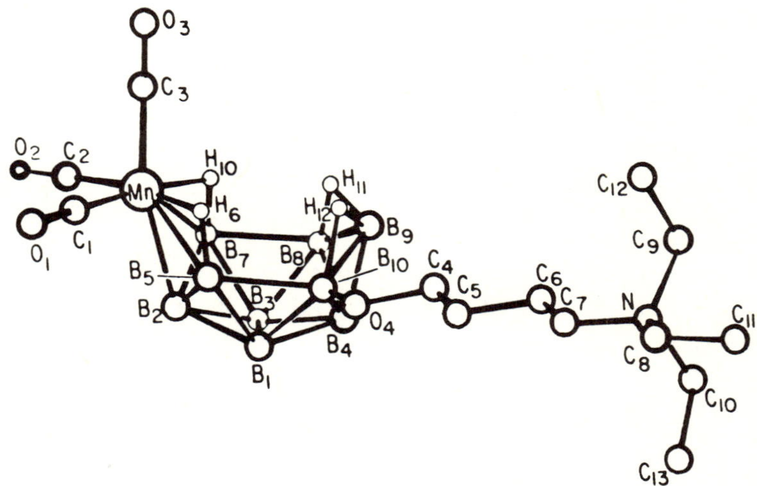

Figure 19. Structure of $8\text{-}(C_2H_5)_3N(CH_2)_4O\text{-}6\text{-}(CO)_3MnB_9H_{12}$ (Ref. 63).

boranes from the reaction of $B_5H_8^-$ ion with $CoCl_2$ and NaC_5H_5 in THF;[41,42] when the reaction was conducted with $C_5(CH_3)_5^-$ in place of $C_5H_5^-$, an apparently analogous complex $[\eta^5\text{-}C_5(CH_3)_5]CoB_9H_{13}$, was produced,[65a] but with the cobalt in the 6 rather than the 5 position. This species [Figure 20(B)], thus, matches the $6\text{-}(CO)_3MnB_9H_{13}^-$ ion in both cage geometry and metal placement. In addition, two dicobalt counterparts, 5,7- and $6,9\text{-}[\eta^5\text{-}C_5(CH_3)_5]_2\text{-}Co_2B_8H_{12}$, were obtained and fully characterized.[65a] In all of these pentamethylcyclopentadienyl cobalt complexes, each metal atom forms Co—H—B bridges and the overall geometry again deviates only slightly (allowing for the larger radius of cobalt versus boron) from that of $B_{10}H_{14}$ itself.

The reaction of cobalt vapor with $B_{10}H_{14}$ and cyclopentadiene[65b] gave $6\text{-}(\eta^5\text{-}C_5H_5)CoB_9H_{13}$, analogous to the pentamentylcyclopentadienyl complex mentioned above, as well as the $5\text{-}(\eta^5\text{-}C_5H_5)CoB_9H_{13}$ species reported earlier.[42,64]

A novel type of metal–boron cluster involving M—H—B bonding is found in the species $(CO)_3MnB_8H_{13}$ (Figure 21) which has been structurally elucidated by X-ray methods.[66] Although this complex is in a strict sense a metal–borane chelate rather than a metallaboron cluster, in that there are no metal–boron bonds, it does feature three Mn—H—B bridges whose spectroscopic and geometric properties closely resemble those found in the metallaboranes previously discussed (for a comparison see Table 2). The most interesting and unique aspect of this structure is the link between manganese and the erstwhile "terminal" hydrogen attached to B(1), a boron *not* on the open face; clearly the Mn—H—B(1) bridge serves to anchor the manganese, complete its octahedral coordination, and satisfy the 18-electron rule.[66]

Other *nido* clusters are known in which metal–hydrogen bonds might have

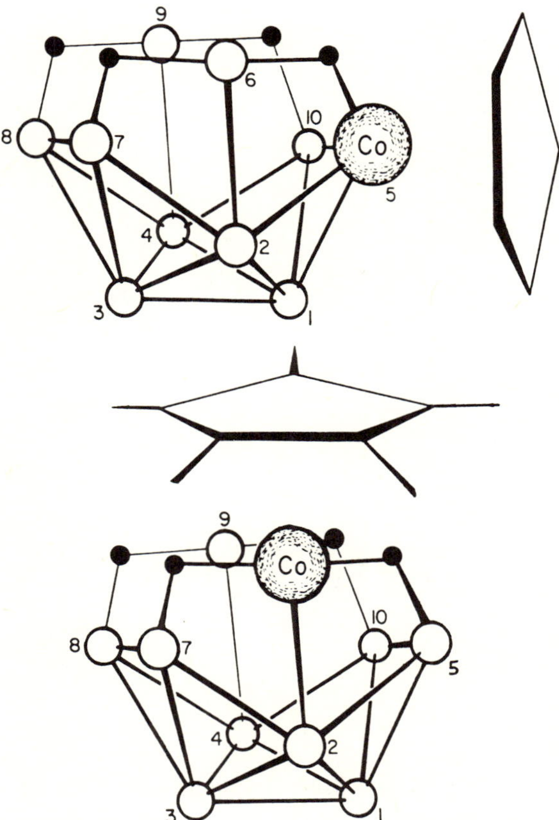

Figure 20.   (A) Structure of $5\text{-}(\eta^5\text{-}C_5H_5)CoB_9H_{13}$ (Ref. 64). (B) Structure of $6\text{-}[\eta^5\text{-}C_5(CH_3)_5]CoB_9H_{13}$ (Ref. 65a).

been anticipated from the presence of metal(s) on the open face and of "extra" (nonterminal) hydrogens, but which in fact have no M—H interactions, the bridging hydrogens preferring to adopt B—H—B locations. A recent example is the 10-vertex species, $9,6\text{-}[(C_6H_5)_3P]_2PtSB_8H_{10}$, formally analogous to the MB_9 and M_2B_8 complexes just described, which has a $B_{10}H_{14}$-like geometry

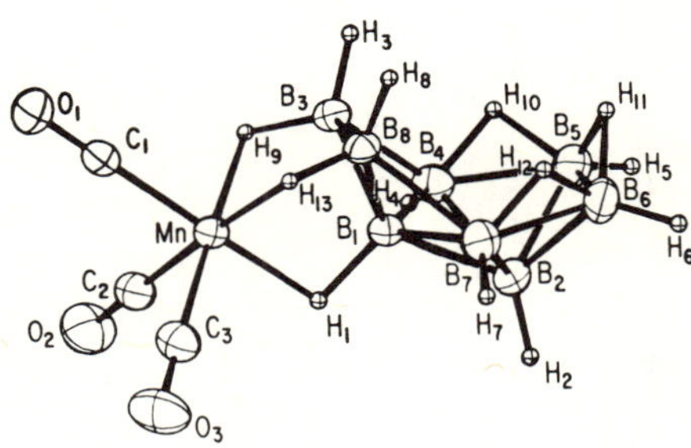

Figure 21.   Structure of $(CO)_3MnB_8H_{13}$
(Ref. 66).

Table 2. Structural and Spectroscopic Data on Clusters Containing Bridging or Face-Capping Hydrogen Atoms[a]

Compound	M–H distance (Å)	B–H(–M) distance (Å)	^{1}H nmr δ_{M-H}(J,Hz)	Solvent	Ref.
Metallaboranes					
Manganese					
2-(CO)$_3$MnB$_5$H$_{10}$	—	—	−11.7(77), −12.3(84)	A	59
1–Br–(CO)$_3$MnB$_5$H$_9$	—	—	−12.0(74)	D	59
(CO)$_3$MnB$_5$H$_9^-$	—	—	−11.6, −12.1(47)	A / I	59
(CO)$_3$MnB$_8$H$_{13}$	1.68(4), 1.81(3)	1.21(4), 1.15(3)	−9.3, −11.2	A	66
8-Et$_3$N(CH$_2$)$_4$O-6–(CO)$_3$MnB$_9$H$_{12}$	1.64, 1.70	1.17, 1.15	−10.8, −11.8	A	63
6-(CO)$_3$MnB$_9$H$_{13}^-$	—	—	−11.4	A	60–62
6-(CO)$_3$MnB$_9$H$_{12}$-2–O(CH$_2$)$_4$	1.75(4)	1.24	−11.9	A	60–62
6-(CO)$_3$MnB$_9$H$_{12}$-5–O(CH$_2$)$_4$	1.83, 1.67	1.27, 1.26	−9.2, −11.9	A	62
6-(CO)$_3$MnB$_9$H$_{12}$-2–O(C$_2$H$_5$)$_2$	—	—	−11.6	A	62
(CO)$_{10}$Mn$_3$B$_2$H$_7$	b	—	−19		50
Rhenium					
6-(CO)$_3$ReB$_9$H$_{13}$	—	—	−8.0	A	62
Iron					
2-(C$_5$H$_5$)FeB$_5$H$_{10}$	—	—	−16.0(70)	H	56,57
1,2-(CO)$_6$Fe$_2$B$_3$H$_7$	1,57(4), 1.58(5)	1.33(6), 1.26(4)	−16.6	A	47
(CO)$_3$FeB$_5$H$_9^-$	—	—	−16.8	A	51,52
(CO)$_3$FeB$_5$H$_8^-$	[1.52]^c	1.15	−16.0	d	52
μ–[(C$_6$H$_5$)$_3$P]$_2$Cu-2–(CO)$_3$FeB$_5$H$_8$^{e}	Fe–H, 1.56(6); Cu–H, 1.96(7)	—	—	—	54
(CO)$_6$Fe$_2$B$_2$H$_6$	—	—	−10.3	A	48

Cobalt					
2–$(C_5H_5)CoB_4H_8$	1.43(5)	1.43(5)	—	—	43
$(C_5H_5)CoB_4H_7^-$			−15.28	E	41,42
2–$(C_5Me_5)CoB_4H_8$	—	—	−15.40	H	45
5–$(C_5H_5)CoB_9H_{13}$	—	—	−14.2(6)	E	65a
	1.49(1)	1.26(1)	—	—	64
6–$(C_5H_5)CoB_9H_{13}$	—	—	−19.2	E	42
6–$(C_5Me_5)CoB_9H_{13}$	—	—	−12.78	CS_2	65
	1.52(5),	1.17(5),	−11.8(35)	E	65b
	1.65(6)	1.34(6)	—	—	—
6,9–$(C_5Me_5)_2Co_2B_8H_{12}$	1.60(5), 1.41(5)	1.30(5), 1.31(5)	−11.65	E	65a
	1.59(5), 1.32(5)	1.38(5), 1.42(5)	—	—	—
5,7–$(C_5Me_5)_2Co_2B_8H_{12}$	1.38(1),	1.33(1),	−21.0	E	65a
	1.37(1)	1.33(1)	—	—	—
5,9–$(C_5Me_5)_2Co_2B_8H_{12}$		—	−10.7,−14.2,	E	65a
			−19.5		
6–Cl–5,7–$(C_5Me_5)_2Co_2B_8H_{11}$	1.48(1),	1.27(1),	−20.0	E	65a
	1.38(1)	1.23(1)	—	—	—
1,2–$(C_5H_5)_2Co_2B_4H_6$	1.62(3),	1.51(3)	—	—	79
	1.48(3)	—	—	—	—
1,2–$(C_5Me_5)_2Co_2B_4H_6$	—	—	−12.58	E	42
3–C_5H_9–1,2–$(C_5H_5)_2Co_2B_4H_5$	—	—	−12.85	E	65a
4–C_5H_9–1,2–$(C_5H_5)_2Co_2B_4H_5$	—	—	−12.60	E	42
$(C_5H_5)_3Co_3B_3H_5$	—	—	−12.67	E	42
1,2–$(C_5Me_5)_2Co_2B_5H_7$	f,g	—	−14.48	E	42
$(C_5Me_5)_2Co_2B_5H_9$	—	—	−14.6	E	65a
	—	—	−12.4, −14.1	E	65a
Copper					
$[(Ph_3P)_2Cu]_2B_{10}H_{10} \cdot CHCl_3$[h,k]	2.08(7)[i]	—	—	—	100
	1.86(6)[j]	—	—	—	—
$[(Ph_3P)_2Cu]_2B_{10}H_{10}$[h]	—	—	+0.85	E	104

(continued overleaf)

Table 2 (*cont.*)

Compound	M–H distance (Å)	B–H(–M) distance (Å)	δ_{M-H} (J,Hz)	Solvent	Ref.
			^{1}H nmr		
Metallacarboranes					
Iron					
$1,2,3-(C_5H_5)FeHC_2B_4H_6$	—	—	−14.40	G	6,7
$1,2,4-(C_5H_5)FeHC_2B_4H_6$	—	—	−7.80	G	7
$1,2,4,5-(C_5H_5)_2FeHCo(CH_3)_2C_2B_3H_3$	f,m	—	−16.5	E	92
$(C_5H_5)Co[(CH_3)_2C_2B_3H_3]FeH_2[(CH_3)_2C_2B_4H_4]$	—	—	−12.7	E	92
$[2,3-(CH_3)_2C_2B_4H_4]_2FeH_2$	f,n	—	−10.44	E	90
Cobalt					
$[2,3-(CH_3)_2C_2B_4H_4]_2CoH$	—	—	−8.7	E	85
$[2,3-(CH_3)_2C_2B_4H_4]CoH[(CH_3)_2C_2B_3H_5]$	f,o	—	−6.3	E	85
$(C_5H_5)Co[(CH_3)_2C_2B_3H_3]CoH[(CH_3)_2C_2B_3H_5]$	—	—	−14.3	E	85
$(C_5H_5)Co[(CH_3)_2C_2B_3H_3]CoH[(CH_3)_2C_2B_3H_3]Co(C_5H_5)$	—	—	−15.8	E	85
$7,8,9-(C_5H_5)CoC_2B_8H_{10}-11-NC_5H_5$	—	—	−11.1	p	69
$[(1,2-C_2B_9H_{11})-3,9'-Co-(7',8'-C_2B_8H_{10}-11'-NC_5H_5]^-$	f,r	—	−9.2	p	69
$7,8,9-(C_5H_5)CoC_2B_8H_{10}-11-NHC_5H_{10}$	—	—	−11.0	q	69
$[(1,2-C_2B_9H_{11})-3,9'-Co-(7',8'-C_2B_8H_{12})]^-$	—	—	−6.10	p	69
$7,8,9-(C_5H_5)CoC_2B_8H_{12}$	—	—	−13.14	E	69
$8,6,7-(C_5H_5)CoC_2B_7H_{11}$	1.40(4)	1.31(4)	—	—	72
	—	—	−18.30	E	71

[a] Abbreviations and codes are as in Table 1. [b] X-ray study revealed symmetrical Mn–H–Mn bridge with Mn–Mn distance of 2.845 (3) Å; Mn–H distance not given. [c] Not refined. [d] $(CD_3)_2$O. [e] Cu–H–B IR stretching band observed at 2292 cm^{-1}. [f] X-ray structure determination did not directly locate bridging H atoms. [g] Ref. 80. [h] Intermolecular B–H–Cu bonds. [i] Cu–H$_{apical}$ distance. [j] Cu–H$_{equatorial}$ distance. [k] Cu–H–B stretching absorptions observed in 2150–2400 cm^{-1} region of IR spectrum (Nujol) as weak, broad bands. [n] Ref. 95. [n] Ref. 91. [o] Ref. 105. [p] CD_3CN. [q] $(CD_3)_2$CO. [r] Ref. 70.

with B—H—B bridges spanning the B(7)—B(8) and B(5)—B(10) edges.[67,68] Such cases as this, in which the bridging protons fail to adopt M—H—B locations despite the opportunity to do so, may eventually help to illuminate more clearly the factors which favor metal-hydrogen bonding. At present, neither the empirical evidence nor theoretical efforts in this area thus far have generated a useful predictive model for dealing with this question.

3.2. Nido-Metallacarboranes

A few examples of metallacarboranes containing M—H—B bridges on an open face have been discovered. Treatment of the 11-vertex *closo*-carborane $1,2,4-(\eta^5-C_5H_5)CoC_2B_8H_{10}$ with pyridine opens the cage (without removal of boron) and generates the *nido*-11 vertex system $(\eta^5-C_5H_5)CoC_2B_8H_{10} \cdot NC_5H_5$, which contains a single hydrogen bridge between Co and B (Figure 22).[69] The cage-opening is a consequence of the addition of two electrons (via pyridine), and the displacement of a terminal hydrogen by pyridine creates a hydrogen bridge on the open face. The structure shown is based on the crystallographically determined geometry[70] of the analogous species $(C_2B_9H_{11})Co(C_2B_8H_{10}) \cdot NC_5H_5^-$ (in which $C_2B_9H_{11}^{2-}$ replaces $C_5H_5^-$), which was similarly obtained by reaction of *closo, closo*-$(C_2B_9H_{11})Co(C_2B_8H_{10})^-$ with pyridine. The bridging hydrogens in these complexes were located from their high-field ^1H nmr signals (Table 2), since the crystallographic study did not directly reveal their positions in that species.

A piperidine-substituted derivative, $(\eta^5-C_5H_5)CoC_2B_8H_{10} \cdot NHC_5H_{10}$, has also been prepared and it too has a Co—H—B bridge;[69] however, in the unsubstituted parent molecule $(C_2B_9H_{11})Co(C_2B_8H_{11})^{2-}$ the bridging proton occupies a B—H—B rather than a Co—H—B location, as shown by nmr evidence. Thus, replacement of a terminal hydrogen by pyridine or piperidine at B(11) causes the bridge proton to move to a Co—H—B location.

The parent species $(C_2B_9H_{11})Co(C_2B_8H_{11})^{2-}$ can be reversibly protonated by treatment with HCl in methanol to give the monoanion shown in Figure 23, which has both B—H—B and Co—H—B bridges.[69] In strongly basic media the Co—H—B proton is selectively removed, a result analogous to the deprotonation of $2-(\eta^5-C_5H_5)CoB_4H_8$ as discussed in Section 3.1.1.

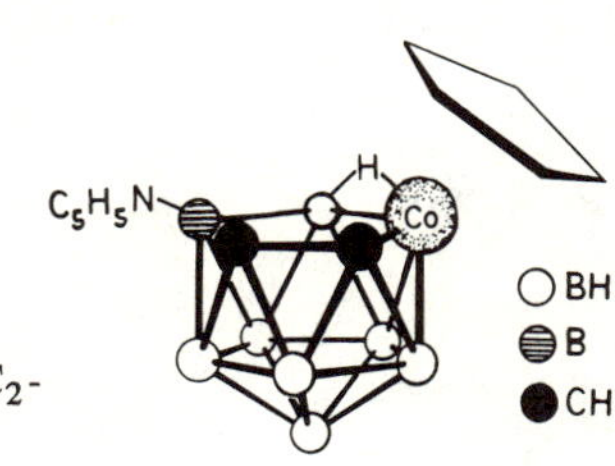

Figure 22. Structure of $4-C_5H_5N-7,1,2-(\eta^5-C_5H_5)CoC_2-B_8H_{10}$ (Ref. 69).

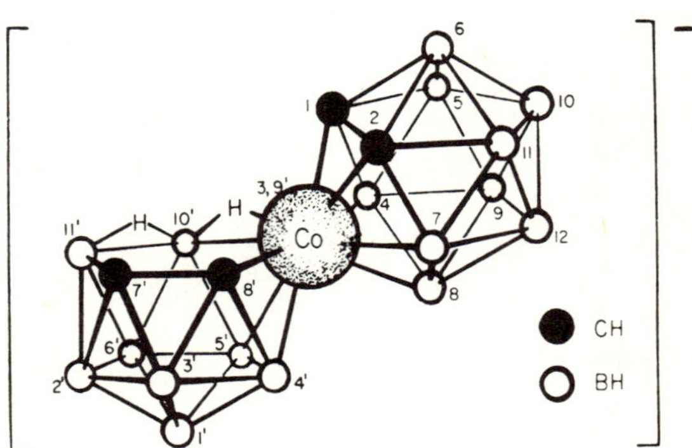

Figure 23. Proposed structure of the $[(C_2B_9H_{11})Co(C_2B_8H_{12})]^-$ ion (Ref. 69).

Similar chemistry has been observed with the analogous cyclopentadienyl-cobalt complex $(\eta^5\text{-}C_5H_5)CoC_2B_8H_{11}^-$, which is obtained by degradation of the icosahedral *closo*-1,2,3-$(\eta^5\text{-}C_5H_5)CoC_2B_9H_{11}$ in strong base. The $(\eta^5\text{-}C_5H_5)CoC_2B_8H_{11}^-$ ion itself has a B—H—B, but no Co—H—B bridge; however, protonation with HCl yields neutral, yellow $(\eta^5\text{-}C_5H_5)CoC_2B_8H_{12}$ which has a structure analogous to that in Figure 23, but with a $C_5H_5^-$ ligand replacing $C_2B_9H_{11}^{2-}$, and does have both Co—H—B and B—H—B hydrogen bridges as judged from nmr evidence.

The species containing "extra" (i.e., bridging) hydrogens described here can be oxidized to *closo*-cobaltacarboranes. On pyrolysis they tend to eliminate H_2, again forming *closo* systems, in a manner analogous to the conversion of *nido*-carboranes to *closo*-carboranes (e.g., $C_2B_9H_{13} \longrightarrow C_2B_9H_{11} + H_2$).[69]

The controlled degradation of *closo*-1,2,4-$(\eta^5\text{-}C_5H_5)CoC_2B_8H_{10}$ with FeCl$_3$ in ethanol[71] yields a *nido*-cobaltacarborane product, 8,6,7-$(\eta^5\text{-}C_5H_5)$-CoC$_2$B$_7$H$_{11}$, whose crystallographically determined structure[72] is shown in Figure 24. The Co—H—B and B—H—B hydrogens in this molecule were both located and refined in the X-ray study, thereby supporting the ^{1}H nmr indications of their presence (which, however, in the case of the Co—H—B proton was not originally recognized as such[71]). This cobalt complex is isoelectronic with *nido*-$C_2B_8H_{12}$ and $B_{10}H_{14}$, and also with a nickel species, $[(C_2H_5)_3P]_2Ni$-$(CH_3)_2C_2B_7H_9$.[73] The nickel compound, however, was obtained by a different rout (insertion of a Ni$[P(C_2H_5)_3]_2$ moiety into *arachno*-$(CH_3)_2C_2B_7H_{11}$) and is structurally unlike the cobalt species; the nickel does occupy a vertex on

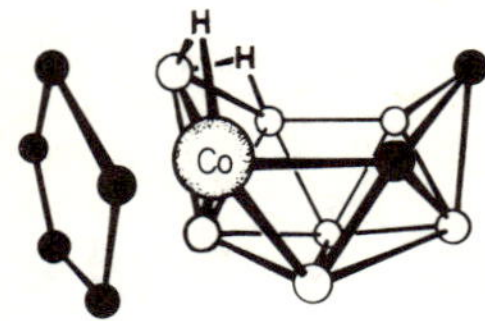

Figure 24. Structure of 8,6,7-$(\eta^5\text{-}C_5H_5)CoC_2B_7H_{11}$ (Ref. 72).

the open face, but the "extra" hydrogens occupy B—H—B and C—H locations, and there are no metal-hydrogen interactions in the molecule.[73]

The analogous platinum complex, $[(C_2H_5)_3P]_2Pt(CH_3)_2C_2B_7H_9$, prepared in a similar manner by reaction of $Pt[P(C_2H_5)_3]_2$(stilbene) with $(CH_3)_2$-$C_2B_7H_{11}$, also contains no M—H—B bridges and presumably is isostructural with its nickel counterpart; interestingly, however, an intermediate species having a Pt—H bond was detected from its doublet resonance at δ-6.24 ($J=$ 22.0 Hz) in the ^1H spectrum during the course of the reaction.[73] This high-field signal eventually disappeared, suggesting that the initial platinacarborane formed has a Pt—H—B bridge from which the proton migrates to another location. Infrared evidence also supports the postulated Pt—H intermediate.[73]

3.3. σ-Metal-Carborane Derivatives with M—H—B Bridge Bonds

A mode of metal-hydrogen-boron bridge bonding that is different from those discussed to this point involves the interaction between an *exo*-polyhedral metal atom and a terminal hydrogen on boron. This can occur when the metal has an incomplete coordination sphere and an orbital of suitable symmetry to interact with a hydrogen attached to the cage proper. Several structurally established examples are known, the first of which is the $Co[1,2-C_2B_{10}H_{10})_2]_2^-$ anion, a complex containing two *bis*-(*o*-carborane) ligands coordinated to cobalt primarily via four Co—C bonds. In this molecule, which was synthesized by reaction of $CoCl_2$ with two mole equivalents of the C,C'-dilithio derivative $(LiCB_{10}H_{10}C-CB_{10}H_{10}Li)$ in ether,[74] the four icosahedral cages are situated in approximately tetrahedral fashion around cobalt (Figure 25); however, the metal is within bonding distance not only of four carbon atoms (one on each

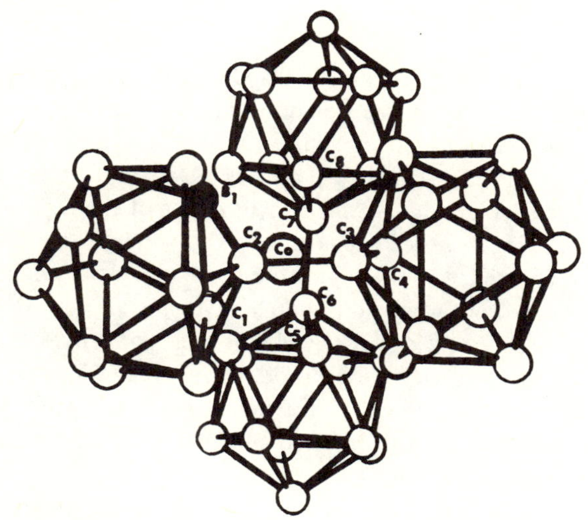

Figure 25. Structure of the $Co[(C_2B_{10}H_{10})_2]_2^-$ ion (Ref. 75).

cage), but also of a hydrogen atom on a nearby boron (B1), so that the actual symmetry of the coordinating carbon atoms is severely distorted from that of a regular tetrahedron.[75] The Co—H—B bridge is highly unsymmetrical with metal–hydrogen and boron–hydrogen distances of 1.82(7) and 1.11(7) Å, respectively, suggesting a weaker Co—H interaction than in *nido*-cobaltaboranes and cobaltacarboranes (Table 2). It is significant, however, that the driving force for Co—H bonding is sufficiently strong to force distortion of the tetrahedral coordination environment around cobalt in order to allow it to occur.

A similar situation occurs in a rhodium(I) complex, $1\text{-}[(C_6H_5)_3P]_2Rh\text{-}2\text{-}C_6H_5\text{-}1,2\text{-}C_2B_{10}H_{10}$, in which the rhodium atom is linked to the cage through an ordinary Rh—C bond and also via a Rh—H—B bridge interaction with one of the BH units.[76a] It is interesting that the atoms involved in the coordination (two phosphorus, carbon, hydrogen, and rhodium) are nearly coplanar, again indicating that the drive favoring metal–hydrogen bonding is a powerful one which can have major structural consequences.

In the molecule $3,9\text{-}[(p\text{-tolyl})_3P]_2H_2IrC_2B_9H_{12}$, the iridium atom is attached to the open C_2B_9 cage *solely* through interaction with two terminal hydrogen atoms on adjacent borons on the open face, forming a pair of Ir—H—B bridges.[76b] Insofar as there are no metal-to-cage links other than the hydrogen bridges, this complex is related to the manganese species $(CO)_3MnB_8H_{13}$ (Figure 21) described in Section 3.1.2. The geometry around the iridium atom is roughly octahedral, with the $P(p\text{-tolyl})_3$ units *trans* to each other, the two bridging hydrogens *cis*, and the remaining (*cis*) locations occupied by terminal hydride ligands (which were not located crystallographically).

4. CLUSTERS CONTAINING FACE-CAPPING HYDROGEN ATOMS

Hydrogen is capable of bonding to as many as three atoms simultaneously; indeed, in certain exotic species coordination numbers as high as six have been observed.[77] A particularly intriguing class of metallaboron clusters consists of polyhedra in which one or more triangular faces is capped (or bridged) by hydrogen. Special importance can be attached to this type of cluster for two reasons: first, hydrogen-capping has become a common feature in metallaborane and metallacarborane chemistry, and second, there is a very close parallel with metal clusters containing hydrogen-capped trimetallic faces.

Before we proceed with discussion of this area, some caveats are in order. Face-bonded hydrogen atoms have been definitely located in only a few cases, owing to the difficulties of observing hydrogen in the vicinity of heavy atoms by X-ray crystallography; neutron diffraction studies are likely to be more successful, but at this writing few have been attempted. Nevertheless, in most of the species to be described there is no serious doubt as to the general positions of the face-capping hydrogen atoms, supported by several arguments: (1) ^{1}H and

^{11}B nmr evidence supports the assignments; (2) hydrogen face-capping is an established feature in metal cluster chemistry, and is, therefore, reasonable to expect in metallaboron clusters; (3) X-ray studies support, and in at least one case definitely prove, the presence of face-bridging hydrogens in certain metallaboranes; and (4) most of the compounds in question are *closo*-polyhedra with no open faces, so that triangular faces offer the most plausible sites for placement of "extra" (nonterminal) hydrogen atoms.

A fine point of discussion centers on the distinction between true capping of a triangular face wherein the hydrogen experiences a bonding interaction with all three facial atoms, and edge-bridging (e.g., M—H—M or M—H—B) in which the hydrogen actually bonds to only two atoms on the face. Presently available evidence cannot in most cases clearly distinguish between these possibilities, and the difference in some cases, such as $1,2\text{-}(\eta^5\text{-}C_5H_5)_2Co_2B_4H_6$, may be rather subtle; moreover, there is also the possibility that a given hydrogen atom may alternate between capping and bridging modes, perhaps adopting different locations in the crystal and in solution. For present purposes we shall simply assume idealized face-capping structures for the systems to be discussed.

4.1. Metallaboranes

At this writing, the only known examples of polyhedral metallaboranes containing face-bridging hydrogens are a family of air-stable octahedral clusters consisting of $1,2\text{-}(\eta^5\text{-}C_5H_5)_2Co_2B_4H_6$, its 3- and 4-$C_5H_9$– substituted derivatives, and $1,2,3\text{-}(\eta^5\text{-}C_5H_5)_3Co_3B_3H_5$.[41,42,78] X-ray studies of the violet Co_2B_4[79] and brown Co_3B_3[80] parent complexes confirmed the *closo* 6-vertex polyhedral structures originally deduced from nmr evidence (Figure 26), and in the case of the Co_2B_4 system, revealed the locations of the two face-capping hydrogen atoms. Refinement of these hydrogen positions by full-matrix least-squares methods[79] established that they are nearly centered over the two

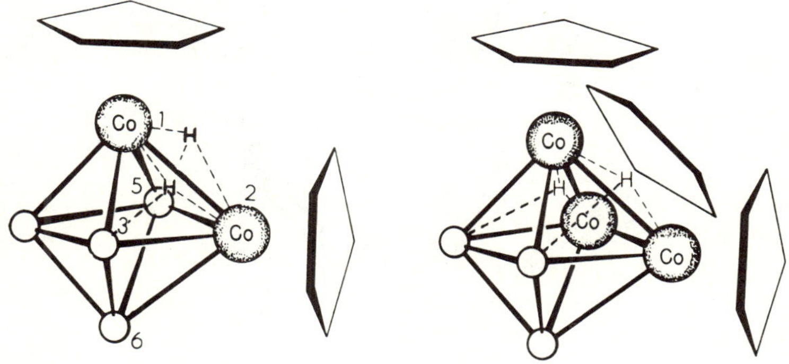

Figure 26. Structures of $(\eta^5\text{-}C_5H_5)_2Co_2B_4H_6$ and $(\eta^5\text{-}C_5H_5)_3Co_3B_3H_5$ (Refs. 42,79,80). The bridging hydrogen locations on the Co_3B_3 cluster were assigned from nmr data, and probably tautomerize over the three Co_2B triangular faces.

equivalent Co–Co–B triangular faces, such that the H–B distance of 1.51(3) Å is the same as the mean H–Co interaction [1.54(3) Å], with the hydrogen 0.78 Å above the Co–Co–B facial plane. These data together with a plot of the anisotropic thermal motion of the hydrogen atom (Figure 27) suggest a basic Co–H–Co bridging interaction with some bonding attraction to the nearby boron atom. The hydrogen bridge in this molecule is markedly similar to that in the tetrahedral metal cluster $HFeCo_3(CO)_9[P(OCH_3)_3]_3$, in which a hydrogen atom caps a triangular Co_3 face with mean H–Co distances of 1.63(15) Å and the hydrogen is situated 0.75 Å above the tricobalt plane.[81]

A conspicuous feature of the $(\eta^5\text{-}C_5H_5)_2Co_2B_4H_6$ structure is the long [2.557(1) Å] cobalt-cobalt distance, which is 0.1 Å (100 standard deviations) longer than the Co–Co bond lengths in $(\eta^5\text{-}C_5H_5)_3Co_3B_3H_5$ and $(\eta^5\text{-}C_5H_5)_3Co_3B_4H_4$. The bond-lengthening in $(\eta^5\text{-}C_5H_5)_2Co_2B_4H_6$ has been attributed to the presence of the bridging hydrogens which tend to lower the metal–metal bond order.[79]

The companion cluster $(\eta^5\text{-}C_5H_5)_3Co_3B_3H_5$ also contains two extra hydrogens, but unfortunately these were not conclusively located in the crystallographic study.[80] From ^{11}B and ^1H nmr evidence[42] it is shown that these hydrogens are associated with cobalt and migrate on the nmr time scale over the polyhedral surface such that the three cobalt atoms are equivalent, producing C_{3v} symmetry; in the crystal, the two hydrogen atoms are evidently disordered on the three Co–Co edges (or Co_2B triangles). Since there is an average of only 2/3 of a hydrogen per Co–Co edge, the failure to establish the hydrogen positions crystallographically is understandable (in the Co_2B_4 complex just described, the corresponding ratio is 2). As expected, the Co–Co distances in the Co_3B_3 cluster [2.472(1)–2.488(1) Å] are substantially shorter than in the Co_2B_4 system, a difference attributed to the larger H/Co–Co ratio in the dicobalt species mentioned previously.[79] All available evidence indicates that the extra

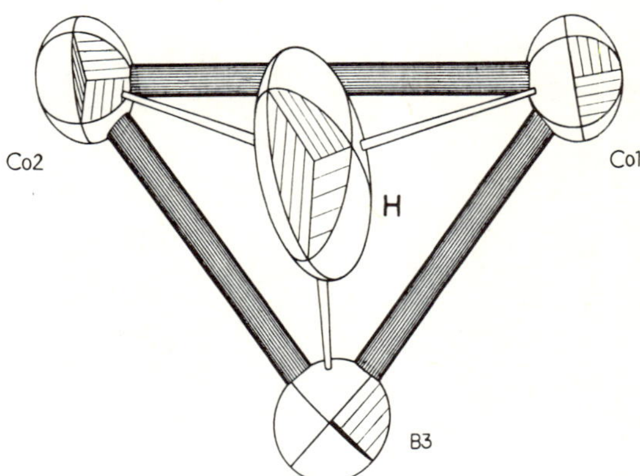

Figure 27. Co–Co–B face of 1,2-$(\eta^5\text{-}C_5H_5)_2Co_2B_4H_6$, showing the vibration ellipsoid of the bridging hydrogen atom (Ref. 79).

hydrogens in $(\eta^5\text{-}C_5H_5)_3Co_3B_3H_5$ cap Co_2B faces in a manner similar to those established in $(\eta^5\text{-}C_5H_5)_2Co_2B_4H_6$.

4.2. Metallacarboranes

The first indication in boron cluster chemistry that bridging hydrogen atoms could be accommodated on a closed polyhedral cage was the structural characterization of the octahedral carborane CB_5H_7, whose nmr spectra[82] suggested that the unique hydrogen tautomerizes through four equivalent B_3 faces or B_2 edges. More recently, microwave evidence[83] pointed to a face-bridging location for the static structure and this conclusion was supported by electron diffraction studies.[84] This molecule is isoelectronic with the octahedral family $C_2B_4H_6$, $B_6H_6^{2-}$, and $CB_5H_6^{-}$, and is, in fact, the conjugate acid of the latter species.

In metallacarborane chemistry, the earliest examples of face-capping hydrogens were found in the isomeric complexes 1,2,3- and 1,2,4-$(\eta^5\text{-}C_5H_5)$-$FeHC_2B_4H_6$ (Figure 28) and several derivatives.[6,7] Subsequently, this feature has turned up frequently in the author's laboratory, primarily in metallacarboranes incorporating formal cobalt(III) and iron(II) (Table 2), and the bridging hydrogens have been associated with some interesting and novel chemistry. Although a number of these compounds have been crystallographically characterized, the face-bridging hydrogen atoms have thus far eluded direct detection. However, the nmr evidence for metal–hydrogen interaction is in most cases clear, consisting of (1) the presence of ^1H nmr signals at high field (above δ–5 ppm) and (2) the absence of resonances attributable to alternative locations for extra hydrogens, such as BH_2 or CH_2 groups. In addition, in some instances it has been possible to demonstrate via decoupling experiments that the capping proton(s) are spin-coupled to both boron and metal atoms.[85]

The *closo* geometries of these clusters are well established from X-ray studies of prototype species, and it is notable that in all cases encountered thus far, the electron-counting rules of Wade[4] are in conformity with the observed structures when the electrons supplied by the extra (capping) hydrogen atoms are included. Thus, in general, these hydrogens permit the achievement of filled bonding molecular orbitals in the skeletal metallacarborane network, and of 18-electron shells for transition metal atoms. In 1,2,3-$(\eta^5\text{-}C_5H_5)FeHC_2B_4H_6$ (Figure 28), for example, the requirement of 16 skeletal electrons $(2n + 2)$ for the 7-vertex cage is met by donations of two from each BH group, three from each CH, one* from $(\eta^5\text{-}C_5H_5)Fe$, and the remaining electron from hydrogen. Effectively, the FeH moiety functions as the equivalent of a Co atom (note the stable existence[86] of the analogous species 1,2,3- and 1,2,4-$(\eta^5\text{-}C_5H_5)CoC_2B_4H_6$, which lack extra hydrogens); similarly, CoH (or FeH_2) provide the electronic equivalent of a nickel atom.

---

*Treating iron as a neutral atom, it contributes one electron to C_5H_5–Fe bonding and stores six electrons in nonbonding orbitals, leaving one for cage donation.

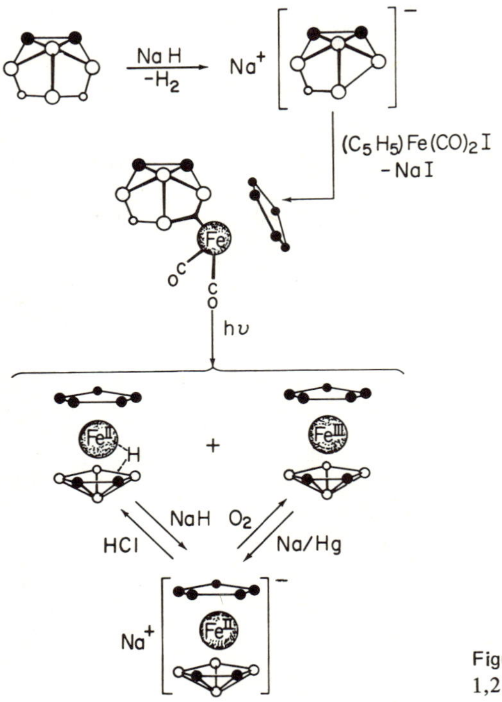

Figure 28. Synthesis and chemistry of $1,2,3-(\eta^5-C_5H_5)FeHC_2B_4H_6$ (Ref. 7).

The syntheses of *closo*-metallacarboranes containing extra hydrogens, in most cases have been based on addition of metal groups to carborane substrates containing bridge hydrogens, as illustrated in Figure 28. Attempts at direct protonation of *closo*-metallacarborane anions have tended to give ambiguous results; thus, treatment of $[(1,2-C_2B_9H_{11})_2Fe(II)]^{2-}$ with strong acid produced a $[(C_2B_9H_{11})Fe(II)(C_2B_9H_{12})]^-$ species which gave no Fe—H signal in the proton nmr spectrum.[87] This result contrasts with the protonation of ferrocene, which occurs readily to generate $(\eta^5-C_5H_5)_2FeH^+$, whose ^1H nmr spectrum exhibits a resonance at $\delta-2.09$ ppm that has been attributed to FeH.[88] However, recent studies[89] have indicated that protonation of the $(\eta^5-C_5H_5)_2Co^-$ ion occurs at a ligand site rather than the metal, and this is evidently true of $1,2,3-(\eta^5-C_5H_5)CoC_2B_9H_{11}$ according to the same investigation. On the other hand, anions derived from hydrogen-capped metallacarboranes by deprotonation with strong bases can, in general, be reprotonated to generate the original metal-hydrogen species, as shown for the iron system[7] in Figure 28.

As these observations suggest, metallaboranes containing face-capping protons tend to be acidic with respect to loss of one such proton in the presence of strong bases, and the process is usually reversible. The tendency to lose specifically a metal-bound proton is shown by the *closo, nido* species $[(CH_3)_2C_2B_3H_5]-Co(III)H[(CH_3)_2C_2B_4H_4]$, depicted in Figure 29, which contains ordinary

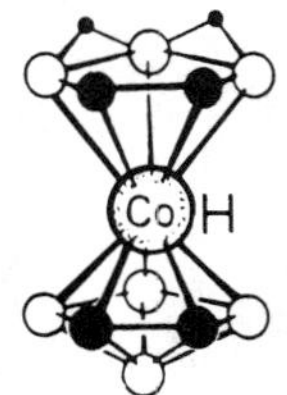

Figure 29.   Structure of $[2,3\text{-}(CH_3)_2C_2B_3H_5]CoH[2,3\text{-}(CH_3)_2C_2B_4H_4]$
(Refs. 85, 105). The presence of a $Co-H$ hydrogen atom, shown schematically, has been established from nmr evidence.

$B-H-B$ bridges as well as a face-capping (metal-bonded) proton. On treatment
with excess NaH or KOH, *only* the capping proton is removed; the $B-H-B$
protons are unaffected.[85]

Complexes that are related to those in Figures 28 and 29 are the *bis*-(di-
carbahexaboranyl) metal species, $[2,3\text{-}[(CH_3)_2C_2B_4H_4]_2Fe(II)H_2$ (Figure
30)[90] and $[2,3\text{-}(CH_3)_2C_2B_4H_4)_2Co(III)H$,[85] which are obtained by reaction
of the $(CH_3)_2C_2B_4H_5^-$ anion with metal halides in THF solution. In these systems there are two *closo* 7-vertex polyhedra joined at a common metal atom,
and, hence, there are four equivalent MB_2 faces* to which the extra hydrogens
might bind. Since nmr evidence indicates the presence of symmetry in both
molecules, the extra hydrogens are assumed to tautomerize through these
equivalent faces. An X-ray study of the iron complex[91] revealed that the
ligands are mutually rotated by 90° and are tilted at an angle of 7.80° such that
two of the $C-CH_3$ units are forced close together with a separation of only
3.5 Å, or 0.5 Å less than the van der Waals distance (Figure 30). This tilting is
attributed to the two hydrogens which are postulated to be disordered over the
four equivalent iron-boron-boron faces.[91]

The face-capping hydrogens in $[(CH_3)_2C_2B_4H_4]_2FeH_2$ and $[(CH_3)_2C_2B_4H_4]_2^-$
CoH are believed to be involved in the oxidative fusion process[2d,85,90-92]
whereby the $[(CH_3)_2C_2B_4H_4]^{2-}$ ligands in these complexes, on exposure to
air or other oxidants, join face-to-face with a net loss of four electrons and ex-
pulsion of the metal to form a neutral four-carbon species, $(CH_3)_4C_4B_8H_8$ (Figure 31). The oxidative fusion reaction has been discussed elsewhere,[2d] but we
reiterate here that it is (1) broad in scope, having been observed in a variety of
metallacarborane complexes, and (2) efficient and facile, occurring at or below
room temperature and often in high yield. While the mechanism has not been
established, it may be significant that every instance of fusion thus far has involved species containing metal-bound hydrogen atoms; thus, there is reason
to believe that these hydrogens may play a central role. One can envision attack
of the oxidizing agent (e.g., O_2) at one of the highly reactive metal–hydrogen
sites, initiating a transfer of electrons from the ligands which could induce linkage at the B–B edges. In the example depicted in Figure 31, the process continues

---

*Faces involving carbon atoms are assumed to be excluded since $C-H_{bridge}$ bonding in car-
borane systems is virtually unknown.

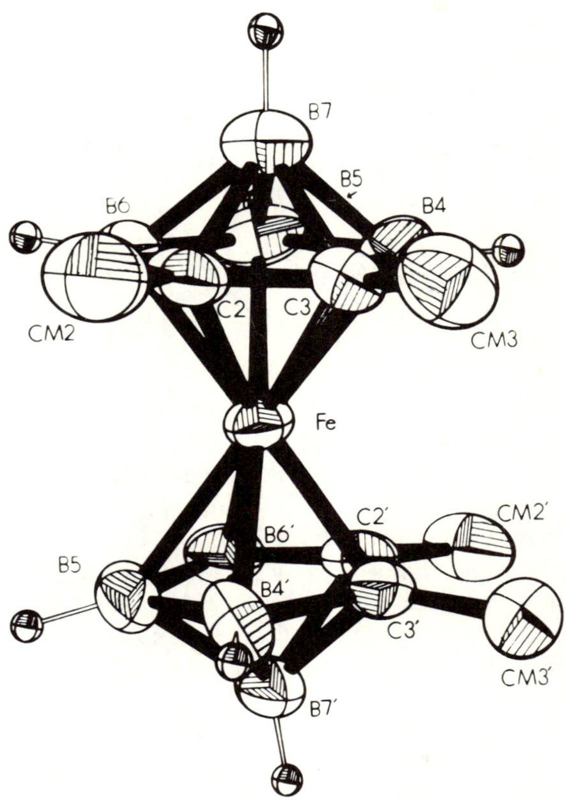

Figure 30. Structure of $[2,3\text{–}(CH_3)_2 C_2 B_4 H_4]_2 FeH_2$ (Ref. 91). The metal-bound hydrogen atoms are not shown, but nmr data indicate that they tautomerize between bridging locations on the four equivalent Fe –B—B faces.

with elimination of the metal and formation of a carbon-carbon link to complete the fusion; in other instances fusion is only partial and the carbons on the two ligands remain mutually separated.[93a]

At this writing, it has been established that fusion of the $R_2 C_2 B_4 H_4{}^{2-}$ pyramidal ligands ($R = CH_3$, $C_2 H_5$, or $C_3 H_7$) occurs intramolecularly, with no dissociation of ligands from the metal.[93b] Thus, oxidation of mixtures of $[(CH_3)_2 C_2 B_4 H_4]_2 FeH_2$ and $[(C_2 H_5)_2 C_2 B_4 H_4]_2 FeH_2$ in THF solution gives only $(CH_3)_4 C_4 B_8 H_8$ and $(C_2 H_5)_4 C_4 B_8 H_8$; no $(CH_3)_2 (C_2 H_5)_2 C_4 B_8 H_8$ is formed. Solutions of red, diamagnetic $[(CH_3)_2 C_2 B_4 H_4]_2 FeH_2$ in THF, on standing at room temperature, slowly convert to a paramagnetic purple species which appears stable as long as air is excluded. On exposure to atmospheric O_2, it rapidly forms the colorless final product, $(CH_3)_4 C_4 B_8 H_8$. X-ray diffraction and Mössbauer analyses on the purple complex reveal the presence of *two* iron atoms which have been identified as low-spin and high-spin Fe(III). The structure

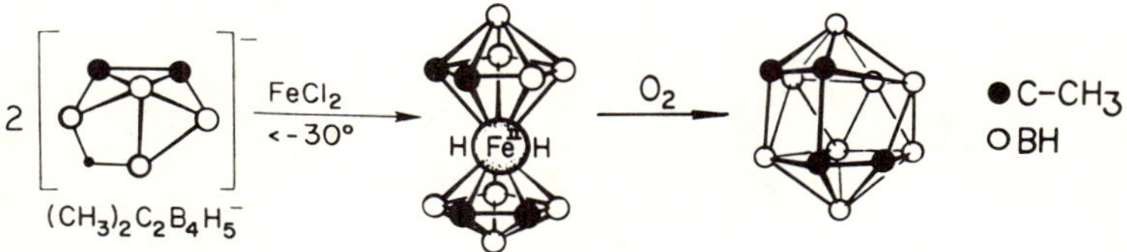

Figure 31. Oxidative fusion of $[2,3-(CH_3)_2C_2B_4H_4]_2FeH_2$ to form $(CH_3)_4C_4B_8H_8$ (Ref. 2d).

consists of a $[(CH_3)_2C_2B_4H_4]_2Fe$ sandwich like that in Figure 31, with a second iron atom wedged between the two ligands and located within normal bonding distance (2.414 Å) of the first iron; solvent molecules $[2THF$ or $(CH_3O)_2-C_2H_4]$ complete the coordination shell of the outer iron. The second iron atom evidently originates from traces of Fe(III) in the $(R_2C_2B_4H_4)_2FeH_2$ solutions, and it has been found[93b] that $FeCl_3$ catalyzes the conversion of red $[(CH_3)_2-C_2B_4H_4]_2FeH_2$ to the purple complex.

The bis(carboranyl)iron(II) dihydrogen complex and its cobalt(III) mono-hydrogen analog undergo a variety of other reactions, many of which involve de-protonation or displacement of the face-capping hydrogens with metal groups. Treatment of red solid $[(CH_3)_2C_2B_4H_4]_2Fe(II)H_2$ with NaH in THF removes one of the metal-bound protons, forming the wine-red $[(CH_3)_2C_2B_4H_4]_2Fe(II)H^-$ anion, which on air oxidation eliminates the other proton and generates the para-magnetic, red–orange $[(CH_3)_2C_2B_4H_4]_2Fe(III)^-$ anion.[90] Reaction of this latter species with dry HCl in dichloroethane regenerates the original complex, $[(CH_3)_2C_2B_4H_4]_2Fe(II)H_2$, together with the previously described tetracarbon metallacarborane, $(CH_3)_4C_4B_8H_8$; evidently the $[(CH_3)_2C_2B_4H_4]_2Fe(III)H$ species (not observed) forms initially and then undergoes auto-oxidation to generate the observed products:

$$[(CH_3)_2C_2B_4H_4]_2Fe(II)H^- \xrightarrow{O_2} [(CH_3)_2C_2B_4H_4]_2Fe(III)^- \xrightarrow{H^+}$$
$$[(CH_3)_2C_2B_4H_4]_2Fe(III)H$$

$$2[(CH_3)_2C_2B_4H_4]_2Fe(III)H \longrightarrow [(CH_3)_2C_2B_4H_4]_2Fe(II)H_2 +$$
$$(CH_3)_4C_4B_8H_8 + Fe$$

The neutral dihydrogen compound reacts with $(\eta^5-C_5H_5)Co(CO)_2$ under ultraviolet light to yield a structurally unique product, dark green $(\eta^5-C_5H_5)-CoFe(CH_3)_4C_4B_8H_8$ (Figure 32).[90] Crystallographic analysis[94] has shown that the cobalt has displaced a BH unit, which in turn replaces the face-capping hydrogens and adopts a wedging position between the pyramidal C_2B_4 and CoC_2B_3 groups. This geometry is a consequence of the electron-hyperdeficiency of the cage system, which arises because the molecule has two electrons fewer than the number required for a "normal" structure, i.e., one having *closo* 7–vertex (FeC_2B_4) and *closo* 8-vertex $(FeCoC_2B_4)$ polyhedra linked by an iron

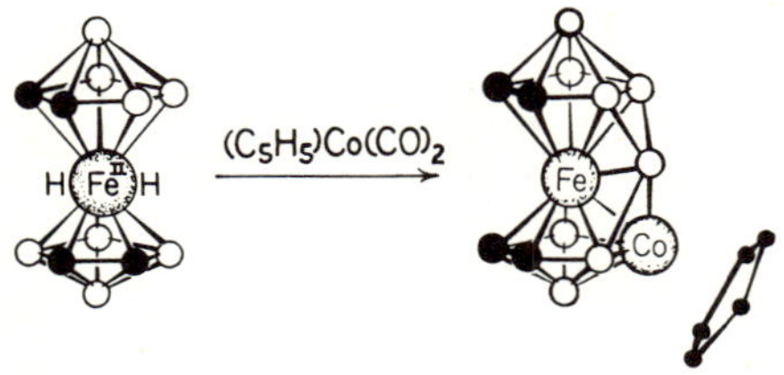

Figure 32.   Insertion of $(\eta^5\text{–}C_5H_5)Co$ into $[2,3\text{–}(CH_3)_2C_2B_4H_4]_2FeH_2$ (Ref. 90).

vertex common to both. Hyperdeficient (or hyper-*closo*) cages are those having fewer than $2n + 2$ framework bonding electrons, and usually adopt capped poly-hedral geometries (for additional examples, see Refs. 80 and 94 and papers cited therein).

Other hyperdeficient metallacarboranes are the germanium–iron and tin–iron species which are generated from the bis(carboranyl)iron hydrogen anion:[92]

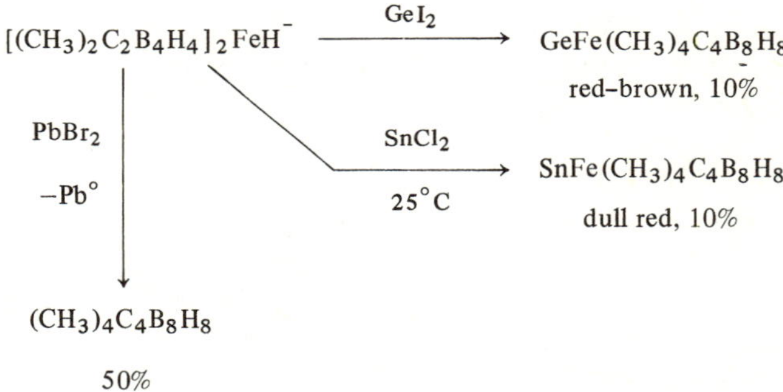

Again, the Ge and Sn atoms effectively replace the capping hydrogens in the iron dihydrogen system; crystal structures are not available, but nmr spectra of these products suggest highly symmetrical arrangements with the group IV atom, rather than BH, occupying the "wedging" position, in contrast to the Co—Fe species described previously. (Such a structure has, in fact, recently been observed[93b] in the purple, paramagnetic $[(CH_3)_2C_2B_4H_4]_2Fe(II)_2(CH_3O)_2\text{-}C_2H_4$ complex mentioned earlier). As shown, $PbBr_2$ failed to effect lead inser-tion, instead functioning as an oxidant.

Treatment of $[(CH_3)_2C_2B_4H_4]_2FeH_2$ with $CoCl_2$ in ethanolic KOH with subsequent work-up in air, produced a complex mixture of products, some of which were iron–cobalt metallacarboranes with adjacent metal centers, and some having capping hydrogens associated with the metals;[92] two products incor-porating both of these features are depicted in Figure 33. Both structures were assigned from ^{11}B and ^1H nmr spectra, and an X-ray study[95] has confirmed that of $(\eta^5\text{–}C_5H_5)_2CoFeH(CH_3)_2C_2B_3H_3$ (the unique hydrogen, however, was not located crystallographically).

Still another remarkable reaction of the iron–dihydrogen complex is that observed with carbon monoxide gas at 200°C, which produces in 76% yield a

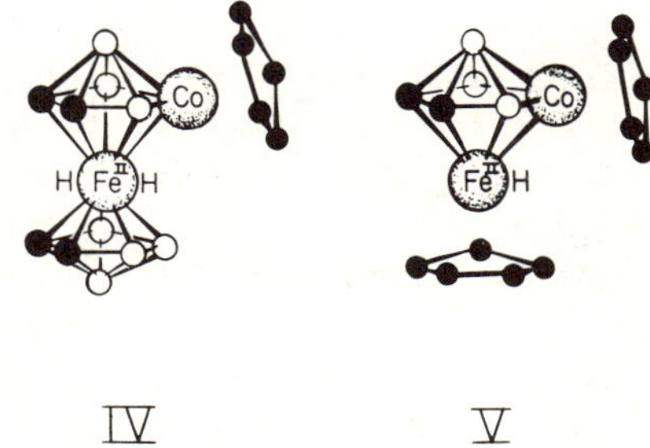

Figure 33. Proposed structure of $[(CH_3)_2-C_2B_4H_4]FeH_2[(\eta^5-C_5H_5)Co(CH_3)_2C_2B_3H_3]$ and established structure of $1,2,4,5-(\eta^5-C_5H_5)_2Fe(H)Co(CH_3)_2C_2B_3H_3$ (Refs. 92 and 95).

10-vertex *closo*-ferracarborane, $2,1,4-(CO)_3Fe(CH_3)_2C_2B_7H_7$.[92] This author and his co-workers have suggested[92] that the mechanism may be initiated by displacement of one or both metal-bound hydrogens by CO, forming $(CO)_3-Fe(CH_3)_2C_2B_4H_4$ which is attacked by a reactive borane fragment formed by thermal decomposition of the original species, to give the observed product. Although this is speculative, it may be noted that carbon monoxide is known to displace hydrogen in metal hydrides;[96] moreover, the high yield of product in this case implies a facile mechanism.

The red cobalt monohydrogen complex, $[2,3-(CH_3)_2C_2B_4H_4]_2Co(III)H$, exhibits a rather different chemistry from the iron–dihydrogen system. In aqueous acid it loses a BH unit and converts to the yellow *nido, closo* species $[2,2-(CH_3)_2C_2B_4H_4]CoH[2,3-(CH_3)_2C_2B_3H_5]$ (Figure 29), retaining the metal-bound "capping" proton in the process.[85] This species contains three "extra" hydrogens, two of the B—H—B bridging type and one assumed to cap a CoB_2 face or faces. Nmr evidence shows that there is no exchange between these hydrogens on the nmr time scale. As mentioned earlier, treatment with strong base affects removal of the capping proton only; the resulting yellow anion can be protonated with HCl, regenerating the neutral compound. The complex $[(CH_3)_2C_2B_4H_4]_2CoH$, like its iron–dihydrogen counterpart, reacts with $(\eta^5-C_5H_5)Co(CO)_2$ to replace one or more BH units with cobalt (Figure 34). The dicobalt product has been structurally confirmed by X-ray analysis,[95]

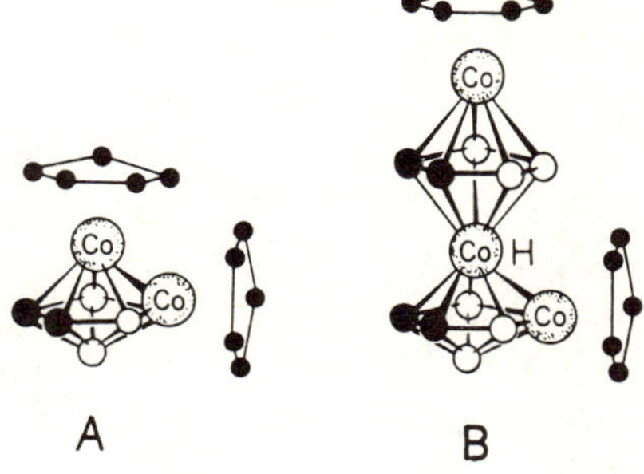

Figure 34. (A) Structure of $1,2,4,5-(\eta^5-C_5H_5)_2Co_2(CH_3)_2C_2B_3H_5$ (Ref. 95). (B) Proposed structure of $(\eta^5-C_5H_5)Co(CH_3)_2-C_2B_3H_3Co(H)(CH_3)_2C_2B_3H_3Co(\eta^5-C_5H_5)$ (Ref. 85).

but the geometry of the tricobalt species (which retains a metal-bound hydrogen) has been proposed from nmr evidence.[85]

The behavior of the face-capping hydrogen atoms in these complexes is not understood in mechanistic detail, but nmr spectra invariably point to direct association with a metal center, and the capping hydrogens are highly reactive. It should be stressed to readers not familiar with the metallacarborane field that the chemistry described above for this class of complexes is unusual in comparison to metallacarboranes in general. For example, viewed against the air-stability of most metallacarboranes, the extreme sensitivity of $[(CH_3)_2C_2B_4H_4]_2FeH_2$ and $[(CH_3)_2C_2B_4H_4]_2CoH$ to oxygen (which converts them to $(CH_3)_4C_4B_8H_8$) is extraordinary. The presence of the capping hydrogens has profound chemical consequences which are just beginning to be explored.

5. INTERMOLECULAR AND INTERCAGE M–H–B BRIDGES

Hydrogen bridging between metal and boron atoms is not restructed to intramolecular bonding, as in the molecules described up to this point. M–H–B interactions can also occur in intermolecular fashion, i.e., between metals and clusters that are not otherwise linked. Chelate complexes of transition metal cations and $B_3H_8^-$ anions (described in Chapter 3) form an important class of compounds of this type, and indeed a number of these, as well as BH_4^- complexes, were known and characterized prior to the discovery of the metallaborane and metallacarborane cluster compounds.[97]

Of the large boron polyhedra, the first in which intermolecular M–H–B bonding was convincingly demonstrated is the *closo*-$B_{10}H_{10}^{2-}$ anion, whose copper salts have been extensively studied. The original X-ray investigation of $Cu_2B_{10}H_{10}$[98,99] revealed the existence of significant covalent interactions between the Cu^{2+} ions and boron atoms in the $B_{10}H_{10}^{2-}$ cluster, but hydrogen bridge bonding was not then considered. More recently, infrared and crystallographic studies[100] of $\{[(C_6H_5)_3P]_2Cu\}_2B_{10}H_{10} \cdot CHCl_3$ have established that the borane cage in this compound is involved in a total of four Cu–H–B bridges, consisting of apical and equatorial BH interactions with each of two copper atoms (Figure 35). The mean Cu–H apical and Cu–H equatorial distances of 2.08(7) and 1.86(6) Å are significantly different, suggesting that the interaction of copper with the apex hydrogen is weaker than that involving the equatorial hydrogen. This observation is not accountable from charge distribution calculations,[101,102] which place greater negative charge on the apex positions, but is in accord with simple geometric considerations.[100]

Detailed studies of the infrared spectra of $Cu_2B_{10}H_{10}$[98,99,103] and $\{[(C_6H_5)_3P]_2Cu\}_2B_{10}H_{10}$[104] in the solid state indicate that Cu–H–B bonding exists in both cases, manifested by characteristic broad weak bands in the region 2100–2400 cm^{-1} that are absent in alkali metal salts of $B_{10}H_{10}^{2-}$. In the case of the *bis*(triphenylphosphine) copper salt, the IR spectrum in CHCl$_3$

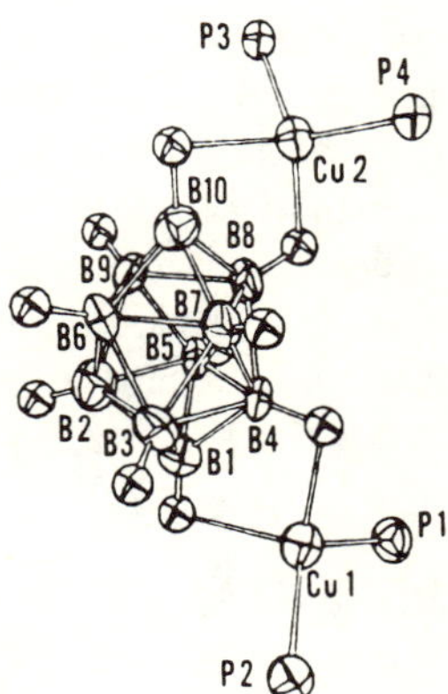

Figure 35. Structure of $\{[(C_6H_5)_3P]_2Cu\}_2B_{10}H_{10}$ (Ref. 100).

solution[104] is very similar to the solid-state spectrum, indicating that $Cu-H-B$ bridging persists in solution; moreover, a ^1H nmr signal at $\delta + 0.85$ has been assigned to the bridging proton. These spectroscopic characteristics are also exhibited by $[(C_6H_5)_3P]_2CuBH_4$ and $[(C_6H_5)_3P]_2CuB_3H_8$, both of which have $Cu-H-B$ bonding, but are *not* found in $\mu-[(C_6H_5)_3P]_2CuB_5H_8$, a complex which has a three-center $Cu-B_2$ covalent bridge bond; hence, it has been concluded that this latter species has no significant $Cu-H$ bonding interaction.[104] The spectroscopic data and $Cu-H$ bond distances indicate that the $Cu-H-B$ bonds in copper-$B_{10}H_{10}^{2-}$ salts are relatively weak in comparison to intramolecular $M-H-B$ or $B-H-B$ bridges, but their existence has been clearly established. At this writing, examples of metal–hydrogen–boron bridging involving other $B_nH_n^{2-}$ polyhedra, or metals other than copper, have not been reported.

A different type of hydrogen-bonded complex is the phosphinorhodacarborane dimer $[(C_6H_5)_3PRhC_2B_9H_{11}]_2$, discussed earlier and shown in Figure 4. In this molecule,[18] identical *closo*-metallacarborane polyhedra are linked by two $Rh-H-B$ bridges as well as by a direct $Rh-Rh$ bond. The entire system is, thus, a single covalently linked molecule, and the hydrogen bridges are not, strictly speaking, intermolecular; we designate this situation (which is at present unique in boron cluster chemistry) as intercage $M-H-B$ bridging.

An example of "pure" intermolecular $M-H-B$ linkage has recently been provided by the synthesis and structure determination of $[(C_6H_5)_3PAgC_2B_8H_{11}]_2$, which crystallizes as a dimer in which two AgC_2B_8 open polyhedra (formally of the *arachno* class) are joined by two $Ag-H-B$ bridges derived from terminal $B-H$ groups.[106] In this instance, in contrast to the rhodium species just described, there is evidently no direct metal-metal interaction and the dimer is held together entirely by $M-H-B$ bridge bonds.

Other species mentioned in this reivew might also be described as intermolecular $M-H-B$ linked complexes, notably $(CO)_3MnB_8H_{13}$ (Section 3.1.2) and $[(p\text{-tolyl})_3P]_2(H)_2(C_2B_9H_{12})Ir$ (Section 3.3). In each of these compounds the metal is linked to the borane or carborane unit solely by hydrogen bridges, as depicted schematically in Figure 1(X).

The propensity for transition metal-hydrogen bonding is such that further examples of intermolecular and intercage M—H—B, and possibly M—H—M, bridging are to be expected. Structures of this type are most likely, of course, when the transition metal center would be coordinatively unsaturated in the absence of bridging, as is the case in the monomer of the rhodium complex discussed above.

REFERENCES

1. R. Bau, ed., *Transition Metal Hydrides, Advances in Chemistry Series*, No. 167, American Chemical Society, Washington, D.C. (1978); E. L. Muetterties, ed., *Transition Metal Hydrides*, Marcel Dekker, New York (1971); W. M. Mueller, J. P. Blackledge, and G. C. Libowitz, eds., *Metal Hydrides*, Academic Press, New York (1968).
2. (a) E. L. Muetterties, ed., *Boron Hydride Chemistry*, Academic Press, New York (1975); (b) R. N. Grimes, in: *Organometallic Reactions and Syntheses, Vol. 6*, E. I. Becker and M. Tsutsui, eds., ch. 2, pp. 63–221, Plenum, New York (1977); (c) N. N. Greenwood and I. M. Ward, *Chem. Soc. Rev.*, **3**, 231 (1974); (d) R. N. Grimes, *Acc. Chem. Res.*, **11**, 420 (1978); (e) R. N. Grimes, *Coord. Chem. Rev.*, **28**, 47 (1979).
3. J. P. Jesson, *Transition Metal Hydrides* (E. L. Muetterties, ed.), pp. 76–77, Marcel Dekker, New York (1971).
4. (a) K. Wade, *Adv. Inorg. Chem. Radiochem.*, **18**, 1 (1976); (b) R. W. Rudolph, *Acc. Chem. Res.*, **9**, 446 (1976); (c) D. M. P. Mingos, *Nat. (London) Phys. Sci.*, **236**, 99 (1972); (d) M. E. O'Neill and K. Wade, *Metal Interactions with Boron Clusters* (R. N. Grimes, ed.), Ch. 1, pp. 1–4, Plenum, New York (198x) (this volume).
5. T. E. Paxson and M. F. Hawthorne, *J. Am. Chem. Soc.*, **96**, 4674 (1974).
6. L. G. Sneddon and R. N. Grimes, *J. Am. Chem. Soc.*, **94**, 7161 (1972).
7. L. G. Sneddon, D. C. Beer, and R. N. Grimes, *J. Am. Chem. Soc.*, **95**, 6623 (1973).
8. L. I. Zakharkin and V. N. Grandberg, *Zh. Obshch. Khim*, **47**, 228 (1977).
9. G. E. Hardy, K. P. Callahan, C. E. Strouse, and M. F. Hawthorne, *Acta Crystallogr., Sec. B*, **B32**, 264 (1976).
10. E. L. Hoel, M. Talebinasab-Savari, and M. F. Hawthorne, *J. Am. Chem. Soc.*, **99**, 4356 (1977).
11. E. H. S. Wong and M. F. Hawthorne, *J. Chem. Soc., Chem. Commun.*, 257 (1976).
12. E. H. S. Wong and M. F. Hawthorne, *Inorg. Chem.*, **17**, 2863 (1978).
13. (a) A. R. Siedle, *J. Organomet. Chem.*, **90**, 249 (1975); (b) M. S. Delaney, C. B. Knobler, and M. F. Hawthorne, *Inorg. Chem.*, **20**, 1341 (1981).
14. L. I. Zakharkin and T. B. Agakhanova, *Izv. Akad. Nauk SSSR, Ser. Khim*, **2151** (1978).
15. L. I. Zakharkin, I. V. Pisareva, and T. B. Agakhanova, *Izv. Akad. Nauk SSSR, Ser. Khim.*, 2389 (1977).
16. L. I. Zakharkin and T. B. Agakhanova, *Izv. Akad. Nauk SSSR, Ser. Khim.*, 2833 (1978).
17. E. L. Hoel and M. F. Hawthorne, *J. Am. Chem. Soc.*, **96**, 4676 (1974).
18. R. T. Baker, R. E. King III, C. Knobler, C. A. O'Con, and M. F. Hawthorne, *J. Am. Chem. Soc.*, **100**, 8266 (1978).
19. W. C. Kalb, R. G. Teller, and M. F. Hawthorne, *J. Am. Chem. Soc.*, **101**, 5417 (1979).
20. Z. Demidowicz, R. G. Teller, and M. F. Hawthorne, *J. Chem. Soc., Chem. Commun.*, 831 (1979).

21. B. A. Sosinsky, W. C. Kalb, R. A. Greuz, V. A. Uski, and M. F. Hawthorne, *J. Am. Chem. Soc.*, **99**, 6768 (1977).

22. E. S. Chandrasekaran, D. A. Thompson, and R. W. Rudolph, *Inorg. Chem.*, **17**, 760 (1978).

23. S. B. Miller and M. F. Hawthorne, *J. Chem. Soc., Chem. Commun.*, 786 (1976).

24. C. W. Jung and M. F. Hawthorne, *J. Chem. Soc., Chem. Commun.*, 499 (1976).

25. A. R. Kane, L. J. Guggenberger, and E. L. Muetterties, *J. Am. Chem. Soc.*, **92**, 2571 (1970).

26. D. A. Thompson and R. W. Rudolph, *J. Chem. Soc., Chem. Commun.*, 770 (1976).

27. B. J. Meneghelli and R. W. Rudolph, *J. Am. Chem. Soc.*, **100**, 4626 (1978).

28. A. R. Siedle, *J. Organomet. Chem.*, **97**, C4 (1975).

29. S. Bresadola, B. Longato, and F. Morandini, *J. Chem. Soc., Chem. Commun*, 510 (1974).

30. B. Longato, F. Morandini, and S. Bresadola, *J. Organomet. Chem.*, **121**, 113 (1976).

31. (a) B. Longato, F. Morandini, and S. Bresadola, *Inorg. Chem.*, **15**, 650 (1976); (b) B. Longato and S. Bresadola, *Inorg. Chim. Acta*, **33**, 189 (1979).

32. B. Longato, F. Morandini, and S. Bresadola, *Proc. Int. Conf. Coord. Chem., 16th*, 2.2b (1974).

33. E. L. Hoel and M. F. Hawthorne, *J. Am. Chem. Soc.*, **97**, 6388 (1975).

34. E. L. Hoel and M. F. Hawthorne, *J. Am. Chem. Soc.*, **95**, 2712 (1973).

35. E. L. Hoel and M. F. Hawthorne, *J. Am. Chem. Soc.*, **96**, 6770 (1974).

36. M. A. Bennett and D. L. Milner, *J. Am. Chem. Soc.*, **91**, 6983 (1969).

37. G. W. Parshall, W. H. Knoth, and R. A. Schunn, *J. Am. Chem. Soc.*, **91**, 4990 (1969).

38. R. W. Marks, S. S. Wreford, and D. D. Traficante, *Inorg. Chem.*, **17**, 756 (1978).

39. G. K. Barker, M. Green, T. P. Onak, F. G. A. Stone, C. B. Ungermann, and A. J. Welch, *J. Chem. Soc., Chem. Commun.*, 169 (1978).

40. S. G. Shore, *Boron Hydride Chemistry* (E. L. Muetterties, ed.), Ch. 3, pp. 79–174, Academic Press, New York (1975).

41. V. R. Miller and R. N. Grimes, *J. Am. Chem. Soc.*, **95**, 5078 (1973).

42. V. R. Miller, R. Weiss, and R. N. Grimes, *J. Am. Chem. Soc.*, **99**, 5646 (1977).

43. L. G. Sneddon and D. Voet, *J. Chem. Soc., Chem. Commun.*, 118 (1976).

44. (a) D. Schwoch, A. B. Burg, and R. A. Beaudet, *Inorg. Chem.*, **16**, 3219 (1977) (microwave study); (b) W. J. Dulmage and W. N. Lipscomb, *Acta Crystallogr.*, **5**, 260 (1962).

45. R. Weiss, J. R. Bowser, and R. N. Grimes, *Inorg. Chem.*, **17**, 1522 (1978).

46. N. N. Greenwood, C. G. Savory, R. N. Grimes, L. G. Sneddon, A. Davison, and S. S. Wreford, *J. Chem. Soc., Chem. Commun.*, 718 (1974).

47. E. L. Anderson, K. J. Haller, and T. P. Fehlner, *J. Am. Chem. Soc.*, **101**, 4390 (1979).

48. E. L. Anderson and T. P. Fehlner, *J. Am. Chem. Soc.*, **100**, 4606 (1978).

49. R. N. Grimes, *Annals N.Y. Acad. Sci.*, **239**, 180 (1974).

50. H. D. Kaesz, W. Fellmann, G. R. Wilkes, and L. F. Dahl, *J. Am. Chem. Soc.*, **87**, 2753 (1965).

51. S. G. Shore, J. D. Ragaini, R. L. Smith, C. E. Cottrell, and T. P. Fehlner, *Inorg. Chem.*, **18**, 670 (1979).

52. T. P. Fehlner, J. Ragaini, and M. Mangion, *J. Am. Chem. Soc.*, **98**, 7085 (1976).

53. (a) A. Davison, D. D. Traficante, and S. S. Wreford, *J. Am. Chem. Soc.*, **96**, 2802 (1974); (b) D. F. Gaines and T. V. Iorns, *J. Am. Chem. Soc.*, **90**, 6617 (1968); (c) C. P. Magee, L. G. Sneddon, D. C. Beer, and R. N. Grimes, *J. Organomet. Chem.*, **86**, 159 (1975); (d) M. L. Thompson and R. N. Grimes, *Inorg. Chem.*, **11**, 1925 (1972); (e) A. Tabereaux, and R. N. Grimes, *Inorg. Chem.*, **12**, 792 (1973); (f) C. G. Savory and M. G. H. Wallbridge, *J. Chem. Soc. Dalton*, 918 (1972); (g) N. N. Greenwood, J. A. Howard, and W. S. McDonald, *J. Chem. Soc., Dalton*, 37 (1977); (h) M. Mangion, W. R. Clayton, O. Hollander, and S. G. Shore, *Inorg. Chem.*, **16**, 2110 (1977); (i) J. C. Calabrese, and L. F. Dahl, *J. Am. Chem. Soc.*, **93**, 6042 (1971); (j) D. F. Gaines, K. M. Coleson, and

J. C. Calabrese, *J. Am. Chem. Soc.*, **101**, 3979 (1979); (k) J. P. Brennan, R. Schaeffer, A. Davison, and S. S. Wreford, *J. Chem. Soc., Chem. Commun.*, 354 (1973); (l) N. S. Hosmane, and R. N. Grimes, *Inorg. Chem.*, **18**, 2886 (1979); (m) D. C. Finster and R. N. Grimes, *Inorg. Chem.*, **20**, 863 (1981).

54. M. Mangion, J. D. Ragaini, T. A. Schmitkons, and S. G. Shore, *J. Am. Chem. Soc.*, **101**, 754 (1979).

55. V. T. Brice and S. G. Shore, *J. Chem. Soc. Chem. Commun.*, 1312 (1970).

56. R. Weiss and R. N. Grimes, *J. Am. Chem. Soc.*, **99**, 8087 (1977).

57. R. Weiss and R. N. Grimes, *Inorg. Chem.*, **18**, 3291 (1979).

58. R. Wilczynski and L. G. Sneddon, *Inorg. Chem.*, **18**, 864 (1979).

59. M. B. Fischer and D. F. Gaines, *Inorg. Chem.*, **18**, 3200 (1979).

60. D. F. Gaines, J. W. Lott, and J. C. Calabrese, *J. Chem. Soc., Chem. Commun.*, 295 (1973).

61. J. W. Lott, D. F. Gaines, H. Shenhav, and J. Schaeffer, *J. Am. Chem. Soc.*, **95**, 3042 (1973).

62. J. W. Lott and D. F. Gaines, *Inorg. Chem.*, **13**, 2261 (1974).

63. D. F. Gaines and J. C. Calabrese, *Inorg. Chem.*, **13**, 2419 (1974).

64. (a) J. R. Pipal and R. N. Grimes, *Inorg. Chem.*, **16**, 3251 (1977).

65. (a) T. L. Venable and R. N. Grimes, *Inorg. Chem.*, **21**, 887 (1982); T. L. Venable, E. Sinn, and R. N. Grimes, *Inorg. Chem.*, **21**, 895 (1982); (b) G. J. Zimmerman, L. W. Hall, and L. G. Sneddon, *Inorg. Chem.*, **19**, 3642 (1980).

66. J. C. Calabrese, M. B. Fischer, D. F. Gaines, and J. W. Lott, *J. Am. Chem. Soc.*, **96**, 6318 (1974).

67. D. A. Thompson, T. K. Hilty, and R. W. Rudolph, *J. Am. Chem. Soc.*, **99**, 6774 (1977).

68. T. K. Hilty, D. A. Thompson, W. M. Butler, and R. W. Rudolph, *Inorg. Chem.*, **18**, 2642 (1979).

69. C. J. Jones, J. N. Francis, and M. F. Hawthorne, *J. Am. Chem. Soc.*, **95**, 7633 (1973).

70. M. R. Churchill and K. Gold, *Inorg. Chem.*, **12**, 1157 (1973).

71. C. J. Jones, J. N. Francis, and M. F. Hawthorne, *J. Am. Chem. Soc.*, **94**, 8391 (1972).

72. K. P. Callahan, F. Y. Lo, C. E. Strouse, A. L. Sims, and M. F. Hawthorne, *Inorg. Chem.*, **13**, 2842 (1974).

73. M. Green, J. A. K. Howard, T. L. Spencer, and F. G. A. Stone, *J. Chem. Soc., Dalton*, 2274 (1975).

74. D. A. Owen and M. F. Hawthorne, *J. Am. Chem. Soc.*, **93**, 873 (1971).

75. R. A. Love and R. Bau, *J. Am. Chem. Soc.*, **94**, 8274 (1972).

76. (a) G. Allegra, M. Calligaris, R. Gurlanetto, G. Nardin, and L. Randaccio, *Cryst. Struct. Commun.*, **3**, 69 (1974); (b) J. A. Doi, R. G. Teller, and M. F. Hawthorne, *J. Chem. Soc., Chem. Commun.*, 80 (1980).

77. D. W. Hart, R. G. Teller, C-Y. Wei, R. Bau, G. Longoni, S. Campanella, P. Chini, and T. F. Koetzle, *Angew. Chem. Int. Ed. Engl.*, **18**, 80 (1979).

78. V. R. Miller and R. N. Grimes, *J. Am. Chem. Soc.*, **98**, 1600 (1976).

79. J. R. Pipal and R. N. Grimes, *Inorg. Chem.*, **18**, 252 (1979).

80. J. R. Pipal and R. N. Grimes, *Inorg. Chem.*, **16**, 3255 (1977).

81. B. T. Huie, C. B. Knobler, and H. D. Kaesz, *J. Am. Chem. Soc.*, **100**, 3059 (1978).

82. (a) T. Onak, R. Drake, and G. B. Dunks, *J. Am. Chem. Soc.*, **87**, 2505 (1965); (b) T. Onak and J. B. Leach, *J. Chem. Soc., Chem. Commun.*, 76 (1971); (c) T. Onak, and E. Wan, *J. Chem. Soc., Dalton*, 665 (1974).

83. G. L. McKown, B. P. Don, R. A. Beaudet, P. J. Vergamini, and L. H. Jones, *J. Am. Chem. Soc.*, **98**, 6909 (1976).

84. E. A. McNeill and F. R. Scholer, *Inorg. Chem.*, **14**, 1081 (1975).

85. W. M. Maxwell, V. R. Miller, and R. N. Grimes, *J. Am. Chem. Soc.*, **98**, 4818 (1976).

86. V. R. Miller and R. N. Grimes, *J. Am. Chem. Soc.*, **95**, 2830 (1973); R. N. Grimes, D. C. Beer, L. G. Sneddon, V. R. Miller and R. Weiss, *Inorg. Chem.*, **13**, 1138 (1974).

87. M. F. Hawthorne, L. F. Warren, Jr., K. P. Callahan, and N. F. Travers, *J. Am. Chem. Soc.*, **93**, 2407 (1971).

88. T. J. Curphey, J. O. Santer, M. Rosenblum, and J. H. Richards, *J. Am. Chem. Soc.*, **82**, 5249 (1960).

89. W. E. Geiger, Jr., W. L. Bowden, and N. El Murr, *Inorg. Chem.*, **18**, 2358 (1979).

90. W. M. Maxwell, V. R. Miller, and R. N. Grimes, *Inorg. Chem.*, **15**, 1343 (1976).

91. J. R. Pipal, and R. N. Grimes, *Inorg. Chem.*, **18**, 263 (1979).

92. W. M. Maxwell, K. S. Wong, and R. N. Grimes, *Inorg. Chem.*, **16**, 3094 (1977).

93. (a) J. R. Pipal and R. N. Grimes, *Inorg. Chem.*, **18**, 1936 (1979); (b) R. N. Grimes, R. B. Maynard, E. Sinn, and G. J. Long, Abstract INOR-10, *Abstracts of Papers, 182nd National Meeting of the American Chemical Society*, American Chemical Society, New York (August 1981).

94. W. M. Maxwell, E. Sinn, and R. N. Grimes, *J. Am. Chem. Soc.*, **98**, 3490 (1976).

95. R. N. Grimes, E. Sinn, and R. B. Maynard, *Inorg. Chem.*, **9**, 2384 (1980).

96. R. A. Schunn, *Transition Metal Hydrides*, (E. L. Muetterties, ed.), Ch. 5, Marcel Dekker, New York (1971).

97. B. A. Frenz and J. A. Ibers, *Transition Metal Hydrides*, (E. L. Muetterties, ed.), ch. 3, pp. 66–69, Marcel Dekker, New York (1971); F. A. Cotton, M. Jeremic, and A. Shaver, *Inorg. Chim. Acta*, **6**, 543 (1972) (and references therein); K. M. Melmed, D. Coucouvanis, and S. J. Lippard, *Inorg. Chem.*, **12**, 232 (1973); K. M. Melmed, T. Li, J. J. Mayerle, and S. J. Lippard, *J. Am. Chem. Soc.*, **96**, 69 (1974); S. J. Lippard and K. M. Melmed, *Inorg. Chem.*, **8**, 2755 (1969); S. J. Lippard, and K. M. Melmed, *J. Am. Chem. Soc.*, **89**, 3929 (1967); S. J. Lippard, and K. M. Melmed, *Inorg. Chem.*, **6**, 2223 (1967).

98. R. D. Dobrott and W. N. Lipscomb, *J. Chem. Phys.*, **37**, 1779 (1962).

99. A. Kaczmarczyk, R. D. Dobrott, and W. N. Lipscomb, *Proc. Nat. Acad. Sci., U. S.*, **48**, 729 (1962).

100. J. T. Gill and S. J. Lippard, *Inorg. Chem.*, **14**, 751 (1975).

101. W. N. Lipscomb, *Boron Hydrides*, Benjamin, New York (1963).

102. W. N. Lipscomb, A. R. Pitochelli, and M. F. Hawthorne, *J. Am. Chem. Soc.*, **81**, 5833 (1959).

103. T. E. Paxson, M. F. Hawthorne, L. D. Brown, and W. N. Lipscomb, *Inorg. Chem.*, **13**, 2772 (1974).

104. C. G. Outterson, V. T. Brice, and S. G. Shore, *Inorg. Chem.*, **15**, 1456 (1976).

105. J. R. Pipal, W. M. Maxwell, and R. N. Grimes, *Inorg. Chem.*, **17**, 1447 (1978).

106. H. M. Colquhoun, T. J. Greenhough, and M. G. H. Wallbridge, *J. Chem. Soc., Chem. Commun.*, 192 (1980).

Index